SIXTH EDITION

STATISTICS

for Engineering and the Sciences

STUDENT SOLUTIONS MANUAL

SIXTH EDITION
STATISTICS
for Engineering and the Sciences
STUDENT SOLUTIONS MANUAL

William M. Mendenhall

Terry L. Sincich

Nancy S. Boudreau

CRC Press
Taylor & Francis Group
Boca Raton   London   New York

CRC Press is an imprint of the
Taylor & Francis Group, an **informa** business

A CHAPMAN & HALL BOOK

CRC Press
Taylor & Francis Group
6000 Broken Sound Parkway NW, Suite 300
Boca Raton, FL 33487-2742

Printed on acid-free paper
Version Date: 20160805

International Standard Book Number-13: 978-1-4987-3182-9 (Paperback)

Visit the Taylor & Francis Web site at
http://www.taylorandfrancis.com

and the CRC Press Web site at
http://www.crcpress.com

Contents

Contents

1

Introduction

1.1   a. The population of interest to the researchers is the population of all young women who recently participated in a STEM program.

      b. The sample is the set of 159 young women who were recruited to complete an on-line survey.

      c. We could infer that approximately 27% of all young women who recently participated in a STEM program felt that participation in the STEM program increased their interest in science.

1.3   There are two populations – male students at Griffin University who were video game players and male students at Griffin University who were not video game players. There were two samples — those male students in the 65 chosen who were video game players and those male students in the 65 chosen who were not video game players.

1.5   a. The experimental units for this study are the earthquakes.

      b. The data from the 15 earthquakes represent a sample. There are many more than 15 earthquakes from around the world. Only 15 of the many were studied.

1.7   a. The variable measured is the level of carbon monoxide gas in the atmosphere. The experimental unit is the atmosphere at the Cold Bay, Alaska, weather station each week.

      b. If we are interested in only the weekly carbon monoxide values at the Cold Bay station for the years 2000-2002, then this data represents the population because all that data were collected.

1.9   a. Sampling method would be qualitative.

      b. Effective stress level would be quantitative.

      c. Damping ratio would be quantitative.

1.11 a. Town where sample was collected is qualitative.

      b. Type of water supply is qualitative.

      c. Acidic level is quantitative.

      d. Turbidity level is quantitative.

      e. Temperature is quantitative.

      f. Number of fecal coliforms per 100 milliliters is quantitative.

      g. Free chlorine-residual is quantitative.

      h. Presence of hydrogen sulfide is qualitative.

1.13 a. The experimental units are the smokers.

      b. Two variables measured on each smoker are screening method and age at which scanning method first detects a tumor.

      c. Screening method is qualitative and age is quantitative.

      d. The inference is which screening method (CT or X-ray) is more effective in pinpointing small tumors.

1.15   Answers will vary. First, we number the wells from 1 to 223. We will use Table 1, Appendix B to select the sample of 5. Start in column 8, row 11, and look at the first 3 digits. We proceed down the column until we select 5 numbers between 1 and 223: 58, 176, 136, 47, and 153. Thus, wells numbered 47, 58, 136, 153, and 176 will be selected.

1.17   Answers will vary. First, we number the weeks from 1 to 590. Using the MINITAB random sample procedure, the following sample is selected:

Weeks	Sample	Weeks	Sample
1	568	9	192
2	584	10	590
3	329	11	81
4	379	12	67
5	54	13	230
6	104	14	56
7	171	15	154
8	439		

The 15 weeks with the numbers listed in the Sample column will be selected.

1.19 a. The population of interest is all computer security personnel at all U.S. corporations and government agencies.

   b. The data-collection method is a survey of 5,412 firms. Only 351 computer security personnel responded. Since this was a survey, the computer security personnel elected to either respond or not. Because only 351 of the 5,412 firms survey responded, there could be a nonresponse bias. In addition, the security personnel chose whether to respond or not.

   c. The variable measured is whether or not unauthorized use of the computer system occurred at the firm during the year. This variable is qualitative because the response would be yes or no.

   d. Because 41% of the sample admitted that there was unauthorized use of their computer system, we can infer that approximately 41% of all firms had unauthorized use of their computer systems during the year.

1.21   First, suppose we number all of the intersections from 1 to 5,000. Then, we will use a random number generator to select 50 numbers between 1 and 5,000. The intersections with the 50 numbers selected will then be used for digging.

   Second, we will number the rows from 1 to 100 and the columns from 1 to 50. We will then use a random number generator to select 50 rows from 1 to 100 (rows can be selected more than once) and 50 columns from 1 to 50 (columns can be selected more than once). We will then combine the rows and columns selected to get the intersections used for the digs. For instance, the first row selected might be row 10 and the first column selected might be 47. Then the intersection of row 10 and column 47 will be the first intersection selected.

1.23 a. The population of interest is the set of all computer components (e.g. the hard disk drives).

 b. The sample is the 100 computer components tested.

 c. The data are quantitative because the lifelength of the component is a numerical value.

 d. The mean lifelength of the computer components tested could be used to estimate the mean lifelength of all computer components.

1.25 a. The experimental units are the 2-ml portions of the newly developed cleaning solution.

 b. The variable measured is the amount of hydrochloric acid necessary to achieve neutrality of 2-ml of the newly developed cleaning solution.

 c. The population of interest is the set of all amounts of hydrochloric acid necessary to neutralize all 2-ml portions of the newly developed cleaning solution.

 d. The sample is the set of amounts of hydrochloric acid necessary to neutralize the five 2-ml portions of the newly developed cleaning solution.

1.27 a. The experimental units are the undergraduate engineering students at Penn State.

 b. The population of interest is the set of all undergraduate engineering at Penn State. The sample is the set of 21 undergraduate engineering students in a first-year, project-based design course.

 c. The data collected are the Perry scores which are quantitative.

 d. We estimate that the mean Perry score for all undergraduate engineering students at Penn State is 3.27.

 e. Answers will vary. First, we number the students from 1 to 21. Using the MINITAB random sample procedure, the following sample is selected:

Aftershock	Sample
1	14
2	3
3	16

The 3 students with the numbers listed in the Sample column will be selected.

1.29 a. The variable of interest is the status of bridges in the United States.

 b. The variable is qualitative with values structurally deficient, functionally obsolete, and safe.

 c. The data set is a population since all of the bridges in the United States were inspected.

d. The data for the study were obtained from the FHWA inspection ratings.
e. Answers will vary. First, number the bridges from 1 to 600,000. Using the MINITAB random sample procedure, the following sample is selected:

Sample	Sample
369,891	69,324
481,030	28,952
58,902	365,481
301,594	187,834
538,562	569,846
255,565	566,258
350,835	250,030
267,191	325,747
470,533	528,693
482,519	400,430
403,882	252,044
202,888	191,159
360,439	

The 25 bridges with the numbers listed in the Sample column will be selected.

2

Descriptive Statistics

2.1   a.   The graph used is a bar chart.
      b.   The variable measured is the type of robotic limbs on social robots.
      c.   The social robot design that is currently used the most is legs only.
      d.   The relative frequencies are found by dividing the frequencies by the sample size, $n = 106$.

Robotic Limbs	Frequency	Relative Frequency
None	15	15/106 = 0.1415
Both	8	8/106 = 0.0755
Legs only	63	63/106 = 0.5943
Wheels only	20	20/106 = 0.1887

      e.   Using MINITAB, the Pareto chart is:

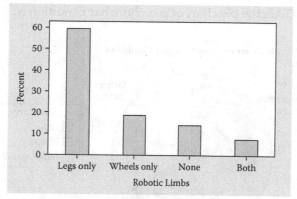

2.3   Using MINITAB, the pie chart is:

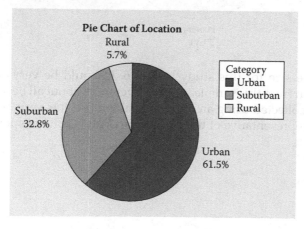

The majority of young women who recently participated in a STEM program are from urban areas (61.5%) and very few are from rural areas (5.7%).

2.5 a. The variable beach condition is qualitative, nearshore bar condition is qualitative, and long-term erosion rate is quantitative.

b. Using MINITAB, the pie chart for beach condition is:

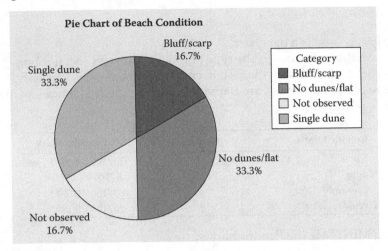

c. Using MINITAB, the pie chart of nearshore bar condition is:

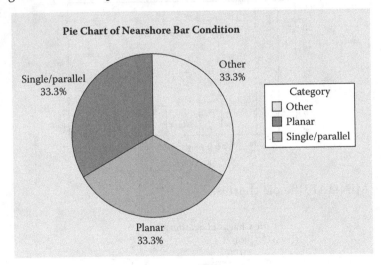

d. The sample size for this study is only 6. It would be very risky to use the information from this sample to make inferences about all beach hotspots. The data were collected using an online questionnaire. It is very doubtful that this sample is representative of the population of all beach hot spots.

2.7   Using MINITAB, pie charts to compare the two ownership sectors of LEO and GEO satellites are:

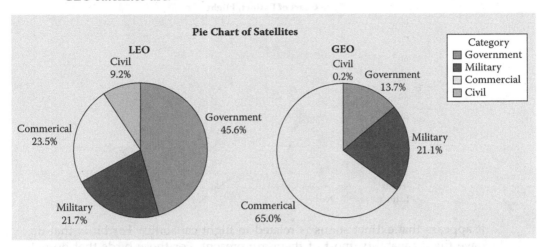

Most LEO satellites are owned by entities in the government (45.6%) while most GEO satellites are owned by entities in the commercial sector (65.0%). The fewest percentage of LEO satellites are owned by entities in the civil sector (9.2%). The fewest percentage of GEO satellites are also owned by entities in the civil sector (0.2%), but the percentage is much smaller than that for the LEO satellites.

2.9   a.   Using MINITAB, the Pareto chart is:

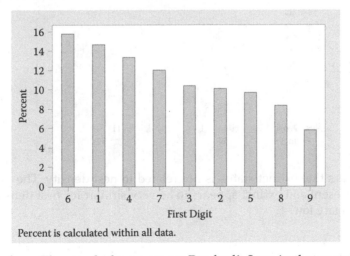

b.   Yes and no. The graph does support Benford's Law in that certain digits are more likely to occur than others. In this set of data, the number 6 occurs first 15.7% of the time while the number 9 occurs first only 5.8% of the time. However, Benford's Law also states that the number 1 is the most likely to occur at 30% of the time. In this set of data, the number 1 is not the most frequent number to occur first, and it also only occurs as the first significant digit 14.7% of the time, not the 30% specified by Benford's Law.

2.11   Using MINITAB, a bar chart for the Extinct status versus flight capability is:

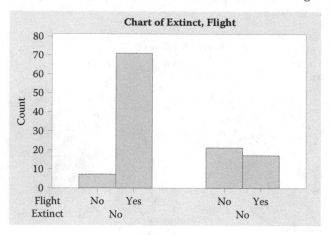

It appears that extinct status is related to flight capability. For birds that do have flight capability, most of them are present. For those birds that do not have flight capability, most are extinct.

The bar chart for Extinct status versus Nest Density is:

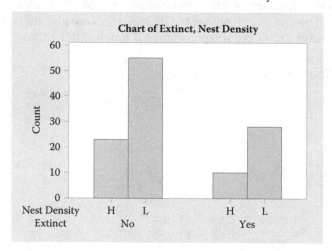

It appears that extinct status is not related to nest density. The proportion of birds present and extinct appears to be very similar for nest density high and nest density low.

The bar chart for Extinct status versus Habitat is:

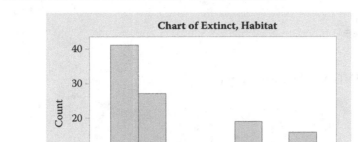

It appears that the extinct status is related to habitat. For those in aerial terrestrial (TA), most species are present. For those in ground terrestrial (TG), most species are extinct. For those in aquatic, most species are present.

2.13 a. The measurement class 10-20 contains the highest proportion of respondents.

   b. The approximate proportion of organizations that reported a percentage monetary loss from malicious insider actions less than 20% is $0.30 + 0.38 = 0.68$.

   c. The approximate proportion of organizations that reported a percentage monetary loss from malicious insider actions greater than 60% is $0.07 + 0.025 + .035 + .045 = 0.175$.

   d. The approximate proportion of organizations that reported a percentage monetary loss from malicious insider actions between 20% and 30% is 0.12. The actual number is approximately $0.12(144) = 17.28$ or approximately 17.

2.15 a. Using MINITAB, the dotplot is:

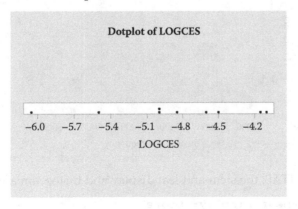

   b. Using MINITAB, the stem-and-leaf display is:

Stem-and-Leaf Display: LOGCES

```
Stem-and-leaf of LOGCES  N = 9
Leaf Unit = 0.10
```

```
 1   -6 0
 2   -5 5
 4   -5 00
(3)  -4 865
 2   -4 11
```

c. Using MINITAB, the histogram is:

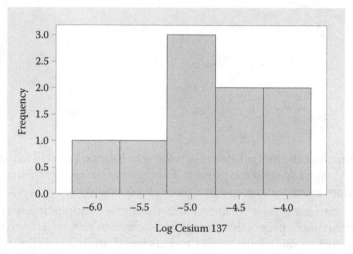

d. Answers may vary. It appears that the histogram is more informative.
e. Four of the nine measurements are −5.00 or less. The proportion is $4/9 = 0.444$.

2.17   Using MINITAB, a histogram of the sound frequencies is:

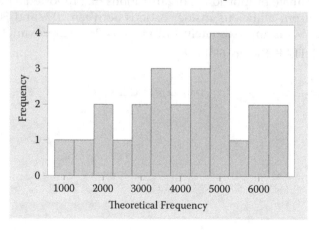

2.19 a. Using MINITAB, the stem-and-leaf display and histogram are:

Stem-and-Leaf Display: Score

```
Stem-and-leaf of Score N = 186
Leaf Unit = 1.0
```

```
 1    6   9
 1    7
 2    7   3
 3    7   4
 4    7   6
 5    7   8
 7    8   11
 8    8   3
11    8   445
20    8   666677777
25    8   99999
36    9   00001111111
54    9   222222333333333333
89    9   444444444444444444555555555555555555
(42)  9   6666666666666666666666777777777777777777777
55    9   888888888888888888999999999999999999999
19   10   0000000000000000000
```

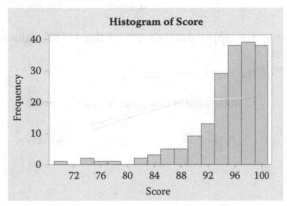

b. Of the 186 scores, only 11 are less than 86. Thus, there are 175 scores that are 86 or higher. The proportion of ships that have an accepted sanitation standard 175/186 = 0.941. The stem-and-leaf display was used because one can identify the actual values.

c. A score of 72 would be located above the number 72 on the histogram and in the 3rd row of the stem-and-leaf display.

2.21 a. Using MINITAB, the histogram is:

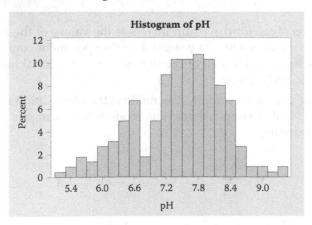

From the histogram, approximately 0.25 of the wells have pH values less than 7.0.

b. Using MINITAB, the histogram of the MTBE values for contaminated wells is:

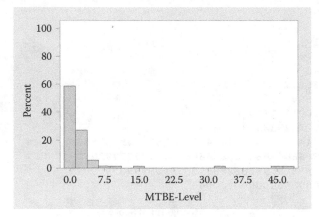

From the histogram, approximately 9% of the MTBE values exceed 5 micrograms per liter.

2.23   Using MINITAB, the histograms are:

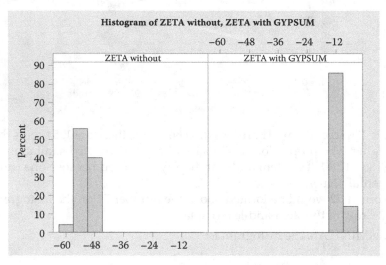

The addition of calcium/gypsum increases the values of the zeta potential of silica. All of the values of zeta potential for the specimens containing calcium/gypsum are greater than all of the values of zeta potential for the specimens without calcium/gypsum.

2.25 a. Assume the data are a sample. The mode is the observation that occurs most frequently. For this sample, there is no mode or all are modes.

The sample mean is:

$$\bar{y} = \frac{\sum y}{n} = \frac{4+3+10+8+5}{5} = \frac{30}{5} = 6$$

The median is the middle number when the data are arranged in order. The data arranged in order are: 3, 4, 5, 8, 10. The middle number is the 3rd number, which is $m = 5$.

b.   Assume the data are a sample. The mode is the observation that occurs most frequently. For this sample, there are 2 modes, 4 and 6.
The sample mean is:

$$\bar{y} = \frac{\sum y}{n} = \frac{9+6+12+4+4+2+5+6}{8} = \frac{48}{8} = 6$$

The median is the middle number when the data are arranged in order. The data arranged in order are: 2, 4, 4, 5, 6, 6, 9, 12. The average of the middle 2 numbers is $m = \frac{5+6}{2} = 5.5$.

2.27 a.   The sample mean is $\bar{y} = \frac{\sum y}{n} = \frac{18.12+19.48+\cdots+16.20}{18} = \frac{296.99}{18} = 16.499$. The average dentary depth of molars is 16.499 mm.
If the largest depth measurement were doubled, then the mean would increase.

b.   The data arranged in order are:
13.25, 13.96, 14.02, 14.04, 15.70, 15.76, 15.83, 15.94, 16.12, 16.20, 16.55, 17.00, 17.83, 18.12, 18.13, 19.36, 19.48, 19.70
There is an even number of observations, so the median is the average of the middle two numbers, $m = \frac{16.12+16.20}{2} = 16.16$. Half of the observations are less than 16.16 and half are greater than 16.16.
If the largest depth measurement were doubled, then the median would not change.

c.   Since no observation occurs more than once, there is either no mode or all of the observations are considered modes.

2.29   The sample mean is $\bar{y} = \frac{\sum y}{n} = \frac{10.94+13.71+\cdots+6.77}{13} = \frac{126.32}{13} = 9.717$. The average rebound length is 9.717 meters.
The data arranged in order are: 4.90, 5.10, 5.44, 5.85, 6.77, 7.26, 10.94, 11.38, 11.87, 11.92, 13.35, 13.71, 17.83
There is an odd number of observations so the median is the middle number or $m = 10.94$. Half of the rebound lengths are less than 10.94 and half are greater than 10.94.

2.31 a.   The sample mean is $\bar{y} = \frac{\sum y}{n} = \frac{3.3+0.5+\cdots+4.0}{16} = \frac{29}{16} = 1.813$.

b.   The data arranged in order are: 0.1, 0.2, 0.2, 0.3, 0.4, 0.5, 0.5, 1.3, 1.4, 2.4, 2.4, 3.3, 4.0, 4.0, 4.0, 4.0. There is an even number of observations so the median is the average of the middle two numbers or $m = \frac{1.3+1.4}{2} = 1.35$.

c.   The mode is the number that occurs the most which is 4.0.

d.   The data arranged in order for the no crude oil present are: 0.1, 0.3, 1.4, 2.4, 2.4, 3.3, 4.0, 4.0, 4.0, 4.0. There is an even number of observations so the median is the average of the middle two numbers or $m = \frac{2.4+3.3}{2} = 2.85$.

e.   The data arranged in order for the crude oil present are: 0.2, 0.2, 0.4, 0.5, 0.5, 1.3. There is an even number of observations so the median is the average of the middle two numbers or $m = \frac{0.4+0.5}{2} = 0.45$.

d.   The median dioxide amount for no crude oil present is 2.85, while the median dioxide amount for crude oil present is 0.45. It appears that dioxide amount is less when crude oil is present.

2.33 a. The average permeability measurement for Group A sandstone is 73.62. The median permeability for Group A is 70.45. Half of the permeability measurements for Group A are less than 70.45 and half are greater than 70.45.

  b. The average permeability measurement for Group B sandstone is 128.54. The median permeability for Group B is 139.30. Half of the permeability measurements for Group B are less than 139.30 and half are greater than 139.30.

  c. The average permeability measurement for Group C sandstone is 83.07. The median permeability for Group C is 78.65. Half of the permeability measurements for Group C are less than 78.65 and half are greater than 78.65.

  d. The mode for Group C is 70.9. Three observations were 70.9. The permeability measurement that occurred the most often from Group C is 70.9.

  e. Group B appears to result in faster decay because all three measures of central tendency for Group B are larger than the corresponding measures for Group C.

2.35 a. Using MINITAB, the descriptive statistics are:

Descriptive Statistics: PRDiff _ outlier

Variable	N	Mean	Median	Mode	N for Mode
PRDiff_outlier	14	-1.091	-0.655	*	0

The average difference is –1.091. The median difference is -0.655. Half of the differences are less than –0.655 and half of the differences are greater than -0.655. No difference occurs more than once, so there is no mode.

  b. The one large difference is -8.11.

  c. Using MINITAB, the descriptive statistics are:

Descriptive Statistics: PRDiff

Variable	N	Mean	Median	Mode	N for Mode
PRDiff	14	-0.519	-0.520	*	0

The mean increases from –1.091 to -0.519 or increases by 0.572. The median increases from -0.655 to –0.520 or increases by 0.135. The mean is much more affected by correcting the outlier than the median.

2.37 a. Using MINITAB, the histogram of the data is:

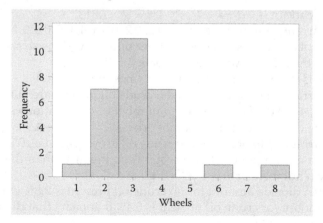

No, the distribution is somewhat mound-shaped but it is not symmetric. The distribution is skewed to the right.

b. The sample mean is $\bar{y} = \dfrac{\sum y}{n} = \dfrac{4+4+\cdots+2}{28} = \dfrac{90}{28} = 3.214$.

The sample variance is $s^2 = \dfrac{\sum y^2 - \dfrac{(\sum y)^2}{n}}{n-1} = \dfrac{340 - \dfrac{90^2}{28}}{28-1} = \dfrac{50.7143}{27} = 1.8783$.

The sample standard deviation is $s = \sqrt{1.8783} = 1.371$.

c. $\bar{y} \pm 2s \Rightarrow 3.214 \pm 2(1.371) \Rightarrow 3.214 \pm 2.742 \Rightarrow (0.472,\ 5.956)$

d. According to Chebyshev's rule, at least ¾ or 75% of the observations will fall in this interval.

e. According to the Empirical Rule, approximately 95% of the observations will fall in this interval.

f. The actual proportion of observations that fall in the interval is $26/28 = 0.929$ or 92.9%. Yes, the Empirical Rule provides a good estimate of the proportion even though the distribution is not perfectly symmetric.

2.39 a. The range is $R = 1.55 - 1.37 = 0.18$.

b. The sample variance is

$$s^2 = \frac{\sum y^2 - \dfrac{\left(\sum y\right)^2}{n}}{n-1} = \frac{17.3453 - \dfrac{11.77^2}{8}}{8-1} = \frac{0.0286875}{7} = 0.00410.$$

c. The sample standard deviation is $s = \sqrt{0.00410} = 0.0640$.

d. The standard deviation for the morning is 1.45 ppm, while the standard deviation for the afternoon is 0.0640. The morning drive-time has more variable ammonia levels.

2.41 a. The range for Group A is 67.20. $R = 122.40 - 55.20 = 67.20$

b. The standard deviation for Group A is 14.48. $s = \sqrt{209.53} = 14.48$

c. Using MINITAB, a histogram of Group A data is:

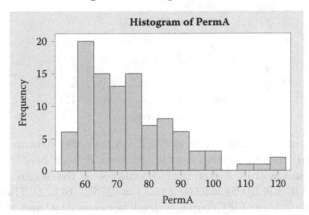

From Exercise 2.33, the mean is 73.62. Because the data are skewed to the right, we will use Chebyshev's rule. At least 8/9 or 88.9% of the observations will fall within 3 standard deviations of the mean. This interval is $\bar{y} \pm 3s \Rightarrow 73.62 \pm 3(14.48) \Rightarrow 73.62 \pm 43.44 \Rightarrow (30.18,\ 117.06)$. Thus, at least 88.8% of the measurements for Group A will fall between 30.18 and 117.06.

d. The range for Group B is 99.60. $R = 150.00 - 50.40 = 99.60$

The standard deviation for Group B is 21.97. $s = \sqrt{482.75} = 21.97$
Using MINITAB, a histogram of Group B data is:

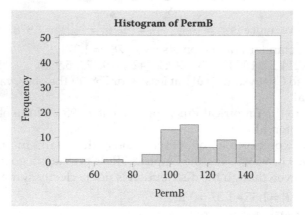

From Exercise 2.33, the mean is 128.54. Because the data are skewed to the left, we will use Chebyshev's rule. At least 8/9 or 88.9% of the observations will fall within 3 standard deviations of the mean. This interval is $\bar{y} \pm 3s \Rightarrow 128.54 \pm 3(21.97) \Rightarrow 128.54 \pm 65.91 \Rightarrow (62.63,\ 194.45)$. Thus, at least 88.8% of the measurements for Group B will fall between 62.63 and 194.45.

e. The range for Group C is 76.80. $R = 129.00 - 52.20 = 76.80$

The standard deviation for Group C is 20.05. $s = \sqrt{401.94} = 20.05$
Using MINITAB, a histogram of Group C data is:

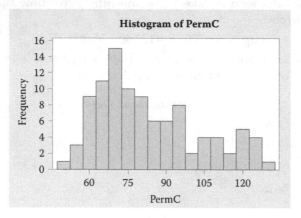

From Exercise 2.33, the mean is 83.07. Because the data are skewed to the left, we will use Chebyshev's rule. At least 8/9 or 88.9% of the observations will fall within 3 standard deviations of the mean. This interval is $\bar{y} \pm 3s \Rightarrow 83.07 \pm 3(20.05) \Rightarrow 83.07 \pm 60.15 \Rightarrow (22.92,\ 143.22)$. Thus, at least 88.8% of the measurements for Group C will fall between 22.92 and 143.22.

f. From all of the analyses, Group B appears to result in higher permeability measurements. The interval of the $\bar{y} \pm 3s$ for Group B is shifted to the right of that for Group C. Also, the histogram for Group B is skewed to the left, while the histogram for Group C is skewed to the right. Most of the observations for Group B are to the right of the observations from Group C.

2.43 a. If we assume that the distributions of scores are mound-shaped, then we know that approximately 95% of all observations are within 2 standard deviations of the mean. For flexed arms, this interval is $\bar{y} \pm 2s \Rightarrow 59 \pm 2(4) \Rightarrow 59 \pm 8 \Rightarrow (51, 67)$. For extended arms, this interval is $\bar{y} \pm 2s \Rightarrow 43 \pm 2(2) \Rightarrow 43 \pm 4 \Rightarrow (39, 47)$. Since these intervals do not overlap, the scores for those with extended arms tend to be smaller than those with flexed arms. Thus, this supports the researchers' theory.

  b. Changing the standard deviations: The interval for flexed arms is $\bar{y} \pm 2s \Rightarrow 59 \pm 2(10) \Rightarrow 59 \pm 20 \Rightarrow (39, 79)$. The interval for extended arms is $\bar{y} \pm 2s \Rightarrow 43 \pm 2(15) \Rightarrow 43 \pm 30 \Rightarrow (13, 73)$. Since these intervals significantly overlap, there is no evidence to support the researchers' theory.

2.45 a. For the private wells, $\bar{y} = 1.00$ and $s = 0.950$. Assuming that the distribution is approximately mound-shaped, approximately 95% of the observations will be within 2 standard deviations of the mean. This interval is $\bar{y} \pm 2s \Rightarrow 1.00 \pm 2(0.95) \Rightarrow 1.00 \pm 1.90 \Rightarrow (-0.90, 2.90)$.

  b. For the public wells, $\bar{y} = 4.56$ and $s = 10.39$. Assuming that the distribution is approximately mound-shaped, approximately 95% of the observations will be within 2 standard deviations of the mean. This interval is $\bar{y} \pm 2s \Rightarrow 4.56 \pm 2(10.39) \Rightarrow 4.56 \pm 20.78 \Rightarrow (-16.22, 25.34)$.

2.47 Using MINITAB, the descriptive statistics are:

Descriptive Statistics: Strength

```
Variable   N   Mean   StDev  Minimum  Median  Maximum
Strength  10  234.74  9.91   215.70   234.55  248.80
```

The mean and standard deviation are $\bar{y} = 234.74$ and $s = 9.91$. Assuming that the data are approximately mound-shaped and symmetric, the interval of the mean plus or minus two standard deviations is $\bar{y} \pm 2s \Rightarrow 234.74 \pm 2(9.91) \Rightarrow 234.74 \pm 19.82 \Rightarrow (214.92, 254.56)$. Approximately 95% of all the observations will be between 214.92 and 254.56.

2.49 a. From the histogram, the approximate 30th percentile would be 10%.
  b. From the histogram, the approximate 95th percentile would be 90%.

2.51 a. Using the Empirical Rule, the 84th percentile would correspond to 1 standard deviation above the mean. Thus, the 84th percentile would be $\$126,417 + \$15,000 = \$141,417$.

  b. Using the Empirical Rule, the 2.5[th] percentile would correspond to 2 standard deviations below the mean. Thus, the 2.5[th] percentile would be $\$126,417 - 2(\$15,000) = \$126,417 - \$30,000 = \$96,417$.

  c. $z = \dfrac{y - \mu}{\sigma} = \dfrac{\$100,000 - \$126,417}{\$15,000} = -1.76$

2.53 a. In the text, it was given that the mean number of sags is 353 and the standard deviation of the number of sags is 30. The z-score for 400 sags is $z = \frac{y - \mu}{\sigma} = \frac{400 - 353}{30} = 1.57$. A value of 400 sags is 1.57 standard deviations above the mean number of sags.

  b. In the text, it was given that the mean number of swells is 184 and the standard deviation of the number of swells is 25. The z-score for 100 swells is $z = \frac{y - \mu}{\sigma} = \frac{100 - 184}{25} = -3.36$. A value of 100 swells is 3.36 standard deviations below the mean number of swells. This would be a very unusual value to observe.

2.55   Using MINITAB, the descriptive statistics are:

Descriptive Statistics: Score

```
Variable   N    Mean   StDev
Score     186  94.441  5.335
```

a. The z-score for the *Nautilus Explorer*'s score of 74 is $z = \frac{y-\bar{y}}{s} = \frac{74-94.441}{5.335} = -3.83$. The *Nautilus Explorer*'s score of 74 is 3.83 standard deviations below the mean sanitation score.

b. The z-score for the *Rotterdam*'s score of 86 is $z = \frac{y-\bar{y}}{s} = \frac{86-94.441}{5.335} = -1.58$. The *Rotterdam*'s score of 86 is 1.58 standard deviations below the mean sanitation score.

2.57   Using MINITAB, the descriptive statistics are:

Descriptive Statistics: ZETA without, ZETA with GYPSUM

```
Variable           N     Mean   StDev
ZETA without      50   -52.070  2.721
ZETA with GYPSUM  50   -10.958  1.559
```

a. The z-score for a zeta potential measurement for solutions prepared without calcium/gypsum of −9.0 is $z = \frac{y-\bar{y}}{s} = \frac{-9.0-(-52.07)}{2.721} = 15.83$

b. The z-score for a zeta potential measurement for solutions prepared with calcium/gypsum of −9.0 is $z = \frac{y-\bar{y}}{s} = \frac{-9.0-(-10.958)}{1.559} = 1.26$.

c. The solution prepared with calcium/gypsum is more likely to have a zeta potential measurement of -9.0 because the z-score of 1.26 is reasonable. The z-score of 15.83 is highly unlikely.

2.59 a. The median is $m = 170$. Half of the clinkers had barium content less than or equal to 170 mg/kg and half of the clinkers had barium content greater than or equal to 170 mg/kg.

b. $Q_L = 115$. 25% of the clinkers had barium content less than or equal to 115 mg/kg and 75% of the clinkers had barium content greater than or equal to 115 mg/kg.

c. $Q_U = 260$. 75% of the clinkers had barium content less than or equal to 260 mg/kg and 25% of the clinkers had barium content greater than or equal to 260 mg/kg.

d. $IQR = Q_U - Q_L = 260 - 115 = 145$.

e. Lower inner fence $= Q_L - 1.5(IQR) = 115 - 1.5(145) = -102.5$.

Upper inner fence $= Q_U + 1.5(IQR) = 260 + 1.5(145) = 477.5$.

f. Because no clinkers had barium content levels beyond the inner fences, there is no evidence of outliers.

2.61 a. The z-score for 400 sags is $z = \frac{y-\mu}{\sigma} = \frac{400-353}{30} = 1.57$. We would not consider 400 sags to be unusual because the z-score is less than 2.

b. The z-score for 100 swells is $z = \frac{y-\mu}{\sigma} = \frac{100-184}{25} = -3.36$. We would consider 100 swells to be unusual because it is more than 3 standard deviations fro the mean.

2.63 The z-score for 1.80% is $z = \frac{y-\mu}{\sigma} = \frac{1.80-2.00}{0.08} = -2.50$. A reading of 1.80% zinc phosphide is 2.5 standard deviations below the mean value. This would be a suspect outlier. There is some evidence to indicate that there is too little zinc phosphide in the day's production.

2.65 Using MINITAB, boxplots for the three groups are:

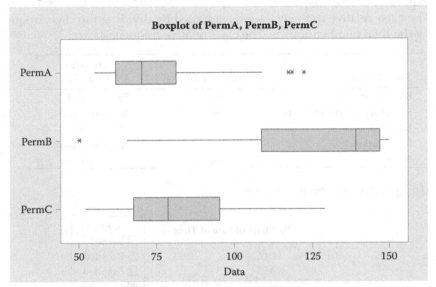

a. There are 3 observations beyond the inner fences for Group A that are suspect outliers. They have values 117.3, 118.5, and 122.4.
b. There is 1 observation beyond the inner fences for Group B that is a suspect outlier. The value is 50.4.
c. There are no observations beyond the inner fences for Group C. There are no suspect outliers for Group C.

2.67 a. By using the cumulative number of barrels collected per day, it looks like BP was collecting more barrels of oil on each successive day, when they were collecting only about an average of 1500 barrels each day.
b. Using MINITAB, the bar chart of the data is:

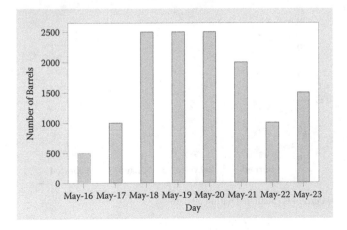

From the graph, we see that the amount of oil collected per day increased from May 16 to May 18, then remained constant for three days, then decreased for two days before increasing again on May 23.

2.69 a. The variable measured for each scrapped tire is the fate of the tire.

b. There are 4 classes or categories: burned for fuel, recycled into new products, exported, or land disposed.

c. The class relative frequencies are computed by dividing the class frequencies by the total number of tires. The class relative frequencies are:

Fate of Tires	Frequency (millions)	Relative Frequency
Burned for Fuel	155	155/300 = 0.517
Recycled into new products	96	96/300 = 0.320
Exported	7	7/300 = 0.023
Land disposed	42	42/300 = 0.140
Totals	300	1.000

d. Using MINITAB, the pie chart is:

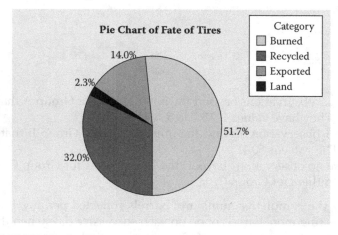

e. Using MINITAB, the Pareto chart is:

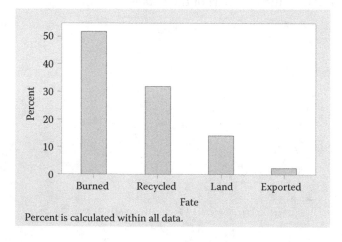

Over half of all scrapped tires are burned for fuel. About a third of scrapped tires are recycled. Only a very small percentage of scrapped tires are exported.

2.71   Using MINITAB, the Pareto diagram is:

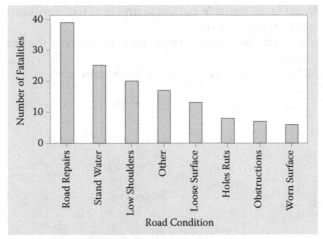

Most of the fatalities are due to road repairs, standing water, and low shoulders. Very few fatalities are due to worn road surface, obstructions without warning, and holes and ruts.

2.73   Using MINITAB, the descriptive statistics are:

Descriptive Statistics: ROUGH

```
Variable   N    Mean    StDev
ROUGH      20   1.881   0.524
```

We know that approximately 95% of all observations will be within 2 standard deviations of the mean. This interval is $\bar{y} \pm 2s \Rightarrow 1.881 \pm 2(0.524) \Rightarrow 1.881 \pm 1.048 \Rightarrow (0.833, 2.929)$

2.75   A driver-head-injury rating of 408 has a z-score of $z = \frac{y - \bar{y}}{s_y} = \frac{408 - 603.7}{185.4} = -1.06$. A head-injury rating of 408 is a little over one standard deviation below the mean. This is not an unusual rating.

2.77   Using MINITAB, the bar graph is:

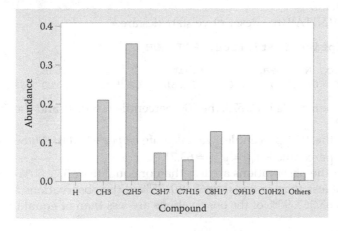

The red dye component with the highest abundance is C_2H_5 with 35.4%. The next highest red dye component is CH_3 with 21.0%. The three components with the least abundance are $C_{10}H_{21}$ (2.5%), H (2.1%), and Others (1.9%).

2.79 a. The population is all possible bulk specimens of Chilean lumpy iron ore in a 35,325-long-ton shipload of ore.

 b. Answers may vary. One possible objective is to estimate the percentage of iron ore in the shipment.

 c. Using MINITAB, the relative frequency histogram is:

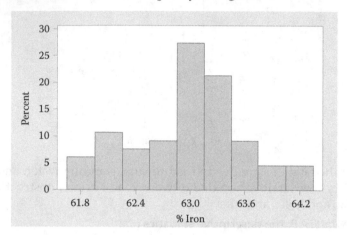

 d. Using MINITAB, the descriptive statistics are:

Descriptive Statistics: PCTIRON

Variable	N	Mean	StDev	Minimum	Q1	Median	Q3	Maximum
PCTIRON	66	62.963	0.609	61.680	62.573	63.010	63.362	64.340

$$\bar{y} = 62.963 \text{ and } s = 0.609$$

 e. $\bar{y} \pm 2s \Rightarrow 62.963 \pm 2(0.609) \Rightarrow 62.963 \pm 1.218 \Rightarrow (61.745, 64.181)$
 64 of the 66 observations or 96.97% of the observations fall in this interval. This does not agree with the Empirical Rule. The Empirical Rule states that approximately 95% of the observations will fall within 2 standard deviations of the mean.

 f. Using MINITAB, the descriptive statistics are:

Descriptive Statistics: PCTIRON

Variable	N	Mean	StDev	Minimum	Q1	Median	Q3	Maximum
PCTIRON	66	62.963	0.609	61.680	62.573	63.010	63.362	64.340

The 25th percentile is 62.573, the 50th percentile is 63.010, and the 75th percentile is 63.362.
To find the 90th percentile, we calculate $i = p(n+1)/100 = 90(66+1)/100 = 60.3$. The 90th percentile is $y_{(i)} = y_{(60)} = 63.71$.
   25% of the observations are less than or equal to 62.573. 50% of the observations are less than or equal to 63.010. 75% of the observations are less than or equal to 63.362. 90% of the observations are less than or equal to 63.71.

2.81   Using MINITAB, the descriptive statistics are:

Descriptive Statistics: SCRAMS

```
Variable  N    Mean   StDev  Minimum    Q1   Median    Q3   Maximum
SCRAMS    56   4.036  3.027    0.000  2.000  3.000  5.750  13.000
```

To find the 95th percentile, we calculate $i = p(n+1)/100 = 95(56+1)/100 = 54.15$. The 95th percentile is the 54th observation, $y_{(i)} = y_{(54)} = 9$. Thus, 95% of all observations are less than or equal to 9. A value of 11 would not be very likely. A score of 11 would be $z = \frac{y - \bar{y}}{s} = \frac{11 - 4.036}{3.027} = 2.30$ standard deviations above the mean. A score greater than 2 standard deviations from the mean is not very likely.

2.83 a. The number of seabirds present and the lengths of the transects are quantitative. Whether the transect was in an oiled area or not is qualitative.

   b. The experimental unit is a transect.

   c. Using MINITAB, the pie chart is:

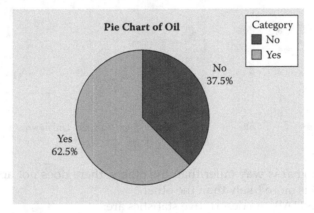

   d. Using MINITAB, a scatterplot is:

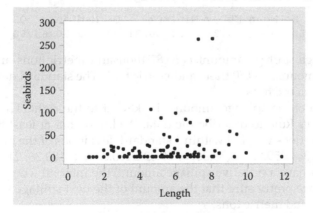

   e. From the output, the means for the two groups are similar as are the medians and standard deviations. It appears that the distributions of seabird densities are similar for transects in oiled and unoiled areas.

f. The data appear to be skewed, so we will use Chebyshev's Rule. At least 75% of the observations will fall within 2 standard deviations of the mean. For unoiled transects, this interval is $\bar{y} \pm 2s \Rightarrow 3.27 \pm 2(6.70) \Rightarrow 3.27 \pm 13.40 \Rightarrow (-10.13, 16.67)$. Since a density cannot be negative, the interval should be $(0, 16.67)$.

g. The data appear to be skewed, so we will use Chebyshev's Rule. At least 75% of the observations will fall within 2 standard deviations of the mean. For oiled transects, this interval is $\bar{y} \pm 2s \Rightarrow 3.495 \pm 2(5.968) \Rightarrow 3.495 \pm 11.936 \Rightarrow (-8.441, 15431)$ Since a density cannot be negative, the interval should be $(0, 15.431)$.

h. It appears that unoiled transects is more likely to have a seabird density of 16 because 16 falls in the interval in part f, but not in part g.

2.85 a. Using MINITAB, a bar chart is:

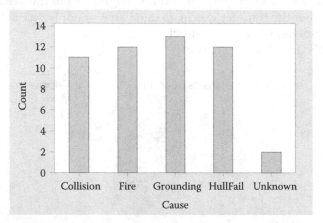

Because no bar is way taller than the others, there does not appear to be one cause that is more likely than the others.

b. Using MINITAB, the descriptive statistics are:

Descriptive Statistics: Spillage

Variable	N	Mean	StDev	Minimum	Q1	Median	Q3	Maximum
Spillage	50	59.82	53.36	21.00	31.00	39.50	63.50	257.00

The average spillage amount is 59.82 thousand metric tons and the median spillage amount is 39.50 thousand metric tons. The standard deviation is 53.36 thousand metric tons.

The graph of the spillage amounts is skewed to the right. Thus, we will use Chebyshev's Rule to describe the data. We know that at least 8/9 or 88.9% of the observations will fall within 3 standard deviations of the mean. This interval is $\bar{y} \pm 3s \Rightarrow 59.82 \pm 3(53.36) \Rightarrow 59.82 \pm 160.08 \Rightarrow (-100.26, 219.90)$. Because we cannot have a negative spillage amount, the interval would be $(0, 219.90)$. Thus, we are pretty sure that the amount of the next spillage will be less than 219.9 thousand metric tons.

2.87 a. The figure portrays quantitative data because diameters are measured using numbers.

  b. A frequency histogram is used to display the data.

  c. There are about 80 observations between 1.0025 and 1.0035 and about 63 observations between 1.0035 and 1.0045. Thus, between 1.0025 and 1.0045, we have about 143 observations. This proportion is $143/500 = 0.286$.

  d. Yes. The shape of the distribution is almost mound-shaped, except for the interval from 0.9995 and 1.0005 and the interval from 0.9985 and 0.9995. The number of observations in the interval 0.9995 and 1.0005 is bigger than what would be expected and the number of observations in the interval 0.9985 and 0.9995 is smaller than what would be expected.

3

Probability

3.1 a. The sample points of the study are: legs only, wheels only, both legs and wheels, and neither legs nor wheels.

b. Reasonable probabilities would be the relative frequencies:

$$P(\text{legs only}) = \frac{63}{106} = 0.5943 \quad P(\text{wheels only}) = \frac{20}{106} = 0.1887$$

$$P(\text{both}) = \frac{8}{106} = 0.0755 \quad P(\text{neither}) = \frac{15}{106} = 0.1415$$

c. $P(\text{wheels}) = P(\text{wheels only}) + P(\text{both}) = 0.1887 + 0.0755 = 0.2642$

d. $P(\text{legs}) = P(\text{legs only}) + P(\text{both}) = 0.5943 + 0.0755 = 0.6698$

3.3 It is more likely that a sound picked up by the acoustical equipment would be a sound from a passing ship than a whale scream because the probability of the sound being from a passing ship is 0.14 while the probability of the sound being from a whale scream is only 0.03.

3.5 a. There are 5 simple events:
A: {Incident occurs in school laboratory}
B: {Incident occurs in transit}
C: {Incident occurs in chemical plant}
D: {Incident occurs in nonchemical plant}
E: {Incident occurs in other}

b. Reasonable probabilities would correspond to the percent of incidents:

$$P(A) = 0.06 \quad P(B) = 0.26 \quad P(C) = 0.21 \quad P(D) = 0.35 \quad P(E) = 0.12$$

c. $P(\text{Incident occurs in school laboratory}) = P(A) = 0.06$

3.7 a. The simple events are:
A: {Private/Bedrock/BelowLimit}
B: {Private/Bedrock/Detect}
C: {Private/Unconsolidated/BelowLimit}
D: {Private/Unconsolidated/Detect}
E: {Public/Bedrock/BelowLimit}
F: {Public/Bedrock/Detect}
G: {Public/Unconsolidated/BelowLimit}
H: {Public/Unconsolidated/Detect}

b. $P(A) = \frac{81}{223} = 0.3632 \quad P(B) = \frac{22}{223} = 0.0987 \quad P(C) = \frac{0}{223} = 0.0000 \quad P(D) = \frac{0}{223} = 0.0000$

$P(E) = \frac{57}{223} = 0.2556 \quad P(F) = \frac{41}{223} = 0.1839 \quad P(G) = \frac{15}{223} = 0.0673 \quad P(H) = \frac{7}{223} = 0.0314$

c. $P(\text{Detect}) = P(B) + P(D) + P(F) + P(H) = 0.0987 + 0.0000 + 0.1839 + 0.0314 = 0.3140$

3.9   a. $P(\text{Beech tree in East Central Europe damaged by fungi}) = \dfrac{49}{188} = 0.2606$

b.   The sample points are:

A: {Area damaged is trunk}
B: {Area damaged is leaf}
C: {Area damaged is branch}

$P(A) = 0.85 \quad P(B) = 0.10 \quad P(C) = 0.05$

3.11 a.   The different purchase order cards possible are

A,Red,1 A,Red,2 A,Red,3 A,Black,1 A,Black,2 A,Black,3
B,Red,1 B,Red,2 B,Red,3 B,Black,1 B,Black,2 B,Black,3

b.   No. If the demand for black TVs is higher than the demand for red TVs, then the probabilities associated with the simple events involving black TVs will be higher than the probabilities associated with the simple events involving red TVs.

3.13 Define the following events:

A: {Incident occurs in school laboratory}
B: {Incident occurs in transit}
C: {Incident occurs in chemical plant}
D: {Incident occurs in nonchemical plant}
E: {Incident occurs in other}

a. $P(C \cup D) = P(C) + P(D) = 0.21 + 0.35 = 0.56$
b. $P(A^c) = 1 - P(A) = 1 - 0.06 = 0.94$

3.15 Define the following events:

A: {Color code 0}
B: {Color code 5}
C: {Color code b}
D: {Color code c}
E: {Model 2}

a. $P(C) = \dfrac{46}{170} = 0.2706$

b. $P(B \text{ or } A) = P(B \cup A) = \dfrac{85}{170} + \dfrac{35}{170} = 0.7059$

c. $P(E \text{ and } A) = P(E \cap A) = \dfrac{15}{170} = 0.0882$

3.17 Define the following events:

A: {Pass inspection with fecal contamination}
B: {Pass inspection without fecal contamination}

a.   The simple events will be:

AAAAA AAAAB AAABA AABAA ABAAA BAAAA
AAABB AABAB ABAAB BAAAB AABBA ABABA
BAABA ABBAA BABAA BBAAA AABBB ABABB
ABBAB ABBBA BAABB BABAB BABBA BBAAB
BBABA BBBAA ABBBB BABBB BBABB BBBAB
BBBBA BBBBB

b. If each of the simple events is equally likely, then each has a probability of 1/32.

P(At least one of the five chickens passes with fecal contamination)

$= 1 - P$(None of the chickens passes inspection with fecal contamination)

$= 1 - P(BBBBB) = 1 - \dfrac{1}{32} = 0.9688$

c. In Exercise 3.8, it says that approximately 1 in 100 slaughtered chickens passes inspection with fecal contamination. We would surely expect that more than 1 in 100 slaughtered chickens passes inspection without fecal contamination. Thus, it is much more likely that event BBBBB occurs than event AAAAA.

3.19 a. The simple events are:

  A: {Caisson, Active} B: {Caisson, Inactive}
  C: {Well Protector, Active} D: {Well Protector, Inactive}
  E: {Fixed Platform, Active} F: {Fixed Platform, Inactive}

  b. Reasonable probabilities would be the relative frequencies for each:

$P(A) = \dfrac{503}{(2,175 + 1,225)} = \dfrac{503}{3,400} = 0.1479 \qquad P(B) = \dfrac{598}{3,400} = 0.1759$

$P(C) = \dfrac{225}{3,400} = 0.0662 \qquad\qquad\qquad P(D) = \dfrac{177}{3,400} = 0.0521$

$P(E) = \dfrac{1,447}{3,400} = 0.4256 \qquad\qquad\qquad P(F) = \dfrac{450}{3,400} = 0.1324$

  c. Define the following events:

  G: {Well is active} H: {Well is caisson}
  J: {Well is well protector} K: {Well is fixed platform}

$P(G) = P(A) + P(C) + P(E) = 0.1479 + 0.0662 + 0.4256 = 0.6397$

  d. $P(J) = P(C) + P(D) = 0.0662 + 0.0521 = 0.1183$

  e. $P(G^C \cap H) = P(B) = 0.1759$

  f. $P(G^C \cup K) = P(B) + P(C) + P(D) + P(F) = 0.1759 + 0.0662 + 0.0521 + 0.1324 = 0.4266$

  g. $P(H^C) = 1 - P(H) = 1 - [P(A) + P(B)] = 1 - [0.1479 + 0.1759] = 0.6762$

3.21 For each drilling location, there are 2 possible outcomes – dry well or oil gusher. If 6 wells are drilled, then the total number of possible outcomes or simple events is $2 \times 2 \times 2 \times 2 \times 2 \times 2 = 64$. If we let D represent a dry well and G represent oil gusher, then a few examples of the 64 simple events are: {DDDDDD}, {DDDGGG}, {DGDGDG}, etc.

  If each simple event is equally likely, then each simple event has a probability of $1/64 = 0.0156$. Thus, the probability of at least one oil gusher

$= 1 - P$(No oil gushers) $= 1 - P(DDDDDD) = 1 - \dfrac{1}{64} = 0.9844.$

3.23   From Exercises 3.1 and 3.12, the simple events and their corresponding probabilities are:

A: {Robots have legs only}
B: {Robots have wheels only}
C: {Robots have legs and wheels}
D: {Robots have neither legs nor wheels}

$$P(A) = 0.5943 \quad P(B) = 0.1887 \quad P(C) = 0.0755 \quad P(D) = 0.1415$$

Define the following events:
L: {Robot has legs} W: {Robot has wheels}
$$P(L) = P(A \cup C) = P(A) + P(C) = 0.5943 + 0.0755 = 0.6698$$
$$P(W) = P(B \cup C) = P(B) + P(C) = 0.1887 + 0.0755 = 0.2642$$
$$P(L \cap W) = P(C) = 0.0755$$

Thus, $P(L \mid W) = \dfrac{P(L \cap W)}{P(W)} = \dfrac{0.0755}{0.2642} = 0.2858$

3.25 $P(A \mid B) = 0$

3.27   From Exercise 3.19, we defined the events:

G: {Well is active} H: {Well is caisson}
J: {Well is well protector} K: {Well is fixed platform}

a. $P(G \mid K) = \dfrac{P(G \cap K)}{P(K)} = \dfrac{1,447/3,400}{1,897/3,400} = 0.7628$

b. $P(J \mid G^C) = \dfrac{P(J \cap G^C)}{P(G^C)} = \dfrac{177/3,400}{1,225/3,400} = 0.1445$

3.29   From Exercise 3.15,

A : {Color code 0} B: {Color code 5} C: {Color code b} D: {Color code c}
E: {Model 2}

a. $P(B \mid E) = \dfrac{50}{75} = 0.6667$

b. $P(E^C \mid B \cup A) = \dfrac{(20 + 35)}{(35 + 85)} = 0.4583$

3.31   Define the following events:

S: {System has high selectivity}
F: {System has high fidelity}
From the Exercise, $P(S) = 0.72$, $P(F) = 0.59$, and $P(S \cap F) = 0.33$.

Thus, $P(S \mid F) = \dfrac{P(S \cap F)}{P(F)} = \dfrac{0.33}{0.59} = 0.5593$.

3.33   Define the following events:
          K: {Lab mouse responds to kaitomone}
          A: {Lab mouse responds to Mups A}
          B: {Lab mouse responds to Mups B}

$$P(A \mid K) = \frac{P(A \cap K)}{P(K)} = \frac{(0.025 + 0.19)}{(0.165 + 0.025 + 0.19 + 0.025)} = \frac{0.215}{0.405} = 0.5309$$

$$P(B \mid K) = \frac{P(B \cap K)}{P(K)} = \frac{(0.025 + 0.19)}{(0.165 + 0.025 + 0.19 + 0.025)} = \frac{0.215}{0.405} = 0.5309$$

3.35   Define the following events:
          A: {Firefighter has no SOP for detecting/monitoring hydrogen cyanide}
          B: {Firefighter has no SOP for detecting/monitoring carbon monoxide}
          From the Exercise, $P(A) = 0.80$, $P(B) = 0.49$, and $P(A \cup B) = 0.94$.
          We know that $P(A \cup B) = P(A) + P(B) - P(A \cap B)$, so $P(A \cap B) = P(A) + P(B) - P(A \cup B)$
          Thus, $P(A \cap B) = P(A) + P(B) - P(A \cup B) = 0.80 + 0.49 - 0.94 = 0.35$

3.37   Define the following events:
          A: {Electrical switching device can monitor the quality of the power running through the device}
          B: {Device is wired to monitor the quality of the power running through the device}
          From the Exercise, $P(A) = 0.90$ and $P(B^C \mid A) = 0.90$. Also,
          $P(B \mid A) = 1 - P(B^C \mid A) = 1 - 0.90 = 0.10$.
          $P(A \cap B) = P(B \mid A)P(A) = 0.10(0.90) = 0.09$

3.39   Define the following events:
          A: {System A sounds alarm}
          B: {System B sounds alarm}
          I: {Intruder}
          N: {No intruder}
     a. $P(A \mid I) = 0.90 \qquad P(B \mid I) = 0.95 \qquad P(A \mid N) = 0.20 \qquad P(B \mid N) = 0.10$
     b.   We are given that the systems work independently. Therefore,
          $P(A \cap B \mid I) = P(A \mid I) \times P(B \mid I) = 0.90(0.95) = 0.855$
     c. $P(A \cap B \mid N) = P(A \mid N) \times P(B \mid N) = 0.20(0.10) = 0.02$
     d. $P(A \cup B \mid I) = P(A \mid I) + P(B \mid I) - P(A \cap B \mid I) = 0.90 + 0.95 - 0.855 = 0.995$

3.41   Define the following events:
          A_t: {Report is OK at time period t}
          A_{t+1}: {Report is OK at time period $t + 1$}
          A_{t+2}: {Report is OK at time period $t + 2$}
          From the Exercise, $P(A_{t+1} \mid A_t) = 0.2$, $P(A_{t+1} \mid A_t^C) = 0.55$. Thus, $P(A_{t+1}^C \mid A_t) = 0.8$
          and $P(A_{t+1}^C \mid A_t^C) = 0.45$

We know that the report was OK at time t. Thus, the possible outcomes for times $t+1$ and $t+2$ are:

$t+1$	$t+2$	probability
A	A	$0.2(0.2) = 0.04$
A	A^C	$0.2(0.8) = 0.16$
A^C	A	$0.8(0.55) = 0.44$
A^C	A^C	$0.8(0.45) = 0.36$

Thus, the probability of an "OK" report in two consecutive time periods $t+1$ and $t+2$ given an "OK" report in time t is 0.04.

3.43   Define the following events:
A: {Seed carries single spikelet}
B: {Seed carries paired spikelet}
C: {Seed produces single spikelet}
D: {Seed produces paired spikelet}
From the Exercise, $P(A) = 0.40$, $P(B) = 0.60$, $P(C \mid A) = 0.29$, $P(D \mid A) = 0.71$, $P(C \mid B) = 0.26$, and $P(D \mid B) = 0.74$.

a. $P(A \cap C) = P(C \mid A)P(A) = 0.29(0.40) = 0.116$

b. $P(D) = P(D \mid A)P(A) + P(D \mid B)P(B) = 0.71(0.40) + 0.74(0.60) = 0.284 + 0.444 = 0.728$

3.45   Define the following events:
A: {Receive erroneous ciphertext}
B: {Error in restoring plaintext}
From the Exercise, $P(A) = \beta$, $P(B \mid A) = 0.5$, and $P(B \mid A^C) = \alpha\beta$.
Thus, $P(B) = P(B \mid A)P(A) + P(B \mid A^C)P(A^C) = 0.5(\beta) + \alpha\beta(1 - \beta) = (0.5 + \alpha - \alpha\beta)\beta$

3.47   a.   The simple events are $D \cap A$, $D \cap A^C$, $D^C \cap A^C$. P_1 is defined as the probability that the effect occurs when factor A is present or not. Thus, $P_1 = P(D)$. Also, P_0 is defined as the probability that the effect occurs when factor A is not present. Thus, $P_0 = P(D \cap A^C)$.
We know that
$$P(D) = P(D \cap A) + P(D \cap A^C) = P_1 \Rightarrow P(D \cap A) + P_0 = P_1 \Rightarrow P(D \cap A) = P_1 - P_0$$
Thus, $P(A \mid D) = \dfrac{P(D \cap A)}{P(D)} = \dfrac{P_1 - P_0}{P_1}$

b.   The simple events are $D \cap A \cap B^C$, $D \cap A^C \cap B$. P_1 is defined as the probability that the effect occurs when factor A is present or $P_1 = P(D \cap A \cap B^C)$ and P_2 is defined as the probability that the effect occurs when factor B is present or $P_2 = P(D \cap A^C \cap B)$. Thus, $P(D) = P_1 + P_2$.
Then $P(A \mid D) = \dfrac{P(D \cap A \cap B^C)}{P(D)} = \dfrac{P_1}{P_1 + P_2}$ and $P(B \mid D) = \dfrac{P(D \cap A^C \cap B)}{P(D)} = \dfrac{P_2}{P_1 + P_2}$

c.   The simple events are $DC\,D^C \cap A^C \cap B^C$, $D \cap A \cap B^C$, $D \cap A^C \cap B$, $D \cap A \cap B$. P_1 is defined as the probability that the effect occurs when factor A is present or $P_1 = P(D \cap A \cap B^C) + P(D \cap A \cap B) = P(D \cap A)$. P2 is defined as the probability that the effect occurs when factor B is present or $P_2 = P(D \cap A^C \cap B) +$

$P(D \cap A \cap B) = P(D \cap B)$. Since the factors A and B affect D independently, $P(D)$
$= P(D \cap A) + P(D \cap B) - P(D \cap A)P(D \cap B) = P_1 + P_2 - P_1 P_2$

Then $P(A|D) = \dfrac{P(D \cap A)}{P(D)} = \dfrac{P_1}{P_1 + P_2 + P_1 P_2}$ and $P(B|D) = \dfrac{P(D \cap B)}{P(D)} = \dfrac{P_2}{P_1 + P_2 + P_1 P_2}$

3.49   From the Exercise, we know $P(D) = 0.5$, $P(S) = 0.2$, and $P(T) = 0.3$. Suppose we
define the event J: {Joystick is pointed straight}. In addition, from the Exercise,
we know $P(J|D) = 0.3$, $P(J|S) = 0.4$, and $P(J|T) = 0.05$.
Then

$$P(D|J) = \frac{P(J|D)P(D)}{P(J|D)P(D) + P(J|S)P(S) + P(J|T)P(T)}$$

$$= \frac{0.3(0.5)}{0.3(0.5) + 0.4(0.2) + 0.05(0.3)} = \frac{0.15}{0.15 + 0.08 + 0.015} = \frac{0.15}{0.245} = 0.6122$$

$$P(S|J) = \frac{P(S|D)P(D)}{P(J|D)P(D) + P(J|S)P(S) + P(J|T)P(T)}$$

$$= \frac{0.4(0.2)}{0.3(0.5) + 0.4(0.2) + 0.05(0.3)} = \frac{0.08}{0.245} = 0.3265$$

$$P(T|J) = \frac{P(T|D)P(D)}{P(J|D)P(D) + P(J|S)P(S) + P(J|T)P(T)}$$

$$= \frac{0.05(0.3)}{0.3(0.5) + 0.4(0.2) + 0.05(0.3)} = \frac{0.015}{0.245} = 0.0612$$

Thus, if the wheelchair user points the joystick straight, his most likely desti-
nation is the door because it has the highest probability.

3.51   Define the following events:
A: {Matched pair is correctly identified}
B: {Similar Distractor pair is correctly identified}
C: {Non-similar Distractor pair is correctly identified}
D: {Participant is an expert}
From the Exercise, $P(A|D) = 0.9212$, $P(A|D^C) = 0.7455$, $P(D) = 0.5$
We want to find $P(D|A^C)$ and $P(D^C|A^C)$.
From above, we know $P(A^C|D) = 1 - P(A|D) = 1 - 0.9212 = 0.0788$
and $P(A^C|D^C) = 1 - P(A|D^C) = 1 - 0.7455 = 0.2545$. Also,
$P(D^C) = 1 - P(D) = 1 - 0.5 = 0.5$.

$$P(D|A^C) = \frac{P(A^C|D)P(D)}{P(A^C|D)P(D) + P(A^C|D^C)P(D^C)} = \frac{0.0788(0.5)}{0.0788(0.5) + 0.2545(0.5)}$$

$$= \frac{0.0394}{0.0394 + 0.12725} = \frac{0.0394}{0.16665} = 0.2364$$

$$P(D^C|A^C) = \frac{P(A^C|D^C)P(D^C)}{P(A^C|D)P(D) + P(A^C|D^C)P(D^C)} = \frac{0.2545(0.5)}{0.0788(0.5) + 0.2545(0.5)}$$

$$= \frac{0.12725}{0.0394 + 0.12725} = \frac{0.12725}{0.16665} = 0.7636$$

Thus, if the participant fails to identify the match, the participant is more likely to be a novice than an expert because the probability of being a novice given the participant fails to identify the match is 0.7636 which is greater than the probability of being an expert given the participant fails to identify the match which is 0.2364.

3.53   From the Exercise, we know $P(E_1) = 0.3$, $P(E_2) = 0.2$, $P(E_3) = 0.5$ and

$$P(error) = P(error \mid E_1)P(E_1) + P(error \mid E_2)P(E_2) + P(error \mid E_3)P(E_3)$$

$$= 0.01(0.3) + 0.03(0.2) + 0.02(0.5) = 0.019$$

a. $P(E_1 \mid error) = \dfrac{P(error \mid E_1)P(E_1)}{P(error)} = \dfrac{0.01(0.3)}{0.019} = \dfrac{0.003}{0.019} = 0.1579$

b. $P(E_2 \mid error) = \dfrac{P(error \mid E_2)P(E_2)}{P(error)} = \dfrac{0.03(0.2)}{0.019} = \dfrac{0.006}{0.019} = 0.3158$

c. $P(E_3 \mid error) = \dfrac{P(error \mid E_3)P(E_3)}{P(error)} = \dfrac{0.02(0.5)}{0.019} = \dfrac{0.01}{0.019} = 0.5263$

d.   Based on the probabilities above, Engineer 3 is most likely responsible for making the serious error because Engineer has the highest probability given a serious error.

3.55   Define the following events:
D: {Sample is dolomite}
S: {Sample is shale}
G: {Gamma ray reading exceeds 60 API units}
From the Exercise we know:

$$P(D) = \frac{476}{771} = 0.6174, \ P(S) = \frac{295}{771} = 0.3826, \ P(G \mid S)$$

$$= \frac{280}{295} = 0.9492, \ P(G \mid D) = \frac{34}{476} = 0.0714$$

We want to mine if the area is abundant in dolomite.

$$P(D \mid G) = \frac{P(G \mid D)P(D)}{P(G \mid D)P(D) + P(G \mid S)P(S)}$$

$$= \frac{\dfrac{34}{476}\left(\dfrac{476}{771}\right)}{\dfrac{34}{476}\left(\dfrac{476}{771}\right) + \dfrac{280}{295}\left(\dfrac{295}{771}\right)} = \frac{\dfrac{34}{771}}{\dfrac{34 + 280}{771}} = \frac{34}{314} = 0.1083$$

Since this probability is small, the area should not be mined.

3.57   a.   If $\frac{P(T \mid E)}{P(T^C \mid E)} < 1$, then $P(T \mid E) < P(T^C \mid E)$. Thus, the probability that more than two bullets were used given the evidence used by the HSCA is greater than the probability that two bullets were used given the evidence used by the HSCA

b. Using Bayes' Theorem, we know

$$P(T \mid E) = \frac{P(E \mid T)P(T)}{P(E \mid T)P(T) + P(E \mid T^C)P(T^C)} \text{ and}$$

$$P(T^C \mid E) = \frac{P(E \mid T^C)P(T^C)}{P(E \mid T)P(T) + P(E \mid T^C)P(T^C)}$$

From the first equation, we get $P(E \mid T)P(T) + P(E \mid T^C)P(T^C) = \dfrac{P(E \mid T)P(T)}{P(T \mid E)}$

From the second equation, we get $P(E \mid T)P(T) + P(E \mid T^C)P(T^C) = \dfrac{P(E \mid T^C)P(T^C)}{P(T^C \mid E)}$

The left sides of both of these equations are the same, so

$$\frac{P(E \mid T)P(T)}{P(T \mid E)} = \frac{P(E \mid T^C)P(T^C)}{P(T^C \mid E)} \Rightarrow \frac{P(T \mid E)}{P(T^C \mid E)} = \frac{P(E \mid T)P(T)}{P(E \mid T^C)P(T^C)}$$

3.59 a. There will be a total of $2 \times 3 \times 3 = 18$ maintenance organization alternatives.
b. The probability that a randomly selected alternative is feasible is P(feasible) = $P(\text{feasible}) = \frac{4}{18} = 0.2222$.

3.61 a. There are 4 color partitions of the partition 3. There are $4 \times 3 = 12$ color partitions of the partition $2 + 1$. There are $4 \times 3 \times 2 = 24$ color partitions of the partition $1 + 1 + 1$. Thus, the total number of color partitions of the number 3 is $4 + 12 + 24 = 40$.
b. There are 7 partitions of the number 5: 5, $4 + 1$, $3 + 2$, $3 + 1 + 1$, $2 + 2 + 1$, $2 + 1 + 1 + 1$, $1 + 1 + 1 + 1 + 1$.
There are 4 color partitions of the partition 5. There are $4 \times 3 = 12$ color partitions of the partition $4 + 1$. There are $4 \times 3 = 12$ color partitions of the partition $3 + 2$. There are $4 \times 3 \times 2 = 24$ color partitions of the partition $3 + 1 + 1$. There are $4 \times 3 \times 2 = 24$ color partitions of the partition $2 + 2 + 1$. There are $4 \times 3 \times 2 \times 1 = 24$ color partitions of the partition $2 + 1 + 1 + 1$. There are no color partitions of the partition $1 + 1 + 1 + 1 + 1$. Thus, the total number of color partitions of the number 5 is $4 + 12 + 12 + +24 + 24 + 24 = 100$.

3.63 a. There are a total of $4 \times 4 = 16$ metal-support combinations possible.
b. There are a total of $P_4^4 = \frac{4!}{(4-4)!} = 4 \cdot 3 \cdot 2 \cdot 1 = 24$ orderings of the four supports.

3.65 a. The total number of different responses is $2 \times 3 \times 3 = 18$.
b. The total number of parameter-part combinations is $2 \times 3 = 6$. The number of different rankings of these six combinations is $P_3^6 = \frac{6!}{(6-3)!} = 6 \cdot 5 \cdot 4 = 120$.

3.67 a. The total number of conditions possible is $2 \times 2 \times 6 \times 7 = 168$.
b. The total number of ways one could select 8 combinations from the 168 is

$$\binom{168}{8} = \frac{168!}{8!(168-8)!}$$

The total number of ways one could select 8 combinations from the 168, where one is the one that can detect the flaw is

$$\binom{1}{1}\binom{167}{7} = \frac{167!}{7!(167-7)!} = \frac{167!}{7!(160)!}.$$

Thus, the probability that the experiment conducted in the study will detect the system flaw is

$$\frac{\binom{1}{1}\binom{167}{7}}{\binom{168}{8}} = \frac{\dfrac{167!}{7!(167-7)!}}{\dfrac{168!}{8!(168-8)!}} = \frac{\dfrac{167!}{7!(160)!}}{\dfrac{168!}{8!(160)!}} = \frac{8}{168} = 0.0476$$

c.   There are 2 conditions in the 8 conditions in the experiment that used steel with a 0.25-inch drill size at a speed of 2,500 rpm. Thus, the probability that the experiment conducted will detect the system flaw is $\frac{2}{8} = 0.25$.

3.69   First, we find the number of combinations of 16 task force members taken 4 at a time or $\binom{16}{4} = \dfrac{16!}{4!(16-4)!} = \dfrac{16 \cdot 15 \cdot 14 \cdots 1}{4 \cdot 3 \cdot 2 \cdot 1 \cdot 12 \cdot 11 \cdots 1} = 1,820$. Therefore, for the first facility, there are 1820 possible assignments. Once the first facility is filled, there are only 12 task force members left to fill the remaining 3 facilities. Therefore, once the first facility has been filled, there are 12 task force members from which to pick 4 to fill the second facility. There are a total of $\binom{12}{4} = \dfrac{12!}{4!(12-4)!} = \dfrac{12 \cdot 11 \cdot 10 \cdots 1}{4 \cdot 3 \cdot 2 \cdot 1 \cdot 8 \cdot 9 \cdots 1} = 495$ ways to pick 4 task members for the second facility. Once the second facility is filled, there are only 8 task force members from which to pick 4 to fill the third facility. There are a total of $\binom{8}{4} = \dfrac{8!}{4!(8-4)!} = \dfrac{8 \cdot 7 \cdot 6 \cdots 1}{4 \cdot 3 \cdot 2 \cdot 1 \cdot 4 \cdot 3 \cdot 2 \cdot 1} = 70$ ways to fill the third facility. Once the third facility is filled, there is only one way to fill the fourth facility. Therefore, the total number of ways to fill the 4 facilities is $1,820 \times 495 \times 70 = 63,063,000$.

3.71   a.   Let A = dealer draws blackjack. In order to draw blackjack, the dealer has to get 1 ace and 1 card that can be a 10, jack, queen, or king. There are a total of 16 cards that have a value of 10. The total number of ways to get 1 ace and 1 card worth 10 points is $\binom{4}{1}\binom{16}{1} = \dfrac{4!}{1!(4-1)!} \dfrac{16!}{1!(16-1)!} = \dfrac{4 \cdot 3 \cdot 2 \cdot 1}{1 \cdot 3 \cdot 2 \cdot 1} \cdot \dfrac{16 \cdot 15 \cdot 14 \cdots 1}{1 \cdot 15 \cdot 14 \cdots 1} = 64$. The total number of ways to draw 2 cards from 52 is $\binom{52}{2} = \dfrac{52!}{2!(52-2)!} = \dfrac{52 \cdot 51 \cdot 50 \cdots 1}{2 \cdot 1 \cdot 50 \cdot 49 \cdots 1} = 1,326$. Thus, $P(A) = \dfrac{64}{1,326} = 0.0483$.

b.   In order for a player to win with blackjack, the player must get blackjack and the dealer cannot get blackjack. Let B = player draws blackjack. Using our notation, then we want to find $P(A^C \cap B)$. We need to find the probability that the player wins with blackjack, $P(B)$, and the probability that the dealer does not draw blackjack given the player does, $P(A^C | B)$. Then the probability that a player wins with blackjack is $P(A^C | B)P(B)$.

The probability that a player draws blackjack is the same as the probability that the dealer draws blackjack or $P(B) = \dfrac{64}{1326} = 0.0483$.

There are 5 scenarios where the dealer will not draw blackjack give the player does. First, the dealer could draw an ace and not a card worth 10. Next, the dealer could draw a card worth 10 and not an ace. Third, the dealer could draw two aces. Fourth, the dealer could draw two cards worth 10 each. Finally, the dealer could draw two cards that are not aces and not worth 10.

The number of ways the dealer could draw an ace and not a card worth 10 given the player draws blackjack is $\dbinom{3}{1}\dbinom{36}{1} = \dfrac{3!}{1!(3-1)!} \cdot \dfrac{32!}{1!(32-1)!} = 96$.

The number of ways the dealer could draw a card worth 10 and not an ace given the player draws blackjack is $\dbinom{15}{1}\dbinom{32}{1} = \dfrac{15!}{1!(15-1)!} \cdot \dfrac{32!}{1!(32-1)!} = 480$.

The number of ways the dealer could draw two aces given the player draws blackjack is $\dbinom{3}{2} = \dfrac{3!}{2!(3-1)!} = 3$.

The number of ways the dealer could draw two cards worth 10 given the player draws blackjack is $\dbinom{15}{2} = \dfrac{15!}{2!(15-1)!} = 105$.

The number of ways the dealer could draw two cards that are not aces and not worth 10 given the player draws blackjack is $\dbinom{32}{2} = \dfrac{32!}{2!(32-2)!} = 496$.

The total number of ways a dealer could draw two cards given the player draws blackjack is $\dbinom{50}{2} = \dfrac{50!}{2!(50-2)!} = 1225$.

Thus, $P(A^C \mid B) = \dfrac{96 + 480 + 3 + 105 + 496}{1225} = \dfrac{1180}{1225}$.

Finally, $P(A^C \cap B) = P(A^C \mid B)P(B) = \dfrac{1180}{1225} \cdot \dfrac{64}{1326} = 0.0465$.

3.73   Define $D = \{CD \text{ is defective}\}$. From the problem, $P(D) = \dfrac{1}{100} = 0.01$. Then $P(D^C) = 1 - P(D) = 1 - 0.01 = 0.99$

$P(\text{at least 1 defective in next 4}) = 1 - P(0 \text{ defectives in next 4})$

$= 1 - P(D^C)P(D^C)P(D^C)P(D^C)$

$= 1 - 0.99(0.99)(0.99)(0.99) = 1 - 0.9606 = 0.0394$

Since this probability is so small, we would infer that the claimed defective rate is too small.

3.75 a. Define H = {antiaircraft shells are fired and strike within 30-foot radius of target}. From the Exercise, $P(H) = 0.45$. Thus, $P(H^C) = 1 - P(H) = 1 - 0.45 = 0.55$. $P(3 \text{ miss}) = P(H^C)P(H^C)P(H^C) = 0.55(0.55)(0.55) = 0.1664$. Since this probability is not small, it would not be reasonable to conclude that in battle conditions, p differs from 0.45.

  b. $P(10 \text{ miss}) = P(H^C)P(H^C)\cdots P(H^C) = P(H^C)^{10} = (0.55)^{10} = 0.0025$. Since this probability is so small, it would be reasonable to conclude that in battle conditions, p differs from 0.45.

3.77 a. The simple events are: Basic browns. True-blue greens, Greenback greens, Sprouts, and Grousers.

  b. The probabilities would be equal to the proportions:
      $P(\text{Basic browns}) = 0.28$, $P(\text{True-blue greens}) = 0.11$, $P(\text{Greenback greens}) = 0.11$, $P(\text{Sprouts}) = 0.26$, $P(\text{Grousers}) = 0.24$

  c. $P(\text{Basic brown or grouser}) = P(\text{Basic browns}) + P(\text{Grousers}) = 0.28 + 0.24 = 0.52$

  d. $P(\text{Supports environmentalism}) = P(\text{True-blue greens}) + P(\text{Greenback greens})$ $+ P(\text{Sprouts}) = 0.11 + 0.11 + 0.26 = 0.48$

3.79   Define the following events:
      A: {Intruder is detected}
      B: {Day is cloudy}
      C: {Day is snowy}

  a. $P(A \mid B) = \dfrac{P(A \cap B)}{P(B)} = \dfrac{228/692}{234/692} = 0.9744$

  b. $P(C \mid A^C) = \dfrac{P(C \cap A^C)}{P(A^C)} = \dfrac{3/692}{25/692} = 0.12$

3.81 a. $P(\text{dragonfly species inhabits a dragonfly hotspot}) = 0.92$

  b. $P(\text{butterfly species inhabits a bird hotspot}) = 1.00$

  c. We know that $P(\text{bird species inhabits a butterfly hotspot}) = 0.70$. Thus, 70% of all British bird species inhabit a butterfly hotspot. Since 70% of the bird species are present, this makes the butterfly hotspots also bird hotspots.

3.83 a. There are 30 lots, 6 tablets from each lot, and 8 measurements for each tablet. The total number of measurements is $30 \times 6 \times 8 = 1,440$.

  b. We will average the readings for the 6 tablets at each time period within each lot. The total number of averages is $30 \times 8 = 240$.

3.85   Let A = {critical-item failure}. From the Exercise, $P(A) = \dfrac{1}{60}$. We know
      $P(A^C) = 1 - P(A) = 1 - \dfrac{1}{60} = \dfrac{59}{60}$.

a. P(at least 1 of the next 8 shuttle flights results in critical-item failure)

     $= 1 - P$(none of the next 8 shuttle flights results in critical-item failure)

$$= 1 - P(A^C)P(A^C)P(A^C)P(A^C)P(A^C)P(A^C)P(A^C)P(A^C)$$

$$= 1 - \left(\frac{59}{60}\right)^8 = 0.1258$$

b. P(at least 1 of the next 40 shuttle flights results in critical-item failure)

     $= 1 - P$(none of the next 40 shuttle flights results in critical-item failure)

$$= 1 - \underbrace{P(A^C)P(A^C)\cdots P(A^C)}_{40}$$

$$= 1 - \left(\frac{59}{60}\right)^{40} = 0.4895$$

3.87   Define the following events:
     A: {First machine breaks down}
     B: {Second machine breaks down}
     From the Exercise, $P(A) = 0.20$, $P(B \mid A) = 0.30$.

a. $P(A \cap B) = P(B \mid A)P(A) = 0.30(0.20) = 0.06$

b.   The probability that the system is working is the probability that the first machine is working plus the probability that the second machine is working and the first is not.

     Reliability $= P(A^C) + P(A \cap B^C) = 1 - P(A) + P(B^C \mid A)P(A)$

$$= 1 - 0.20 + (1 - 0.30)(0.2) = 0.8 + 0.14 = 0.94$$

3.89   Define the following events:
     A: {System shuts down}
     B: {System suffers hardware failure}
     C: {System suffers software failure}
     D: {System suffers power failure}
     From the Exercise,
     $P(A \mid B) = 0.73$, $P(A \mid C) = 0.12$, $P(A \mid D) = 0.88$, $P(B) = 0.01$, $P(C) = 0.05$, $P(D) = 0.02$
     $P(A) = P(A \mid B)P(B) + P(A \mid C)P(C) + P(A \mid D)P(D)$

$$= 0.73(0.01) + 0.12(0.05) + 0.88(0.02) = 0.0309$$

$$P(B \mid A) = \frac{P(B \cap A)}{P(A)} = \frac{P(A \mid B)P(B)}{P(A)} = \frac{0.73(0.01)}{0.0309} = 0.2362$$

$$P(C \mid A) = \frac{P(C \cap A)}{P(A)} = \frac{P(A \mid C)P(C)}{P(A)} = \frac{0.12(0.05)}{0.0309} = 0.1942$$

$$P(D \mid A) = \frac{P(D \cap A)}{P(A)} = \frac{P(A \mid D)P(D)}{P(A)} = \frac{0.88(0.02)}{0.0309} = 0.5696$$

3.91   We need to find how many ways to choose 2 suppliers from 5, 3 suppliers from 5, 4 suppliers from 5 and 5 suppliers from 5:
The total number of choices is

$$\binom{5}{2}+\binom{5}{3}+\binom{5}{4}+\binom{5}{5}=\frac{5!}{2!(5-2)!}+\frac{5!}{3!(5-3)!}+\frac{5!}{4!(5-4)!}+\frac{5!}{5!(5-5)!}$$

$$=\frac{5\cdot4\cdot3\cdot2\cdot1}{2\cdot1\cdot3\cdot2\cdot1}+\frac{5\cdot4\cdot3\cdot2\cdot1}{3\cdot2\cdot1\cdot2\cdot1}+\frac{5\cdot4\cdot3\cdot2\cdot1}{4\cdot3\cdot2\cdot1\cdot1}+\frac{5\cdot4\cdot3\cdot2\cdot1}{5\cdot4\cdot3\cdot2\cdot1\cdot1}=10+10+5+1=26$$

3.93 a.   Starting at a randomly selected point in Table 1, select consecutive sets of 7 digits, assuming that any combination of 7 digits represent phone numbers.

   b.   Answers will vary. Suppose we start in row 10, column 9 and go down the column. The 10 7-digit numbers are:
0815817, 9010631, 5218020, 3001508, 0151126, 9773585, 4944253, 0118865, 7158585, 2349564

   c.   Answers will vary. If the first three digits have to be 373, then we just need to generate five 4-digit numbers. Suppose we start in row 52, column 4 and go down the column. The 5 7-digit numbers will be:
3739196, 3738763, 3734932, 3731442, 3739827

3.95 a.   Since no company can be awarded more than 1 DOT contract, the total number of ways the bids can be awarded is $P_3^5 = \frac{5!}{(5-3)!} = \frac{5\cdot4\cdot3\cdot2\cdot1}{2\cdot1} = 60.$

   b.   The number of ways 2 additional companies can be awarded a contract is $P_2^4 = \frac{4!}{(4-2)!} = \frac{4\cdot3\cdot2\cdot1}{2\cdot1} = 12.$ Company 2 can be awarded any of the 3 jobs, so the total number of ways company 2 can be awarded a job is $3\times12 = 36.$
The probability that company 2 is awarded a bid is $\frac{36}{60} = 0.60.$

   c.   The number of ways companies 4 and 5 can be positioned is $P_1^3 = \frac{3!}{(3-2)!} = \frac{3\cdot2\cdot1}{1} = 6.$ There are 3 additional companies that can fill the 3rd bid. Thus, the total number of ways companies 4 and 5 can be awarded a contract is $6\times3 = 18$
The probability that companies 4 and 5 are awarded a bid is $\frac{18}{60} = 0.30.$

3.97 a. The total number of ways to draw 5 cards from 52 is

$$\binom{52}{5} = \frac{52!}{5!(52-5)!} = \frac{52 \cdot 51 \cdot 50 \cdots 1}{5 \cdot 4 \cdot 3 \cdot 2 \cdot 1 \cdot 47 \cdot 46 \cdot 45 \cdots 1} = 2,598,960.$$

The total number of ways to draw 5 cards of the same suit is

$$\binom{13}{5} = \frac{13!}{5!(13-5)!} = \frac{13 \cdot 12 \cdot 11 \cdots 1}{5 \cdot 4 \cdot 3 \cdot 2 \cdot 1 \cdot 8 \cdot 7 \cdot 6 \cdots 1} = 1,287.$$

Since there are 4 suits, the total number of ways to draw a flush is $4 \times 1,287 = 5,148$.

The probability of drawing a flush is $P(A) = \dfrac{5,148}{2,598,960} = 0.0019808.$

b. To get a straight, one only needs 5 consecutive cards of any suit. Thus, the number of ways to get 5 cards in a sequence of any suit is $4 \times 4 \times 4 \times 4 \times 4 = 1,024$. Now, there are 10 starting positions for a straight – Ace, 2, 3, 4, 5, 6, 7, 8, 9, or 10. Thus, the total number of ways to draw a straight is $1,024 \times 10 = 10,240$.

The probability of drawing a straight is $P(B) = \dfrac{10,240}{2,598,960} = 0.00394.$

c. To get a straight flush, the cards must all be of the same suit and in sequence. For one suit, the number of ways to get 5 cards in sequence is $1 \times 1 \times 1 \times 1 \times 1 = 1$. Again, there are 10 starting positions for a straight. Thus, the total number of ways to get a straight in one suit is $10 \times 1 = 10$. There are four suits, so the total number of ways to get a straight flush is $10 \times 4 = 40$.

The probability of drawing a straight flush is $P(A \cap B) = \dfrac{40}{2,598,960} = 0.00001539.$

3.99 Define D = {Chip produces incorrect result when dividing 2 numbers}.

Thus, $P(D) = \dfrac{1}{9,000,000,000} = 1.111111 \times 10^{-10}$

Now, a heavy SAS user will make about 1 billion divisions in a short period of time. The probability of at least 1 incorrect division in 1 billion divisions is

$P(\text{at least 1 incorrect division}) = 1 - P(0 \text{ incorrect divisions})$

$$= 1 - \left(\frac{1}{9,000,000,000} \right)^{1,000,000,000} = 1 - 0 = 1$$

So, for heavy SAS users, the probability of at least one incorrect division in 1 billion divisions is 1. The flawed chip will create definite problems for heavy SAS users.

4

Discrete Random Variables

4.1   a.   The number of solar energy cells manufactured in China is a countable number: 0, 1, 2, 3, 4 or 5.

   b. $p(0) = \dfrac{5!(0.35)^0(0.65)^{5-0}}{0!(5-0)!} = \dfrac{5 \cdot 4 \cdot 3 \cdot 2 \cdot 1}{1 \cdot 5 \cdot 4 \cdot 3 \cdot 2 \cdot 1}(0.35)^0(0.65)^5 = 0.1160$

   $p(1) = \dfrac{5!(0.35)^1(0.65)^{5-1}}{1!(5-1)!} = \dfrac{5 \cdot 4 \cdot 3 \cdot 2 \cdot 1}{1 \cdot 4 \cdot 3 \cdot 2 \cdot 1}(0.35)^1(0.65)^4 = 0.3124$

   $p(2) = \dfrac{5!(0.35)^2(0.65)^{5-2}}{2!(5-2)!} = \dfrac{5 \cdot 4 \cdot 3 \cdot 2 \cdot 1}{2 \cdot 1 \cdot 3 \cdot 2 \cdot 1}(0.35)^2(0.65)^3 = 0.3364$

   $p(3) = \dfrac{5!(0.35)^3(0.65)^{5-3}}{3!(5-3)!} = \dfrac{5 \cdot 4 \cdot 3 \cdot 2 \cdot 1}{3 \cdot 2 \cdot 1 \cdot 2 \cdot 1}(0.35)^3(0.65)^2 = 0.1811$

   $p(4) = \dfrac{5!(0.35)^4(0.65)^{5-4}}{4!(5-4)!} = \dfrac{5 \cdot 4 \cdot 3 \cdot 2 \cdot 1}{4 \cdot 3 \cdot 2 \cdot 1 \cdot 1}(0.35)^4(0.65)^1 = 0.0488$

   $p(5) = \dfrac{5!(0.35)^5(0.65)^{5-5}}{5!(5-5)!} = \dfrac{5 \cdot 4 \cdot 3 \cdot 2 \cdot 1}{5 \cdot 4 \cdot 3 \cdot 2 \cdot 1 \cdot 1}(0.35)^5(0.65)^0 = 0.0053$

   c.   The properties of a discrete probability distribution are:

   i. $0 \le p(y) \le 1$

   ii. $\sum_{\text{all } y} p(y) = 1$

   All of the probabilities above are between 0 and 1. The sum of these probabilities is $0.1160 + 0.3121 + 0.3361 + 0.1811 + 0.0488 + 0.0053 = 1$. Thus both properties are met.

   d. $P(Y \ge 4) = p(4) + p(5) = 0.0488 + 0.0053 = 0.0541$

4.3   a.   To find the probabilities, divide each of the frequencies by the total number of observations, which is 100. The probability distribution of Y is:

y	1	2	3	4
$p(y)$	0.40	0.54	0.02	0.04

   b. $P(Y \ge 3) = p(3) + p(4) = 0.02 + 0.04 = 0.06$

4.5   a. $p(1) = (0.23)(0.77)^{1-1} = 0.23$ The probability that the first cartridge sampled is contaminated is 0.23.

     b. $p(5) = (0.23)(0.77)^{5-1} = 0.081$ The probability that the 5th cartridge sampled is the first one that is contaminated is 0.081.

     c. $P(Y \geq 2) = 1 - p(1) = 1 - (0.23)(0.77)^{1-1} = 1 - 0.23 = 0.77$ The probability that the first cartridge sampled is not contaminated is 0.77.

4.7   a.   There are two links in the system. Therefore, the number of free links can be 0, 1, or 2.

     b.   The probability that link $A \leftrightarrow B$ is free is the probability that both A and B are free. We are given that the probability that any point in the system is free is 0.5 and that the points are independent. Therefore, P (both A and B are free) $= 0.5(0.5) = 0.25$. Also, P(at least one of A or B is not free) $= 1 - 0.25 = 0.75$. Similarly, P(both B and C are free) $= 0.5(0.5) = 0.25$ and P(at least one of B or C is not free) $= 1 - 0.25 = 0.75$

$P(Y = 2) = P(A \leftrightarrow B$ is free and $B \leftrightarrow C$ is free)

     $= P$(both A and B are free) $\cdot P$(both B and C are free) $= 0.25(0.25) = 0.0625$

$P(Y = 0) = P(A \leftrightarrow B$ is not free and $B \leftrightarrow C$ is not free)

     $= P$(at least one of A or B is not free) $\cdot P$(at least one of B or C is not free)

     $= 0.75(0.75) = 0.5625$

$P(Y = 1) = P(A \leftrightarrow B$ is free and $B \leftrightarrow C$ is not free or $A \leftrightarrow B$ is not free
         and $B \leftrightarrow C$ is free)
     $= P(A$ and B are free) $\cdot P$(at least one of B or C is not free)
      $+ P$(at least one of A or B is not free) $\cdot P(B$ and C are free)
     $= 0.25(0.75) + 0.75(0.25) = 0.3750$

The probability distribution of Y is:

y	0	1	2
$p(y)$	0.5625	0.3750	0.0625

4.9   a.   Section 1: All probabilities are between 0 and 1.

$$\sum_{\text{all } y} p(y) = 0.05 + 0.25 + 0.25 + 0.45 = 1.00$$

      Section 2: All probabilities are between 0 and 1.

$$\sum_{\text{all } y} p(y) = 0.10 + 0.25 + 0.35 + 0.30 = 1.00$$

      Section 3: All probabilities are between 0 and 1.

$$\sum_{\text{all } y} p(y) = 0.15 + 0.20 + 0.30 + 0.35 = 1.00$$

b. The variable y can take on the values 30, 40, 50, and 60. Because the freeway is equally divided into three sections, each section has probability of 1/3.

$P(Y = 30) = P(Y = 30 \mid \text{section 1})P(\text{section 1}) + P(Y = 30 \mid \text{section 2})P(\text{section 2})$

$\qquad + P(Y = 30 \mid \text{section 3})P(\text{section 3})$

$\qquad = (0.05)\left(\dfrac{1}{3}\right) + (0.10)\left(\dfrac{1}{3}\right) + (0.15)\left(\dfrac{1}{3}\right) = 0.10$

$P(Y = 40) = P(Y = 40 \mid \text{section 1})P(\text{section 1}) + P(Y = 40 \mid \text{section 2})P(\text{section 2})$

$\qquad + P(Y = 40 \mid \text{section 3})P(\text{section 3})$

$\qquad = (0.25)\left(\dfrac{1}{3}\right) + (0.25)\left(\dfrac{1}{3}\right) + (0.20)\left(\dfrac{1}{3}\right) = 0.2333$

$P(Y = 50) = P(Y = 50 \mid \text{section 1})P(\text{section 1}) + P(Y = 50 \mid \text{section 2})P\left(\text{section 2}\right)$

$\qquad + P(Y = 50 \mid \text{section 3})P(\text{section 3})$

$\qquad = (0.25)\left(\dfrac{1}{3}\right) + (0.35)\left(\dfrac{1}{3}\right) + (0.30)\left(\dfrac{1}{3}\right) = 0.3$

$P(Y = 60) = P(Y = 60 \mid \text{section 1})P(\text{section 1}) + P(Y = 60 \mid \text{section 2})P(\text{section 2})$

$\qquad + P(Y = 60 \mid \text{section 3})P(\text{section 3})$

$\qquad = (0.45)\left(\dfrac{1}{3}\right) + (0.30)\left(\dfrac{1}{3}\right) + (0.35)\left(\dfrac{1}{3}\right) = 0.3667$

The probability distribution of y is:

y	30	40	50	60
$p(y)$	0.1000	0.2333	0.3000	0.3667

c. $P(Y \geq 50) = P(Y = 50) + P(Y = 60) = 0.3000 + 0.3667 = 0.6667$

4.11   For this problem, Y can take on values 1, 2, or 3. The probability that the first firing pin tested is defective is $P(Y = 1) = \dfrac{3}{5} = 0.6$.

The probability that the second firing pin tested is the first defective pin is
$P(Y = 2) = \dfrac{2}{5} \cdot \dfrac{3}{4} = 0.3$.

The probability that the third firing pin tested is the first defective pin is
$P(Y = 3) = \dfrac{2}{5} \cdot \dfrac{1}{4} \cdot \dfrac{3}{3} = 0.1$.

The probability distribution for y is:

y	1	2	3
$p(y)$	0.6	0.3	0.1

The graph is:

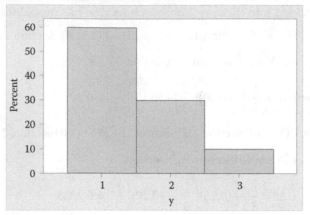

4.13   a. $\mu = E(Y) = \sum yp(y) = 0(0.09) + 1(0.30) + 2(0.37) + 3(0.20) + 4(0.04) = 1.8$

On average, for every 4 homes tested, 1.8 will have dust mite levels that exceed $2\,\mu g/g$.

b. $\sigma^2 = E[(Y - \mu)^2] = \sum (y - \mu)^2 p(y)$

$\quad = (0 - 1.8)^2 (0.09) + (1 - 1.8)^2 (0.30) + (2 - 1.8)^2 (0.37)$

$\quad\quad + (3 - 1.8)^2 (0.20) + (4 - 1.8)^2 (0.04) = 0.98$

$\quad\quad \sigma = \sqrt{\sigma^2} = \sqrt{0.98} = 0.990$

c. $\mu \pm 2\sigma \Rightarrow 1.8 \pm 2(0.99) \Rightarrow 1.8 \pm 1.98 \Rightarrow (-0.18,\ 3.78)$

$P(-0.18 < Y < 3.78) = p(0) + p(1) + p(2) + p(3) = 0.09 + 0.30 + 0.37 + 0.20 = 0.96$

This probability is very close to the Empirical Rule (approximately 0.95) and agrees with Chebyshev's Rule (at least 0.75).

4.15 $P(Y \le 14) = p(13) + p(14) = 0.04 + 0.25 = 0.29$

4.17   a. $\mu = E(Y) = \sum yp(y) = 1(0.10) + 2(0.25) + 3(0.40) + 4(0.15) + 5(0.10) = 2.9$

The average number of training units needed to master the program is 2.9. The median number of training units needed to master the program is 3. At least half of the students need 3 or fewer training units to master the program.

b. $P(Y \le 3) = p(1) + p(2) + p(3) = 0.10 + 0.25 + 0.40 = 0.75$.

If the firm wants to ensure at least 75% of students master the program, then 3 training units must be administered.
$P(Y \le 4) = p(1) + p(2) + p(3) + p(4) = 0.10 + 0.25 + 0.40 + 0.15 = 0.90$. If the firm wants to ensure at least 90% of students master the program, then 4 training units must be administered.

c. $\mu = E(Y) = \sum yp(y) = 1(0.25) + 2(0.35) + 3(0.40) = 2.15$

By changing the training program, the average number of training units necessary to master the program has been reduced from 2.9 to 2.15.
$P(Y \le 3) = p(1) + p(2) + p(3) = 0.25 + 0.35 + 0.40 = 1.00$

By changing the training program, the firm would have to administer 3 training units to ensure both 75% and 90% of the students master the program.

4.19 Let the total cost be $C = \$2,000Y$. From Exercise 4.13, $\mu = E(Y) = 1.8$. Then $\mu = E(C) = E(\$2,000Y) = \$2,000E(Y) = \$2,000(1.8) = \$3,600$.

$$\sigma^2 = E(C^2) - \mu_C^2 = [0(2000)]^2(0.09) + [1(2000)]^2(0.30) + [2(2000)]^2(0.37)$$

$$+ [3(2000)]^2(0.20) + [4(2000)]^2(0.04) - 3,600^2$$

$$= 16,880,000 - 12,960,000 = 3,920,000$$

$$\sigma = \sqrt{\sigma^2} = \sqrt{3,920,000} = 1979.899$$

We would expect most of the observations to fall within 2σ of the mean cost. This range would be

$\mu \pm 2\sigma \Rightarrow \$3,600 \pm 2(\$1,9979.899) \Rightarrow \$3,600 \pm \$3,959.798 \Rightarrow (-\$359.798,$

$\$7,559.798)$ or $(0, \$7,559.798)$.

4.21 From Exercise 4.12, $\mu = 4.655$.

$$\sigma^2 = E(X^2) - \mu^2 = 0^2(0.17) + 1^2(0.10) + 2^2(0.11) + \cdots + 20^2(0.005) - 4.655^2$$

$$= 41.525 - 4.655^2 = 19.856$$

This agrees with Exercise 4.12.

4.23 $$E(c) = \sum_{all\,c} cp(c) = c \sum_{all\,c} p(c) = c(1) = c$$

4.25 $$E[g_1(y) + g_2(y) + \cdots g_k(y)] = \sum_{all\,y} [g_1(y) + g_2(y) + \cdots g_k(y)]p(y)$$

$$= \sum_{all\,y} \left[g_1(y)p(y) + g_2(y)p(y) + \cdots g_k(y)p(y) \right]$$

$$= \sum_{all\,y} g_1(y)p(y) + \sum_{all\,y} g_2(y)p(y) + \cdots + \sum_{all\,y} g_k(y)p(y)$$

$$= E[g_1(y)] + E[g_2(y)] + \cdots + E[g_k(y)]$$

4.27 a. The experiment consists of $n = 100$ trials. Each trial results in an S (lab mouse responds positively to chemically produced cat Mups) or an F (lab mouse does not respond positively to chemically produced cat Mups). The probability of success, p, is 0.40 and $q = 1 - p = 1 - 0.40 = 0.60$. We assume the trials are independent. Therefore, Y = number of lab mice who respond positively to chemically produced Mups in 100 trials and Y has a binomial distribution with $n = 100$ and $p = 0.40$.

   b. $\mu = E(y) = np = 100(0.40) = 40$ In random samples of size 100 lab mice exposed to cat Mups, on average 40 will react positively.

   c. $\sigma^2 = npq = 100(0.40)(0.60) = 24$

   d. Since p is close to 0.5, the distribution of the sample proportion is approximately symmetric. Therefore, most of the observations will

fall within 2 standard deviations of the mean or within the interval
$\mu \pm 2\sigma \Rightarrow 40 \pm 2\sqrt{24} \Rightarrow 40 \pm 9.798 \Rightarrow (30.202, 49.798)$.

4.29   a.   The experiment consists of $n = 5$ trials. Each trial results in an S (water in bottle is tap water) or an F (water in bottle is not tap water). The probability of success, p, is 0.25 and $q = 1 - p = 1 - 0.25 = 0.75$. We assume the trials are independent. Therefore, Y = the number of bottles of water that contain tap water in 5 trials and Y has a binomial distribution with $n = 5$ and $p = 0.25$. The probability distribution for Y is:

$$P(Y = y) = \binom{5}{y}(0.25)^y (0.75)^{5-y}, \ y = 0, 1, 2, 3, 4, 5$$

   b. $P(Y = 2) = \binom{5}{2}(0.25)^2 (0.75)^{5-2} = \dfrac{5!}{2!3!}(0.25)^2 (0.75)^3 = 0.2637$

   c. $P(Y \leq 1) = p(0) + p(1) = \binom{5}{0}(0.25)^0 (0.75)^{5-0} + \binom{5}{1}(0.25)^1 (0.75)^{5-1}$

$$= \frac{5!}{0!5!}(0.25)^0 (0.75)^5 + \frac{5!}{1!4!}(0.25)^1 (0.75)^4 = 0.2373 + 0.3955 = 0.6328$$

4.31   The experiment consists of $n = 150$ trials. Each trial results in an S (packet is detected) or an F (packet is not detected). The probability of success, p, is 0.001 and $q = 1 - p = 1 - 0.001 = 0.999$. We assume the trials are independent. Therefore, Y = the number of packets detected in 150 trials and Y has a binomial distribution with $n = 150$ and $p = 0.001$.

$$P(Y \geq 1) = 1 - P(Y = 0) = 1 - p(0) = 1 - \binom{150}{0}(0.001)^0 (0.999)^{150-0}$$

$$= 1 - 0.8606 = 0.1394$$

4.33   The experiment consists of $n = 5$ trials. Each trial results in an S (chicken passes inspection with fecal contamination) or an F (chicken does not pass inspection with fecal contamination). The probability of success, p, is 0.01 and $q = 1 - p = 1 - 0.01 = 0.99$. We assume the trials are independent. Therefore, Y = the number of chickens that pass inspection with fecal contamination in 5 trials and Y has a binomial distribution with $n = 5$ and $p = 0.01$.

$$P(Y \geq 1) = 1 - P(Y = 0) = 1 - p(0) = 1 - \binom{5}{0}(0.01)^0 (0.99)^{5-0}$$

$$= 1 - 0.9510 = 0.0490$$

4.35   The experiment consists of $n = 4$ trials. Each trial results in an S (bytes differ on the two strings) or an F (bytes on the two strings match). The probability of success, p, is 0.5 and $q = 1 - p = 1 - 0.5 = 0.5$. We assume the trials are independent. Therefore, Y = the number of bytes that differ on the two strings in 4 bytes and Y has a binomial distribution with $n = 4$ and $p = 0.5$.

4.37 a.   The experiment consists of $n = 15$ trials. Each trial results in an S (child develops the neurological disorder) or an F (child develops the neurological disorder). The probability of success, p, is $1/5 = 0.2$ and $q = 1 - p = 1 - 0.2 = 0.8$. We assume the trials are independent. Therefore, Y = the number of children who develop the neurological disorder in 15 trials and Y has a binomial distribution with $n = 15$ and $p = 0.2$.

$$P(Y > 8) = 1 - P(Y \le 8) = 1 - 0.9992 = 0.0008$$

    b.   Now, y has a binomial distribution with $n = 10,000$ and $p = 0.2$.

$$\mu = E(y) = np = 10,000(0.2) = 2,000$$

$$\sigma^2 = npq = 10,000(0.2)(0.8) = 1,600 \qquad \sigma = \sqrt{\sigma^2} = \sqrt{1,600} = 40$$

We would expect most sample of size 10,000 to have the number of children developing the neurological disorder to fall in the interval $\mu \pm 2\sigma \Rightarrow 2,000 \pm 2(40) \Rightarrow 2,000 \pm 80 \Rightarrow (1920, 2,080)$.

Since this interval is entirely below 3,000, it is extremely likely that fewer than 3,000 will have children that develop the disorder.

4.39 $$E[Y(Y-1)] = \sum_{y=0}^{n} y(y-1)p(y) = \sum_{y=0}^{n} y(y-1)\binom{n}{y}p^y q^{n-y} = \sum_{y=0}^{n} y(y-1)\frac{n!}{y!(n-y)!}p^y q^{n-y}$$

$$= \sum_{y=2}^{n} \frac{n!}{(y-2)!(n-y)!}p^y q^{n-y} = n(n-1)p^2 \sum_{y=2}^{n} \frac{(n-2)!}{(y-2)!(n-y)!}p^{y-2} q^{n-y}$$

Now, let $z = y - 2$. Then we can rewrite the above as

$$n(n-1)p^2 \sum_{z=0}^{n-2} \frac{(n-2)!}{(z)!(n-2-z)!}p^z q^{n-2-z} = n(n-1)p^2$$

Now we can rewrite the above as

$$n(n-1)p^2 = np^2(n-1) = np(np-p) = np(np-1+q) = n^2 p^2 - np + npq$$

$$= \mu^2 - \mu + npq = npq + \mu^2 - \mu$$

Thus, $E[Y(Y-1)] = npq + \mu^2 - \mu$

4.41   From Theorem 4.4, we know $\sigma^2 = E(Y^2) - \mu^2$.

From Exercise 4.40, we showed that $E(Y^2) = npq + \mu^2$.

Thus, $\sigma^2 = E(Y^2) - \mu^2 = npq + \mu^2 - \mu^2 = npq$.

4.43 a. This experiment consists of 50 identical trials. There are three possible outcomes on each trial with the probabilities indicated in the table below. Assuming the trials are independent, this is a multinomial experiment with $n = 50$, $k = 3$, $p_1 = 0.25$, $p_2 = 0.10$, and $p_3 = 0.65$.

Subcarriers	Proportion
Pilot	0.25
Null	0.10
Data	0.65

The number of subcarriers we would expect to be pilot subcarriers is
$\mu_1 = E(Y_1) = np_1 = 50(0.25) = 12.5$.
The number of subcarriers we would expect to be null subcarriers is
$\mu_2 = E(Y_2) = np_2 = 50(0.10) = 5$.
The number of subcarriers we would expect to be data subcarriers is
$\mu_3 = E(Y_3) = np_3 = 50(0.65) = 32.5$.

   b. $P(10,10,30) = \dfrac{50!}{10!10!30!}(0.25)^{10}\left(0.10\right)^{10}(0.65)^{30} = 0.0020$

   c. $\sigma_1^2 = np_1q_1 = 50(0.25)(0.75) = 9.375 \qquad \sigma_1 = \sqrt{9.375} = 3.062$

   We would expect most of the observations to fall within 2 standard deviations of the mean. For pilot subcarriers, this interval would be $\mu_1 \pm 2\sigma_1 \Rightarrow 12.5 \pm 2(3.062) \Rightarrow 12.5 \pm 6.124 \Rightarrow (6.376, 18.624)$. Thus, it would be extremely unlikely to observe more than 25 pilot subcarriers.

4.45 a. This experiment consists of 20 identical trials. There are six possible outcomes on each trial with the probabilities indicated in the table below. Assuming the trials are independent, this is a multinomial experiment with $n = 20$, $k = 6$, $p_1 = 0.30$, $p_2 = 0.10$, $p_3 = 0.07$, $p_4 = 0.07$, $p_5 = 0.08$, and $p_6 = 0.38$.

Industry	Proportion
Wood/Paper	0.30
Grain/Foodstuffs	0.10
Metal	0.07
Power Generation	0.07
Plastics/Mining/Textile	0.08
Miscellaneous	0.38

$P(7,5,2,0,1,5) = \dfrac{20!}{7!5!2!0!1!5!}(0.30)^7(0.10)^5(0.07)^2(0.07)^0(0.08)^1(0.38)^5 = 0.000114$

   b. Let $Y_1 =$ number of explosions in the wood/paper industry. Then Y_1 has a binomial distribution with $n = 20$ and $p = 0.3$.
   $P(Y_1 < 3) = P(Y_1 \le 2) = 0.0355$ using Table 2, Appendix B.

4.47 a. This experiment consists of 8 identical trials. There are three possible outcomes on each trial with the probabilities indicated in the table below. Assuming the trials are independent, this is a multinomial experiment with $n = 8$, $k = 3$, $p_1 = 0.65$, $p_2 = 0.15$, and $p_3 = 0.20$.

Orientation	Proportion
Brighter side up	0.65
Darker side up	0.15
Brighter & darker side aligned	0.20

$$P(8,0,0) = \frac{8!}{8!0!0!}(0.65)^8(0.15)^0(0.20)^0 = 0.0319$$

b. $P(4,3,1) = \dfrac{8!}{4!3!1!}(0.65)^4(0.15)^3(0.20)^1 = 0.0337$

c. $\mu_1 = E(Y_1) = np_1 = 8(0.65) = 5.2$

4.49 This experiment consists of n identical trials. There are three possible outcomes on each trial with the probabilities indicated in the table below. Assuming the trials are independent, this is a multinomial experiment with n, $k = 3$, p_1, p_2, and p_3.

# Defects	Proportion
Zero	p_0
One	p_1
More than one	p_2

Because $p_1 + p_2 < 1$, Y_1 and Y_2 are independent.

$$E(C) = E(4Y_1 + Y_2) = E(4Y_1) + E(Y_2) = 4np_1 + np_2 = n(4p_1 + p_2)$$

4.51 $[a + (b+c)]^2 = \begin{pmatrix} 2 \\ 0 \end{pmatrix} a^2 + \begin{pmatrix} 2 \\ 1 \end{pmatrix} a^1(b+c)^1 + \begin{pmatrix} 2 \\ 2 \end{pmatrix}(b+c)^2$

$$= \begin{pmatrix} 2 \\ 0 \end{pmatrix} a^2 + \begin{pmatrix} 2 \\ 1 \end{pmatrix} a^1 b^1 + \begin{pmatrix} 2 \\ 1 \end{pmatrix} a^1 c^1 + \begin{pmatrix} 2 \\ 2 \end{pmatrix}\left[\begin{pmatrix} 2 \\ 0 \end{pmatrix} b^2 + \begin{pmatrix} 2 \\ 1 \end{pmatrix} b^1 c^1 + \begin{pmatrix} 2 \\ 2 \end{pmatrix} c^2 \right]$$

$$= \frac{2!}{0!2!} a^2 + \frac{2!}{1!1!} a^1 b^1 + \frac{2!}{1!1!} a^1 c^1 + \frac{2!}{2!0!}\left[\frac{2!}{0!2!} b^2 + \frac{2!}{1!1!} b^1 c^1 + \frac{2!}{0!2!} c^2 \right]$$

$$= \frac{2!}{2!0!0!} a^2 b^0 c^0 + \frac{2!}{1!1!0!} a^1 b^1 c^0 + \frac{2!}{1!0!1!} a^1 b^0 c^1 + \frac{2!}{0!2!0!} a^0 b^2 c^0$$

$$+ \frac{2!}{0!1!1!} a^0 b^1 c^1 + \frac{2!}{0!0!2!} a^0 b^0 c^2$$

Now, substituting $a = p_1$, $b = p_2$, $c = p_3$ yields:

$$P(2,0,0) + P(1,1,0) + P(1,0,1) + P(0,2,0) + P(0,1,1) + P(0,0,2) = 1$$

4.53 a. Let S = observed slug is *Milax rusticus* and F = observed slug is not *Milax rus-ticus*. If we let Y = the number of tests until the 10th *Milax rusticus* slug is collected, then the probability distribution of Y is a negative binomial and the formula is:

$$p(y) = \binom{y-1}{r-1} p^r q^{y-r} = \binom{y-1}{10-1}(0.2)^{10}(0.8)^{y-10}$$

b. $\mu = E(Y) = \dfrac{r}{p} = \dfrac{10}{0.2} = 50$

On average, we would have to observe 50 slugs before observing 10 *Milax rusticus* slugs.

c. $P(Y = 25) = \binom{25-1}{10-1}(0.2)^{10}(0.8)^{25-10} = \dfrac{24!}{9!15!}(0.2)^{10}(0.8)^{15} = 0.0047$

4.55 a. Let S = lab mice cell responds positively to cat Mups and F = lab mice cell does not respond positively to cat Mups. If we let Y = the number of lab mice cells tested until one responds positively to cat Mups, then the probability distribution of Y is a geometric and the formula is:

$$p(y) = pq^{y-1} = (0.4)(0.6)^{y-1}$$

b. $\mu = E(Y) = \dfrac{1}{p} = \dfrac{1}{0.4} = 2.5$

On average, when testing lab mice cells, it will take 2.5 cells until the first responds positively to cat Mups.

c. $\sigma^2 = \dfrac{q}{p^2} = \dfrac{0.6}{0.4^2} = 3.75$

d. $\sigma = \sqrt{3.75} = 1.936$

Most observations will fall within 2 standard deviations of the mean. This interval would be
$\mu \pm 2\sigma \Rightarrow 2.5 \pm 2(1.936) \Rightarrow 2.5 \pm 3.872 \Rightarrow (-1.372, 6.372)$

4.57 a. $E(Y) = \sum_{all\ y} yp(y) = 1(0.25) + 2(0.20) + 3(0.125) + 4(0.125) + 5(0.09) + 6(0.05)$

$+ 7(0.04) + 8(0.045) + 9(0.04) + 10(0.025) + 11(0.01) + 12(0.0075) + 13(0.005)$

$+ 14(0.000) + 15(0.000)$

$= 3.79$

b. For the geometric distribution, $\mu = E(Y) = \dfrac{1}{p}$. Therefore,

$$\mu = E(Y) = \frac{1}{p} = 3.79 \Rightarrow p = 0.264.$$

c. $P(Y = 7) = pq^{7-1} = (0.264)(0.736)^{7-1} = 0.042$

4.59   Let Y = number of shuttle flights until a "critical item" fails. Then Y is a geometric random variable with $p = \dfrac{1}{63}$.

a. $\mu = E(Y) = \dfrac{1}{p} = \dfrac{1}{1/63} = 63$

b. $\sigma^2 = \dfrac{q}{p^2} = \dfrac{62/63}{(1/63)^2} = 3906$ $\sigma = \sqrt{3906} = 62.498$

c. Approximately 0.95 of the observations will fall within 2 standard deviations f the mean. This interval would be $\mu \pm 2\sigma \Rightarrow 63 \pm 2(62.498) \Rightarrow 63 \pm 124.996 \Rightarrow (-61.996, 187.996) \Rightarrow (0, 188)$.

4.61 a. Let S = drilling location hits oil and F = drilling location does not hit oil. If we let Y = the number of drilling locations until hitting oil, then the probability distribution of Y is a geometric and the formula is:

$$p(y) = pq^{y-1} = (0.3)(0.7)^{y-1}$$

$$P(Y \le 3) = p(1) + p(2) + p(3) = (0.3)(0.7)^{1-1} + (0.3)(0.7)^{2-1} + (0.3)(0.7)^{3-1}$$

$$= 0.3 + 0.21 + 0.147 = 0.657$$

b. $\mu = E(Y) = \dfrac{1}{p} = \dfrac{1}{0.3} = 3.3333$

$$\sigma^2 = \frac{q}{p^2} = \frac{0.7}{(0.3)^2} = 7.7778 \quad \sigma = \sqrt{7.7778} = 2.7889$$

c. No. The value of 10 would be $z = \dfrac{x - \mu}{\sigma} = \dfrac{10 - 3.3333}{2.7889} = 2.39$ standard deviations from the mean. It would be unlikely for observations to be more than 2.39 standard deviations above the mean.

d. Let S = drilling location hits oil and F = drilling location does not hit oil. If we let Y = the number of drilling locations until the second success occurs, then the probability distribution of Y is a negative binomial and the formula is:

$$p(y) = \binom{y-1}{r-1} p^r q^{y-r} = \binom{y-1}{2-1}(0.3)^2(0.7)^{y-2}$$

$P(Y \le 7) = p(2) + p(3) + p(4) + p(5) + p(6) + p(7)$

$$= \binom{2-1}{2-1}(0.3)^2(0.7)^{2-2} + \binom{3-1}{2-1}(0.3)^2(0.7)^{3-2} + \binom{4-1}{2-1}(0.3)^2(0.7)^{4-2} +$$

$$\binom{5-1}{2-1}(0.3)^2(0.7)^{5-2} + \binom{6-1}{2-1}(0.3)^2(0.7)^{6-2} + \binom{7-1}{2-1}(0.3)^2(0.7)^{7-2}$$

$$= \frac{1!}{1!0!}(0.3)^2(0.7)^0 + \frac{2!}{1!1!}(0.3)^2(0.7)^1 + \frac{3!}{1!2!}(0.3)^2(0.7)^2 + \frac{4!}{1!3!}(0.3)^2(0.7)^3$$

$$+ \frac{5!}{1!4!}(0.3)^2(0.7)^4 + \frac{6!}{1!5!}(0.3)^2(0.7)^5$$

$$= 0.09 + 0.126 + 0.1323 + 0.1235 + 0.1080 + 0.0908 = 0.6706$$

4.63 a. This probability distribution should not be approximated by the binomial distribution because the sampling is done without replacement. In addition, $N = 106$ is not large and $n/N = 10/106 = 0.09$ is not less than 0.05.

   b. We are drawing $n = 10$ robots without replacement from a total of $N = 106$, of which $r = 15$ have neither legs nor wheels and $N - r = 106 - 15 = 91$ have either legs or wheels or both. The ratio of $n/N = 10/106 = 0.09$ is greater than 0.05. We let $Y =$ number of robots with neither legs nor wheels observed in $n = 10$ trials.

   c. $\mu = \dfrac{nr}{N} = \dfrac{10(15)}{106} = 1.4151$

   $\sigma^2 = \dfrac{r(N-r)n(N-n)}{N^2(N-1)} = \dfrac{15(106-15)10(106-10)}{106^2(106-1)} = 1.1107$, $\sigma = \sqrt{1.1107} = 1.0539$

   d. $P(X = 2) = \dfrac{\binom{r}{x}\binom{N-r}{n-x}}{\binom{N}{n}} = \dfrac{\binom{15}{2}\binom{106-15}{10-2}}{\binom{106}{10}} = \dfrac{\dfrac{15!}{2!13!} \cdot \dfrac{91!}{8!83!}}{\dfrac{106!}{10!96!}} = 0.2801$

4.65   Let $Y =$ the number of committee members who are from the Department of Engineering Physics. Then Y has a hypergeometric distribution with $N = 10$, $r = 4$, and $n = 3$. The formula for the distribution is

$$p(y) = \frac{\binom{r}{y}\binom{N-r}{n-y}}{\binom{N}{n}} = \frac{\binom{4}{y}\binom{10-4}{3-y}}{\binom{10}{3}}.$$

4.67 Let Y = the number of clean cartridges chosen in 5 trials. Then Y has a hypergeometric distribution with $N = 158$, $r = 158 - 36 = 122$, and $n = 5$.

$$P(Y=5) = \frac{\binom{r}{y}\binom{N-r}{n-y}}{\binom{N}{n}} = \frac{\binom{122}{5}\binom{158-122}{5-5}}{\binom{158}{5}} = \frac{\frac{122!}{5!117!} \cdot \frac{36!}{0!36!}}{\frac{158!}{5!153!}} = 0.2693$$

4.69 a. Let Y = the number of bird species selected that are extinct in 10 trials. Then Y has a hypergeometric distribution with $N = 132$, $r = 38$, and $n = 10$.

$$P(Y=5) = \frac{\binom{r}{y}\binom{N-r}{n-y}}{\binom{N}{n}} = \frac{\binom{38}{5}\binom{132-38}{10-5}}{\binom{132}{10}} = \frac{\frac{38!}{5!33!} \cdot \frac{94!}{5!89!}}{\frac{132!}{10!122!}} = 0.0883$$

b. $$P(Y \le 1) = p(0) + p(0) = \frac{\binom{38}{0}\binom{132-38}{10-0}}{\binom{132}{10}} + \frac{\binom{38}{1}\binom{132-38}{10-1}}{\binom{132}{10}}$$

$$= \frac{\frac{38!}{0!38!} \cdot \frac{94!}{10!84!}}{\frac{132!}{10!122!}} + \frac{\frac{38!}{1!37!} \cdot \frac{94!}{9!85!}}{\frac{132!}{10!122!}} = 0.02897 + 0.12953 = 0.1585$$

4.71 a. Let Y = the number of packets containing genuine cocaine in 4 trials. Then Y has a hypergeometric distribution with $N = 496$, $r = 331$, and $n = 4$.

$$P(Y=4) = \frac{\binom{r}{y}\binom{N-r}{n-y}}{\binom{N}{n}} = \frac{\binom{331}{4}\binom{496-331}{4-4}}{\binom{496}{4}} = \frac{\frac{331!}{4!327!} \cdot \frac{165!}{0!165!}}{\frac{496!}{4!492!}} = 0.1971$$

b. Let Y = the number of packets containing genuine cocaine in 2 trials. Then Y has a hypergeometric distribution with $N = 492$, $r = 327$, and $n = 2$.

$$P(Y=0) = \frac{\binom{r}{y}\binom{N-r}{n-y}}{\binom{N}{n}} = \frac{\binom{327}{0}\binom{492-327}{2-0}}{\binom{492}{2}} = \frac{\frac{327!}{0!327!} \cdot \frac{165!}{2!163!}}{\frac{492!}{2!490!}} = 0.1120$$

c.   Let Y = the number of packets containing genuine cocaine in 2 trials. Then Y has a hypergeometric distribution with $N = 492$, $r = 396$, and $n = 2$.

$$P(Y=0)=\frac{\binom{r}{y}\binom{N-r}{n-y}}{\binom{N}{n}}=\frac{\binom{396}{0}\binom{492-396}{2-0}}{\binom{492}{2}}=\frac{\dfrac{396!}{0!396!}\cdot\dfrac{96!}{2!94!}}{\dfrac{492!}{2!490!}}=0.0378$$

4.73   For this problem, Y has a Poisson distribution with $\lambda = 0.5$.

$$P(Y\geq 3)=1-P(Y\leq 2)=1-p(0)-p(1)-p(2)=1-\frac{0.5^0 e^{-0.5}}{0!}-\frac{0.5^1 e^{-0.5}}{1!}-\frac{0.5^2 e^{-0.5}}{2!}$$

$$=1-0.6065-0.3033-0.0758=0.0144$$

4.75   For this problem, Y has a Poisson distribution with $\lambda = 0.8$.

$$P(Y>0)=1-P(Y=0)=1-\frac{0.8^0 e^{-0.8}}{0!}=1-0.4493=0.5507.$$

We assumed that the probability that a flaw occurs in a given 4-meter length of wire is the same for all 4-meter lengths of wire and the number of flaws that occur in a 4-meter length of wire is independent of the number of flaws that occur in any other 4-meter lengths of wire.

4.77 a.   For this problem, Y has a Poisson distribution with $\lambda = 1.6$.

$$P(Y=0)=\frac{1.6^0 e^{-1.6}}{0!}=0.2019$$

b. $$P(Y=1)=\frac{1.6^1 e^{-1.6}}{1!}=0.3230$$

c. $E(Y)=\mu=\lambda=1.6$

$\sigma=\sqrt{\lambda}=\sqrt{1.6}=1.2649$

4.79 a. $\sigma=\sqrt{\lambda}=\sqrt{4}=2$

b. $P(Y>10)=1-P(Y\leq 10)=1-0.9972=0.0028$
Because the probability of getting a value greater than 10 is so small, it is very unlikely that the value of Y would exceed 10.

c.   We must assume that the probability of one part per million emissions is the same for all million parts and the amount of emissions in one million parts is independent of the amount of emissions for any other one million parts.

4.81 a. $P(Y\leq 20)=p(0)+p(1)+p(2)+\cdots+p(20)$

$$=\frac{18^0 e^{-18}}{0!}+\frac{18^1 e^{-18}}{1!}+\frac{18^2 e^{-18}}{2!}+\cdots+\frac{18^{20} e^{-18}}{20!}=0.7307$$

b. $P(5 \le Y \le 10) = p(5) + p(6) + p(7) + \cdots + p(10)$

$$= \frac{18^5 e^{-18}}{5!} + \frac{18^6 e^{-18}}{6!} + \frac{18^7 e^{-18}}{7!} + \cdots + \frac{18^{10} e^{-18}}{10!} = 0.0303$$

c. $\sigma^2 = \lambda = 18$; $\sigma = \sqrt{\lambda} = \sqrt{18} = 4.2426$

We would expect y to fall within $\mu \pm 2\sigma \Rightarrow 18 \pm 2(4.2426) \Rightarrow 18 \pm 8.4852 \Rightarrow$ (9.51, 26.49)

d. The trend would indicate that the numbers of occurrences are not independent of each other. This cast doubts on the independence characteristic of the Poisson random variable.

4.83 a. N = number of batches arriving in a specific time period and N has a Poisson distribution with $\lambda = 1.1$. $P(Y = 0) = P(N = 0) = \dfrac{1.1^0 e^{-1.1}}{0!} = 0.3329$.

b. $P(Y = 1) = P(X_1 = 1 | N = 1)P(N = 1) = (0.4)\dfrac{1.1^1 e^{-1.1}}{1!} = 0.1465$

c. $P(Y = 2) = P(X_1 = 2 | N = 1)P(N = 1) + P(X_1 + X_2 = 2 | N = 2)P(N = 2)$

$$= (0.6)\frac{1.1^1 e^{-1.1}}{1!} + \binom{2}{2}(0.4)^2(0.6)^{2-2}\frac{1.1^2 e^{-1.1}}{2!} = 0.2197 + 0.0322 = 0.2519$$

d. $P(Y = 3) = P(X_1 + X_2 = 3 | N = 2)P(N = 2) + P(X_1 + X_2 + X_3 = 3 | N = 3)P(N = 3)$

$$= \binom{2}{1}(0.4)(0.6)\frac{1.1^2 e^{-1.1}}{2!} + \binom{3}{3}(0.4)^3(0.6)^{3-3}\frac{1.1^3 e^{-1.1}}{3!} = 0.0967 + 0.0047 = 0.1014$$

4.85 We know that $\sigma^2 = E(Y - \mu)^2 = E(Y^2) - \mu^2$. We know $\mu = E(Y) = \lambda$ and from Exercise 4.84, we know that $E(Y^2) = \lambda^2 + \lambda$.

Thus, $\sigma^2 = E(Y - \mu)^2 = E(Y^2) - \mu^2 = \lambda^2 + \lambda - \lambda^2 = \lambda$.

4.87 $\mu_1 = \mu = \dfrac{dm(t)}{dt}\bigg]_{t=0} = \lambda e^{\lambda(e^t - 1) + t}\bigg]_{t=0} = \lambda e^{\lambda(1-1)+0} = \lambda$

$\mu_2' = \dfrac{d^2 m(t)}{dt^2}\bigg]_{t=0} = \lambda \dfrac{d}{dt}\left(e^{\lambda(e^t-1)+t}\right)\bigg]_{t=0} = \lambda \dfrac{d}{dt}\left(e^{\lambda(e^t-1)}e^t\right)\bigg]_{t=0}$

$= \lambda e^{\lambda(e^t-1)}e^t + \lambda e^t \lambda e^{\lambda(e^t-1)+t}\bigg]_{t=0} = \lambda e^{\lambda(1-1)}e^0 + \lambda^2 e^0 e^{\lambda(1-1)+0} = \lambda + \lambda^2$

$\sigma^2 = E(Y^2) - \mu^2 = \mu_2' - \left(\mu_1'\right)^2 = \lambda + \lambda^2 - \lambda^2 = \lambda$

4.89   The experiment consists of $n = 3$ trials. Each trial results in an S (accident caused by engineering and design) or an F (accident not caused by engineering and design). The probability of success, p, is 0.32 and $q = 1 - p = 1 - 0.32 = 0.68$. Even though we are sampling without replacement, we assume the trials are independent because we are only sampling 3 accidents from all industrial accidents caused by management system failures. Therefore, Y = number of accidents caused by engineering and design in 3 trials and Y has a binomial distribution with $n = 3$ and $p = 0.32$.

The probability distribution of Y is

$$p(y) = \binom{n}{y} p^y q^{n-y} = \binom{3}{y}(0.32)^y(0.68)^{3-y}, \ y = 0,1,2,3$$

$$p(0) = \binom{3}{0}(0.32)^0(0.68)^{3-0} = \frac{3!}{0!3!}0.68^3 = 0.3144$$

$$p(1) = \binom{3}{1}(0.32)^1(0.68)^{3-1} = \frac{3!}{1!2!}(0.32)^1(0.68)^2 = 0.4439$$

$$p(0) = \binom{3}{2}(0.32)^2(0.68)^{3-2} = \frac{3!}{2!1!}(0.32)^2(0.68)^1 = 0.2089$$

$$p(0) = \binom{3}{3}(0.32)^3(0.68)^{3-3} = \frac{3!}{3!0!}(0.32)^3 = 0.0328$$

Using MINITAB, the graph is:

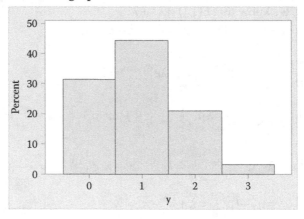

4.91 a. Let Y = number of consumers sampled until the first environmentalist is found. Then Y is a geometric random variable with $p = 0.11 + 0.11 + 0.26 = 0.48$.

$P(Y = 1) = 0.48(0.52)^{1-1} = 0.48$ $P(Y = 2) = 0.48(0.52)^{2-1} = 0.2496$

$P(Y = 3) = 0.48(0.52)^{3-1} = 0.1298$ $P(Y = 4) = 0.48(0.52)^{4-1} = 0.0675$

$P(Y = 5) = 0.48(0.52)^{5-1} = 0.0351$ $P(Y = 6) = 0.48(0.52)^{6-1} = 0.0182$

$P(Y = 7) = 0.48(0.52)^{7-1} = 0.0095$ $P(Y = 8) = 0.48(0.52)^{8-1} = 0.0049$

$P(Y = 9) = 0.48(0.52)^{9-1} = 0.0026$ $P(Y = 10) = 0.48(0.52)^{10-1} = 0.0013$

$P(Y = 11) = 0.48(0.52)^{11-1} = 0.0007$ $P(Y = 12) = 0.48(0.52)^{12-1} = 0.0004$

$P(Y = 13) = 0.48(0.52)^{13-1} = 0.0002$ $P(Y = 14) = 0.48(0.52)^{14-1} = 0.0001$

$P(Y = 15) = 0.48(0.52)^{15-1} = 0.0001$ $P(Y = 16) = 0.48(0.52)^{16-1} = 0.0000$

The probability distribution is:

y	$p(y)$	y	$p(y)$
1	0.4800	9	0.0026
2	0.2496	10	0.0013
3	0.1298	11	0.0007
4	0.0675	12	0.0004
5	0.0351	13	0.0002
6	0.0182	14	0.0001
7	0.0095	15	0.0001
8	0.0049	16	0.0000

b. $p(y) = pq^{y-1} = (0.48)(0.52)^{y-1}$, $y = 1, 2, \ldots$

c. $\mu = \dfrac{1}{p} = \dfrac{1}{0.48} = 2.0833$; $\sigma^2 = \dfrac{q}{p^2} = \dfrac{0.52}{(0.48)^2} = 2.2569$; $\sigma = \sqrt{\sigma^2} = \sqrt{2.2569} = 1.5023$

d. We would expect most observation to fall within 2 standard deviations of the mean, or in the interval

$\mu \pm 2\sigma \Rightarrow 2.0833 \pm 2(1.5023) \Rightarrow 2.0833 \pm 3.0046 \Rightarrow (-0.9213, 5.0879)$

or (1, 5.09)

4.93 a. If the number of respondents with symptoms does not depend on the daily amount of water consumed, then the probabilities for each of the categories would be 0.25.

b. This experiment consists of 40 identical trials. There are four possible outcomes on each trial with the probabilities of each equal to 0.25. Assuming the trials are independent, this is a multinomial experiment with $n = 40$, $k = 4$, $p_1 = p_2 = p_3 = p_4 = 0.25$.

$P(6,11,13,10) = \dfrac{40!}{6!11!13!10!}(0.25)^6(0.25)^{11}(0.25)^{13}(0.25)^{10} = 0.0010$

4.95 a. Let Y = the number of engineers chosen with experience in the design of steam turbine power plants in 2 trials. Then Y has a hypergeometric distribution with $N = 5$, $r = 2$, and $n = 2$.

$$P(Y=2)=\frac{\binom{r}{y}\binom{N-r}{n-y}}{\binom{N}{n}}=\frac{\binom{2}{2}\binom{5-2}{2-2}}{\binom{5}{2}}=\frac{\frac{2!}{2!0!}\cdot\frac{3!}{0!3!}}{\frac{5!}{2!3!}}=\frac{1}{10}=0.1$$

   b. $$P(Y\geq1)=1-P(Y=0)=1-\frac{\binom{2}{0}\binom{5-2}{2-0}}{\binom{5}{2}}=1-\frac{\frac{2!}{2!0!}\cdot\frac{3!}{2!1!}}{\frac{5!}{2!3!}}=1-\frac{3}{10}=0.7$$

4.97 a. Let Y = the number of bird species selected that inhabit butterfly hotspots in 4 trials. Then Y has a hypergeometric distribution with $N = 10$, $r = 7$, and $n = 4$.

$$P(Y=2)=\frac{\binom{r}{y}\binom{N-r}{n-y}}{\binom{N}{n}}=\frac{\binom{7}{2}\binom{10-7}{4-2}}{\binom{10}{4}}=\frac{\frac{7!}{2!5!}\cdot\frac{3!}{2!1!}}{\frac{10!}{4!6!}}=\frac{21(3)}{210}=0.3$$

   b. We know that 7 of the 10 species inhabit a butterfly habitat. Therefore, only 3 species do not inhabit a butterfly habitat. If we sample 4 bird species, then we will sample at least one species that inhabits a butterfly hotspot with $P(Y\geq1)=1$.

4.99 a. $$P(Y\geq3)=1-p(0)-p(1)-p(2)=1-\frac{1^{0}e^{-1}}{0!}-\frac{1^{1}e^{-1}}{1!}-\frac{1^{2}e^{-1}}{2!}$$

   $$=1-0.3679-0.3679-0.1839=0.0803$$

   b. $$P(Y>3)=1-p(0)-p(1)-p(2)-p(3)=1-\frac{1^{0}e^{-1}}{0!}-\frac{1^{1}e^{-1}}{1!}-\frac{1^{2}e^{-1}}{2!}-\frac{1^{3}e^{-1}}{3!}$$

   $$=1-0.3679-0.3679-0.1839-0.0613=0.0190$$

   Yes. The probability of more than 3 arrivals in 1 minute is very small.

4.101 a. The distribution of Y is Poisson with $\lambda = 1.57$.

   $$\mu=\lambda=1.57; \qquad \sigma=\sqrt{\lambda}=\sqrt{1.57}=1.253$$

   b. $$P(Y\geq3)=1-p(0)-p(1)-p(2)=1-\frac{1.57^{0}e^{-1.57}}{0!}-\frac{1.57^{1}e^{-1.57}}{1!}-\frac{1.57^{2}e^{-1.57}}{2!}$$

   $$=1-0.2080-0.3266-0.2564=0.2090$$

4.103 a. The experiment consists of $n = 20$ trials. Each trial results in an S (beech tree has been damaged by fungi) or an F (beech tree has not been damaged by fungi). The probability of success, p, is 0.25 and $q = 1 - p = 1 - 0.25 = 0.75$. We assume the trials are independent. Therefore, Y = number beech trees damaged by fungi in 20 trials and Y has a binomial distribution with $n = 20$ and $p = 0.25$.
$P(Y < 10) = P(Y \le 9) = 0.9861$ using computer software

b. $P(Y > 15) = 1 - P(Y \le 15) = 1 - 1.0000 = 0$ using computer software

c. $E(y) = \mu = np = 20(0.25) = 5$

4.105 a. For this problem, Y has a Poisson distribution with $\lambda = 1.1$.
$$P(Y > 2) = 1 - P(Y \le 2) = 1 - p(0) - p(1) - p(2)$$

$$= 1 - \frac{1.1^0 e^{-1.1}}{0!} - \frac{1.1^1 e^{-1.1}}{1!} - \frac{1.1^2 e^{-1.1}}{2!} = 1 - 0.3329 - 0.3662 - 0.2014 = 0.0995$$

b. $P(Y = 3) = \dfrac{1.1^3 e^{-1.1}}{3!} = 0.0738$

4.107 a. $P(Y > 1) = 1 - p(0) - p(0) = 1 - \dfrac{5^0 e^{-5}}{0!} - \dfrac{5^1 e^{-5}}{1!} = 1 - 0.0067 - 0.0337 = 0.9596$

b. $P(Y > 1) = 1 - p(0) - p(0) = 1 - \dfrac{2.5^0 e^{-2.5}}{0!} - \dfrac{2.5^1 e^{-2.5}}{1!} = 1 - 0.0821 - 0.2052 = 0.7127$

c. Y has a binomial distribution with $n = 55$ and $p = 0.12$. If we use the Poisson approximation, $\lambda = \mu = np = 55(0.12) = 6.6$.

$$P(Y = 0) = \frac{6.6^0 e^{-6.6}}{0!} = 0.00136$$

The exact probability using the binomial distribution is:

$$P(Y = 0) = \binom{55}{0}(0.12)^0 (0.88)^{55-0} = \frac{55!}{0!55!}(0.88)^{55} = 0.00088$$

The difference between the exact probability and the approximate probability is $0.00088 - 0.00136 = -0.00048$.

d. Using Table 4, Appendix B, the value of $\lambda = \mu$ that yields $P(Y \le 1) \approx 0.88$ is somewhere between 0.5 and 1.0. If we use $\lambda = 0.6$,

$$P(Y \le 1) = p(0) + p(1) = \frac{0.6^0 e^{-0.6}}{0!} + \frac{0.6^1 e^{-0.6}}{1!} = 0.5488 + 0.3293 = 0.8781$$

If we use $\lambda = 0.59$,

$$P(Y \le 1) = p(0) + p(1) = \frac{0.59^0 e^{-0.59}}{0!} + \frac{0.59^1 e^{-0.59}}{1!} = 0.5543 + 0.3271 = 0.8814$$

Thus, $\mu = \lambda = 0.59$.

4.109 a. $m(t) = \dfrac{1}{5}e^t + \dfrac{2}{5}e^{2t} + \dfrac{2}{5}e^{3t}$

$$\mu = \mu_1' = \dfrac{dm(t)}{dt}\Bigg]_{t=0} = \dfrac{d\left[\dfrac{1}{5}e^t + \dfrac{2}{5}e^{2t} + \dfrac{2}{5}e^{3t}\right]}{dt}\Bigg]_{t=0}$$

$$= \dfrac{1}{5}e^t + \dfrac{2}{5}e^{2t}(2) + \dfrac{2}{5}e^{3t}(3)\Bigg]_{t=0} = \dfrac{1}{5}e^t + \dfrac{4}{5}e^{2t} + \dfrac{6}{5}e^{3t}\Bigg]_{t=0} = \dfrac{1}{5} + \dfrac{4}{5} + \dfrac{6}{5} = \dfrac{11}{5} = 2.2$$

b. $\sigma^2 = \mu_2' - (\mu_1')^2$

$$\mu_2' = \dfrac{d^2m(t)}{dt^2}\Bigg]_{t=0} = \dfrac{d\left[\dfrac{1}{5}e^t + \dfrac{4}{5}e^{2t} + \dfrac{6}{5}e^{3t}\right]}{dt}\Bigg]_{t=0}$$

$$= \dfrac{1}{5}e^t + \dfrac{4}{5}e^{2t}(2) + \dfrac{6}{5}e^{3t}(3)\Bigg]_{t=0} = \dfrac{1}{5}e^t + \dfrac{8}{5}e^{2t} + \dfrac{18}{5}e^{3t}\Bigg]_{t=0} = \dfrac{1}{5} + \dfrac{8}{5} + \dfrac{18}{5} = 5.4$$

$\sigma^2 = \mu_2' - (\mu_1')^2 = 5.4 - 2.2^2 = 0.56$

4.111 a. $E(t^Y) = \displaystyle\sum_{y=0}^{\infty} \dfrac{(\lambda t)^y e^{-\lambda}}{y!} = e^{\lambda(t-1)} \displaystyle\sum_{y=0}^{\infty} \dfrac{(\lambda t)^y e^{-\lambda t}}{y!} = e^{\lambda(t-1)}$

$E(Y) = \dfrac{dP(t)}{dt}\Bigg]_{t=1} = \dfrac{d[e^{\lambda(t-1)}]}{dt}\Bigg]_{t=1} = e^{\lambda(t-1)}\lambda\Big]_{t=1} = \lambda$

$E[Y(Y-1)] = \dfrac{d^2P(t)}{dt^2}\Bigg]_{t=1} = \dfrac{d[\lambda e^{\lambda(t-1)}]}{dt}\Bigg]_{t=1} = \lambda e^{\lambda(t-1)}\lambda\Big]_{t=1}$

$$= \lambda^2 \Rightarrow E(Y^2) = \lambda^2 + E(Y) = \lambda^2 + \lambda$$

$\sigma^2 = E(Y^2) - \mu^2 = \lambda^2 + \lambda - \lambda^2 = \lambda$

5

Continuous Random Variables

5.1   a. We know

$$\int_0^2 cy^2\, dy = 1$$

$$\Rightarrow \frac{1}{3}cy^3 \bigg]_0^2 = 1$$

$$\Rightarrow \frac{1}{3}c(2)^3 - \frac{1}{3}c(0)^3 = 1$$

$$\Rightarrow \frac{8}{3}c = 1 \Rightarrow c = \frac{3}{8}$$

b. $F(y) = \int_0^y f(t)\, dt = \int_0^y \frac{3}{8}t^2\, dt = \frac{t^3}{8}\bigg]_0^y = \frac{y^3}{8} - \frac{0^3}{8} = \frac{y^3}{8}$

c. $F(1) = \frac{1^3}{8} = \frac{1}{8}$

d. $F(0.5) = \frac{(0.5)^3}{8} = \frac{0.125}{8} = 0.0156$

e. $P(1 \le Y \le 1.5) = \int_1^{1.5} \frac{3}{8}y^2\, dy = \frac{y^3}{8}\bigg]_1^{1.5} = \frac{1.5^3}{8} - \frac{1^3}{8} = 0.2969$

5.3   a. We know

$$\int_{-1}^0 (c+y)\, dy + \int_0^1 (c-y)\, dy = 1$$

$$\Rightarrow \left(cy + \frac{y^2}{2} \right)\bigg]_{-1}^0 + \left(cy - \frac{y^2}{2} \right)\bigg]_0^1 = 1$$

$$\Rightarrow \left(c(0) + \frac{0^2}{2} \right) - \left(c(-1) + \frac{(-1)^2}{2} \right) + \left(c(1) - \frac{1^2}{2} \right) - \left(c(0) - \frac{0^2}{2} \right) = 1$$

$$\Rightarrow c - \frac{1}{2} + c - \frac{1}{2} = 1$$

$$\Rightarrow 2c - 1 = 1 \Rightarrow 2c = 2 \Rightarrow c = 1$$

b. For $-1 \le y < 0$:

$$F(y) = \int_{-1}^{y} f(t)\,dt = \int_{-1}^{y} (1+t)\,dt = t + \frac{t^2}{2}\Bigg]_{-1}^{y} = y + \frac{y^2}{2} - (-1) - \frac{(-1)^2}{2} = \frac{1}{2} + y + \frac{y^2}{2}$$

For $0 \le y \le 1$:

$$F(y) = \frac{1}{2} + \int_{0}^{y} f(t)\,dt = \frac{1}{2} + \int_{0}^{y} (1-t)\,dt = \frac{1}{2} + \left(t - \frac{t^2}{2} \right) \Bigg]_{0}^{y}$$

$$= \frac{1}{2} + y - \frac{y^2}{2} - (0) - \frac{(0)^2}{2} = \frac{1}{2} + y - \frac{y^2}{2}$$

$$F(y) = \begin{cases} \dfrac{1}{2} + y + \dfrac{y^2}{2} & -1 \le y < 0 \\[4mm] \dfrac{1}{2} + y - \dfrac{y^2}{2} & 0 \le y \le 1 \end{cases}$$

c. $F(-0.5) = \dfrac{1}{2} + (-0.5) + \dfrac{(-0.5)^2}{2} = 0.125$

d. $P(0 \le Y \le 0.5) = \displaystyle\int_{0}^{0.5} (1-y)\,dy = \left(y - \frac{y^2}{2} \right) \Bigg]_{0}^{0.5} = \left(0.5 - \frac{(0.5)^2}{2} \right) - \left(0 - \frac{(0)^2}{2} \right) = 0.375$

5.5 a. We know

$$\int_{-5}^{5} \frac{c}{500} (25 - y^2)\,dy = 1$$

$$\Rightarrow \frac{cy}{20} - \frac{cy^3}{1500} \Bigg]_{-5}^{5} = 1$$

$$\Rightarrow \left(\frac{5c}{20} - \frac{5^3 c}{1500} \right) - \left(\frac{-5c}{20} - \frac{(-5)^3 c}{1500} \right) = 1$$

$$\Rightarrow \left(\frac{c}{4} - \frac{125c}{1500} \right) - \left(\frac{-c}{4} - \frac{-125c}{1500} \right) = 1$$

$$\Rightarrow \frac{c}{2} - \frac{c}{6} = 1 \Rightarrow \frac{2c}{6} = 1 \Rightarrow c = 3$$

b. $F(y) = \int_{-5}^{y} f(t)\,dt = \int_{-5}^{y} \frac{3}{500}(25 - t^2)\,dt = \frac{3}{500}\left(25t - \frac{t^3}{3}\right)\Big]_{-5}^{y}$

$$= \frac{3}{500}\left[\left(25y - \frac{y^3}{3}\right) - \left(25(-5) - \frac{(-5)^3}{3}\right)\right]$$

$$= \frac{3}{500}\left(25y - \frac{y^3}{3} + 125 - \frac{125}{3}\right) = \frac{75y - y^3}{500} + \frac{1}{2}$$

c. $P(Y \le 3) = F(3) = \dfrac{75(3) - 3^3}{500} + \dfrac{1}{2} = 0.396 + 0.5 = 0.896$

5.7 a. Show $f(y) \ge 0$. For $c > 0$, $y > 0$, show $f(y) = ce^{-cy} \ge 0$. If $c > 0$ and $y > 0$, then $e^{-cy} \ge 0$. Thus, $f(y) = ce^{-cy} \ge 0$.
Show

$$\int_{-\infty}^{\infty} ce^{-cy}\,dy = F(\infty) = 1$$

$$\Rightarrow -e^{-cy}\Big]_0^{\infty} = -e^{-\infty} + e^0 = 1$$

Show

$$P(a < Y < b) = \int_a^b ce^{-cy}\,dy = -e^{-cy}\Big]_a^b = -e^{-cb} + e^{-ca}$$

b. $F(y) = \int_0^y 0.04e^{-0.04t}\,dt = -e^{-0.04t}\Big]_0^y = -e^{-0.04y} + e^0 = 1 - e^{-0.04y}$

c. $R(5) = 1 - F(5) = 1 - (1 - e^{-0.04(5)}) = 0.8187$
The earthquake system reliability at time 5 is 0.8187.

5.9 a. $F(y) = \int_0^y f(t)\,dt = \int_0^y \dfrac{t}{2}\,dt = \dfrac{t^2}{4}\Big]_0^y = \dfrac{y^2}{4} - \dfrac{0^2}{4} = \dfrac{y^2}{4}$

b. $\bar{F}(x + y) = 1 - F(x + y) = 1 - \dfrac{(x + y)^2}{4} = \dfrac{4 - x^2 - 2xy - y^2}{4}$

$$\bar{F}(x)\bar{F}(y) = (1 - F(x))(1 - F(y)) = \left(1 - \frac{x^2}{4}\right)\left(1 - \frac{y^2}{4}\right) = 1 - \frac{x^2}{4} - \frac{y^2}{4} + \frac{x^2 y^2}{16}$$

If we let $x = 1$ and $y = 1/2$, then

$$\bar{F}(x + y) = \bar{F}(1 + 1/2) = \frac{4 - (1)^2 - 2(1)(1/2) - (1/2)^2}{4} = 0.4375$$

$$\bar{F}(x)\bar{F}(y) = \bar{F}(1)\bar{F}(1/2) = 1 - \frac{(1)^2}{4} - \frac{(1/2)^2}{4} + \frac{(1)^2(1/2)^2}{16} = 0.7031$$

Since $\bar{F}(x + y) \le \bar{F}(x)\bar{F}(y)$, the "life" distribution is NBU.

5.11 a. $\mu = E(Y) = \int_0^1 yf(y)dy = \int_0^1 6y^2(1-y)dy = \int_0^1 (6y^2 - 6y^3)dy$

$= 2y^3 - \frac{6}{4}y^4 \Big]_0^1 = \left(2(1)^3 - \frac{6}{4}(1)^4\right) - \left(2(0)^3 - \frac{6}{4}(0)^4\right) = 2 - \frac{3}{2} = \frac{1}{2}$

The average acceleration in sea level rise (standardized between 0 and 1) is 0.5.

b. $E(Y^2) = \int_0^0 y^2 f(y)dy = \int_0^1 6y^3(1-y)dy = \int_0^1 (6y^3 - 6y^4)dy =$

$= \frac{6}{4}y^4 - \frac{6}{5}y^5 \Big]_0^1 = \left(\frac{6}{4}(1)^4 - \frac{6}{5}(1)^5\right) - \left(\frac{6}{4}(0)^4 - \frac{6}{5}(0)^5\right) = \frac{30}{20} - \frac{24}{20} = \frac{6}{20} = 0.3$

$\sigma^2 = E(Y^2) - \mu^2 = 0.3 - 0.5^2 = 0.05$

c. From the Empirical Rule, approximately 0.95 of the observations will fall within 2 standard deviations of the mean. Thus, from the Empirical Rule, $P(\mu - 2\sigma < Y < \mu + 2\sigma) \approx 0.95$.

$\sigma = \sqrt{0.05} = 0.2236$

$P(\mu - 2\sigma < Y < \mu + 2\sigma) = P(0.5 - 2(0.2236) < Y < 0.5 + 2(0.2236))$

$= P(0.0528 < Y < 0.9472)$

$= \int_{0.0528}^{0.9472} (6y - 6y^2)dy = (3y^2 - 2y^3)\Big]_{0.0528}^{0.9472}$

$= [3(0.9472)^2 - 2(0.9472)^3] - [3(0.0528)^2 - 2(0.0528)^3]$

$= 0.9919 - 0.0081 = 0.9838$

This is somewhat larger than the 0.95 found using the Empirical Rule.

5.13 a. $E(Y) = \int_0^\infty 0.04ye^{-0.04y} dy = e^{-0.04y}(-y-25)\Big]_0^\infty$

$= \lim_{y\to\infty}(-ye^{-0.04y}) - 25e^{-\infty} - e^{-0.04(0)}(0-25) = 0 - 0 + 25 = 25$

b. $E(Y^2) = \int_0^\infty 0.04y^2e^{-0.04y} dy = e^{-0.04y}((-y-50)y - 1250)\Big]_0^\infty$

$= \lim_{y\to\infty}(e^{-0.04y}(-y-50)y) - e^{-0.04(\infty)}(1250) - e^{-0.04(0)}((-0-50)0 - 1250)$

$= 0 - 0 + 1250 = 1250$

Thus, $\sigma^2 = E(Y^2) - \mu^2 = 1250 - 25^2 = 1250 - 625 = 625$

c. By the Empirical Rule, $P(\mu - 2\sigma < Y < \mu + 2\sigma) \approx 0.95$

d. $\sigma = \sqrt{\sigma^2} = \sqrt{625} = 25$; $\mu - 2\sigma = 25 - 2(25) = -25$ and $\mu + 2\sigma = 25 + 2(25) = 75$

$P(\mu - 2\sigma < Y < \mu + 2\sigma) = P(-25 < Y < 75) = P(0 < Y < 75)$

$$= \int_0^{75} 0.04 e^{-0.04y} \, dy = -e^{-0.04y} \Big]_0^{75} = -e^{-0.04(75)} + e^{-0.04(0)}$$

$$= -0.0498 + 1 = 0.9502$$

This is very close to the approximation in part c.

5.15   Show $E(c) = c$

$$E(c) = \int_{-\infty}^{\infty} cf(y) \, dy = c \int_{-\infty}^{\infty} f(y) \, dy = c(1) = c$$

Show $E(cy) = cE(y)$

$$E(cy) = \int_{-\infty}^{\infty} cyf(y) \, dy = c \int_{-\infty}^{\infty} yf(y) \, dy = cE(y)$$

Show $E[g_1(y) + g_2(y) + \cdots + g_k(y)] = E[g_1(y)] + E[g_2(y)] + \cdots + E[g_k(y)]$

$$E[g_1(y) + g_2(y) + \cdots + g_k(y)] = \int_{-\infty}^{\infty} [g_1(y) + g_2(y) + \cdots + g_k(y)] f(y) \, dy$$

$$= \int_{-\infty}^{\infty} [g_1(y)f(y) + g_2(y)f(y) + \cdots + g_k(y)f(y)] \, dy$$

$$= \int_{-\infty}^{\infty} g_1(y)f(y) \, dy + \int_{-\infty}^{\infty} g_2(y)f(y) \, dy + \cdots + \int_{-\infty}^{\infty} g_k(y)f(y) \, dy$$

$$= E[g_1(y)] + E[g_2(y)] + \cdots E[g_k(y)]$$

5.17 a. $f(y) = \begin{cases} \dfrac{1}{b-a} = \dfrac{1}{3-1} = \dfrac{1}{2} & 1 \le y \le 3 \\ 0 & \text{otherwise} \end{cases}$

$$E(Y) = \int_1^3 y \frac{1}{2} \, dy = \frac{y^2}{4} \Big]_1^3 = \frac{3^2}{4} - \frac{1^2}{4} = \frac{9}{4} - \frac{1}{4} = \frac{8}{4} = 2$$

b. $P(2 < Y < 2.5) = (2.5 - 2)\dfrac{1}{2} = \dfrac{0.5}{2} = 0.25$

c. $P(Y \le 1.75) = (1.75 - 1)\dfrac{1}{2} = \dfrac{0.75}{2} = 0.375$

5.19 $f(y) = \begin{cases} \dfrac{1}{b-a} = \dfrac{1}{115-100} = \dfrac{1}{15} & 100 \le y \le 115 \\ 0 & \text{otherwise} \end{cases}$

$P(Y > L) = 0.1 \Rightarrow (115 - L)\dfrac{1}{15} = 0.1 \Rightarrow 115 - L = 1.5 \Rightarrow L = 113.5$

5.21 $\mu = \dfrac{a+b}{2} = \dfrac{0+1}{2} = 0.5$

$\sigma^2 = \dfrac{(b-a)^2}{12} = \dfrac{(1-0)^2}{12} = \dfrac{1}{12} = 0.0833 \, ; \quad \sigma = \sqrt{0.0833} = 0.2887$

$f(y) = \begin{cases} \dfrac{1}{b-a} = \dfrac{1}{1-0} = \dfrac{1}{1} = 1 & 0 \le y \le 1 \\ 0 & \text{otherwise} \end{cases}$

Let k be the 10th percentile. $P(Y < k) = 0.1 \Rightarrow (k-0)1 = 0.1 \Rightarrow k = 0.1$
Let h be the lower quartile: $P(Y < h) = 0.25 \Rightarrow (h-0)1 = 0.25 \Rightarrow h = 0.25$
Let j be the lower quartile: $P(Y < j) = 0.75 \Rightarrow (j-0)1 = 0.75 \Rightarrow j = 0.75$

5.23   Let Y = distance of gouge from one end of the spindle. The density function of Y is

$f(y) = \begin{cases} \dfrac{1}{b-a} = \dfrac{1}{18-0} = \dfrac{1}{18} & 0 \le y \le 18 \\ 0 & \text{otherwise} \end{cases}$

In order to get at least 14 consecutive inches without a gouge, the gouge must be within 4 inches of either end. Thus, we must find:

$P(Y < 4) + P(Y > 14) = (4-0)\dfrac{1}{18} + (18-14)\dfrac{1}{18} = \dfrac{4}{18} + \dfrac{4}{18} = \dfrac{8}{18} = 0.4444$

5.25 a. $f(y) = \begin{cases} 1 & 0 \le y \le 1 \\ 0 & \text{otherwise} \end{cases}$

Let $w = by$. The density function of w is

$f(w) = \begin{cases} \dfrac{1}{b} & 0 \le w \le b \\ 0 & \text{otherwise} \end{cases}$

This implies that w is a uniform random variable on the interval from 0 to b.

b.   Let $z = a + (b-a)y$. The density function of z is

$f(z) = \begin{cases} \dfrac{1}{b-a} & a \le z \le b \\ 0 & \text{otherwise} \end{cases}$

5.27   For the uniform distribution on the interval $(0,1)$,

$$f(y) = \begin{cases} 1 & 0 \le y \le 1 \\ 0 & \text{otherwise} \end{cases} \quad \text{and} \quad F(y) = \begin{cases} 0 & y < 0 \\ y & 0 \le y \le 1 \\ 1 & y > 1 \end{cases}$$

To be NBU, $\bar{F}(x+y) \le \bar{F}(x)\bar{F}(y)$

Let $x = \dfrac{1}{4}, y = \dfrac{1}{2}$

$$\bar{F}(x+y) = \bar{F}\left(\frac{3}{4}\right) = 1 - F\left(\frac{3}{4}\right) = 1 - \frac{3}{4} = \frac{1}{4}$$

$$\bar{F}(x)\bar{F}(y) = \left(1 - F\left(\frac{1}{4}\right)\right)\left(1 - F\left(\frac{1}{2}\right)\right) = \left(1 - \frac{1}{4}\right)\left(1 - \frac{1}{2}\right) = \left(\frac{3}{4}\right)\left(\frac{1}{2}\right) = \frac{3}{8}$$

$\Rightarrow \bar{F}(x+y) \le \bar{F}(x)\bar{F}(y) \Rightarrow$ The uniform distribution on the interval $(0,1)$ is NBU.

5.29 a. $P(Y > 120) = P\left(Z > \dfrac{120 - 105.3}{8}\right) = P(Z > 1.84)$

$\qquad\qquad = 0.5 - P(0 < Z < 1.84) = 0.5 - 0.4671 = 0.0329$

(using Table 5, Appendix B)

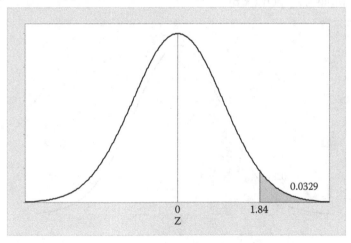

b. $P(100 < Y < 110) = P\left(\dfrac{100 - 105.3}{8} < Z < \dfrac{110 - 105.3}{8}\right)$

$\qquad\qquad = P(-0.66 < Z < 0.59) = P(-0.66 < Z < 0) + P(0 < Z < 0.59)$

$\qquad\qquad = 0.2454 + 0.2224 = 0.4678$

(using Table 5, Appendix B)

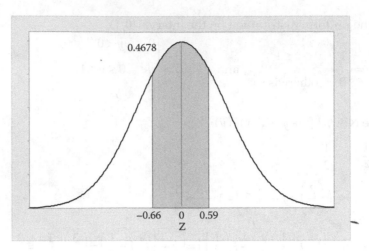

c. $P(Y < a) = 0.25 \Rightarrow P\left(Z < \dfrac{a - 105.3}{8}\right) = P(Z < z_0) = 0.25$

$A_1 = 0.5 - 0.25 = 0.2500$. Looking up area 0.2500 in Table 5 gives $z_0 = -0.67$

$\Rightarrow z_0 = \dfrac{a - 105.3}{8} = -0.67 \Rightarrow a - 105.3 = -5.36 \Rightarrow a = 99.94$

(using Table 5, Appendix B)

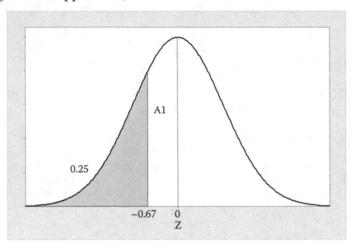

5.31　　Let Y = transmission delay. Then Y is normally distributed with $\mu = 48.5$ and $\sigma = 8.5$.

a. $P(Y < 57) = P\left(Z < \dfrac{57 - 48.5}{8.5}\right) = P(Z < 1) = 0.5 + P(0 < Z < 1)$

$= 0.5 + 0.3413 = 0.8413$

(using Table 5, Appendix B)

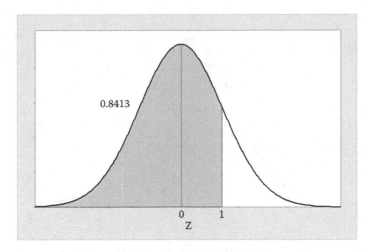

b. $P(40 < Y < 60) = P\left(\dfrac{40-48.5}{8.5} < Z < \dfrac{60-48.5}{8.5}\right)$

$= P(-1 < Z < 1.35) = P(-1 < Z < 0) + P(0 < Z < 1.35)$

$= 0.3413 + 0.4115 = 0.7528$

(using Table 5, Appendix B)

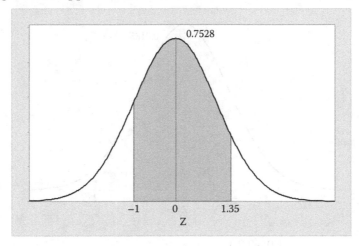

5.33 $P(Y > a) = 0.70 \Rightarrow P\left(Z > \dfrac{a-0.5}{0.1}\right) = P(Z > z_0) = 0.70$

$A_1 = 0.7 - 0.5 = 0.2000$. Looking up area 0.2000 in Table 5

gives $z_0 = -0.52$

$\Rightarrow z_0 = \dfrac{a-0.5}{0.1} = -0.52 \Rightarrow a - 0.5 = -0.052 \Rightarrow a = 0.448$

(using Table 5, Appendix B)

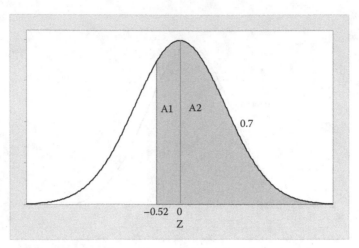

5.35   Let Y = tip resistance. Then Y is normally distributed with $\mu = 2.2$ and $\sigma = 0.9$.

a. $P(1.3 < Y < 4.0) = P\left(\dfrac{1.3 - 2.2}{0.9} < Z < \dfrac{4.0 - 2.2}{0.9} \right)$

$$= P(-1 < Z < 2) = P(-1 < Z < 0) + P(0 < Z < 2)$$

(using Table 5, Appendix B)

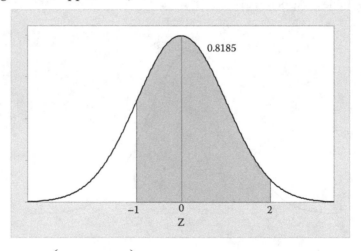

b. $P(Y > 1.0) = P\left(Z > \dfrac{1.0 - 2.2}{0.9} \right) = P(Z > -1.33)$

$$= P(-1.33 < Z < 0) + 0.5 = 0.4082 + 0.5 = 0.9082$$

(using Table 5, Appendix B)

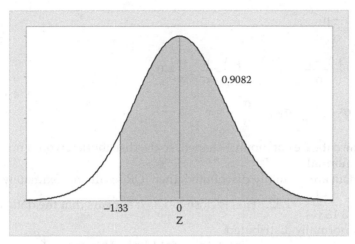

5.37 $\quad P(Y > 3) = P\left(Z > \dfrac{3 - 2.2}{0.5} \right) = P(Z > 1.6) = 0.5 - P(0 < Z < 1.6)$

$\quad\quad\quad = 0.5 - 0.4452 = 0.0548$

(using Table 5, Appendix B)

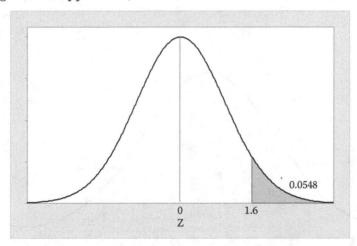

5.39 a. Let Y = fill of container. Then Y is normally distributed with $\mu = 10$ and $\sigma = 0.2$.

$\quad P(Y < 10) = P\left(Z < \dfrac{10 - 10}{0.2} \right) = P(Z < 0) = 0.5$

  b. Profit = Price − cost − reprocessing fee = \$230 − \$20(10.6) − \$10 = \$230 − \$212 − \$10 = \$8

  c. If the probability of underfill is approximately 0, then Profit = Price − Cost.

$\quad E(\text{Profit}) = E(\text{Price} - \text{Cost}) = \$230 - E(\text{Cost}) = \$230 - \$20E(X)$

$\quad\quad\quad = \$230 - \$20(10.5) = \$230 - \$210 = \$20$

5.41 $Z = \dfrac{Y - \mu}{\sigma}$

$E(Z) = E\left(\dfrac{Y - \mu}{\sigma}\right) = E\left(\dfrac{Y}{\sigma} - \dfrac{\mu}{\sigma}\right) = \dfrac{\mu}{\sigma} - \dfrac{\mu}{\sigma} = 0$

$\sigma_Z^2 = \sigma_{\frac{Y-\mu}{\sigma}}^2 = \dfrac{1}{\sigma^2}\sigma_{Y-\mu}^2 = \dfrac{\sigma^2}{\sigma^2} - = 1$

5.43   No. The data are not mound-shaped, so the distribution would not be approximately normal.

5.45   a.   If the data are normally distributed, then *IQR/s* will approximately equal 1.3.

$\dfrac{IQR}{s} = \dfrac{4.84}{3.18344} = 1.52$. This is close to 1.3. It appears that the data are approximately normally distributed.

   b.   The data form an approximately straight line. This suggests that the data are approximately normally distributed.

5.47   Using MINITAB, the histogram of the data is:

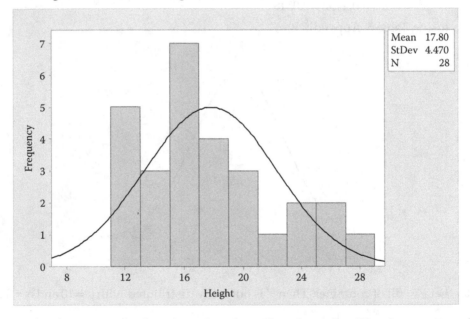

The data appear to be somewhat skewed to the right. The data may not be normally distributed.

The descriptive statistics are:

Descriptive Statistics: Height

Variable	N	Mean	StDev	Minimum	Q1	Median	Q3	Maximum	IQR
Height	28	17.796	4.470	12.400	13.850	16.750	19.825	27.300	5.975

If the data are normally distributed, then *IQR/s* will approximately equal 1.3.

$\dfrac{IQR}{s} = \dfrac{5.975}{4.470} = 1.34$. This is close to 1.3. It appears that the data may be approximately normally distributed.

The normal probability plot is:

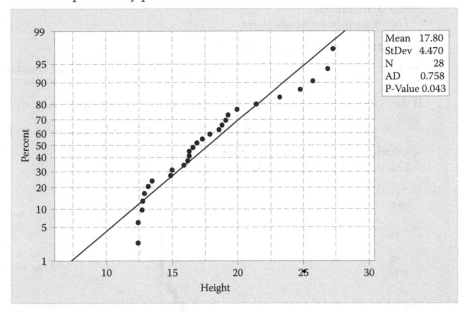

The data do not form a very straight line. This indicates the data may not be normally distributed.

From the three checks, two indicate the data may not be normally distributed.

5.49   Using MINITAB, the histogram of the data is:

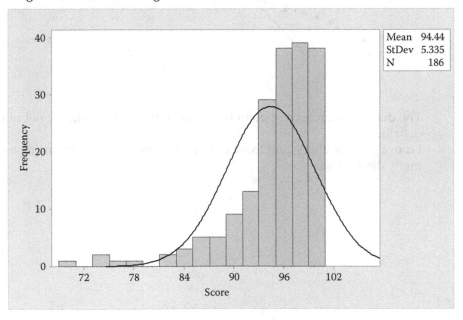

The data appear to be skewed to the left. The data do not appear to be normally distributed.

The descriptive statistics are:

Descriptive Statistics: Score

Variable	N	Mean	StDev	Minimum	Q1	Median	Q3	Maximum	IQR
Score	186	94.441	5.335	69.000	93.000	96.000	98.000	100.000	5.000

If the data are normally distributed, then IQR/s will approximately equal 1.3.

$\dfrac{IQR}{s} = \dfrac{5}{5.335} = 0.94$. This is not close to 1.3. It appears that the data may not be approximately normally distributed.

The normal probability plot is:

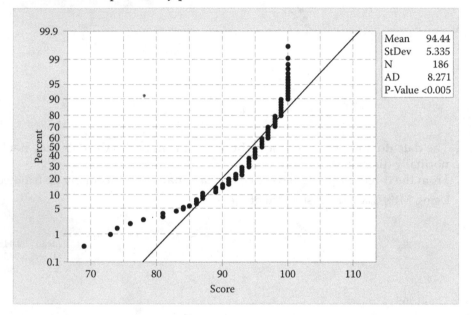

The data do not form a straight line. This indicates the data are not normally distributed.

From the three checks, all three indicate the data are not approximately normally distributed.

5.50   **ZETA without Gypsum**
Using MINITAB, the histogram of the data is:

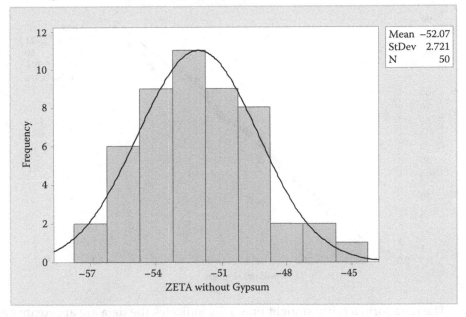

The data appear to be fairly mound-shaped. The data appear to be approximately normally distributed.

The descriptive statistics are:

```
Descriptive Statistics: ZETA without, ZETA with GYPSUM

Variable                 N    Mean  StDev       Q1   Median       Q3    IQR
ZETA without Gypsum     50  -52.070  2.721  -53.900  -52.250  -50.200  3.700
```

If the data are normally distributed, then IQR/s will approximately equal 1.3.

$\dfrac{IQR}{s} = \dfrac{3.7}{2.721} = 1.36$. This is fairly close to 1.3. It appears that the data may be approximately normally distributed.

The normal probability plot is:

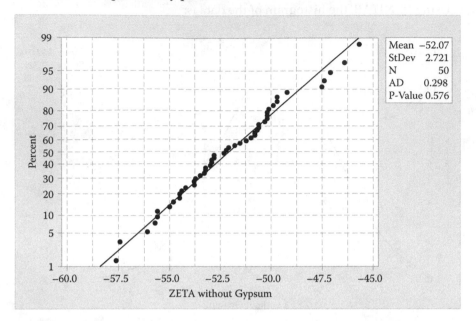

The data form a fairly straight line. This indicates the data are approximately normally distributed.

From the three checks, all three indicate the data are approximately normally distributed.

ZETA with Gypsum
Using MINITAB, the histogram of the data is:

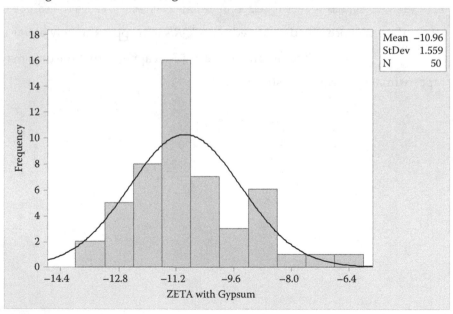

The data appear to be somewhat skewed to the right. The data may not be approximately normally distributed.

The descriptive statistics are:

```
Descriptive Statistics: ZETA without, ZETA with GYPSUM

Variable              N     Mean   StDev      Q1   Median       Q3     IQR
ZETA with GYPSUM     50  -10.958   1.559  -12.100  -11.300  -10.075   2.025
```

If the data are normally distributed, then IQR/s will approximately equal 1.3.

$\dfrac{IQR}{s} = \dfrac{2.025}{1.559} = 1.299$. This is close to 1.3. It appears that the data may be approximately normally distributed.

The normal probability plot is:

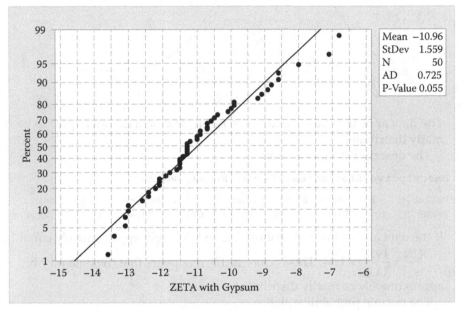

The data vary from a straight line. This indicates the data may not be approximately normally distributed.

From the three checks, two of the three indicate the data may not be approximately normally distributed.

Of the two groups, the data for ZETA without Gypsum is better approximated by the normal distribution.

5.51   For **Group A**:
Using MINITAB, the histogram of the data is:

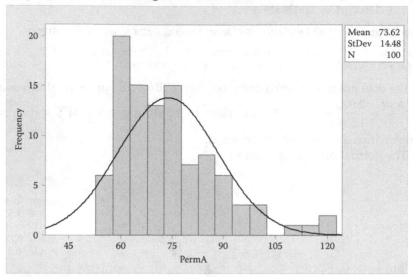

The data appear to be skewed to the right. The data do not appear to be normally distributed.

The descriptive statistics are:

Descriptive Statistics: PermA, PermB, PermC

Variable	N	Mean	StDev	Minimum	Q1	Median	Q3	Maximum	IQR
PermA	100	73.62	14.48	55.20	62.00	70.45	81.42	122.40	19.42

If the data are normally distributed, then IQR/s will approximately equal 1.3.

$\dfrac{IQR}{s} = \dfrac{19.42}{14.48} = 1.34$. This is close to 1.3. It appears that the data may be approximately normally distributed.

The normal probability plot is:

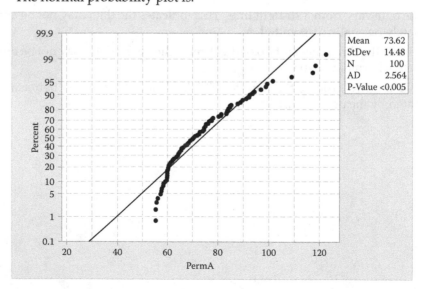

The data do not form a straight line. This indicates the data are not normally distributed.

From the three checks, two of the three indicate the data are not approximately normally distributed.

For **Group B**:
Using MINITAB, the histogram of the data is:

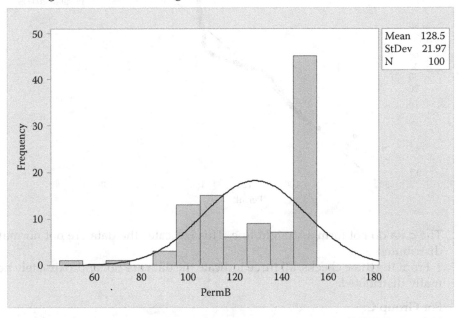

The data appear to be skewed to the left. The data do not appear to be normally distributed.

The descriptive statistics are:

Descriptive Statistics: PermA, PermB, PermC

Variable	N	Mean	StDev	Minimum	Q1	Median	Q3	Maximum	IQR
PermB	100	128.54	21.97	50.40	108.65	139.30	147.02	150.00	38.37

If the data are normally distributed, then IQR/s will approximately equal 1.3.

$\dfrac{IQR}{s} = \dfrac{38.37}{21.97} = 1.75$. This is not close to 1.3. It appears that the data may not be approximately normally distributed.

The normal probability plot is:

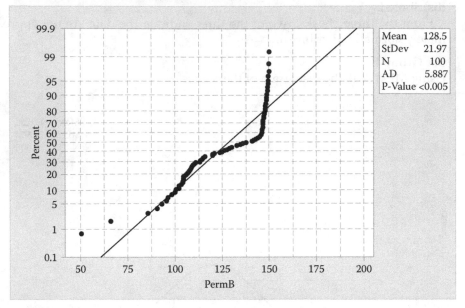

The data do not form a straight line. This indicates the data are not normally distributed.

From the three checks, all three indicate the data are not approximately normally distributed.

For **Group C:**
Using MINITAB, the histogram of the data is:

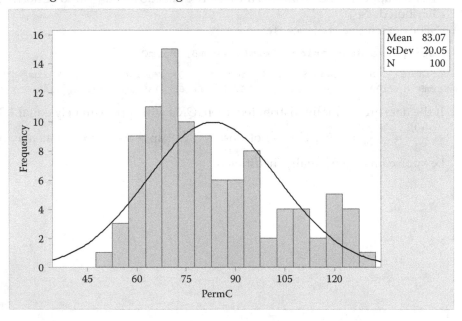

The data appear to be skewed to the right. The data do not appear to be normally distributed.

The descriptive statistics are:

```
Descriptive Statistics: PermA, PermB, PermC
Variable    N    Mean  StDev  Minimum       Q1  Median      Q3  Maximum     IQR
PermC     100   83.07  20.05    52.20    67.72   78.65   95.35   129.00   27.63
```

If the data are normally distributed, then IQR/s will approximately equal 1.3.

$\dfrac{IQR}{s} = \dfrac{27.63}{20.05} = 1.38$. This is close to 1.3. It appears that the data may be approximately normally distributed.

The normal probability plot is:

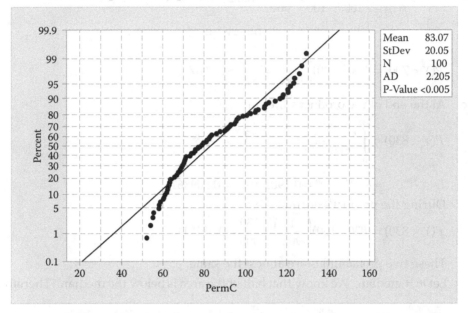

The data do not form a straight line. This indicates the data are not normally distributed.

From the three checks, two of the three indicate the data are not approximately normally distributed.

5.53 a. $P(Y > 2) = \displaystyle\int_{2}^{\infty} \frac{e^{-y/2.5}}{2.5}\,dy = -e^{-y/2.5}\Big]_{2}^{\infty} = -e^{-\infty/2.5} + e^{-2/2.5}$

$= -e^{-\infty} + e^{-0.8} = 0 + 0.449329 = 0.449329$ (using Table 3, Appendix B)

b. $P(Y < 5) = \displaystyle\int_{0}^{5} \frac{e^{-y/2.5}}{2.5}\,dy = -e^{-y/2.5}\Big]_{0}^{5} = -e^{-5/2.5} + e^{-0/2.5}$

$= -e^{-2} + e^{0} = -0.135335 + 1 = 0.864665$ (using Table 3, Appendix B)

5.55 a. At the end of the product's lifetime, the time till failure follows a gamma distribution with $\alpha = 1$ and $\beta = 500$. This is the same as an exponential distribution with $\beta = 500$.

$$P(Y < 700) = \int_0^{700} \frac{e^{-y/500}}{500} dy = -e^{-y/500} \Big]_0^{700} = -e^{-700/500} + e^{-0/500}$$

$$= -e^{-1.4} + e^0 = 0.246597 + 1 = 0.753403 \quad \text{(using Table 3, Appendix B)}$$

b. During the product's normal life, the time till failure is uniformly distributed over the range 100 thousand to 1 million hours. Thus,

$$f(y) = \begin{cases} \dfrac{1}{1000 - 100} = \dfrac{1}{900} & 100 \le y \le 1000 \\ 0 & \text{otherwise} \end{cases}$$

$$P(Y < 700) = (700 - 100)\left(\frac{1}{900}\right) = \frac{600}{900} = 0.6667$$

c. At the end of the product's lifetime,

$$P(Y < 830) = \int_0^{830} \frac{e^{-y/500}}{500} dy = -e^{-y/500} \Big]_0^{830} = -e^{-830/500} + e^{-0/500}$$

$$= -e^{-1.66} + e^0 = 0.190139 + 1 = 0.809861$$

During the product's normal life,

$$P(Y < 830) = (830 - 100)\left(\frac{1}{900}\right) = \frac{730}{900} = 0.81111$$

These two probabilities are almost the same.

5.57   Let m = median. We know that half of the area is below the median. Therefore,

$$\int_0^m \frac{e^{-y/\beta}}{\beta} = 0.5 \Rightarrow -e^{-y/\beta} \Big]_0^m = 0.5 \Rightarrow -e^{-m/\beta} + e^0 = 0.5 \Rightarrow 1 - e^{-m/\beta} = 0.5$$

$$\Rightarrow e^{-m/\beta} = 0.5 \Rightarrow \frac{-m}{\beta} = \ln(0.5) \Rightarrow \frac{-m}{\beta} = -0.693147 \Rightarrow m = 0.693147\beta$$

5.59 a. $P(Y_1 > 1) = \int_1^\infty \frac{e^{-y/1}}{1} dy = -e^{-y} \Big]_1^\infty = -e^{-\infty} + e^{-1} = 0 + 0.3679 = 0.3679$

b. $P(Y_2 > 1) = \int_1^\infty \frac{e^{-y/2}}{2} dy = -e^{-y/2} \Big]_1^\infty = -e^{-\infty} + e^{-1/2} = 0 + 0.6065 = 0.6065$

c. Machine 3: $P(Y_3 > 1) = \int_1^\infty \frac{e^{-y/0.5}}{0.5} dy = -e^{-2y} \Big]_1^\infty = -e^{-\infty} + e^{-2} = 0 + 0.1353 = 0.1353$

Machine 4: $P(Y_4 > 1) = \int\limits_1^\infty \dfrac{e^{-y/0.5}}{0.5} dy = -e^{-2y} \Big]_1^\infty = -e^{-\infty} + e^{-2} = 0 + 0.1353 = 0.1353$

d. $P(\text{Repair time exceeds 1 hour}) = P(Y_1 > 1)P(Y_2 > 1)P(Y_3 > 1)P(Y_4 > 1)$

$$= 0.3679(0.6065)(0.1353)(0.1353) = 0.0041$$

5.61 a. $R(t) = P(Y > t) = \int\limits_t^\infty \dfrac{e^{-y/25,000}}{25,000} dy = -e^{-y/25,000} \Big]_t^\infty = -e^{-\infty} + e^{-t/25,000} = e^{-t/25,000}$

b. $R(8,760) = e^{-8,760/2,5000} = e^{-0.3504} = 0.7044$

c. $S(t) = P(\text{at least one drive has a lifelength that exceeding } t \text{ hours})$
$= 1 - P(\text{neither exceeds } t \text{ hours})$

$S(t) = 1 - [1 - R(t)]^2 = 1 - [1 - e^{-t/25,000}]^2$

d. $S(8,760) = 1 - [1 - e^{-8,760/25,000}]^2 = 1 - [1 - e^{-0.3504}]^2$

$$= 1 - (1 - 0.7044)^2 = 1 - 0.0874 = 0.9126$$

e. The probability that at least one CD-ROM drive system has a lifelength exceeding 8,760 hours is greater than the probability that one CD-ROM drive system has a lifelength exceeding 8,760 hours.

5.63 $F(a) = P(Y \le a) = \int\limits_0^a \dfrac{e^{-y/\beta}}{\beta} dy = -e^{-y/\beta} \Big]_0^a = -e^{-a/\beta} + e^0 = 1 - e^{-a/\beta}$

Then, $P(Y > a) = 1 - P(Y \le a) = 1 - F(a) = 1 - (1 - e^{-a/\beta}) = e^{-a/\beta}$

5.65 Show $\Gamma(\alpha) = (\alpha - 1)\Gamma(\alpha - 1)$.
By definition, $\Gamma(\alpha) = (\alpha - 1)!$
$\Gamma(\alpha) = (\alpha - 1)! = (\alpha - 1)(\alpha - 2)! = (\alpha - 1)\Gamma(\alpha - 1)$

5.67 We know $\int\limits_0^\infty cy^2 e^{-y/2} dy = 1$

We know for a gamma type random variable, $\int\limits_0^\infty \dfrac{y^{\alpha-1}e^{-y/\beta}}{\beta^2\Gamma(\alpha)} dy = 1$
Thus, $\beta = 2$, $\alpha - 1 = 2 \Rightarrow \alpha = 3$

Therefore, $c = \dfrac{1}{2^3\Gamma(3)} = \dfrac{1}{8(2)!} = \dfrac{1}{16}$.

5.69 a. $\mu = \beta^{1/\alpha}\Gamma\left(\dfrac{\alpha+1}{\alpha}\right) = 1800^{1/6}\Gamma\left(\dfrac{6+1}{6}\right) = 1800^{1/6}\Gamma(1.17) = 3.48775(0.92670) = 3.232$

$\sigma^2 = \beta^{2/\alpha}\left[\Gamma\left(\dfrac{\alpha+2}{\alpha}\right) - \Gamma^2\left(\dfrac{\alpha+1}{\alpha}\right)\right] = 1800^{2/6}\left[\Gamma\left(\dfrac{6+2}{6}\right) - \Gamma^2\left(\dfrac{6+1}{6}\right)\right]$

$= 1800^{1/3}[\Gamma(1.33) - \Gamma^2(1.17)] = 12.1644(0.89338 - 0.92670^2)$

$= 12.1644(0.89338 - 0.85877289) = 0.42097$

b.   Using the Empirical Rule, ≈ 0.95 of fracture toughness values lie within 2 standard deviations of the mean.

c. $\sigma = \sqrt{0.42097} = 0.6488$;

$\mu \pm 2\sigma \Rightarrow 3.232 \pm 2(0.6488) \Rightarrow 3.232 \pm 1.2976 \Rightarrow (1.9344,\ 4.5296)$

From Example 5.15, $F(y) = 1 - e^{-y^\alpha/\beta}$.

$P(1.9344 < Y < 4.5296) = F(4.5296) - F(1.9344) = (1 - e^{-4.5296^6/1800}) - (1 - e^{-1.9344^6/1800})$

$= (1 - e^{-4.7983}) - (1 - e^{-0.0291}) = (1 - 0.0082) - (1 - 0.9713) = 0.9918 - 0.0287 = 0.9631$

5.71 $P(Y < 50,000) = F(50,000)$. From Example 5.15, $F(y) = 1 - e^{-y^\alpha/\beta}$

$F(50,000) = 1 - e^{-50,000^1/100,000} = 1 - e^{-0.5} = 1 - 0.6065 = 0.3935$

5.73 a.   For $\alpha = 2$,

$\mu = \beta^{1/\alpha}\Gamma\left(\dfrac{\alpha+1}{\alpha}\right) = \beta^{1/2}\Gamma\left(\dfrac{2+1}{2}\right) = \beta^{1/2}\Gamma(1.5) = \beta^{1/2}(0.88623) = 0.88623\sqrt{\beta}$

b. $\sigma^2 = \beta^{2/\alpha}\left[\Gamma\left(\dfrac{\alpha+2}{\alpha}\right) - \Gamma^2\left(\dfrac{\alpha+1}{\alpha}\right)\right] = \beta^{2/2}\left[\Gamma\left(\dfrac{2+2}{2}\right) - \Gamma^2\left(\dfrac{2+1}{2}\right)\right]$

$= \beta[\Gamma(2) - \Gamma^2(1.5)] = \beta(1 - 0.88623^2)$

$= (1 - 0.7854)\beta = 0.2146\beta$

c. $P(Y > C) = 1 - P(Y \le C) = 1 - F(C)$

From Example 5.15, $F(y) = 1 - e^{-y^\alpha/\beta}$

$P(Y > C) = 1 - F(C) = 1 - (1 - e^{-C^2/\beta}) = e^{-C^2/\beta}$

5.75 $P(Y < C) = 0.05 \Rightarrow F(C) = 0.05 \Rightarrow 1 - e^{-C^\alpha/\beta} = 0.05 \Rightarrow 1 - e^{-C^2/60} = 0.05 \Rightarrow -e^{-C^2/60} = 0.95$

$\Rightarrow -C^2/60 = \ln(0.95) \Rightarrow -C^2/60 = -0.05129 \Rightarrow C^2 = 3.0774 \Rightarrow C = \sqrt{3.0774} = 1.7543$

5.77   If Y has a Weibull distribution, then

$$f(y) = \begin{cases} \dfrac{\alpha}{\beta} y^{\alpha-1} e^{-y^\alpha/\beta} & 0 \leq y < \infty \\ 0 & \text{elsewhere} \end{cases}$$

$$E(Y) = \int_0^\infty y \frac{\alpha}{\beta} y^{\alpha-1} e^{-y^\alpha/\beta} dy$$

Let $Z = Y^\alpha$. Then $dz = \alpha y^{\alpha-1} dy$. $y = 0 \Rightarrow z = 0^\alpha = 0$ and $y = \infty \Rightarrow z = \infty^\alpha = \infty$

Thus, $E(Y) = \displaystyle\int_0^\infty y \frac{\alpha}{\beta} y^{\alpha-1} e^{-y^\alpha/\beta} dy = \int_0^\infty \frac{1}{\beta} z^{1/\alpha} e^{-z/\beta} dz$

We know from the Gamma distribution that $\displaystyle\int_0^\infty y^{\alpha-1} e^{-y/\beta} dy = \Gamma(\alpha)\beta^\alpha$

So, $E(Y) = \displaystyle\int_0^\infty \frac{1}{\beta} z^{1/\alpha} e^{-z/\beta} dz = \frac{1}{\beta} \int_0^\infty z^{1/\alpha} e^{-z/\beta} dz = \frac{1}{\beta} \Gamma\left(\frac{1}{\alpha}+1\right)\beta^{\frac{1}{\alpha}+1} = \Gamma\left(\frac{\alpha+1}{\alpha}\right)\beta^{\frac{1}{\alpha}}$

5.79   The density function for the Beta distribution is

$$f(y) = \begin{cases} \dfrac{y^{\alpha-1}(1-y)^{\beta-1}}{B(\alpha,\beta)} & \text{if } 0 \leq y \leq 1; \alpha>0; \beta>0 \\ 0 & \text{elsewhere} \end{cases}$$

where $B(\alpha,\beta) = \dfrac{\Gamma(\alpha)\Gamma(\beta)}{\Gamma(\alpha+\beta)}$

For $\alpha = 3$ and $\beta = 2$, $B(3,2) = \dfrac{\Gamma(3)\Gamma(2)}{\Gamma(3+2)} = \dfrac{2!1!}{4!} = \dfrac{2}{24} = \dfrac{1}{12}$

Thus, $P(Y < 0.5) = \displaystyle\int_0^{0.5} \frac{y^{3-1}(1-y)^{2-1}}{1/12} dy = \int_0^{0.5} 12y^2(1-y)dy = 12\int_0^{0.5}(y^2 - y^3)dy$

$$= 12\int_0^{0.5}(y^2 - y^3)dy = 12\left(\frac{y^3}{3} - \frac{y^4}{4}\right)\Bigg|_0^{0.5}$$

$$= 12\left[\left(\frac{0.5^3}{3} - \frac{0.5^4}{4}\right) - \left(\frac{0^3}{3} - \frac{0^4}{4}\right)\right]$$

$$= 12(0.04167 - 0.15625) = 0.31254$$

5.81 a. $E(Y) = \mu = \dfrac{\alpha^*}{\alpha^* + \beta^*} = \dfrac{\alpha\beta}{\alpha\beta + \beta(1-\alpha)} = \dfrac{\alpha}{\alpha + (1-\alpha)} = \alpha$

   b. $\sigma^2 = \dfrac{\alpha^*\beta^*}{(\alpha^* + \beta^*)^2(\alpha^* + \beta^* + 1)} = \dfrac{\alpha\beta[\beta(1-\alpha)]}{(\alpha\beta + \beta(1-\alpha))^2(\alpha\beta + \beta(1-\alpha) + 1)}$

   $= \dfrac{\alpha\beta^2(1-\alpha)}{\beta^2(\beta+1)} = \dfrac{\alpha(1-\alpha)}{(\beta+1)}$

5.83 a. $E(Y) = \mu = \dfrac{\alpha}{\alpha + \beta} = \dfrac{1}{1+25} = \dfrac{1}{26} = 0.0385$

   $\sigma^2 = \dfrac{\alpha\beta}{(\alpha+\beta)^2(\alpha+\beta+1)} = \dfrac{1(25)}{(1+25)^2(1+25+1)} = \dfrac{25}{(26)^2(27)} = 0.00137$

   b. From Section 5.9, for cases where α and β are integers,

   $P(Y \leq p) = F(p) = \displaystyle\sum_{y=\alpha}^{n} p(y)$ where $p(y)$ is a binomial probability distribution with

   parameters p and $n = (\alpha + \beta - 1)$. Thus, for $n = (\alpha + \beta - 1) = 1 + 25 - 1 = 25$, $p = 0.01$,

   $P(Y > 0.01) = 1 - P(Y \leq 0.01) = 1 - F(0.01) = 1 - \displaystyle\sum_{y=1}^{25} p(y) = \sum_{y=0}^{0} p(y) = 0.7778$

   using Table 2, Appendix B

5.85 a. From Section 5.9, for cases where α and β are integers,

   $P(Y \leq p) = F(p) = \displaystyle\sum_{y=\alpha}^{n} p(y)$ where $p(y)$ is a binomial probability distribution with

   parameters p and $n = (\alpha + \beta - 1)$. Thus, for $n = (\alpha + \beta - 1) = 5 + 6 - 1 = 10$, $p = 0.60$,

   $P(Y \leq 0.60) = F(0.60) = \displaystyle\sum_{y=5}^{10} p(y) = 1 - \sum_{y=0}^{4} p(y) = 1 - 0.1662 = 0.8338$

   using Table 2, Appendix B

   b. For $n = (\alpha + \beta - 1) = 5 + 6 - 1 = 10$, $p = 0.80$

   $P(Y \geq 0.80) = 1 - P(Y \leq 0.80) = 1 - F(0.80) = 1 - \displaystyle\sum_{y=5}^{10} p(y) = \sum_{y=0}^{4} p(y) = 0.0064$

   using Table 2, Appendix B

5.87 We know $\displaystyle\int_{0}^{1} \dfrac{y^{\alpha-1}(1-y)^{\beta-1}}{\dfrac{\Gamma(\alpha)\Gamma(\beta)}{\Gamma(\alpha+\beta)}}\, dy = 1$

Thus, $\alpha - 1 = 5 \Rightarrow \alpha = 6$

$\beta - 1 = 2 \Rightarrow \beta = 3$

Therefore, $c = \dfrac{\Gamma(\alpha+\beta)}{\Gamma(\alpha)\Gamma(\beta)} = \dfrac{\Gamma(6+3)}{\Gamma(6)\Gamma(3)} = \dfrac{8!}{5!2!} = 168$

5.89　We know $f(y) = \begin{cases} \dfrac{y^{\alpha-1}(1-y)^{\beta-1}}{\dfrac{\Gamma(\alpha)\Gamma(\beta)}{\Gamma(\alpha+\beta)}} & 0 \le y \le 1 \\ \\ 0 & \text{elsewhere} \end{cases}$

If $\alpha = 1$ and $\beta = 1$, then $f(y) = \begin{cases} \dfrac{y^{1-1}(1-y)^{1-1}}{\dfrac{\Gamma(1)\Gamma(1)}{\Gamma(1+1)}} = \dfrac{y^0(1-y)^0}{\dfrac{0!0!}{1!}} = 1 & 0 \le y \le 1 \\ \\ 0 & \text{elsewhere} \end{cases}$

This is the density function of a uniform random variable on the interval $(0,1)$.

5.91　From the Key Formulas at the end of the chapter, the moment generating function of the normal random variable is $m(t) = e^{\mu t + (t^2\sigma^2/2)}$.

$$\mu_1' = \dfrac{dm(t)}{dt}\bigg]_{t=0} = \dfrac{d[e^{\mu t+(t^2\sigma^2/2)}]}{dt}\bigg]_{t=0} = e^{\mu t+(t^2\sigma^2/2)}(\mu + 2t\sigma^2/2)\bigg]_{t=0} = e^0(\mu + 2(0)\sigma^2/2) = \mu$$

$$\mu_2' = \dfrac{d^2m(t)}{dt^2}\bigg]_0 = \dfrac{d\left[e^{\mu t+(t^2\sigma^2/2)}(\mu + 2t\sigma^2/2)\right]}{dt}\bigg]_{t=0}$$

$$= e^{\mu t+(t^2\sigma^2/2)}(2\sigma^2/2) + (\mu + 2t\sigma^2/2)e^{\mu t+(t^2\sigma^2/2)}(\mu + 2t\sigma^2/2)\bigg]_{t=0}$$

$$= e^0\sigma^2 + (\mu + 2(0)\sigma^2/2)e^0(\mu + 2(0)\sigma^2/2) = \sigma^2 + \mu^2$$

$\mu = \mu_1' = \mu$

$\sigma^2 = \mu_2' - (\mu_1')^2 = \sigma^2 + \mu^2 - \mu^2 = \sigma^2$

5.93　$m(t) = E(e^{tY}) = \displaystyle\int_a^b \dfrac{e^{ty}}{b-a}\, dy = \dfrac{1}{b-a} \cdot \dfrac{e^{ty}}{t}\bigg]_a^b = \dfrac{1}{b-a}\left(\dfrac{e^{tb}}{t} - \dfrac{e^{ta}}{t}\right) = \dfrac{e^{tb} - e^{ta}}{t(b-a)}$

5.95　a.　Let Y = choice score with flexed arm. Then Y is normally distributed with $\mu = 59$ and $\sigma = 5$.

$$P(Y > 59) = P\left(Z > \dfrac{60-59}{5}\right) = P(Z > 0.2)$$

$$= 0.5 - P(0 < Z < 0.2) = 0.5 - 0.0793 = 0.4207$$

(using Table 5, Appendix B)

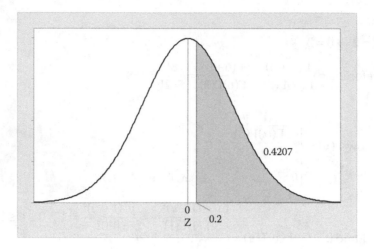

b. Let Y = choice score with extended arm. Then Y is normally distributed with $\mu = 43$ and $\sigma = 5$.

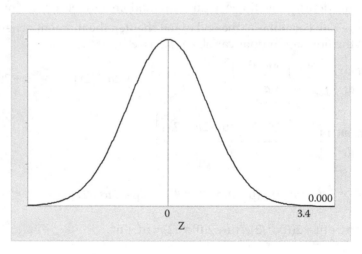

$$P(Y > 59) = P\left(Z > \frac{60-43}{5}\right) = P(Z > 3.4)$$

$$= 0.5 - P(0 < Z < 3.4) \approx 0.5 - 0.5 = 0$$

(using Table 5, Appendix B)

5.97 a. $\mu = \dfrac{a+b}{2} = \dfrac{6.5+7.5}{2} = 7$

$$\sigma^2 = \frac{(b-a)^2}{12} = \frac{(7.5-6.5)^2}{12} = \frac{1}{12} = 0.083333; \quad \sigma = \sqrt{0.083333} = 0.2887$$

$\mu \pm \sigma \Rightarrow 7 \pm 0.2887 \Rightarrow (6.71, 7.29); \mu \pm 2\sigma \Rightarrow 7 \pm 2(0.2887) \Rightarrow 7 \pm 0.5774 \Rightarrow (6.42, 7.58)$

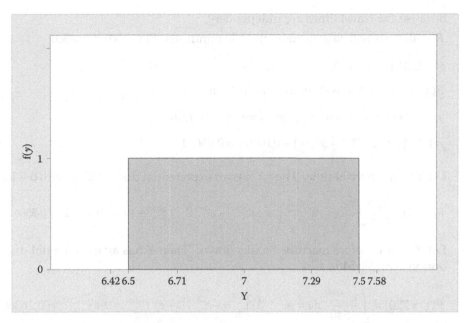

b. $P(Y > 7.2) = (7.5 - 7.2)(1) = 0.3$

5.99 a.   Let Y = length of connector module. Then the random variable Y has a normal distribution with $\mu = 0.3015$ and $\sigma = 0.0016$.

$P(Y < 0.304$ or $Y > 0.322) = P(Y < 0.304) + P(y > 0.322)$

$$= P\left(Z < \frac{0.304 - 0.3015}{0.0016} \right) + P\left(Z > \frac{0.322 - 0.3015}{0.0016} \right) = P(Z < 1.56) + P(Z > 12.81)$$

$$= 0.5 + P(0 < Z < 1.56) + 0.5 - P(0 < Z < 12.81) = 0.5 + 0.4406 + 0.5 - 0.5 = 0.9406$$

(using Table 5, Appendix B)

b.   Let Y = length of connector module. Then the random variable Y has a normal distribution with $\mu = 0.3146$ and $\sigma = 0.0030$.

$P\left(Y < 0.304$ or $Y > 0.322\right) = P\left(Y < 0.304\right) + P\left(Y > 0.322\right)$

$$= P\left(Z < \frac{0.304 - 0.3146}{0.0030} \right) + P\left(Z > \frac{0.322 - 0.3149}{0.0030} \right) = P(Z < -3.53) + P(Z > 2.47)$$

$$= 0.5 - P(-3.53 < Z < 0) + 0.5 - P(0 < Z < 2.47) = 0.5 - 0.5 + 0.5 - 0.4932 = 0.0068$$

(using Table 5, Appendix B)

5.101   Let Y = travel times of successive taxi trips. Then Y has an exponential distribution with $\beta = 20$.

a. $\mu = \beta = 20$

b. $P(Y > 30) = \int_{30}^{\infty} \frac{e^{-y/20}}{20} dy = -e^{-y/20} \Big]_{30}^{\infty} = -e^{-\infty/20} + e^{-30/20} = -e^{-\infty} + e^{-1.5} = 0.223130$

c. Because the travel times are independent,

$P(\text{both taxis will be gone more than 30 minutes}) = P(Y > 30)P(Y > 30)$

$= 0.22313^2 = 0.04979$

$P(\text{at least one taxi will return within 30 minutes})$

$= 1 - P(\text{both taxis will be gone more than 30 minutes})$

$= 1 - P(Y > 30)P(Y > 30) = 1 - 0.04979 = 0.95021$

5.103 a. Let Y = interarrival time. Then Y has an exponential distribution with $\beta = 1.25$.

$$P(Y < 1) = \int_0^1 \frac{e^{-y/1.25}}{1.25}\, dy = -e^{-y/1.25}\Big]_0^1 = -e^{-1/1.25} + e^{-0/1.25} = -e^{-0.8} + 1 = 1 - 0.449329 = 0.550671$$

b. Let Y = time before machine breaks down. Then Y has an exponential distribution with $\beta = 540$.

$$P(Y > 720) = \int_{720}^{\infty} \frac{e^{-y/540}}{540}\, dy = -e^{-y/540}\Big]_{720}^{\infty} = -e^{-\infty/540} + e^{-720/540} = 0 + e^{-1.3333} = 0.26360$$

c. Let Y = repair time. Then Y has a gamma distribution with $\alpha = 2$ and $\beta = 30$.

$\mu = \alpha\beta = 2(30) = 60$

$\sigma^2 = \alpha\beta^2 = 2(30)^2 = 1{,}800$

The average repair time is 60 minutes.

d. $$P(Y > 120) = \int_{120}^{\infty} \frac{y^{2-1}e^{-y/30}}{30^2\Gamma(2)}\, dy = \int_{120}^{\infty} \frac{ye^{-y/30}}{900}\, dy = -\frac{y}{30}e^{-y/30}\Big]_{120}^{\infty} + \int_{120}^{\infty} \frac{e^{-y/30}}{30}\, dy$$

$$= \lim_{y\to\infty}\left(-\frac{y}{30}e^{-y/30}\right) + \frac{120}{30}e^{-120/30} + (-e^{-y/30})\Big]_{120}^{\infty} = 0 + 4e^{-4} - e^{-\infty/30} + e^{-120/30}$$

$$= 0.07326 - 0 + e^{-4} = 0.07326 + 0.01832 = 0.09158$$

5.105 a. Let Y = service life. Then Y has a Weibull distribution with $\alpha = 1.5$ and $\beta = 110$.
From Example 5.15, $F(y) = 1 - e^{-y^\alpha/\beta}$.

$$P(Y < 12.2) = F(12.2) = (1 - e^{-12.2^{1.5}/110}) = 1 - e^{-0.38739} = 1 - 0.6788 = 0.3212$$

b. $$P(Y < 12.2) = \int_0^{12.2} \frac{e^{-y/110}}{110}\, dy = -e^{-y/110}\Big]_0^{12.2} = -e^{-12.2/110} + e^{-0/110} = 1 - e^{-0.1109} = 1 - 0.8950 = 0.1050$$

5.107 Let Y = number of facies bodies required to satisfactorily estimate P. Then Y has an approximate normal distribution with $\mu = 99$ and $\sigma = 4.3$.

$$P(Y < a) = 0.99 \Rightarrow P\left(Z < \frac{a-99}{4.3}\right) = P(Z < z_0) = 0.99$$

$A_1 = 0.99 - 0.50 = 0.4900$. Looking up area 0.4900 in Table 5 gives $z_0 = 2.33$

$$\Rightarrow z_0 = \frac{a-99}{4.3} = 2.33 \Rightarrow a - 99 = 10.019 \Rightarrow a = 109.019$$

(using Table 5, Appendix B)

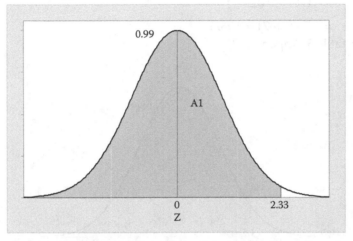

5.109 a. The histogram of the data shown is much less spread out than the normal histogram that is superimposed over the data. The normal distribution does not do an adequate job of modeling the data.

b. The interval $\mu \pm 2s$ will contain more than 95% of the 400 elevation differences. We would expect 95% of the data to fall in this interval for a normal distribution. Because the spread of the elevation difference data is less than that of a normal distribution, we would expect a higher percentage in the interval $\mu \pm 2s$.

5.111 Let Y = time of accident. Then Y has a uniform distribution on the interval from 0 to 30.

$$P(Y > 25) = (30 - 25)\left(\frac{1}{30}\right) = \frac{5}{30} = \frac{1}{6}$$

5.113 a. $8 = \alpha - 1 \Rightarrow \alpha = 9$ and $1 = \beta - 1 \Rightarrow \beta = 2$

b. $\mu = \dfrac{\alpha}{\alpha + \beta} = \dfrac{9}{9+2} = \dfrac{9}{11} = 0.8182$

$$\sigma^2 = \frac{\alpha\beta}{(\alpha+\beta)^2(\alpha+\beta+1)} = \frac{9(2)}{(9+2)^2(9+2+1)} = \frac{18}{(11)^2(12)} = 0.0124$$

c. From Section 5.9, for cases where α and β are integers,

$P(Y \leq p) = F(p) = \sum_{y=\alpha}^{n} p(y)$ where $p(y)$ is a binomial probability distribution with

parameters p and $n = (\alpha + \beta - 1)$. Thus, for $n = (\alpha + \beta - 1) = 9 + 2 - 1 = 10$, $p = 0.80$,

$P(Y > 0.80) = 1 - P(Y \leq 0.80) = 1 - F(0.80) = 1 - \sum_{y=9}^{10} p(y) = \sum_{y=0}^{8} p(y) = 0.6242$

(using Table 2, Appendix B)

5.115 a. $P(Y \geq 11) = P\left(Z \geq \dfrac{11 - 13.6}{\sqrt{2}} \right) = P(Z \geq -1.84)$

$= 0.5 + 0.4671 = 0.9671$

(using Table 5, Appendix B)

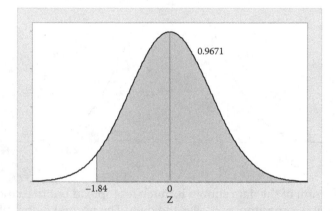

b. $P(Y \geq 11) = P\left(Z \geq \dfrac{11 - 10.1}{\sqrt{2}} \right) = P(Z \geq 0.64)$

$= 0.5 - P(0 < Z < 0.64) = 0.5 - 0.2389 = 0.2611$

(using Table 5, Appendix B)

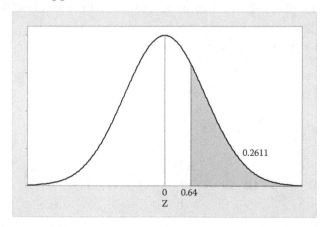

c. No. To make the probability in part a larger, C would have to be smaller than 11. To make the probability in part a smaller, C would have to be larger than 11.

5.117 a. $\displaystyle\int_0^\infty ce^{-y}\,dy = 1 \Rightarrow -ce^{-y}\Big]_0^\infty = -ce^{-\infty} + ce^{-0} = 0 + c = 1 \Rightarrow c = 1$

b. $\displaystyle F(y) = \int_0^y f(t)\,dt = \int_0^y e^{-t}\,dt = -e^{-t}\Big]_0^y = -e^{-y} + e^{-0} = 1 - e^{-y}$

c. $F(2.6) = 1 - e^{-2.6} = 1 - 0.07427 = 0.92573$

d. $F(0) = 1 - e^{-0} = 1 - 1 = 0 \qquad\qquad F(\infty) = 1 - e^{-\infty} = 1 - 0 = 1$

e. $P(1 \le Y \le 5) = F(5) - F(1) = (1 - e^{-5}) - (1 - e^{-1}) = (1 - 0.00674) - (1 - 0.36788)$

$\qquad\qquad = 0.99326 - 0.63212 = 0.36114$

5.119 $\displaystyle\int_0^1 f(y)\,dy = \int_0^\infty [wg(y,\lambda_1) + (1-w)g(1-y,\lambda_2)]\,dy = \int_0^\infty \left[w\,\frac{(1-y)e^{\lambda_1 y}}{\displaystyle\int_0^1 (1-y)e^{\lambda_1 y}\,dy} + (1-w)\,\frac{ye^{\lambda_2 y}}{\displaystyle\int_0^1 ye^{\lambda_2 y}\,dy} \right] dy$

Now, $\displaystyle\int_0^1 (1-y)e^{\lambda_1 y}\,dy$ is a constant with respect to y once it has been evaluated.

Similarly, $\displaystyle\int_0^1 ye^{\lambda_2 y}\,dy$ is a constant with respect to y.

Thus, $\displaystyle\int_0^\infty \left[w\,\frac{(1-y)e^{\lambda_1 y}}{\displaystyle\int_0^1 (1-y)e^{\lambda_1 y}\,dy} + (1-w)\,\frac{ye^{\lambda_2 y}}{\displaystyle\int_0^1 ye^{\lambda_2 y}\,dy} \right] dy$

$\displaystyle = \frac{w\displaystyle\int_0^1 (1-y)e^{\lambda_1 y}\,dy}{\displaystyle\int_0^1 (1-y)e^{\lambda_1 y}\,dy} + \frac{(1-w)\displaystyle\int_0^1 ye^{\lambda_2 y}\,dy}{\displaystyle\int_0^1 ye^{\lambda_2 y}\,dy} = w + (1-w) = 1$

6

Bivariate Probability Distributions and Sampling Distributions

6.1   a.   The properties for a discrete bivariate probability distribution are:

        1. $0 \le p(x,y) \le 1$ for all values of X and Y

        2. $\displaystyle\sum_y \sum_x p(x,y) = 1$

From the table of probabilities, it is clear that $0 \le p(x,y) \le 1$

$$\sum_y \sum_x p(x,y) = 0 + 0.050 + 0.025 + 0 + 0.025 + 0 + 0.200 + 0.050 + 0 + 0.300$$

$$+ 0 + 0 + 0.100 + 0 + 0 + 0 + 0.100 + 0.150 = 1$$

b.   To find the marginal probability distribution for X, we need to find $P(X = 0)$, $P(X = 1)$, $P(X = 2)$, $P(X = 3)$, $P(X = 4)$, $P(X = 5)$. Since $X = 0$ can occur when $Y = 0,1$ or 2 occurs, then $P(X = 0) = p_1(0)$ is calculated by summing the probabilities of 3 mutually exclusive events:

$$P(X = 0) = p_1(0) = p(0,0) + p(0,1) + p(0,2) = 0 + 0.200 + 0.100 = 0.300$$

Similarly,

$$P(X = 1) = p_1(1) = p(1,0) + p(1,1) + p(1,2) = 0.050 + 0.050 + 0 = 0.100$$

$$P(X = 2) = p_1(2) = p(2,0) + p(2,1) + p(2,2) = 0.025 + 0 + 0 = 0.025$$

$$P(X = 3) = p_1(3) = p(3,0) + p(3,1) + p(3,2) = 0 + 0.300 + 0 = 0.300$$

$$P(X = 4) = p_1(4) = p(4,0) + p(4,1) + p(4,2) = 0.025 + 0 + 0.100 = 0.125$$

$$P(X = 5) = p_1(5) = p(5,0) + p(5,1) + p(5,2) = 0 + 0 + 0.150 = 0.150$$

The marginal distribution $p_1(x)$ is given as:

x	0	1	2	3	4	5
$p_1(x)$	0.300	0.100	0.025	0.300	0.125	0.150

c.   The marginal probability distribution for Y is found as in part a.

$$P(Y = 0) = p_2(0) = p(0,0) + p(1,0) + p(2,0) + p(3,0) + p(4,0) + p(5,0)$$

$$= 0 + 0.050 + 0.025 + 0 + 0.025 + 0 = 0.100$$

$$P(Y = 1) = p_2(1) = p(0,1) + p(1,1) + p(2,1) + p(3,1) + p(4,1) + p(5,1)$$

$$= 0.200 + 0.050 + 0 + 0.300 + 0 + 0 = 0.550$$

$$P(Y = 2) = p_2(2) = p(0,2) + p(1,2) + p(2,2) + p(3,2) + p(4,2) + p(5,2)$$

$$= 0.100 + 0 + 0 + 0 + 0.100 + 0.150 = 0.350$$

The marginal distribution $p_2(y)$ is given as:

y	0	1	2
$p_2(y)$	0.100	0.550	0.350

d.   The conditional probability of X given Y is $p_1(x \mid y) = \dfrac{p(x,y)}{p_2(y)}$

Since there are 3 levels of Y, there are 3 conditional probability distributions for X.

When $Y = 0$, $p_1(x \mid 0) = \dfrac{p(x,0)}{p_2(0)}$

$$p_1(0 \mid 0) = \frac{p(0,0)}{p_2(0)} = \frac{0}{0.100} = 0 \qquad\qquad p_1(1 \mid 0) = \frac{p(1,0)}{p_2(0)} = \frac{0.050}{0.100} = 0.500$$

$$p_1(2 \mid 0) = \frac{p(2,0)}{p_2(0)} = \frac{0.025}{0.100} = 0.250 \qquad\qquad p_1(3 \mid 0) = \frac{p(3,0)}{p_2(0)} = \frac{0}{0.100} = 0$$

$$p_1(4 \mid 0) = \frac{p(4,0)}{p_2(0)} = \frac{0.025}{0.100} = 0.250 \qquad\qquad p_1(5 \mid 0) = \frac{p(5,0)}{p_2(0)} = \frac{0}{0.100} = 0$$

The conditional probability distribution of X given $Y = 0$ is given in the table:

x	0	1	2	3	4	5
$p_1(x \mid 0)$	0	0.500	0.250	0	0.250	0

When $Y = 1$, $p_1(x \mid 1) = \dfrac{p(x,1)}{p_2(1)}$

$$p_1(0 \mid 1) = \frac{p(0,1)}{p_2(1)} = \frac{0.200}{0.550} = 0.364 \qquad\qquad p_1(1 \mid 1) = \frac{p(1,1)}{p_2(1)} = \frac{0.050}{0.550} = 0.091$$

$$p_1(2 \mid 1) = \frac{p(2,1)}{p_2(1)} = \frac{0}{0.550} = 0 \qquad\qquad p_1(3 \mid 1) = \frac{p(3,1)}{p_2(1)} = \frac{0.300}{0.550} = 0.545$$

$$p_1(4 \mid 1) = \frac{p(4,1)}{p_2(1)} = \frac{0}{0.550} = 0 \qquad\qquad p_1(5 \mid 1) = \frac{p(5,1)}{p_2(1)} = \frac{0}{0.550} = 0$$

The conditional probability distribution of X given $Y = 1$ is given in the table:

x	0	1	2	3	4	5
$p_1(x \mid 1)$	0.364	0.091	0	0.545	0	0

When $Y = 2$, $p_1(x \mid 2) = \dfrac{p(x,2)}{p_2(2)}$

$$p_1(0 \mid 2) = \frac{p(0,2)}{p_2(2)} = \frac{0.100}{0.350} = 0.286 \qquad\qquad p_1(1 \mid 2) = \frac{p(1,2)}{p_2(2)} = \frac{0}{0.350} = 0$$

$$p_1(2 \mid 2) = \frac{p(2,2)}{p_2(2)} = \frac{0}{0.350} = 0 \qquad\qquad p_1(3 \mid 2) = \frac{p(3,2)}{p_2(2)} = \frac{0}{0.350} = 0$$

$$p_1(4\,|\,2)=\frac{p(4,2)}{p_2(2)}=\frac{0.100}{0.350}=0.286 \qquad p_1(5\,|\,2)=\frac{p(5,2)}{p_2(2)}=\frac{0.150}{0.350}=0.429$$

The conditional probability distribution of X given $Y=2$ is given in the table:

x	0	1	2	3	4	5
$p_1(x\,|\,2)$	0.286	0	0	0	0.286	0.429

e. Similar to part d, the conditional probability distribution of Y given X is $p_2(y\,|\,x)=\dfrac{p(x,y)}{p_1(x)}$. Since there are 6 levels of X, there are 6 conditional distributions of Y.

When $X=0$, $p_2(y\,|\,0)=\dfrac{p(0,y)}{p_1(0)}$

$$p_2(0\,|\,0)=\frac{p(0,0)}{p_1(0)}=\frac{0}{0.300}=0 \qquad p_2(1\,|\,0)=\frac{p(0,1)}{p_1(0)}=\frac{0.200}{0.300}=0.667$$

$$p_2(2\,|\,0)=\frac{p(0,2)}{p_1(0)}=\frac{0.100}{0.300}=0.333$$

The conditional probability distribution of Y given $X=0$ is given in the table:

y	0	1	2
$p_2(y\,|\,0)$	0	0.667	0.333

When $X=1$, $p_2(y\,|\,1)=\dfrac{p(1,y)}{p_1(1)}$

$$p_2(0\,|\,1)=\frac{p(1,0)}{p_1(1)}=\frac{0.050}{0.100}=0.500 \qquad p_2(1\,|\,1)=\frac{p(1,1)}{p_1(1)}=\frac{0.050}{0.100}=0.500$$

$$p_2(2\,|\,1)=\frac{p(1,2)}{p_1(1)}=\frac{0}{0.100}=0$$

The conditional probability distribution of Y given $X=1$ is given in the table:

y	0	1	2
$p_2(y\,|\,1)$	0.500	0.500	0

When $X=2$, $p_2(y\,|\,2)=\dfrac{p(2,y)}{p_1(2)}$

$$p_2(0\,|\,2)=\frac{p(2,0)}{p_1(2)}=\frac{0.025}{0.025}=1.000 \qquad p_2(1\,|\,2)=\frac{p(2,1)}{p_1(2)}=\frac{0}{0.025}=0$$

$$p_2(2\,|\,2)=\frac{p(2,2)}{p_1(2)}=\frac{0}{0.025}=0$$

The conditional probability distribution of Y given $X = 2$ is given in the table:

y	0	1	2
$p_2(y\mid 2)$	0.1000	0	0

When $X = 3$, $p_2(y\mid 3) = \dfrac{p(3, y)}{p_1(3)}$

$$p_2(0\mid 3) = \frac{p(3,0)}{p_1(3)} = \frac{0}{0.300} = 0 \qquad\qquad p_2(1\mid 3) = \frac{p(3,1)}{p_1(3)} = \frac{0.300}{0.300} = 1.000$$

$$p_2(2\mid 3) = \frac{p(3,2)}{p_1(3)} = \frac{0}{0.300} = 0$$

The conditional probability distribution of Y given $X = 3$ is given in the table:

y	0	1	2
$p_2(y\mid 3)$	0	1.0000	0

When $X = 4$, $p_2(y\mid 4) = \dfrac{p(4, y)}{p_1(4)}$

$$p_2(0\mid 4) = \frac{p(4,0)}{p_1(4)} = \frac{0.025}{0.125} = 0.200 \qquad\qquad p_2(1\mid 4) = \frac{p(4,1)}{p_1(4)} = \frac{0}{0.125} = 0$$

$$p_2(2\mid 4) = \frac{p(4,2)}{p_1(4)} = \frac{0.100}{0.125} = 0.800$$

The conditional probability distribution of Y given $X = 4$ is given in the table:

y	0	1	2
$p_2(y\mid 4)$	0.200	0	0.800

When $X = 5$, $p_2(y\mid 5) = \dfrac{p(5, y)}{p_1(5)}$

$$p_2(0\mid 5) = \frac{p(5,0)}{p_1(5)} = \frac{0}{0.150} = 0 \qquad\qquad p_2(1\mid 5) = \frac{p(5,1)}{p_1(5)} = \frac{0}{0.150} = 0$$

$$p_2(2\mid 5) = \frac{p(5,2)}{p_1(5)} = \frac{0.150}{0.150} = 1.000$$

The conditional probability distribution of Y given $X = 5$ is given in the table:

y	0	1	2
$p_2(y\mid 5)$	0	0	1.000

6.3   a.   For this example, X = number of genuine balls selected and Y = number of blue balls selected when 2 balls are drawn. Then $X = 0, 1,$ or 2 and $Y = 0, 1,$ or 2. There is a combination of 10 balls taken 2 at a time or

$$\binom{10}{2} = \frac{10!}{2!8!} = \frac{10 \cdot 9 \cdot 8!}{2!8!} = \frac{10 \cdot 9}{2} = 45 \text{ total ways to draw 2 balls from 10.}$$

Of the 3 yellow balls, 1 is genuine and 2 are not. Let Y_1 designate the genuine yellow ball and Y_2 and Y_3 designate the non-genuine yellow balls. Of the 2 blue balls, 1 is genuine (B_1) and 1 is non-genuine (B_2). The one red ball is genuine (R_1). Of the 3 purple balls, 1 is genuine and 2 are not. Let P_1 designate the genuine purple ball and P_2 and P_3 designate the non-genuine purple balls. The one orange ball is genuine (O_1).

The event $(0,0)$ is the event that neither of the balls selected is genuine and neither of the balls is blue. There are 2 yellow balls and 2 purple balls that are neither genuine nor blue. There are 6 ways to get 2 of these balls – $(Y_2, Y_3), (Y_2, P_2), (Y_2, P_3), (Y_3, P_2), (Y_3, P_3), (P_2, P_3)$.

Thus, $p(0,0) = \dfrac{6}{45}$.

The event $(0,1)$ is the event that neither of the balls selected is genuine and one ball is blue. There are 4 ways to get this combination – $(Y_2, B_2), (Y_3, B_2), (P_2, B_2), (P_3, B_2)$.

Thus, $p(0,1) = \dfrac{6}{45}$.

The event $(0,2)$ is the event that neither of the balls selected is genuine and two balls are blue. There are no ways to get this combination.

Thus, $p(0,2) = \dfrac{0}{45} = 0$.

The event $(1,0)$ is the event that one ball selected is genuine and neither ball is blue. There are 16 ways to get this combination – $(Y_1, Y_2), (Y_1, Y_3), (Y_1, P_2), (Y_1, P_3), (Y_2, R_1), (Y_2, P_1), (Y_2, O_1),$

$(Y_3, R_1), (Y_3, P_1), (Y_3, O_1), (R_1, P_2), (R_1, P_3), (P_1, P_2), (P_1, P_3), (P_2, O_1), (P_3, O_1)$.

Thus, $p(1,0) = \dfrac{16}{45}$.

The event $(1,1)$ is the event that one ball selected is genuine and one ball is blue. There are 8 ways to get this combination – $(Y_1, B_2), (Y_2, B_1), (Y_3, B_1), (B_1, P_2), (B_1, P_3), (B_2, R_1), (B_2, P_1), (B_2, O_1)$.

Thus, $p(1,1) = \dfrac{8}{45}$.

The event $(1,2)$ is the event that one ball selected is genuine and two balls are blue. There is 1 way to get this combination – (B_1, B_2).

Thus, $p(1,2) = \dfrac{1}{45}$.

The event $(2,0)$ is the event that two balls selected are genuine and neither ball is blue. There are 6 ways to get this combination – $(Y_1, R_1), (Y_1, P_1), (Y_1, O_1), (R_1, P_1), (R_1, O_1), (P_1, O_1)$.

Thus, $p(2,0) = \dfrac{6}{45}$.

The event $(2,1)$ is the event that two balls selected are genuine and one ball is blue. There are 4 ways to get this combination – $(Y_1, B_1), (B_1, R_1), (B_1, P_1), (B_1, O_1)$.

Thus, $p(2,1) = \dfrac{4}{45}$.

The event $(2,2)$ is the event that two balls selected are genuine and two balls are blue. There are 0 ways to get this combination.

Thus, $p(2,2) = \dfrac{0}{45}$.

In table form, the bivariate probability distribution $p(x,y)$ is:

		x		
		0	1	2
	0	6/45	16/45	6/45
y	1	4/45	8/45	4/45
	2	0	0	0

b. $P(X = 0) = p_1(0) = p(0,0) + p(0,1) + p(0,2) = \dfrac{6}{45} + \dfrac{4}{45} + 0 = \dfrac{10}{45}$

$P(X = 1) = p_1(1) = p(1,0) + p(1,1) + p(1,2) = \dfrac{16}{45} + \dfrac{8}{45} + \dfrac{1}{45} = \dfrac{25}{45}$

$P(X = 2) = p_1(2) = p(2,0) + p(2,1) + p(2,2) = \dfrac{6}{45} + \dfrac{4}{45} + 0 = \dfrac{10}{45}$

In table form, the marginal probability distribution $p_1(x)$ is:

x	0	1	2
$p_1(x)$	10/45	25/45	10/45

c. $P(Y = 0) = p_2(0) = p(0,0) + p(1,0) + p(2,0) = \dfrac{6}{45} + \dfrac{16}{45} + \dfrac{6}{45} = \dfrac{28}{45}$

$P(Y = 1) = p_2(1) = p(0,1) + p(1,1) + p(2,1) = \dfrac{4}{45} + \dfrac{8}{45} + \dfrac{4}{45} = \dfrac{16}{45}$

$P(Y = 2) = p_2(2) = p(0,2) + p(1,2) + p(2,2) = 0 + \dfrac{1}{45} + 0 = \dfrac{1}{45}$

In table form, the marginal distribution $p_2(y)$ is given as:

y	0	1	2
$p_2(y)$	28/45	16/45	1/45

d. $p_2(2) = \dfrac{1}{45}$

6.5 a. The probabilities in the problem are the probabilities of a speed given a particular section. Thus, $p_2(y \mid x)$.

b. If the sections are of equal length, a vehicle will be equally likely to be in any section. Thus, $p_1(1) = p_1(2) = p_1(3) = \dfrac{1}{3}$.

c. The probabilities given in the problem are $p_2(y \mid x)$. We want to find $p(x, y)$. We know that $p_2(y \mid x) = \dfrac{p(x, y)}{p_1(x)} \Rightarrow p(x, y) = p_2(y \mid x) p_1(x)$. In part b, we found $p_1(x) = \dfrac{1}{3}$ for all values of x. Therefore, to find $p(x, y)$, we multiply all the probabilities given in the problem by $\dfrac{1}{3}$

The bivariate probability distribution $p(x, y)$ in table form is:

		\multicolumn{3}{c}{x}		
		1	2	3
	30	0.0200	0.0333	0.0500
	40	0.0800	0.0800	0.0600
y	50	0.0800	0.1200	0.1000
	60	0.1533	0.1000	0.1233

6.7 a. The marginal probability distribution of Y is:

$P(Y = 0) = p_2(0) = p(0,0) + p(1,0) + p(2,0) + p(3,0) = 0.01 + 0.02 + 0.07 + 0.01 = 0.11$
$P(Y = 1) = p_2(1) = p(0,1) + p(1,1) + p(2,1) + p(3,1) = 0.03 + 0.06 + 0.10 + 0.06 = 0.25$
$P(Y = 2) = p_2(2) = p(0,2) + p(1,2) + p(2,2) + p(3,2) = 0.05 + 0.12 + 0.15 + 0.08 = 0.40$
$P(Y = 3) = p_2(3) = p(0,3) + p(1,3) + p(2,3) + p(3,3) = 0.02 + 0.09 + 0.08 + 0.05 = 0.24$

In table form, the distribution is:

y	0	1	2	3
$p_2(y)$	0.11	0.25	0.40	0.24

b. The conditional probability distribution of Y given $X = 2$ is $p_2(y \mid 2) = \dfrac{p(2, y)}{p_1(2)}$.

$P(X = 2) = p_1(2) = p(2,0) + p(2,1) + p(2,2) + p(2,3) = 0.07 + 0.10 + 0.15 + 0.08 = 0.40$

When $X = 2$, $p_2(y \mid 2) = \dfrac{p(2, y)}{p_1(2)}$

$p_2(0 \mid 2) = \dfrac{p(2,0)}{p_1(2)} = \dfrac{0.07}{0.40} = 0.175$ $\qquad$ $p_2(1 \mid 2) = \dfrac{p(2,1)}{p_1(2)} = \dfrac{0.10}{0.40} = 0.25$

$p_2(2 \mid 2) = \dfrac{p(2,2)}{p_1(2)} = \dfrac{0.15}{0.40} = 0.375$ $\qquad$ $p_2(3 \mid 2) = \dfrac{p(2,3)}{p_1(2)} = \dfrac{0.08}{0.40} = 0.20$

The conditional probability distribution of Y given $X = 2$ is given in the table:

y	0	1	2	3
$p_2(y \mid 2)$	0.175	0.250	0.375	0.200

6.9   a.   The probabilities are defined as $p(x, y) = \dfrac{1}{n}$ if the ranked pair is contained in the sample, 0 if not. The observed sample contains the similarity values X and Y as follows:

$$(X_1 = 75,\ Y_1 = 60),\ (X_2 = 30,\ Y_2 = 80),\ (X_3 = 15,\ Y_3 = 5)$$

Observation one has ranks $X_{(1)} = 1$ and $Y_{(2)} = 2$.
Observation two has ranks $X_{(2)} = 2$ and $Y_{(1)} = 1$.
Observation one has ranks $X_{(3)} = 3$ and $Y_{(3)} = 3$.
The Copula distribution $p(x, y)$ is shown in the table:

			X	
		1	2	3
	1	0	1/3	0
Y	2	1/3	0	0
	3	0	0	1/3

b.   If each of the algorithms agrees on the signature match, then the rankings for each of the observations will be the same. We would get the following Copula distribution:

			X	
		1	2	3
	1	1/3	0	0
Y	2	0	1/3	0
	3	0	0	1/3

6.11   a.   Verify $F_1(a) = \displaystyle\sum_{x \le a} \sum_{y} p(x, y)$

By definition, $\displaystyle\sum_{y} p(x, y) = p_1(x)$

Thus, $F_1(a) = \displaystyle\sum_{x \le a} \sum_{y} p(x, y) = \sum_{x \le a} p_1(x) = P(X \le a) = F_1(a)$

b.   Verify $F_1(a \mid y) = \dfrac{\displaystyle\sum_{x \le a} p(x, y)}{p_2(y)}$

By definition, $p_1(x \mid y) = \dfrac{p(x, y)}{p_2(y)}$

Given $F_1(a \mid y) = P(X \le a \mid y)$

Thus, $F_1(a \mid y) = P(X \le a \mid y) = \dfrac{\sum\limits_{x \le a} p(x, y)}{p_2(y)}$

6.13 a. $\displaystyle\int\limits_0^\infty \int\limits_{80}^{120} \frac{e^{-x}}{40}\, dy\, dx = \frac{1}{40} \int\limits_0^\infty e^{-x} y \Big]_{80}^{120}\, dx = \frac{1}{40} \int\limits_0^\infty e^{-x}(120 - 80)\, dx$

$$= \int\limits_0^\infty e^{-x}\, dx = -e^{-x} \Big]_0^\infty = -e^{-\infty} + e^0 = 1$$

b. $f_1(x) = \displaystyle\int\limits_{80}^{120} \frac{e^{-x}}{40}\, dy = \frac{1}{40} e^{-x} y \Big]_{80}^{120} = \frac{1}{40} e^{-x}(120 - 80) = e^{-x}$

This is an exponential distribution with $\beta = 1$.

c. $f_2(y) = \displaystyle\int\limits_0^\infty \frac{e^{-x}}{40}\, dx = -\frac{1}{40} e^{-x} \Big]_0^\infty = -\frac{1}{40}(e^{-\infty} - e^0) = \frac{1}{40}$

This is a uniform distribution on the interval $(80, 120)$.

6.15 a. To verify that $f(x, y)$ is a bivariate joint probability distribution function, we must show:

1. $f(x, y) \ge 0$ for all x, y

2. $\displaystyle\int\limits_{-\infty}^\infty \int\limits_{-\infty}^\infty f(x, y)\, dy\, dx = 1$

Show that $f(x, y) \ge 0$

For $x < 0, -\infty < y < \infty$ or $x > 2, -\infty < y < \infty$, $f(x, y) = 0$

For $y < 0, -\infty < x < \infty$ or $y > 2, -\infty < x < \infty$, $f(x, y) = 0$

For $0 \le x \le 1, 0 \le y \le 1$, $f(x, y) = xy \ge 0$

For $1 \le x \le 2, 0 \le y \le 1$, $f(x, y) = (2 - x) \ge 0$

For $0 \le x \le 1, 1 \le y \le 2$, $f(x, y) = x(2 - y) \ge 0$

For $1 \le x \le 2, 1 \le y \le 2$, $f(x, y) = (2 - x)(2 - y) \ge 0$

Thus, $f(x, y) \ge 0$.

Show $\displaystyle\int\limits_{-\infty}^\infty \int\limits_{-\infty}^\infty f(x, y)\, dx\, dy = 1$

$\displaystyle\int\limits_{-\infty}^\infty \int\limits_{-\infty}^\infty f(x, y)\, dx\, dy$

$= \displaystyle\int\limits_0^1 \int\limits_0^1 xy\, dx\, dy + \int\limits_0^1 \int\limits_1^2 (2 - x)y\, dx\, dy + \int\limits_1^2 \int\limits_0^1 x(2 - y)\, dx\, dy + \int\limits_1^2 \int\limits_1^2 (2 - x)(2 - y)\, dx\, dy$

$$= \int_0^1 y\frac{x^2}{2}\Big]_0^1 dy + \int_0^1 \left(2yx - y\frac{x^2}{2}\right)\Big]_1^2 dy + \int_1^2 \left(x^2 - y\frac{x^2}{2}\right)\Big]_0^1 dy + \int_1^2 \left(4x - x^2 - 2yx + y\frac{x^2}{2}\right)\Big]_1^2 dy$$

$$= \int_0^1 y\frac{1^2}{2} dy + \int_0^1 \left(2y(2) - y\frac{2^2}{2} - 2y(1) + y\frac{1^2}{2}\right) dy + \int_1^2 \left(1^2 - y\frac{1^2}{2}\right) dy$$

$$+ \int_1^2 \left(4(2) - 2^2 - 2y(2) + y\frac{2^2}{2} - 4(1) + 1^2 + 2y(1) - y\frac{1^2}{2}\right) dy$$

$$= \int_0^1 \frac{y}{2} dy + \int_0^1 \frac{y}{2} dy + \int_1^2 \left(1 - \frac{y}{2}\right) dy + \int_1^2 \left(1 - \frac{y}{2}\right) dy$$

$$= \frac{y^2}{4}\Big]_0^1 + \frac{y^2}{4}\Big]_0^1 + \left(y - \frac{y^2}{4}\right)\Big]_1^2 + \left(y - \frac{y^2}{4}\right)\Big]_1^2$$

$$= \frac{1^2}{4} + \frac{1^2}{4} + \left(2 - \frac{2^2}{4} - 1 + \frac{1^2}{4}\right) + \left(2 - \frac{2^2}{4} - 1 + \frac{1^2}{4}\right)$$

$$= \frac{1}{4} + \frac{1}{4} + \frac{1}{4} + \frac{1}{4} = 1$$

Thus, $f(x, y)$ is a bivariate joint probability distribution function.

b. $P(X > 0.8, Y > 0.8) = \int_{0.8}^1 \int_{0.8}^1 xy\, dx\, dy + \int_{0.8}^1 \int_1^2 (2 - x)y\, dx\, dy + \int_1^2 \int_{0.8}^1 x(2 - y)\, dx\, dy$

$$+ \int_1^2 \int_1^2 (2 - x)(2 - y)\, dx\, dy$$

$$= \int_{0.8}^1 y\frac{x^2}{2}\Big]_{0.8}^1 dy + \int_{0.8}^1 \left(2yx - y\frac{x^2}{2}\right)\Big]_1^2 dy + \int_1^2 \left(x^2 - y\frac{x^2}{2}\right)\Big]_{0.8}^1 dy + \int_1^2 \left(4x - x^2 - 2yx + y\frac{x^2}{2}\right)\Big]_1^2 dy$$

$$= \int_{0.8}^1 y\left(\frac{1^2}{2} - \frac{0.8^2}{2}\right) dy + \int_{0.8}^1 \left(2y(2) - y\frac{2^2}{2} - 2y(1) + y\frac{1^2}{2}\right) dy + \int_1^2 \left(1^2 - y\frac{1^2}{2} - 0.8^2 + y\frac{0.8^2}{2}\right) dy$$

$$+ \int_1^2 \left(4(2) - 2^2 - 2y(2) + y\frac{2^2}{2} - 4(1) + 1^2 + 2y(1) - y\frac{1^2}{2}\right) dy$$

$$= \int_{0.8}^1 0.18y\, dy + \int_{0.8}^1 \frac{y}{2} dy + \int_1^2 (0.36 - 0.18y)\, dy + \int_1^2 \left(1 - \frac{y}{2}\right) dy$$

$$= 0.09y^2 \Big]_{0.8}^{1} + \frac{y^2}{4}\Big]_{0.8}^{1} + (0.36y - 0.09y^2)\Big]_{1}^{2} + \left(y - \frac{y^2}{4}\right)\Big]_{1}^{2}$$

$$= 0.09(1^2 - 0.8^2) + \frac{1}{4}(1^2 - 0.8^2) + \left(0.36(2) - 0.09(2)^2 - 0.36(1) + 0.09(1)^2\right) + \left(2 - \frac{2^2}{4} - 1 + \frac{1^2}{4}\right)$$

$$= 0.0324 + 0.09 + 0.09 + 0.25 = 0.4624$$

6.17 a. We know that $\displaystyle\int_{-\infty}^{\infty}\int_{-\infty}^{\infty} f(x,y)\,dx\,dy = 1$

Therefore, $\displaystyle\int_{-\infty}^{\infty}\int_{-\infty}^{\infty}(x+cy)\,dx\,dy = \int_{0}^{1}\int_{1}^{2}(x+cy)\,dx\,dy = \int_{0}^{1}\left(\frac{x^2}{2}+cyx\right)\Big]_{1}^{2}\,dy$

$$= \int_{0}^{1}\left(\frac{2^2}{2}+cy(2)-\frac{1^2}{2}-cy(1)\right)dy = \int_{0}^{1}\left(\frac{3}{2}+cy\right)dy = \frac{3}{2}y + \frac{cy^2}{2}\Big]_{0}^{1}$$

$$= \frac{3}{2}(1) + \frac{c(1)^2}{2} - \frac{3}{2}(0) - \frac{c(0)^2}{2} = \frac{3}{2} + \frac{c}{2} = 1 \Rightarrow 3 + c = 2 \Rightarrow c = -1$$

b. $\displaystyle f_2(y) = \int_{-\infty}^{\infty} f(x,y)\,dx = \int_{1}^{2}(x-y)\,dx = \left(\frac{x^2}{2}-xy\right)\Big]_{1}^{2} = \frac{2^2}{2} - 2y - \frac{1^2}{2} + y = \frac{3}{2} - y$

$$\int_{-\infty}^{\infty} f_2(y)\,dy = \int_{0}^{1}\left(\frac{3}{2}-y\right)dy = \left(\frac{3}{2}y - \frac{y^2}{2}\right)\Big]_{0}^{1} = \frac{3}{2}(1) - \frac{1^2}{2} - \frac{3}{2}(0) + \frac{0^2}{2} = 1$$

c. $\displaystyle f_1(x\,|\,y) = \frac{f(x,y)}{f_2(y)} = \frac{x-y}{\dfrac{3}{2} - y}$

6.19 a. $\displaystyle F(-\infty,-\infty) = P(X \le -\infty, Y \le -\infty) = \int_{-\infty}^{-\infty}\int_{-\infty}^{-\infty} f(x,y)\,dx\,dy = 0$

$$F(-\infty,y) = P(X \le -\infty, Y \le y) = \int_{-\infty}^{y}\int_{-\infty}^{-\infty} f(x,y)\,dx\,dy = \int_{-\infty}^{y} 0\,dy = 0$$

$$F(x,-\infty) = P(X \le x, Y \le -\infty) = \int_{-\infty}^{x}\int_{-\infty}^{-\infty} f(x,y)\,dy\,dx = \int_{-\infty}^{x} 0\,dx = 0$$

b. $\displaystyle F(\infty,\infty) = P(X \le \infty, Y \le \infty) = \int_{-\infty}^{\infty}\int_{-\infty}^{\infty} f(x,y)\,dx\,dy = 1$ by definition

c. Show $F(a_2,b_2) - F(a_1,b_2) \ge F(a_2,b_1) - F(a_1,b_1)$ where $a_2 \ge a_1$, $b_2 \ge b_1$

$$F(a_2, b_2) - F(a_1, b_2) = P(X \le a_2, Y \le b_2) - P(X \le a_1, Y \le b_2) = P(a_1 \le X \le a_2, Y \le b_2)$$

and

$$F(a_2, b_1) - F(a_1, b_1) = P(X \le a_2, Y \le b_1) - P(X \le a_1, Y \le b_1) = P(a_1 \le X \le a_2, Y \le b_1)$$

Since $b_2 \ge b_1$, $P(Y \le b_2) \ge P(Y \le b_1)$. Therefore,

$$P(a_1 \le X \le a_2, Y \le b_2) \ge P(a_1 \le X \le a_2, Y \le b_1) \Rightarrow F(a_2, b_2) - F(a_1, b_2) \ge F(a_2, b_1) - F(a_1, b_1)$$

6.21 a. From Exercise 6.5, the marginal probability distribution of X is

$p_1(1) = p_1(2) = p_1(3) = \dfrac{1}{3}$. Thus,

$$E(X) = \sum_x xp(x) = 1\left(\frac{1}{3}\right) + 2\left(\frac{1}{3}\right) + 3\left(\frac{1}{3}\right) = \frac{1}{3} + \frac{2}{3} + \frac{3}{3} = \frac{6}{3} = 2$$

The average section is section 2.

b. From Exercise 6.5, the bivariate probability distribution $p(x, y)$ in table form is:

		x		
		1	2	3
	30	0.0200	0.0333	0.0500
	40	0.0800	0.0800	0.0600
y	50	0.0800	0.1200	0.1000
	60	0.1533	0.1000	0.1233

The marginal probability distribution of Y is

$$P(Y = 30) = p_2(30) = p(1, 30) + p(2, 30) + p(3, 30) = 0.0200 + 0.0333 + 0.0500 = 0.1033$$
$$P(Y = 40) = p_2(40) = p(1, 40) + p(2, 40) + p(3, 40) = 0.0800 + 00.080 + 0.0600 = 0.2200$$
$$P(Y = 50) = p_2(50) = p(1, 50) + p(2, 50) + p(3, 50) = 0.0800 + 0.1200 + 0.1000 = 0.3000$$
$$P(Y = 50) = p_2(50) = p(1, 60) + p(2, 60) + p(3, 60) = 0.1533 + 0.1000 + 0.1233 = 0.3766$$

$$E(Y) = \sum_y yp(y) = 30(0.1033) + 40(0.2200) + 50(0.3000) + 60(0.3766)$$

$$= 3.099 + 8.800 + 15.000 + 22.596 = 49.495$$

The average speed is 49.5 mph.

6.23 a. From Exercise 6.12, $E(Y) = \mu = \beta = 10$. Thus, $E(Y - 10) = E(Y) - 10 = 10 - 10 = 0$

b. $E(3Y) = 3E(Y) = 3(10) = 30$

6.25 a. From Exercise 6.17, $f(x, y) = x - y$, $1 \le x \le 2; 0 \le y \le 1$.

$$f_1(x) = \int_0^1 (x - y) \, dy = xy - \frac{y^2}{2}\Bigg]_0^1 = x(1) - \frac{1^2}{2} - x(0) + \frac{0^2}{2} = x - \frac{1}{2}, \quad 1 \le x \le 2$$

Thus,

$$E(X) = \int_1^2 \left(x^2 - \frac{x}{2} \right) dx = \frac{x^3}{3} - \frac{x^2}{4} \Bigg|_1^2 = \frac{2^3}{3} - \frac{2^2}{4} - \frac{1^3}{3} + \frac{1^2}{4} = \frac{8}{3} - \frac{4}{4} - \frac{1}{3} + \frac{1}{4}$$

$$= \frac{7}{3} - 1 + \frac{1}{4} = \frac{28}{12} - \frac{12}{12} + \frac{3}{12} = \frac{19}{12}$$

b. From Exercise 6.17, $f_2(y) = \frac{3}{2} - y, \ 0 \le y \le 1$

Thus, $E(Y) = \int_0^1 \left(\frac{3y}{2} - y^2 \right) dy = \frac{3y^2}{4} - \frac{y^3}{3} \Bigg|_0^1 = \frac{3(1)^2}{4} - \frac{1^3}{3} - \frac{3(0)^2}{4} + \frac{0^3}{3} = \frac{3}{4} - \frac{1}{3} = \frac{5}{12}$

c. $E(X+Y) = E(X) + E(Y) = \frac{19}{12} + \frac{5}{12} = \frac{24}{12} = 2$

d. $E(XY) = \int_1^2 \int_0^1 xy(x-y) \, dy \, dx = \int_1^2 \int_0^1 (x^2 y - xy^2) \, dy \, dx = \int_1^2 \left(\frac{x^2 y^2}{2} - \frac{xy^3}{3} \right) \Bigg|_0^1 dx$

$$= \int_1^2 \left(\frac{x^2 1^2}{2} - \frac{x1^3}{3} - \frac{x^2 0^2}{2} + \frac{x0^3}{3} \right) dx = \int_1^2 \left(\frac{x^2}{2} - \frac{x}{3} \right) dx = \left(\frac{x^3}{6} - \frac{x^2}{6} \right) \Bigg|_1^2$$

$$= \frac{2^3}{6} - \frac{2^2}{6} - \frac{1^3}{6} + \frac{1^2}{6} = \frac{8}{6} - \frac{4}{6} - \frac{1}{6} + \frac{1}{6} = \frac{4}{6} = \frac{2}{3}$$

6.27 $E(c) = \sum_y \sum_x c \, p(x,y) = c \sum_y \sum_x p(x,y) = c(1) = c$

$E[cg(X,Y)] = \sum_y \sum_x cg(x,y) p(x,y) = c \sum_y \sum_x g(x,y) p(x,y) = cE[g(X,Y)]$

$E[g_1(X,Y) + g_2(X,Y) + \cdots + g_k(X,Y)] = \sum_y \sum_x [g_1(x,y) + g_2(x,y) + \cdots + g_k(x,y)] p(x,y)$

$= \sum_y \sum_x g_1(x,y) p(x,y) + \sum_y \sum_x g_2(x,y) p(x,y) + \cdots + \sum_y \sum_x g_k(x,y) p(x,y)$

$= E[g_1(X,Y)] + E[g_2(X,Y)] + \cdots + E[g_k(X,Y)]$

6.29 X and Y are independent if $p(x,y) = p_1(x) p_2(y)$

From Exercise 6.1, $p(0,0) = 0$, $p_1(0) = 0.300$, and $p_2(0) = 0.100$.

$p(0,0) \overset{?}{=} p_1(0) p_2(0) \Rightarrow 0 \ne 0.300(0.100)$. Thus, X and Y are not independent.

6.31 X and Y are independent if $p(x,y) = p_1(x) p_2(y)$

From Exercise 6.3, $p(0,0) = \frac{6}{45} = 0.1333$, $p_1(0) = \frac{10}{45} = 0.2222$, and

$p_2(0) = \frac{28}{45} = 0.6222$

$$p(0,0) \overset{?}{=} p_1(0)p_2(0) \Rightarrow 0.1333 \overset{?}{=} 0.2222(0.6222) \Rightarrow 0.1333 \neq 0.1383$$

Thus, X and Y are not independent.

6.33   We must assume that $p_1(1) = p_1(2) = 0.5$

The probabilities given in the table are the conditional probabilities $p_2(y \mid x)$. To find the bivariate probability distribution $p(x, y)$ we multiply $p_2(y \mid x)$ by $p_1(x)$.

$p(1,0) = p_2(0 \mid 1)p_1(1) = 0.01(0.5) = 0.005$ $p(2,0) = p_2(0 \mid 2)p_1(2) = 0.002(0.5) = 0.001$

$p(1,12) = p_2(12 \mid 1)p_1(1) = 0.02(0.5) = 0.010$ $p(2,35) = p_2(35 \mid 2)p_1(2) = 0.002(0.5) = 0.001$

$p(1,24) = p_2(24 \mid 1)p_1(1) = 0.02(0.5) = 0.010$ $p(2,70) = p_2(70 \mid 2)p_1(2) = 0.996(0.5) = 0.498$

$p(v) = p_2(36 \mid 1)p_1(1) = 0.95(0.5) = 0.475$

6.35 a. $f_1(x) = \dfrac{1}{5}e^{-x/5} \quad 0 \leq x \leq \infty$

$f_2(y) = \dfrac{1}{5}e^{-y/5} \quad 0 \leq y \leq \infty$

Because X and Y are independent, $f(x, y) = f_1(x)f_2(y) = \dfrac{1}{25}e^{-x/5}e^{-y/5} = \dfrac{1}{25}e^{-(x+y)/5}$

   b. $E(X + Y) = E(X) + E(Y) = 5 + 5 = 10$

6.37   From Exercise 6.15,

$$f(x, y) = \begin{cases} xy & \text{if } 0 \leq x \leq 1; \ 0 \leq y \leq 1 \\ (2-x)y & \text{if } 1 \leq x \leq 2; \ 0 \leq y \leq 1 \\ x(2-y) & \text{if } 0 \leq x \leq 1; \ 1 \leq y \leq 2 \\ (2-x)(2-y) & \text{if } 1 \leq x \leq 2; \ 1 \leq y \leq 2 \end{cases}$$

From Exercise 6.36, the theorem indicates that X and Y are independent if we can write $f(x, y) = g(x)h(y)$ where $g(x)$ is a nonnegative function of X only and $h(y)$ is a nonnegative function of Y only.

For each region above, $f(x, y)$ can be written as $g(x)h(y)$.

For $xy = g(x)h(y)$ where $g(x) = x$ and $h(y) = y$. Both $g(x)$ and $h(y)$ are nonnegative functions of X and Y respectively.

For $(2-x)y = g(x)h(y)$ where $g(x) = 2-x$ and $h(y) = y$. Both $g(x)$ and $h(y)$ are nonnegative functions of X and Y respectively.

For $xy = g(x)h(y)$ where $g(x) = x$ and $h(y) = 2-y$. Both $g(x)$ and $h(y)$ are nonnegative functions of X and Y respectively.

For $(2-x)(2-y) = g(x)h(y)$ where $g(x) = 2-x$ and $h(y) = 2-y$. Both $g(x)$ and $h(y)$ are nonnegative functions of X and Y respectively.

Thus, X and Y are independent.

6.39   From Exercise 6.17, $f(x, y) = x - y$ $1 \le x \le 2;\ 0 \le y \le 1$ and $f_2(y) = \dfrac{3}{2} - y$

$$f_1(x) = \int_0^1 (x - y)\,dy = xy - \frac{y^2}{2}\Bigg]_0^1 = x(1) - \frac{1^2}{2} - x(0) + \frac{0^2}{2} = x - \frac{1}{2}$$

$$f_1(x)f_2(y) = \left(x - \frac{1}{2}\right)\left(\frac{3}{2} - y\right) \ne x - y = f(x, y)$$

Therefore, X and Y are not independent.

6.41 $Cov(X, Y) = E(XY) - \mu_X \mu_Y$

$$E(XY) = \sum_{x=0}^5 \sum_{y=0}^2 xyp(x, y) = 0(0)(0) + 0(1)(0.200) + 0(2)(0.100) + 1(0)(0.050) + 1(1)(0.050)$$

$$+ 1(2)(0) + 2(0)(0.025) + 2(1)(0) + 2(2)(0) + 3(0)(0) + 3(1)(0.300) + 3(2)(0)$$

$$+ 4(0)(0.025) + 4(1)(0) + 4(2)(0.100) + 5(0)(0) + 5(1)(0) + 5(2)(0.150)$$

$$= 0 + 0 + 0 + 0 + 0.050 + 0 + 0 + 0 + 0 + 0 + 0.900 + 0 + 0 + 0 + 0.800 + 0 + 0 + 1.500 = 3.25$$

$$E(X) = \mu_X = \sum_{x=0}^5 xp_1(x) = 0(0.300) + 1(0.100) + 2(0.025) + 3(0.300) + 4(0.125) + 5(0.150)$$

$$= 0 + 0.100 + 0.050 + 0.900 + 0.50 + 0.750 = 2.3$$

$$E(Y) = \mu_Y = \sum_{y=0}^2 yp_2(y) = 0(0.100) + 1(0.550) + 2(0.350) = 0 + 0.550 + 0.700 = 1.25$$

$$Cov(X, Y) = E(XY) - \mu_X \mu_Y = 3.25 - 2.3(1.25) = 3.25 - 2.875 = 0.375$$

6.43 $Cov(X, Y) = E(XY) - \mu_X \mu_Y$

$$E(XY) = \sum_x \sum_y xyp(x, y) = 1(30)(0.0200) + 1(40)(0.0800) + 1(50)(0.0800) + 1(60)(0.1533)$$

$$+ 2(30)(0.0333) + 2(40)(0.0800) + 2(50)(0.1200) + 2(60)(0.1000) + 3(30)(0.0500)$$

$$+ 3(40)(0.0500) + 3(50)(0.1000) + 3(60)(0.1233)$$

$$= 0.6 + 3.2 + 4 + 9.198 + 1.998 + 6.4 + 12 + 12 + 4.5 + 7.2 + 15 + 22.194 = 98.29$$

$$E(X) = \mu_X = \sum_{x=1}^3 xp_1(x) = 1(0.3333) + 2(0.3333) + 3(0.3333) = 0.3333 + 0.6666 + 0.9999 = 2$$

$$E(Y) = \mu_Y = \sum_y yp_2(y) = 30(0.1033) + 40(0.2200) + 50(0.3000) + 60(0.3766) = 49.495$$

$$Cov(X, Y) = E(XY) - \mu_X \mu_Y = 98.29 - 2(49.495) = 98.29 - 98.99 = -0.7$$

$$E(X^2) = \sum_{x=1}^{3} xp_1(x) = 1^2(0.3333) + 2^2(0.3333) + 3^2(0.3333) = 4.6667$$

$$\sigma_X^2 = E(X^2) - \mu_X^2 = 4.6667 - 2^2 = 0.6667 \; ; \quad \sigma_X = \sqrt{0.6667} = 0.8165$$

$$E(Y^2) = \sum_y y^2 p_2(y) = (30)^2(0.1033) + (40)^2(0.2200) + (50)^2(0.3000) + (60)^2(0.3767) = 2551$$

$$\sigma_Y^2 = E(Y^2) - \mu_Y^2 = 2551 - 49.5^2 = 100.75 \; ; \quad \sigma_Y = \sqrt{100.75} = 10.0374$$

$$\rho = \frac{Cov(x,y)}{\sigma_X \sigma_Y} = \frac{-0.7}{0.8165(10.0374)} = -0.0854$$

6.45 $Cov(X,Y) = E(XY) - \mu_X \mu_Y$

$$f_1(x) = \int_0^x 4x^2 \, dy = 4x^2 y \Big]_0^x = 4x^2(x-0) = 4x^3$$

$$f_2(y) = \int_0^1 4x^2 \, dx = \frac{4x^3}{3} \Big]_0^1 = \frac{4(1)^3}{3} - \frac{4(0)^3}{3} = \frac{4}{3}$$

$$E(X) = \int_0^1 4x^4 \, dx = \frac{4x^5}{5} \Big]_0^1 = \frac{4(1)^5}{5} - \frac{4(0)^5}{5} = \frac{4}{5}$$

$$E(Y) = \int_0^1 \frac{4}{3} y \, dy = \frac{4y^2}{6} \Big]_0^1 = \frac{4(1)^2}{6} - \frac{4(0)^2}{6} = \frac{4}{6} = \frac{2}{3}$$

$$E(XY) = \int_0^1 \int_0^x 4x^3 y \, dy \, dx = \int_0^1 2x^3 y^2 \Big]_0^x \, dx = \int_0^1 2x^3(x^2 - 0^2) \, dx$$

$$= \int_0^1 2x^5 \, dx = \frac{1}{3} x^6 \Big]_0^1 = \frac{1}{3}(1^6 - 0^6) = \frac{1}{3}$$

$$Cov(X,Y) = E(XY) - \mu_X \mu_Y = \frac{1}{3} - \frac{4}{5}\left(\frac{2}{3}\right) = \frac{1}{3} - \frac{8}{15} = -\frac{3}{15} = -\frac{1}{5}$$

6.47 a.   From Exercise 6.16, $f(x,y) = 4xy$, $f_1(x) = 2x$, and $f_2(y) = 2y$

From Exercise 6.24, $E(X) = \frac{2}{3}$, $E(Y) = \frac{2}{3}$, and $E(XY) = \frac{4}{9}$

$$Cov(X,Y) = E(XY) - \mu_X \mu_Y = \frac{4}{9} - \frac{2}{3}\left(\frac{2}{3}\right) = \frac{4}{9} - \frac{4}{9} = 0$$

b.   Thus, $\rho = \dfrac{Cov(x,y)}{\sigma_X \sigma_Y} = \dfrac{0}{\sigma_X \sigma_Y} = 0$

6.49 $\quad Cov(X,Y) = E(XY) - E(X)E(Y) = \sum_x \sum_y xyp(x,y) - \sum_x xp_1(x) \sum_y yp_2(y)$

If X and Y are independent, then $p(x,y) = p_1(x)p_2(y)$

Thus, $Cov(X,Y) = \sum_x \sum_y xyp_1(x)p_2(y) - \sum_x xp_1(x) \sum_y yp_2(y)$

$\qquad = \sum_x xp_1(x) \sum_y yp_2(y) - \sum_x xp_1(x) \sum_y yp_2(y) = 0$

6.51 $\quad Cov(X,Y) = E(XY) - E(X)E(Y)$

$E(XY) = \sum_x \sum_y xyp(x,y) = (-1)(-1)\left(\frac{1}{12}\right) + (-1)(0)\left(\frac{2}{12}\right)$

$\qquad + (-1)(1)\left(\frac{1}{12}\right) + (0)(-1)\left(\frac{2}{12}\right) + (0)(0)(0) + (0)(1)\left(\frac{2}{12}\right)$

$\qquad + (1)(-1)\left(\frac{1}{12}\right) + (1)\left(\frac{2}{12}\right)(0) + (1)(1)\left(\frac{1}{12}\right)$

$\qquad = \frac{1}{12} + 0 - \frac{1}{12} + 0 + 0 + 0 - \frac{1}{12} + \frac{1}{12} = 0$

To find $E(X)$ and $E(Y)$, we must find the marginal distributions of X and Y.

$P(X = -1) = p_1(-1) = p(-1,-1) + p(-1,0) + p(-1,0) = \frac{1}{12} + \frac{2}{12} + \frac{1}{12} = \frac{4}{12} = \frac{1}{3}$

$P(X = 0) = p_1(0) = p(0,-1) + p(0,0) + p(0,0) = \frac{2}{12} + 0 + \frac{2}{12} = \frac{4}{12} = \frac{1}{3}$

$P(X = 1) = p_1(1) = p(1,-1) + p(1,0) + p(1,0) = \frac{1}{12} + \frac{2}{12} + \frac{1}{12} = \frac{4}{12} = \frac{1}{3}$

$E(X) = \sum_x xp_1(x) = (-1)\left(\frac{1}{3}\right) + (0)\left(\frac{1}{3}\right) + (1)\left(\frac{1}{3}\right) = 0$

$P(Y = -1) = p_2(-1) = p(-1,-1) + p(0,-1) + p(1,-1) = \frac{1}{12} + \frac{2}{12} + \frac{1}{12} = \frac{4}{12} = \frac{1}{3}$

$P(Y = 0) = p_2(0) = p(-1,0) + p(0,0) + p(1,0) = \frac{2}{12} + 0 + \frac{2}{12} = \frac{4}{12} = \frac{1}{3}$

$P(Y = 1) = p_2(1) = p(-1,1) + p(0,1) + p(1,1) = \frac{1}{12} + \frac{2}{12} + \frac{1}{12} = \frac{4}{12} = \frac{1}{3}$

$E(Y) = \sum_y yp_1(y) = (-1)\left(\frac{1}{3}\right) + (0)\left(\frac{1}{3}\right) + (1)\left(\frac{1}{3}\right) = 0$

$Cov(X,Y) = E(XY) - E(X)E(Y) = 0 - 0(0) = 0$

To show that X and Y are not independent, we must show that $p(x,y) \neq p_1(x)p_2(y)$ for at least one pair (x,y). Let $x = -1$ and $y = -1$.

$$p(-1,-1) = \frac{1}{12}, \; p_1(-1) = \frac{1}{3}, \text{ and } p_2(-1) = \frac{1}{3}$$

$$p(x,y) \overset{?}{=} p_1(x)p_2(y) \Rightarrow \frac{1}{12} \neq \frac{1}{3}\left(\frac{1}{3}\right)$$

Thus, X and Y are not independent even though $Cov(X,Y) = 0$.

6.53   From Exercise 6.2, $p(x,y) = \frac{1}{36}, \; p_1(x) = \frac{1}{6} \quad x = 1,2,3,4,5,6,$

$p_2(y) = \frac{1}{6} \quad y = 1,2,3,4,5,6$

$$E(X) = \sum_{x=1}^{6} xp_1(x) = 1\left(\frac{1}{6}\right) + 2\left(\frac{1}{6}\right) + 3\left(\frac{1}{6}\right) + 4\left(\frac{1}{6}\right) + 5\left(\frac{1}{6}\right) + 6\left(\frac{1}{6}\right) = \frac{21}{6} = 3.5$$

$$E(Y) = \sum_{y=1}^{6} yp_2(y) = 1\left(\frac{1}{6}\right) + 2\left(\frac{1}{6}\right) + 3\left(\frac{1}{6}\right) + 4\left(\frac{1}{6}\right) + 5\left(\frac{1}{6}\right) + 6\left(\frac{1}{6}\right) = \frac{21}{6} = 3.5$$

$E(X+Y) = E(X) + E(Y) = 3.5 + 3.5 = 7$

$V(X+Y) = \sigma_1^2 + \sigma_2^2 + 2Cov(X,Y)$

$$E(X^2) = \sum_{x=1}^{6} x^2 p_1(x) = 1^2\left(\frac{1}{6}\right) + 2^2\left(\frac{1}{6}\right) + 3^2\left(\frac{1}{6}\right) + 4^2\left(\frac{1}{6}\right) + 5^2\left(\frac{1}{6}\right) + 6^2\left(\frac{1}{6}\right) = \frac{91}{6} = 15.1667$$

$\sigma_X^2 = E(X^2) - \mu_X^2 = 15.1667 - 3.5^2 = 2.9167$

$$E(Y^2) = \sum_{y=1}^{6} y^2 p_1(y) = 1^2\left(\frac{1}{6}\right) + 2^2\left(\frac{1}{6}\right) + 3^2\left(\frac{1}{6}\right) + 4^2\left(\frac{1}{6}\right) + 5^2\left(\frac{1}{6}\right) + 6^2\left(\frac{1}{6}\right) = \frac{91}{6} = 15.1667$$

$\sigma_Y^2 = E(Y^2) - \mu_Y^2 = 15.1667 - 3.5^2 = 2.9167$

$Cov(X,Y) = E(XY) - E(X)E(Y)$

$$E(XY) = \sum_{x=1}^{6}\sum_{y=1}^{6} xy\, p(x,y) = 1(1)\left(\frac{1}{36}\right) + 1(2)\left(\frac{1}{36}\right) + 1(3)\left(\frac{1}{36}\right) + \cdots 6(6)\left(\frac{1}{36}\right) = \frac{441}{36} = 12.25$$

$Cov(X,Y) = E(XY) - E(X)E(Y) = 12.25 - 3.5(3.5) = 0$

Thus, $V(X+Y) = \sigma_X^2 + \sigma_Y^2 + 2Cov(X,Y) = 2.9167 + 2.9167 + 2(0) = 5.8334$

6.55   From Exercise 6.14, $f(x,y) = 2e^{-x^2} \quad 0 \leq y \leq x < \infty$

$$E(X-Y) = \int_{0}^{\infty}\int_{0}^{x} (x-y)2e^{-x^2} \, dy \, dx = 2\int_{0}^{\infty}\left\{ \int_{0}^{x} xe^{-x^2} \, dy - \int_{0}^{x} ye^{-x^2} \, dy \right\} dx$$

$$= 2\int_{0}^{\infty}\left\{ \left(xe^{-x^2} y\right)\Big]_{0}^{x} - \left(e^{-x^2}\frac{y^2}{2}\right)\Big]_{0}^{x} \right\} dx = 2\int_{0}^{\infty}\left(x^2 e^{-x^2} - e^{-x^2}\frac{x^2}{2}\right) dx$$

$$= 2\int_0^\infty \frac{x^2 e^{-x^2}}{2}\,dx = \int_0^\infty x^2 e^{-x^2}\,dx = -\frac{xe^{-x^2}}{2}\Bigg]_0^\infty - \int_0^\infty -\frac{e^{-x^2}}{2}\,dx = (0-0) + \frac{1}{2}\int_0^\infty e^{-x^2}\,dx$$

$$= \frac{1}{2}\left(\frac{\sqrt{\pi}}{2}\right) = \frac{\sqrt{\pi}}{4}$$

$$V(X-Y) = E(X-Y)^2 - \left\{E(X-Y)\right\}^2$$

$$E(X-Y)^2 = E(X^2 - 2XY + Y^2) = \int_0^\infty \int_0^x (x^2 - 2xy + y^2)2e^{-x^2}\,dy\,dx$$

$$= 2\int_0^\infty \left\{\int_0^x x^2 e^{-x^2}\,dy - \int_0^x 2xye^{-x^2}\,dy + \int_0^x y^2 e^{-x^2}\,dy\right\}dx$$

$$= 2\int_0^\infty \left\{x^2 e^{-x^2} y\Big]_0^x - xe^{-x^2}y^2\Big]_0^x + e^{-x^2}\frac{y^3}{3}\Big]_0^x\right\}dx$$

$$= 2\int_0^\infty \left(x^3 e^{-x^2} - x^3 e^{-x^2} + \frac{x^3 e^{-x^2}}{3}\right)dx = 2\int_0^\infty \frac{x^3 e^{-x^2}}{3}\,dx = \frac{2}{3}\int_0^\infty x^3 e^{-x^2}\,dx$$

$$= -\frac{x^2 e^{-x^2}}{2}\Bigg]_0^\infty - \int_0^\infty -\frac{2xe^{-x^2}}{2}\,dx = 0 + \int_0^\infty xe^{-x^2}\,dx = \frac{-e^{-x^2}}{2}\Bigg]_0^\infty = 0 - \left(-\frac{1}{2}\right) = \frac{1}{2}$$

$$V(X-Y) = E(X-Y)^2 - \{E(X-Y)\}^2 = \frac{1}{2} - \left(\frac{\sqrt{\pi}}{4}\right)^2 = \frac{1}{2} - \frac{\pi}{16} = \frac{8-\pi}{16}$$

6.57 $\hat{p} = a_1 Y$ where $a_1 = \dfrac{1}{n}$

$$E(\hat{p}) = E(a_1 Y) = a_1 E(Y) = a_1 np = \frac{1}{n}np = p$$

$$V(\hat{p}) = V(a_1 Y) = a_1^2 V(Y) = a_1^2 npq = \left(\frac{1}{n}\right)^2 npq = \frac{pq}{n}$$

6.59 If $C = 3Y + 2$, then $Y = \dfrac{C-2}{3}$.

$$f(y) = \frac{1}{5}e^{-y/5} \quad 0 \le y < \infty \quad \text{Thus, } \int_0^\infty \frac{1}{5}e^{-y/5}\,dy = 1.$$

If $Y = \dfrac{C-2}{3}$, then $dy = \dfrac{dc}{3}$. Also, if $0 \le y < \infty$, then $2 \le c < \infty$.

Thus, $\int_0^\infty \frac{1}{5} e^{-y/5} \, dy = 1 \Rightarrow \int_2^\infty \frac{1}{5} e^{-(c-2)/15} \frac{dc}{3} = 1 \Rightarrow \int_2^\infty \frac{1}{15} e^{-(c-2)/15} \, dc = 1$

The density function of C is $f(c) = \begin{cases} \frac{1}{15} e^{-(c-2)/15} & 2 \le c < \infty \\ 0 & \text{elsewhere} \end{cases}$.

6.61   If $W = Y^2$, then $Y = W^{1/2}$.

$f(y) = \frac{y}{\mu} e^{-y^2/2\mu}$ Thus, $\int_0^\infty \frac{y}{\mu} e^{-y^2/2\mu} \, dy = 1$.

If $Y = W^{1/2}$, then $dy = \frac{1}{2} w^{-1/2} dw$. Also, if $y > 0$, then $w > 0$.

Thus, $\int_0^\infty \frac{y}{\mu} e^{-y^2/2\mu} \, dy = 1 \Rightarrow \int_0^\infty \frac{w^{1/2}}{\mu} e^{-w/2\mu} \frac{1}{2} w^{-1/2} \, dw = 1 \Rightarrow \int_0^\infty \frac{1}{2\mu} e^{-w/2\mu} \, dw = 1$

The density function of W is $f(w) = \begin{cases} \frac{1}{2\mu} e^{-w/2\mu} & w > 0 \\ 0 & \text{elsewhere} \end{cases}$.

Thus, W has an exponential distribution with a mean of 2μ.

6.63   The density function for the beta distribution with $\alpha = 2$ and $\beta = 1$ is

$f(y) = \begin{cases} \dfrac{\Gamma(3)}{\Gamma(2)\Gamma(1)} y & 0 \le y \le 1 \\ 0 & \text{elsewhere} \end{cases}$ where $\Gamma(3) = 2! = 2, \Gamma(2) = 1$, and $\Gamma(1) = 1$.

Thus, $f(y) = \begin{cases} 2y & 0 \le y \le 1 \\ 0 & \text{elsewhere} \end{cases}$

The cumulative distribution function is $F(y) = \int_0^y 2t \, dt = t^2 \Big]_0^y = y^2$.

If we let $W = F(y) = y^2$, then Theorem 6.7 tells us that W has a uniform density function over the interval $0 \le w \le 1$.

   Answers will vary. To draw a random number Y from this function, we first randomly draw a value W from the uniform distribution. This can be done by drawing a random number from Table I of Appendix B or using a computer. Suppose we draw the random number 91646 (1st number in column 6). This corresponds to the random selectin of the value $w_1 = 0.91646$ from a uniform distribution over the interval $0 \le w \le 1$. Substituting this value of w_1 into the formula $W = F(y)$ and solving for Y, we obtain

$w_1 = F(y) = y^2 \Rightarrow 0.91646 = y_1^2 \Rightarrow y_1 = \sqrt{0.91646} = 0.957$.

We continue this for 4 additional random numbers:

Random number 89198, $\Rightarrow w_2 = F(y) = y^2 \Rightarrow 0.89198 = y_2^2 \Rightarrow y_2 = \sqrt{0.89198} = 0.944$

Random number 64809, $\Rightarrow w_3 = F(y) = y^2 \Rightarrow 0.64809 = y_3^2 \Rightarrow y_3 = \sqrt{0.64809} = 0.805$

Random number 16376, $\Rightarrow w_4 = F(y) = y^2 \Rightarrow 0.16376 = y_4^2 \Rightarrow y_4 = \sqrt{0.16376} = 0.405$

Random number 91782, $\Rightarrow w_5 = F(y) = y^2 \Rightarrow 0.91782 = y_5^2 \Rightarrow y_5 = \sqrt{0.91782} = 0.958$

6.65 $E(/) = E\left(\dfrac{1}{2}Y_1 - Y_2 + 2Y_3\right) = \dfrac{1}{2}E(Y_1) - E(Y_2) + 2E(Y_3) = \dfrac{1}{2}(0) - (-1) + 2(5) = 11$

$V(/) = V\left(\dfrac{1}{2}Y_1 - Y_2 + 2Y_3\right) = \left(\dfrac{1}{2}\right)^2 V(Y_1) + (-1)^2 V(Y_2) + 2^2 V(Y_3)$

$\qquad + 2\left(\dfrac{1}{2}\right)(-1)Cov(Y_1, Y_2) + 2\left(\dfrac{1}{2}\right)(2)Cov(Y_1, Y_3) + 2(-1)(2)Cov(Y_2, Y_3)$

$\qquad = \dfrac{1}{4}(2) + 3 + 4(9) - 1 + 2(4) - 4(-2) = 54.5$

6.67   A gamma distribution with $\alpha = n$ and $\beta = \dfrac{2}{n}$ has

$E(Y) = \mu = \alpha\beta = n\left(\dfrac{2}{n}\right) = 2 \text{ and } \sigma^2 = \alpha\beta^2 = n\left(\dfrac{2}{n}\right)^2 = \dfrac{4}{n}.$

If $Y_1, Y_1, \ldots, Y_n$ are gamma random variables with $\alpha = 1$ and $\beta = 2$, then

$E(Y_i) = \mu_i = \alpha\beta = 1(2) = 2 \text{ and } \sigma_i^2 = \alpha\beta^2 = 1(2)^2 = 4$

Then

$E(\bar{Y}) = E\left(\dfrac{\sum\limits_{i=1}^{n} Y_i}{n}\right) = \dfrac{1}{n}E(Y_1) + \dfrac{1}{n}E(Y_2) + \cdots + \dfrac{1}{n}E(Y_n) = \dfrac{1}{n}(2) + \dfrac{1}{n}(2) + \cdots + \dfrac{1}{n}(2) = \dfrac{1}{n}(2n) = 2$

$\sigma_{\bar{Y}}^2 = V(\bar{Y}) = V\left(\dfrac{\sum\limits_{i=1}^{n} Y_i}{n}\right) = \left(\dfrac{1}{n}\right)^2 V(Y_1) + \left(\dfrac{1}{n}\right)^2 V(Y_2) + \cdots + \left(\dfrac{1}{n}\right)^2 V(Y_n)$

$\qquad = \dfrac{1}{n^2}(4) + \dfrac{1}{n^2}(4) + \cdots + \dfrac{1}{n^2}(4) = \dfrac{1}{n^2}(4n) = \dfrac{4}{n}$

6.69 a.   If $W = Y^2$, then $Y = W^{1/2}$.

$f(y) = 2y \quad \text{Thus, } \displaystyle\int_0^1 2y\, dy = 1.$

If $Y = W^{1/2}$, then $dy = \dfrac{1}{2} w^{-1/2} dw$. Also, if $0 \le y \le 1$, then $0 \le w \le 1$.

Thus, $\displaystyle\int_0^1 2y\, dy = 1 \Rightarrow \int_0^1 2w^{1/2} \dfrac{1}{2} w^{-1/2}\, dw = 1 \Rightarrow \int_0^1 1\, dw = 1$

The density function of W is $f(w) = \begin{cases} 1 & 0 \le w \le 1 \\ 0 & \text{elsewhere} \end{cases}$.

b.   If $W = 2Y - 1$, then $Y = \dfrac{W+1}{2}$.

$f(y) = 2y$ Thus, $\displaystyle\int_0^1 2y\, dy = 1$.

If $Y = \dfrac{W+1}{2}$, then $dy = \dfrac{w}{2}\, dw$. Also, if $0 \le y \le 1$, then $-1 \le w \le 1$.

Thus, $\displaystyle\int_0^1 2y\, dy = 1 \Rightarrow \int_{-1}^1 2\dfrac{(w+1)}{2}\dfrac{1}{2}\, dw = 1 \Rightarrow \int_{-1}^1 \dfrac{(w+1)}{2}\, dw = 1$

The density function of W is $f(w) = \begin{cases} \dfrac{(w+1)}{2} & -1 \le w \le 1 \\ 0 & \text{elsewhere} \end{cases}$.

c.   If $W = 1/Y$, then $Y = 1/W$.

$f(y) = 2y$ Thus, $\displaystyle\int_0^1 2y\, dy = 1$.

If $Y = 1/W = W^{-1}$, then $dy = -w^{-2} dw$. Also, if $0 \le y \le 1$, then $w \ge 1$.

If $Y = 0$, then $W = \infty$. If $Y = 1$, then $W = 1$.

Thus, $\displaystyle\int_0^1 2y\, dy = 1 \Rightarrow \int_{-1}^1 2w^{-1}(-w^{-2})\, dw = 1 \Rightarrow \int_{-1}^1 2w^{-3}\, dw = 1$

The density function of W is $f(w) = \begin{cases} 2w^{-3} & 1 \le w \le \infty \\ 0 & \text{elsewhere} \end{cases}$.

6.73   a. $\hat{p} = \dfrac{Y}{n}$ where $Y \sim B(n, p) \sim B(106, 0.4)$

$\mu_{\hat{p}} = E(\hat{p}) = E\left(\dfrac{Y}{n}\right) = \dfrac{1}{n} E(Y) \dfrac{1}{n}(np) = p = 0.4$

$$\sigma_{\hat{p}}^2 = V(\hat{p}) = V\left(\frac{Y}{n}\right) = \left(\frac{1}{n}\right)^2 V(Y) = \frac{1}{n^2}(npq) = \frac{pq}{n} = \frac{0.4(0.6)}{106} = 0.002264$$

$$\sigma_{\hat{p}} = \sqrt{0.002264} = 0.0476$$

b. The sampling distribution of $\hat{p}$ is approximately normal.

c. $P(\hat{p} > 0.59) = P\left(Z > \dfrac{0.59 - 0.4}{0.0476}\right) = P(Z > 3.992) = 0.5 - P(0 < Z < 3.992) \approx 0.5 - 0.5 = 0$

d. Yes. $\hat{p} = \dfrac{Y}{n} = \dfrac{63}{106} = 0.59$. We found the probability of observing $\hat{p} > 0.59$ to be essentially 0. Thus, it would be very unlikely to observe 63 out of 106 robots with legs but no wheels if the population proportion is 0.40.

6.75 $\mu_Y = 293$ and $\sigma_Y = 847$. $n = 50$

a. $\mu_{\bar{Y}} = \mu_Y = 293$; $\sigma_{\bar{Y}} = \dfrac{\sigma_Y}{\sqrt{n}} = \dfrac{847}{\sqrt{50}} = 119.7939$

b. The sampling distribution of $\bar{Y}$ is approximately normal with a mean of 293 and a standard deviation of 119.7939. A sketch of the sampling distribution is:

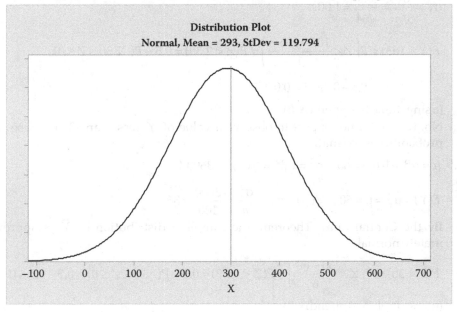

Distribution Plot
Normal, Mean = 293, StDev = 119.794

c. $P(\bar{Y} \geq 550) = P\left(Z \geq \dfrac{550 - 293}{119.7939}\right) = P(Z \geq 2.15) = 0.5 - P(0 \leq Z \leq 2.15)$

$$= 0.5 - 0.4842 = 0.0158$$

6.77 a.   We will stretch and use the Central Limit Theorem $(n = 20)$ to state the sampling distribution of $\bar{Y}$ is approximately normal with a mean of $\mu_{\bar{Y}} = \mu = 1.8$ and a standard deviation of $\sigma_{\bar{Y}} = \dfrac{\sigma_Y}{\sqrt{n}} = \dfrac{0.5}{\sqrt{20}} = 0.1118$.

$$P(\bar{Y} > 1.85) = P\left(Z > \frac{1.85 - 1.8}{0.1118}\right) = P(Z > 0.45) = 0.5 - P(0 \le Z \le 0.45)$$

$$= 0.5 - 0.1736 = 0.3264$$

(using Table 5, Appendix B)

  b. $\bar{Y} = \dfrac{\sum Y_i}{n} = \dfrac{37.62}{20} = 1.881$

  c.   Based on the probability found in part a, it would not be unreasonable to find a sample mean that exceeded 1.85. We have no reason to doubt the assumptions made in part a.

6.79   By the Central Limit Theorem, the sampling distribution of $\bar{Y}$ is approximately normal with a mean of $\mu_{\bar{Y}} = \mu = 105.3$ and a standard deviation of $\sigma_{\bar{Y}} = \dfrac{\sigma_Y}{\sqrt{n}} = \dfrac{8.0}{\sqrt{64}} = 1.00$.

$$P(\bar{Y} < 103) = P\left(Z < \frac{103 - 105.3}{1}\right) = P(Z < 2.30) = 0.5 - P(-2.30 \le Z \le 0)$$

$$= 0.5 - 0.4893 = 0.0107$$

(using Table 5, Appendix B)

No, we would not expect to observe a value of $\bar{Y}$ less than 103 because the probability is so small.

6.81 a. $\mu = \alpha\beta = 1(60) = 60$; $\sigma^2 = \alpha\beta^2 = 1(60)^2 = 3600$

$$E(\bar{Y}) = \mu_{\bar{Y}} = \mu = 60; \quad V(\bar{Y}) = \sigma_{\bar{Y}}^2 = \frac{\sigma_Y^2}{n} = \frac{3600}{100} = 36$$

  b.   By the Central Limit Theorem, the sampling distribution of $\bar{Y}$ is approximately normal.

  c. $P(\bar{Y} < 30) = P\left(Z < \dfrac{30 - 60}{6}\right) = P(Z < -5.0) = 0.5 - P(-5.0 \le Z \le 0) \approx 0.5 - 0.5 = 0$

(using Table 5, Appendix B)

6.83 $P\left(\displaystyle\sum_{i=1}^{45} Y_i \le 10,000\right) = P\left(\dfrac{\displaystyle\sum_{i=1}^{45} Y_i}{n} \le \dfrac{10,000}{n}\right) = P\left(\bar{Y} \le \dfrac{10,000}{45}\right) = P(\bar{Y} \le 222.22)$

By the Central Limit Theorem, the sampling distribution of $\bar{Y}$ is approximately normal with a mean of $\mu_{\bar{Y}} = \mu = 200$ and a standard deviation of

$$\sigma_{\bar{Y}} = \frac{\sigma_Y}{\sqrt{n}} = \frac{55}{\sqrt{45}} = 8.1989.$$

$$P(\bar{Y} < 222.22) = P\left(Z < \frac{222.22 - 200}{8.1989}\right) = P(Z < 2.71) = 0.5 + P(0 < Z < 2.71)$$

$$= 0.5 + 0.4966 = 0.9966$$

(using Table 5, Appendix B)

6.85 $Y_1 \sim B(n_1, p_1)$ and $Y_2 \sim B(n_2, p_2)$

Let $\hat{p}_1 - \hat{p}_2 = a_1 Y_1 - a_2 Y_2 = \dfrac{1}{n_1} Y_1 - \dfrac{1}{n_2} Y_2$

$$E(\hat{p}_1 - \hat{p}_2) = \frac{1}{n_1} E(Y_1) - \frac{1}{n_2} E(Y_2) = \frac{1}{n_1}(n_1 p_1) - \frac{1}{n_2}(n_2 p_2) = p_1 - p_2$$

$$V(\hat{p}_1 - \hat{p}_2) = \left(\frac{1}{n_1}\right)^2 V(Y_1) + \left(\frac{1}{n_2}\right)^2 V(Y_2) + 2\left(\frac{1}{n_1}\right)\left(-\frac{1}{n_2}\right) Cov(Y_1, Y_2)$$

$$= \frac{1}{n_1^2}(n_1 p_1 q_1) + \frac{1}{n_2^2}(n_2 p_2 q_2) - 2\left(\frac{1}{n_1}\right)\left(\frac{1}{n_2}\right) 0 = \frac{p_1 q_1}{n_1} + \frac{p_2 q_2}{n_2}$$

The sampling distribution of $\hat{p}_1 - \hat{p}_2$ has an approximate normal distribution for large values of n_1 and n_2 with mean $\mu_{\hat{p}_1 - \hat{p}_2} = p_1 - p_2$ and standard deviation $\sigma_{\hat{p}_1 - \hat{p}_2} = \sqrt{\dfrac{p_1 q_1}{n_1} + \dfrac{p_2 q_2}{n_2}}$.

Thus, $Z = \dfrac{(\hat{p}_1 - \hat{p}_2) - (p_1 - p_2)}{\sqrt{\dfrac{p_1 q_1}{n_1} + \dfrac{p_2 q_2}{n_2}}}$ has an approximate standard normal distribution.

6.87 The normal approximation can be used if both $np \geq 4$ and $nq \geq 4$.

a. $np = 100(0.01) = 1 \not\geq 4$; $nq = 100(0.99) = 99 \geq 4$

The first condition is not met, so the normal approximation should not be used.

b. $np = 100(0.50) = 50 \geq 4$; $nq = 100(0.50) = 50 \geq 4$

Both conditions are met, so the normal approximation can be used.

c. $np = 100(0.90) = 90 \geq 4$; $nq = 100(0.10) = 10 \geq 4$

Both conditions are met, so the normal approximation can be used.

6.89 $Y_1 \sim B(n, p) \sim B(300, 0.60)$; $\mu = np = 300(0.60) = 180$;

$$\sigma = \sqrt{npq} = \sqrt{300(0.60)(0.40)} = \sqrt{72} = 8.4853$$

The normal approximation can be used if both $np \geq 4$ and $nq \geq 4$.

$np = 300(0.60) = 180 \geq 4$; $nq = 300(0.40) = 120 \geq 4$

Thus, the normal approximation can be used.

$$P(Y < 100) = P\left(Z < \frac{99.5 - 108}{8.4853}\right) = P(Z < -9.49) \approx 0 \text{ (using Table 5, Appendix B)}$$

6.91 $Y \sim B(n, p) \sim B(65, 0.25)$; $\mu = np = 65(0.25) = 16.25$

$$\sigma = \sqrt{npq} = \sqrt{65(0.25)(0.75)} = \sqrt{12.1875} = 3.4911$$

$$P(Y \geq 20) = P\left(Z \geq \frac{19.5 - 16.25}{3.4911}\right) = P(Z \geq 0.93) = 0.5 - P(0 \leq Z \leq 0.93)$$

$$= 0.5 - 0.3238 = 0.1762$$

(using Table 5, Appendix B)

6.93 $Y \sim B(n, p) \sim B(1,000, 0.92)$; $\mu = np = 1,000(0.92) = 920$

$$\sigma = \sqrt{npq} = \sqrt{1,000(0.92)(0.08)} = \sqrt{73.6} = 8.5790$$

a. $\hat{p} = \dfrac{Y}{n}$; $E(\hat{p}) = p = 0.92$

b.   Using the normal approximation to the binomial,

$$P(Y < 900) = P\left(Z < \frac{899.5 - 920}{8.579}\right) = P(Z \geq -2.39) = 0.5 - P(-2.39 < Z < 0)$$

$$= 0.5 - 0.4916 = 0.0084$$

(using Table 5, Appendix B)

6.95   From Theorem 6.11, $\chi^2 = \dfrac{(n-1)S^2}{\sigma^2}$ has a chi-square distribution with $v = n - 1 = 10 - 1 = 9$ degrees of freedom.

a. $P(S^2 > 14.4) = P\left(\chi^2 > \dfrac{(10-1)14.4}{9}\right) = P(\chi^2 > 14.4) \approx 0.100$

b. $P(S^2 > 33.3) = P\left(\chi^2 > \dfrac{(10-1)33.3}{9}\right) = P(\chi^2 > 33.3) < 0.005$

c. $P(S^2 > 16.7) = P\left(\chi^2 > \dfrac{(10-1)16.7}{9}\right) = P(\chi^2 > 16.7) \approx 0.050$

6.97 a.   The sampling distribution of $T = \dfrac{\sqrt{n}\left(\bar{Y} - \mu\right)}{S}$ is a t-distribution with $v = n - 1 = 10 - 1 = 9$ degrees of freedom.

b.   The sampling distribution of $\chi^2 = \dfrac{(n-1)S^2}{\sigma^2}$ is a chi-square distribution with $v = n - 1 = 10 - 1 = 9$ degrees of freedom.

6.99 a. The sampling distribution of $T = \dfrac{\sqrt{n}\left(\bar{Y} - \mu\right)}{S} = \dfrac{\sqrt{16}\left(\bar{Y} - 0.5\right)}{S} = \dfrac{4\left(\bar{Y} - 0.5\right)}{S}$ is a

t-distribution with $v = n - 1 = 16 - 1 = 15$ degrees of freedom.

b. $P\left(\bar{Y} < 0.52\right) = P\left(T < \dfrac{4(0.52 - 0.5)}{0.015}\right) = P\left(T < 5.333\right) = 0.999958$ (using MINITAB)

6.101 $\dfrac{(n_1 - 1)S_1^2}{\sigma_1^2}$ has a chi-square distribution with $v_1 = n_1 - 1$ degrees of freedom.

$\dfrac{(n_2 - 1)S_2^2}{\sigma_2^2}$ has a chi-square distribution with $v_2 = n_2 - 1$ degrees of freedom.

$F = \dfrac{\dfrac{(n_1 - 1)S_1^2}{\sigma_1^2} \Big/ (n_1 - 1)}{\dfrac{(n_2 - 1)S_2^2}{\sigma_2^2} \Big/ (n_2 - 1)} = \left(\dfrac{S_1^2}{S_2^2}\right)\left(\dfrac{\sigma_2^2}{\sigma_1^2}\right)$ has an F-distribution with $v_1 = n_1 - 1$ and

$v_2 = n_2 - 1$ degrees of freedom.

6.103 Let $A = \dfrac{\left(\bar{Y}_1 - \bar{Y}_2\right) - \left(\mu_1 - \mu_2\right)}{\sqrt{\sigma^2\left(\dfrac{1}{n_1} + \dfrac{1}{n_2}\right)}}$. Then A has a standard normal distribution.

Let $B\dfrac{\left(n_1 + n_2 - 2\right)S_p^2}{\sigma^2}$. Then B has a χ^2 distribution with $v = n_1 + n_2 - 2$ degrees of freedom.

Let $T = \dfrac{A}{\sqrt{B/(n_1 + n_2 - 2)}} = \dfrac{\dfrac{\left(\bar{Y}_1 - \bar{Y}_2\right) - \left(\mu_1 - \mu_2\right)}{\sqrt{\sigma^2\left(\dfrac{1}{n_1} + \dfrac{1}{n_2}\right)}}}{\sqrt{\dfrac{\left(n_1 + n_2 - 2\right)S_p^2}{\sigma^2} \Big/ (n_1 + n_2 - 2)}}$

$= \dfrac{\left(\bar{Y}_1 - \bar{Y}_2\right) - \left(\mu_1 - \mu_2\right)}{\sqrt{\sigma^2\left(\dfrac{1}{n_1} + \dfrac{1}{n_2}\right)}\sqrt{\dfrac{S_p^2}{\sigma^2}}} = \dfrac{\left(\bar{Y}_1 - \bar{Y}_2\right) - \left(\mu_1 - \mu_2\right)}{\sqrt{S_p^2\left(\dfrac{1}{n_1} + \dfrac{1}{n_2}\right)}}$

But $T = \dfrac{Z}{\sqrt{\chi^2/v}} \Rightarrow T$ has a t-distribution with $v = n_1 + n_2 - 2$ degrees of freedom.

6.105 a. By the Central Limit Theorem, the sampling distribution of $\overline{Y}$ is approximately normal with mean $\mu_{\overline{Y}} = \mu = 43$ and standard deviation $\sigma_{\overline{Y}} = \dfrac{\sigma}{\sqrt{n}} = \dfrac{7}{\sqrt{40}} = 1.1068$.

b. By the Central Limit Theorem, the sampling distribution of $\overline{Y}$ is approximately normal with mean $\mu_{\overline{Y}} = \mu = 1{,}050$ and standard deviation $\sigma_{\overline{Y}} = \dfrac{\sigma}{\sqrt{n}} = \dfrac{376}{\sqrt{40}} = 59.4508$.

c. By the Central Limit Theorem, the sampling distribution of $\overline{Y}$ is approximately normal with mean $\mu_{\overline{Y}} = \mu = 24$ and standard deviation $\sigma_{\overline{Y}} = \dfrac{\sigma}{\sqrt{n}} = \dfrac{98}{\sqrt{40}} = 15.4952$.

6.107 a. $f_1(x) = \displaystyle\int_0^1 (x+y)\,dy = xy + \dfrac{y^2}{2}\Bigg]_0^1 = x(1) + \dfrac{1^2}{2} - x(0) - \dfrac{0^2}{2} = x + \dfrac{1}{2}$

$f_2(y) = \displaystyle\int_0^1 (x+y)\,dx = \dfrac{x^2}{2} + xy\Bigg]_0^1 = \dfrac{1^2}{2} - y(1) - \dfrac{0^2}{2} - y(0) = y + \dfrac{1}{2}$

b. $\displaystyle\int_{-\infty}^{\infty}\left(x+\dfrac{1}{2}\right)dx = \int_0^1\left(x+\dfrac{1}{2}\right)dx = \dfrac{x^2}{2} + \dfrac{x}{2}\Bigg]_0^1 = \dfrac{1^2}{2} + \dfrac{1}{2} - \dfrac{0^2}{2} - \dfrac{0}{2} = 1$

$\displaystyle\int_{-\infty}^{\infty}\left(y+\dfrac{1}{2}\right)dy = \int_0^1\left(y+\dfrac{1}{2}\right)dy = \dfrac{y^2}{2} + \dfrac{y}{2}\Bigg]_0^1 = \dfrac{1^2}{2} + \dfrac{1}{2} - \dfrac{0^2}{2} - \dfrac{0}{2} = 1$

c. $f_1(x\,|\,y) = \dfrac{f(x,y)}{f_2(y)} = \dfrac{x+y}{y+\dfrac{1}{2}}$ $\qquad$ $f_2(y\,|\,x) = \dfrac{f(x,y)}{f_1(x)} = \dfrac{x+y}{x+\dfrac{1}{2}}$

d. $\displaystyle\int_{-\infty}^{\infty}\dfrac{x+y}{y+\dfrac{1}{2}}\,dx = \int_0^1\dfrac{x+y}{y+\dfrac{1}{2}}\,dx = \dfrac{1}{y+\dfrac{1}{2}}\int_0^1 (x+y)\,dx = \dfrac{1}{y+\dfrac{1}{2}}\left(\dfrac{x^2}{2} + xy\right)\Bigg]_0^1$

$= \dfrac{1}{y+\dfrac{1}{2}}\left(\dfrac{1^2}{2} + (1)y - \dfrac{0^2}{2} + (0)y\right) = \dfrac{1}{y+\dfrac{1}{2}}\left(\dfrac{1}{2} + y\right) = 1$

$\displaystyle\int_{-\infty}^{\infty}\dfrac{x+y}{x+\dfrac{1}{2}}\,dx = \int_0^1\dfrac{x+y}{x+\dfrac{1}{2}}\,dy = \dfrac{1}{x+\dfrac{1}{2}}\int_0^1 (x+y)\,dy = \dfrac{1}{x+\dfrac{1}{2}}\left(xy + \dfrac{y^2}{2}\right)\Bigg]_0^1$

$= \dfrac{1}{x+\dfrac{1}{2}}\left((1)x + \dfrac{1^2}{2} - (0)x - \dfrac{0^2}{2}\right) = \dfrac{1}{x+\dfrac{1}{2}}\left(x + \dfrac{1}{2}\right) = 1$

e. $E(X) = \int_0^1 x\left(x + \dfrac{1}{2}\right)dx = \int_0^1 \left(x^2 + \dfrac{x}{2}\right)dx = \dfrac{x^3}{3} + \dfrac{x^2}{4}\Bigg]_0^1 = \dfrac{1^3}{3} + \dfrac{1^2}{4} - \dfrac{0^3}{3} - \dfrac{0^2}{4} = \dfrac{1}{3} + \dfrac{1}{4} = \dfrac{7}{12}$

$E(Y) = \int_0^1 y\left(y + \dfrac{1}{2}\right)dy = \int_0^1 \left(y^2 + \dfrac{y}{2}\right)dx = \dfrac{y^3}{3} + \dfrac{y^2}{4}\Bigg]_0^1 = \dfrac{1^3}{3} + \dfrac{1^2}{4} - \dfrac{0^3}{3} - \dfrac{0^2}{4} = \dfrac{1}{3} + \dfrac{1}{4} = \dfrac{7}{12}$

$E(XY) = \int_0^1 \int_0^1 xy(x + y)\,dx\,dy = \int_0^1 \int_0^1 (x^2 y + xy^2)\,dx\,dy = \int_0^1 \left(\dfrac{x^3 y}{3} + \dfrac{x^2 y^2}{2}\right)\Bigg]_0^1 dy$

$= \int_0^1 \left(\dfrac{1^3 y}{3} + \dfrac{1^3 y^2}{2} - \dfrac{0^3 y}{3} - \dfrac{0^2 y^2}{2}\right)dy = \int_0^1 \left(\dfrac{y}{3} + \dfrac{y^2}{2}\right)dy = \dfrac{y^2}{6} + \dfrac{y^3}{6}\Bigg]_0^1$

$= \dfrac{1^2}{6} + \dfrac{1^3}{6} - \dfrac{0^2}{6} - \dfrac{0^3}{6} = \dfrac{1}{3}$

$Cov(X, Y) = E(XY) - E(X)E(Y) = \dfrac{1}{3} - \left(\dfrac{7}{12}\right)\left(\dfrac{7}{12}\right) = \dfrac{1}{3} - \dfrac{49}{144} = -\dfrac{1}{144}$

Since $Cov(X, Y) \neq 0$, X and Y are correlated. Thus, X and Y are not independent.

f. $D = 1 - \dfrac{X + Y}{2} = 1 - \dfrac{1}{2}(X + Y)$

$E(D) = E\left(1 - \dfrac{1}{2}(X + Y)\right) = 1 - \dfrac{1}{2}\{E(X) + E(Y)\} = 1 - \dfrac{1}{2}\left(\dfrac{7}{12} + \dfrac{7}{12}\right) = 1 - \dfrac{14}{24} = \dfrac{10}{24} = \dfrac{5}{12}$

$V(D) = V\left(1 - \dfrac{1}{2}(X + Y)\right) = \left(-\dfrac{1}{2}\right)^2 \{V(X) + V(Y) + 2Cov(X, Y)\}$

$E(X^2) = \int_0^1 x^2\left(x + \dfrac{1}{2}\right)dx = \int_0^1 \left(x^3 + \dfrac{x^2}{2}\right)dx = \dfrac{x^4}{4} + \dfrac{x^3}{6}\Bigg]_0^1 = \dfrac{1^4}{4} + \dfrac{1^3}{6} - \dfrac{0^4}{4} - \dfrac{0^3}{6} = \dfrac{1}{4} + \dfrac{1}{6} = \dfrac{5}{12}$

$V(X) = E(X^2) - (E(X))^2 = \dfrac{5}{12} - \left(\dfrac{7}{12}\right)^2 = \dfrac{5}{12} - \dfrac{49}{144} = \dfrac{11}{144}$

$E(Y^2) = \int_0^1 y^2\left(y + \dfrac{1}{2}\right)dy = \int_0^1 \left(y^3 + \dfrac{y^2}{2}\right)dyy = \dfrac{y^4}{4} + \dfrac{y^3}{6}\Bigg]_0^1 = \dfrac{1^4}{4} + \dfrac{1^3}{6} - \dfrac{0^4}{4} - \dfrac{0^3}{6} = \dfrac{1}{4} + \dfrac{1}{6} = \dfrac{5}{12}$

$V(Y) = E(Y^2) - (E(Y))^2 = \dfrac{5}{12} - \left(\dfrac{7}{12}\right)^2 = \dfrac{5}{12} - \dfrac{49}{144} = \dfrac{11}{144}$

Thus,

$V(D) = \left(-\dfrac{1}{2}\right)^2 \{V(X) + V(Y) + 2Cov(X, Y)\} = \dfrac{1}{4}\left(\dfrac{11}{144} + \dfrac{11}{144} + 2\left(-\dfrac{1}{144}\right)\right) = \dfrac{1}{4}\left(\dfrac{20}{144}\right) = \dfrac{5}{144}$

We would expect D to fall within 3 standard deviations of the mean:

$$\mu \pm 3\sigma \Rightarrow \frac{5}{12} \pm 3\sqrt{\frac{5}{144}} \Rightarrow 0.417 \pm 0.559 \Rightarrow (-0.142, 0.976)$$

6.109 a. $Y \sim B(n,p) \sim B(330, 0.54)$; $\mu = np = 330(0.54) = 178.2$

$$\sigma = \sqrt{npq} = \sqrt{330(0.54)(0.46)} = \sqrt{81.972} = 9.054$$

Using the normal approximation to the binomial:

$$P(Y < 100) = P\left(Z < \frac{99.5 - 178.2}{9.0538}\right) = P(Z < -8.69) = 0.5 - P(-8.69 < Z < 0) \approx 0.5 - 0.5 = 0$$

(using Table 5, Appendix B)

b. $P(Y \geq 200) = P\left(Z \geq \frac{199.5 - 178.2}{9.0538}\right) = P(Z \geq 2.35) = 0.5 - P(0 \leq Z \leq 2.35)$

$$= 0.5 - 0.4906 = 0.0094$$

(using Table 5, Appendix B)

6.111   If $W = 10Y - 2$, then $Y = \frac{W + 2}{10}$.

For $0 \leq y \leq 1$, $f(y) = \frac{y}{2}$ and for $1 \leq y \leq 2.5$, $f(y) = \frac{1}{2}$

If $Y = \frac{W + 2}{10}$, then $dy = \frac{1}{10}dw$.

Also, if $0 \leq y \leq 1$, then $-2 \leq w \leq 8$ and if $1 \leq y \leq 2.5$, then $8 \leq w \leq 23$

Thus,

$$\int_0^1 \frac{y}{2} dy + \int_1^{2.5} \frac{1}{2} dy = 1 \Rightarrow \int_{-2}^8 \frac{1}{2}\left(\frac{w+2}{10}\right)\frac{1}{10} dw + \int_8^{23}\left(\frac{1}{2}\right)\frac{1}{10} dw = 1 \Rightarrow \int_{-2}^8 \frac{w+2}{200} dw + \int_8^{23} \frac{1}{20} dw$$

The density function of W is $f(w) = \begin{cases} \dfrac{w+2}{200} & -2 \leq w \leq 8 \\ \dfrac{1}{20} & 8 \leq w \leq 23. \\ 0 & \text{elsewhere} \end{cases}$

6.113 a. By the Central Limit Theorem, the sampling distribution of $\bar{Y}$ is approximately normal with $\mu_{\bar{Y}} = \mu = 121.74$ and $\sigma_{\bar{Y}} = \frac{\sigma}{\sqrt{n}} = \frac{27.52}{\sqrt{32}} = 4.8649$.

b. $P(118 < \bar{Y} < 130) = P\left(\frac{118 - 121.74}{4.8649} < Z < \frac{130 - 121.74}{4.8694}\right) = P(-0.77 < Z < 1.70)$

$$P(-0.77 < Z < 0) + P(0 < Z < 1.70) = 0.2794 + 0.4554 = 0.7348$$

(using Table 5, Appendix B)

6.115   By the Central Limit Theorem, the sampling distribution of $\bar{Y}$ is approximately

normal with $\mu_{\bar{Y}} = \mu = 406$ and $\sigma_{\bar{Y}} = \dfrac{\sigma}{\sqrt{n}} = \dfrac{10.1}{\sqrt{36}} = 1.6833$.

$$P(\bar{Y} \leq 400.8) = P\left(Z \leq \frac{400.8 - 406}{1.6833}\right) = P(Z \leq -3.09) = 0.5 - P(-3.09 \leq Z \leq 0)$$

$$= 0.5 - 0.4990 = 0.0010$$

The first operator is correct. Because the probability of observing a sample mean of 400.8 or less is so small (0.0010) when the true mean is 406, there is evidence that the true mean is not 406 but something less than 406.

6.117 $Y \sim B(n, p) \sim B(2{,}000, 0.16)$; $\mu = np = 2{,}000(0.16) = 320$

$$\sigma = \sqrt{npq} = \sqrt{2{,}000(0.16)(0.84)} = \sqrt{2.68.8} = 16.3951$$

Using the normal approximation to the binomial:

$$P(Y \leq 280) = P\left(Z \leq \frac{280.5 - 320}{16.3951}\right) = P(Z \leq -2.41) = 0.5 - P(-2.41 < Z < 0)$$

$$= 0.5 - 0.4920 = 0.0080$$

(using Table 5, Appendix B)

6.119   By the Central Limit Theorem, the sampling distribution of $\bar{Y}$ is approxi-

mately normal with $\mu_{\bar{Y}} = \mu = 2.5$ and $\sigma_{\bar{Y}} = \dfrac{\sigma}{\sqrt{n}} = \dfrac{\sqrt{2.5}}{\sqrt{35}} = 0.2673$.

$$P(\bar{Y} \leq 2.1) = P\left(Z \leq \frac{2.1 - 2.5}{0.2673}\right) = P(Z \geq -1.50) = 0.5 + P(-1.50 \leq Z \leq 0)$$

$$= 0.5 + 0.4332 = 0.9332$$

(using Table 5, Appendix B)

6.123 $f_2(y \mid x) f_1(x) = \dfrac{f(x,y)}{f_1(x)} = f(x,y)$ $f_1(x \mid y) f_2(y) = \dfrac{f(x,y)}{f_2(y)} f_2(y) = f(x,y)$

Thus, $f_2(y \mid x) f_1(x) = f(x,y) = f_1(x \mid y) f_2(y)$

6.125 a. $\displaystyle\int_0^1 \int_0^2 \int_0^\infty c(y_1 + y_2) e^{-y_3}\, dy_3\, dy_2\, dy_1 = \int_0^1 \int_0^2 c(y_1 + y_2)(-e^{-y_3})\Big]_0^\infty dy_2\, dy_1$

$$= \int_0^1 \int_0^2 c(y_1 + y_2)(-e^{-\infty} + e^0)\, dy_2\, dy_1 = \int_0^1 \int_0^2 c(y_1 + y_2)\, dy_2\, dy_1 = \int_0^1 \left(cy_1 y_2 + \frac{cy_2^2}{2}\right)\Big]_0^2 dy_1$$

$$= \int_0^1 (2cy_1 + 2c)\, dy_1 = cy_1^2 + 2cy_1 \Big]_0^1 = \left(c(1)^2 + 2c(1)\right) - \left(c(0)^2 + 2c(0)\right)$$

$$= 3c = 1 \Rightarrow c = \frac{1}{3}$$

b. $f_1(y_1) = \int_0^2 \int_0^\infty \dfrac{1}{3}(y_1 + y_2)e^{-y_3}\, dy_3\, dy_2 = \int_0^2 \dfrac{1}{3}(y_1 + y_2)(-e^{-y_3})\Big]_0^\infty dy_2$

$$= \int_0^2 \dfrac{1}{3}(y_1 + y_2)(-e^{-\infty} + e^0)\, dy_2 = \int_0^2 \dfrac{1}{3}(y_1 + y_2)\, dy_2 = \dfrac{1}{3}\left(y_1 y_2 + \dfrac{y_2^2}{2}\right)\Big]_0^2$$

$$= \dfrac{1}{3}\left[\left(y_1(2) + \dfrac{2^2}{2}\right) - \left(y_1(0) + \dfrac{0^2}{2}\right)\right] = \dfrac{1}{3}(2y_1 + 2)$$

$f_2(y_2) = \int_0^1 \int_0^\infty \dfrac{1}{3}(y_1 + y_2)e^{-y_3}\, dy_3\, dy_1 = \int_0^1 \dfrac{1}{3}(y_1 + y_2)(-e^{-y_3})\Big]_0^\infty dy_1$

$$= \int_0^1 \dfrac{1}{3}(y_1 + y_2)(-e^{-\infty} + e^0)\Big]_0^\infty dy_1 = \int_0^1 \dfrac{1}{3}(y_1 + y_2)\, dy_1 = \dfrac{1}{3}\left(\dfrac{y_1^2}{2} + y_1 y_2\right)\Big]_0^1$$

$$= \dfrac{1}{3}\left(\dfrac{1^2}{2} + (1)y_2\right) - \left(\dfrac{0^2}{2} + (0)y_2\right) = \dfrac{1}{3}\left(y_2 + \dfrac{1}{2}\right)$$

$f_3(y_3) = \int_0^1 \int_0^2 \dfrac{1}{3}(y_1 + y_2)e^{-y_3}\, dy_2\, dy_1 = \int_0^1 \dfrac{1}{3}e^{-y_3}\left(y_1 y_2 + \dfrac{y_2^2}{2}\right)\Big]_0^2 dy_1$

$$= \int_0^1 \dfrac{1}{3}e^{-y_3}\left[\left(y_1(2) + \dfrac{2^2}{2}\right) - \left(y_1(0) + \dfrac{0^2}{2}\right)\right] dy_1 = \int_0^1 \dfrac{1}{3}e^{-y_3}(2y_1 + 2)\, dy_1$$

$$= \dfrac{1}{3}e^{-y_3}\left(y_1 + 2y_1\right)\Big]_0^1 = \dfrac{1}{3}e^{-y_3}\left[(1 + 2(1)) - (0 + 2(0))\right] = e^{-y_3}$$

$f(y_1, y_2, y_3) = f_1(y_1)f_2(y_2)f_3(y_3)$

$f_1(y_1)f_2(y_2)f_3(y_3) = \dfrac{1}{3}(2y_1 + 2)\dfrac{1}{3}\left(y_2 + \dfrac{1}{2}\right)e^{-y_3} = \dfrac{1}{9}(2y_1 + 2)\left(y_2 + \dfrac{1}{2}\right)e^{-y_3} \neq f(y_1, y_2, y_3)$

Therefore, the 3 variables are not independent.

6.127   Since Y_1 and Y_2 are independent,

$$f(y_1, y_2) = f_1(y_1)f_2(y_2) = \left(\dfrac{1}{\beta}e^{-y_1/\beta}\right)\left(\dfrac{1}{\beta}e^{-y_2/\beta}\right) = \dfrac{1}{\beta^2}e^{-(y_1 + y_2)/\beta}$$

Thus, $f(y_1, y_2) = \begin{cases} \dfrac{1}{\beta^2}e^{-(y_1 + y_2)/\beta} & y_1 > 0,\ y_2 > 0 \\[2mm] 0 & \text{elsewhere} \end{cases}$

For $W = Y_1 + Y_2$,

$$P(w \le w_0) = P(0 < Y_2 \le w - Y_1, 0 \le Y_1 < w)$$

$$= \int_0^{w_0} \int_0^{w_0 - y_1} \frac{1}{\beta^2} e^{-(y_1 + y_2)/\beta} dy_2 \, dy_1$$

$$= \int_0^{w_0} \left(-\frac{1}{\beta} e^{-y_1/\beta} e^{-y_2/\beta} \right) \Bigg]_0^{w_0 - y_1} dy_1 \int_0^{w_0} \left(-\frac{1}{\beta} e^{-y_1/\beta} e^{-(w_0 - y_1)/\beta} \right) + \left(\frac{1}{\beta} e^{-y_1/\beta} e^0 \right) \Bigg]_0^{w_0 - y_1} dy_1$$

$$= \int_0^{w_0} \frac{1}{\beta} \left(e^{-y_1/\beta} - e^{-w_0/\beta} \right) dy_1 = \left(\left(-e^{-y_1/\beta} \right) - \frac{y_1}{\beta} e^{-w_0/\beta} \right) \Bigg]_0^{w_0}$$

$$= -e^{-w_0/\beta} - \frac{w_0}{\beta} e^{-w_0/\beta} + e^0 + 0 = 1 - e^{-w_0/\beta} - \frac{w_0}{\beta} e^{-w_0/\beta}$$

To find the density function of W, we take the derivative of the cumulative distribution function with respect to w.

$$F(w) = 1 - e^{-w/\beta} - \frac{w}{\beta} e^{-w/\beta}$$

$$\frac{dF(w)}{dw} = \frac{d\left(1 - e^{-w/\beta} - \frac{w}{\beta} e^{-w/\beta} \right)}{dw} = \frac{1}{\beta} e^{-w/\beta} - \frac{1}{\beta} e^{-w/\beta} + \frac{w}{\beta^2} e^{-w/\beta} = \frac{w}{\beta^2} e^{-w/\beta} \text{ for } w > 0$$

Thus, the distribution of W is gamma with $\alpha = 2$ and unknown β.

6.129 If $W = Y^2$, then $Y = W^{1/2}$.

$$(y) = \left(\frac{2y}{\beta} \right) e^{-y^2/\beta} \qquad y > 0 \quad \text{Thus, } \int_0^\infty \left(\frac{2y}{\beta} \right) e^{-y^2/\beta} = 1.$$

If $Y = W^{1/2}$, then $dy = \frac{1}{2} w^{-1/2} dw$. Also, if $y \ge 0$, then $w \ge 0$.

Thus, $\int_0^\infty \left(\frac{2y}{\beta} \right) e^{-y^2/\beta} = 1 \Rightarrow \int_0^\infty \frac{2w^{1/2}}{\beta} e^{-w/\beta} \frac{1}{2} w^{-1/2} dw = 1 \Rightarrow \int_0^\infty \frac{e^{-w/\phi}}{\beta} dw = 1$

The density function of W is $f(w) = \begin{cases} \dfrac{e^{-w/\beta}}{\beta} & w \ge 0 \\ 0 & \text{elsewhere} \end{cases}$.

Thus, W has an exponential distribution.

6.133   To use theorem 6.7, we must first find the cumulative distribution function of Y.

$$F(y) = \int_0^y 2te^{-t^2}\,dt = -e^{-t^2}\bigg]_0^y = -e^{-y^2} + e^0 = 1 - e^{-y^2}$$

Let $W = F(y) = 1 - e^{-Y^2}$. By Theorem 6.7, W has a uniform distribution over the interval $0 \le w \le 1$.

$$W = 1 - e^{-Y^2} \Rightarrow e^{-Y^2} = 1 - W \Rightarrow -Y^2 = \ln(1 - W) \Rightarrow Y = \sqrt{-\ln(1 - W)}$$

By selecting random numbers from the random number table and substituting them in for W, we will get random selections for Y.

Suppose we select 5 random numbers from Table 1, Appendix B. Starting with the number in column 9, row 15 and going down the column, the 5 random numbers are:

97735, 49442, 01188, 71585, 23495. These numbers correspond to 0.97735, 0.49442, 0.01188, 0.71585, 0.23495 from a uniform distribution on the interval from 0 to 1.

For $w_1 = 0.97735$, $y_1 = \sqrt{-\ln(1 - 0.97735)} = 1.9462$

For $w_2 = 0.49442$, $y_2 = \sqrt{-\ln(1 - 0.49442)} = 0.8259$

For $w_3 = 0.01188$, $y_3 = \sqrt{-\ln(1 - 0.01188)} = 0.1093$

For $w_4 = 0.71585$, $y_4 = \sqrt{-\ln(1 - 0.71585)} = 1.1217$

For $w_5 = 0.23495$, $y_5 = \sqrt{-\ln(1 - 0.23495)} = 0.5175$

7

Estimation Using Confidence Intervals

7.1   a. $E(\hat{\theta}_1) = E(\bar{y}) = E\left(\dfrac{y_1 + y_2 + y_3}{3}\right) = \dfrac{1}{3}E(y_1 + y_2 + y_3) = \dfrac{1}{3}\left[E(y_1) + E(y_2) + E(y_3)\right]$

$= \dfrac{1}{3}(\theta + \theta + \theta) = \theta$

$E(\hat{\theta}_2) = E(y_1) = \theta$

$E(\hat{\theta}_3) = E\left(\dfrac{y_1 + y_2}{2}\right) = \dfrac{1}{2}E(y_1 + y_2) = \dfrac{1}{2}\left[E(y_1) + E(y_2)\right] = \dfrac{1}{2}(\theta + \theta) = \theta$

b. $V(\hat{\theta}_1) = V(\bar{y}) = V\left(\dfrac{y_1 + y_2 + y_3}{3}\right) = \dfrac{1}{3^2}\left[V(y_1) + V(y_2) + V(y_3)\right]$

$= \dfrac{1}{9}\left[\theta^2 + \theta^2 + \theta^2\right] = \dfrac{\theta^2}{3}$

(Since the y_i's are independent, the covariances are 0.)

$E\left(\hat{\theta}_2\right) = V(y_1) = \theta^2$

$V(\hat{\theta}_3) = V\left(\dfrac{y_1 + y_2}{2}\right) = \dfrac{1}{2^2}\left[V(y_1) + V(y_2)\right] = \dfrac{1}{4}(\theta^2 + \theta^2) = \dfrac{\theta^2}{2}$

Thus, $\hat{\theta}_1$ has the smallest variance.

7.3   a.   The mean of a binominal distribution is $\mu = E(Y) = np$.

$E(\hat{p}) = E\left(\dfrac{Y}{n}\right) = \dfrac{1}{n}E(Y) = \dfrac{1}{n}np = p$

b.   The variance of a binomial distribution is $\sigma^2 = npq$.

$V(\hat{p}) = V\left(\dfrac{Y}{n}\right) = \dfrac{1}{n^2}V(Y) = \dfrac{1}{n^2}npq = \dfrac{pq}{n}$

7.5 $E\left[\left(\hat{\theta} - \theta\right)^2\right] = E\left\{\left[\hat{\theta} - E(\hat{\theta})\right] + \left[E(\hat{\theta}) - \theta\right]\right\}^2$

$= E\left\{\left[\hat{\theta} - E(\hat{\theta})\right]^2 + 2\left[\hat{\theta} - E(\hat{\theta})\right]\left[E(\hat{\theta}) - \theta\right] + \left[E(\hat{\theta}) - \theta\right]^2\right\}$

$= V(\hat{\theta}) + 2\left[E(\hat{\theta}) - \theta\right]E\left[\hat{\theta} - E(\hat{\theta})\right] + \left[E(\hat{\theta}) - \theta\right]^2$

$= V(\hat{\theta}) + 2\left[E(\hat{\theta}) - \theta\right]0 + \left[E(\hat{\theta}) - \theta\right]^2 = V(\hat{\theta}) + b^2(\theta)$

7.7   From Theorem 6.11, we know that when sampling from a normal distribution,

$$\frac{(n-1)S^2}{\sigma^2} = \chi^2$$

where χ^2 is a chi-square random variable with $v = (n-1)$ degrees of freedom. Rearranging terms yields:

$$S^2 = \frac{\sigma^2}{(n-1)}\chi^2$$

We know from Section 5.7 that $E(\chi^2) = v = n-1$ and $V(\chi^2) = 2v = 2(n-1)$.

Thus, $V(S^2) = V\left(\frac{\sigma^2}{(n-1)}\chi^2\right) = \frac{\sigma^4}{(n-1)^2}V(\chi^2) = \frac{\sigma^4}{(n-1)^2}2(n-1) = \frac{2\sigma^4}{(n-1)}$.

7.9 a. If $y_1, y_2, \ldots y_n$ is a random sample of n observations from a Poisson distribution, then the likelihood function is:

$$L = p(y_1)p(y_2)\cdots p(y_n) = \left(\frac{e^{-\lambda}\lambda^{y_1}}{y_1!}\right)\left(\frac{e^{-\lambda}\lambda^{y_2}}{y_2!}\right)\cdots\left(\frac{e^{-\lambda}\lambda^{y_n}}{y_n!}\right) = \frac{e^{-n\lambda}\lambda^{\Sigma y_i}}{\prod\limits_{i=1}^{n}y_i!}$$

Then $\ln L = -n\lambda + \sum y_i \ln\lambda - \ln\prod\limits_{i=1}^{n}y_i!$

The derivative of $\ln L$ with respect of λ is:

$$\frac{d\ln L}{d\lambda} = \frac{d\left[-n\lambda + \sum y_i \ln\lambda - \ln\prod\limits_{i=1}^{n}y_i!\right]}{d\lambda} = -n + \frac{\sum y_i}{\lambda}$$

Setting this equal to 0 and solving, we get:

$$-n + \frac{\sum y_i}{\hat{\lambda}} = 0 \Rightarrow n = \frac{\sum y_i}{\hat{\lambda}} \Rightarrow \hat{\lambda} = \frac{\sum y_i}{n} = \bar{y}$$

b. To determine if the maximum likelihood estimator is unbiased, we must find its expected value. We know $E(Y) = \lambda$ if Y has a Poisson distribution with parameter λ.

$$E(\hat{\lambda}) = E(\bar{y}) = E\left(\frac{y_1 + y_2 + \cdots + y_n}{n}\right) = \frac{1}{n}E(y_1 + y_2 + \cdots + y_n) = \frac{1}{n}(\lambda + \lambda + \cdots + \lambda) = \lambda$$

Therefore, $\hat{\lambda}$ is an unbiased estimator of λ.

7.11 a. Since there is only one unknown parameter, β, to estimate, the moment estimator is found by setting the first population moment, $E(Y)$, equal to the first sample moment, $\bar{y}$. For the gamma distribution with $\alpha = 2$, $E(Y) = \alpha\beta = 2\beta$.

Thus, the moment estimator is $2\hat{\beta} = \bar{y} \Rightarrow \hat{\beta} = \frac{\bar{y}}{2}$

b. $E(\hat{\beta}) = E\left(\frac{\bar{y}}{2}\right) = \frac{1}{2}E(\bar{y}) = \frac{1}{2}E\left(\frac{y_1 + y_2 + \cdots y_n}{n}\right) = \frac{1}{2n}\left[E(y_1) + E(y_2) + \cdots + E(y_n)\right]$

$$= \frac{1}{2n}\left[2\beta + 2\beta + \cdots + 2\beta\right] = \frac{1}{2n}2n\beta = \beta$$

$$V(\hat{\beta}) = V\left(\frac{\bar{y}}{2}\right) = \frac{1}{2^2}V(\bar{y}) = \frac{1}{4}V\left(\frac{y_1 + y_2 + \cdots y_n}{n}\right) = \frac{1}{4n^2}\left[V(y_1) + V(y_2) + \cdots + V(y_n)\right]$$

$$= \frac{1}{4n^2}\left[2\beta^2 + 2\beta^2 + \cdots + 2\beta^2\right] = \frac{1}{4n^2}2n\beta^2 = \frac{\beta^2}{2n}$$

Since the y_i's are a random sample, they are independent of each other. Therefore, all the covariances are equal to 0.

7.13   a.   Since there is only one unknown parameter, β, to estimate, the moment estimator is found by setting the first population moment, $E(Y)$, equal to the first sample moment, $\bar{y}$. For the exponential distribution, $E(Y) = \beta$.

Thus, the moment estimator is $\hat{\beta} = \bar{y}$

b. $E(\hat{\beta}) = E(\bar{y}) = E\left(\frac{y_1 + y_2 + \cdots + y_n}{n}\right) = \frac{1}{n}\left[E(y_1) + E(y_2) + \cdots + E(y_n)\right]$

$$= \frac{1}{n}\left[\beta + \beta + \cdots + \beta\right] = \frac{1}{n}n\beta = \beta$$

c. $V(\hat{\beta}) = V(\bar{y}) = V\left(\frac{y_1 + y_2 + \cdots + y_n}{n}\right) = \frac{1}{n^2}\left[V(y_1) + V(y_2) + \cdots + V(y_n)\right]$

$$= \frac{1}{n^2}\left[\beta^2 + \beta^2 + \cdots + \beta^2\right] = \frac{1}{n^2}n\beta^2 = \frac{\beta^2}{n}$$

The variance of the exponential distribution is σ^2. Since the y_i's are a random sample, they are independent of each other. Therefore, all the covariances are equal to 0.

7.15   Using degrees of freedom $v = \infty$, we find from Table 7, Appendix B:

$$t_{0.05} = 1.645 = z_{0.05}$$
$$t_{0.025} = 1.96 = z_{0.025}$$
$$t_{0.01} = 2.326 = z_{0.01}$$

7.17   By Theorem 7.2, the sampling distribution of $\bar{y}$ is approximately normal with mean $\mu_{\bar{y}} = \mu = \lambda$ and standard deviation $\sigma_{\bar{y}} = \sigma/\sqrt{n} = \sqrt{\lambda/n}$.

Thus, $Z = \frac{\bar{y} - \lambda}{\sqrt{\lambda/n}}$ has an approximate standard normal distribution.

Using Z as the pivotal statistic, the confidence interval for λ is:

$$P\left(-z_{\alpha/2} \leq Z \leq z_{\alpha/2}\right) = P\left(-z_{\alpha/2} \leq \frac{\bar{y} - \lambda}{\sqrt{\lambda/n}} \leq z_{\alpha/2}\right) = 1 - \alpha$$

Now, substitute $\bar{y}$ for λ in the denominator. (We know the maximum likelihood estimator of λ is $\bar{y}$ from Exercise 7.9.)

$$P\left(-z_{\alpha/2} \leq \frac{\bar{y}-\lambda}{\sqrt{\bar{y}/n}} \leq z_{\alpha/2}\right) = P\left(-z_{\alpha/2}\sqrt{\bar{y}/n} \leq \bar{y}-\lambda \leq z_{\alpha/2}\sqrt{\bar{y}/n}\right)$$

$$= P\left(-\bar{y}-z_{\alpha/2}\sqrt{\bar{y}/n} \leq -\lambda \leq -\bar{y}+z_{\alpha/2}\sqrt{\bar{y}/n}\right)$$

$$= P\left(\bar{y}+z_{\alpha/2}\sqrt{\bar{y}/n} \geq \lambda \geq \bar{y}-z_{\alpha/2}\sqrt{\bar{y}/n}\right)$$

$$= P\left(\bar{y}-z_{\alpha/2}\sqrt{\bar{y}/n} \leq \lambda \leq \bar{y}+z_{\alpha/2}\sqrt{\bar{y}/n}\right) = 1-\alpha$$

The $100(1-\alpha)\%$ confidence interval for λ is $\bar{y} \pm z_{\alpha/2}\sqrt{\bar{y}/n}$.

7.19   By Theorem 6.9, the sampling distribution of $\bar{y}_1 - \bar{y}_2$ has an approximate normal distribution with mean $\mu_{\bar{y}_1-\bar{y}_2} = \mu_1 - \mu_2$ and standard deviation $\sigma_{\bar{y}_1-\bar{y}_2} = \sqrt{\dfrac{\sigma_1^2}{n_1} + \dfrac{\sigma_2^2}{n_2}}$. Thus, $Z = \dfrac{(\bar{y}_1 - \bar{y}_2)-(\mu_1 - \mu_2)}{\sqrt{\frac{\sigma_1^2}{n_1} + \frac{\sigma_2^2}{n_2}}}$ has a standard normal distribution.

Using Z as the pivotal statistic, the confidence interval for $\mu_1 - \mu_2$ is:

$$P(-z_{\alpha/2} \leq Z \leq z_{\alpha/2}) = P\left(-z_{\alpha/2} \leq \frac{(\bar{y}_1 - \bar{y}_2)-(\mu_1 - \mu_2)}{\sqrt{\frac{\sigma_1^2}{n_1} + \frac{\sigma_2^2}{n_2}}} \leq z_{\alpha/2}\right) = 1-\alpha$$

Now, substitute s_1^2 and s_2^2 (the maximum likelihood estimates) for σ_1^2 and σ_2^2 in the denominator.

$$P\left(-z_{\alpha/2} \leq \frac{(\bar{y}_1 - \bar{y}_2)-(\mu_1 - \mu_2)}{\sqrt{\frac{s_1^2}{n_1} + \frac{s_2^2}{n_2}}} \leq z_{\alpha/2}\right)$$

$$= P\left(-z_{\alpha/2}\sqrt{\frac{s_1^2}{n_1} + \frac{s_2^2}{n_2}} \leq (\bar{y}_1 - \bar{y}_2)-(\mu_1 - \mu_2) \leq z_{\alpha/2}\sqrt{\frac{s_1^2}{n_1} + \frac{s_2^2}{n_2}}\right)$$

$$= P\left((-\bar{y}_1 - \bar{y}_2)-z_{\alpha/2}\sqrt{\frac{s_1^2}{n_1} + \frac{s_2^2}{n_2}} \leq -(\mu_1 - \mu_2) \leq -(\bar{y}_1 - \bar{y}_2)+z_{\alpha/2}\sqrt{\frac{s_1^2}{n_1} + \frac{s_2^2}{n_2}}\right)$$

$$= P\left((\bar{y}_1 - \bar{y}_2)+z_{\alpha/2}\sqrt{\frac{s_1^2}{n_1} + \frac{s_2^2}{n_2}} \geq (\mu_1 - \mu_2) \geq (\bar{y}_1 - \bar{y}_2)-z_{\alpha/2}\sqrt{\frac{s_1^2}{n_1} + \frac{s_2^2}{n_2}}\right)$$

$$= P\left((\bar{y}_1 - \bar{y}_2)-z_{\alpha/2}\sqrt{\frac{s_1^2}{n_1} + \frac{s_2^2}{n_2}} \leq (\mu_1 - \mu_2) \leq (\bar{y}_1 - \bar{y}_2)+z_{\alpha/2}\sqrt{\frac{s_1^2}{n_1} + \frac{s_2^2}{n_2}}\right) = 1-\alpha$$

Thus, the $100(1-\alpha)\%$ confidence interval for $\mu_1 - \mu_2$ is $(\bar{y}_1 - \bar{y}_2) \pm z_{\alpha/2}\sqrt{\dfrac{s_1^2}{n_1} + \dfrac{s_2^2}{n_2}}$.

7.21　If χ_1^2 and χ_2^2 are independent chi-square random variables with $v_1 = n_1 - 1$ and $v_2 = n_2 - 1$ degrees of freedom, respectively, then, by Theorem 6.12,

$$\chi^2 = \chi_1^2 + \chi_2^2 = \frac{(n_1 - 1)s_1^2}{\sigma^2} + \frac{(n_2 - 1)s_2^2}{\sigma^2} = \frac{(n_1 - 1)s_1^2 + (n_2 - 1)s_2^2}{\sigma^2}$$

is a chi-square random variable with $v_1 + v_2 = (n_1 - 1) + (n_2 - 1) = n_1 + n_2 - 2$ degrees of freedom.

7.23　Using the pivotal statistic $T = \dfrac{(\bar{y}_1 - \bar{y}_2) - (\mu_1 - \mu_2)}{s_p\sqrt{\dfrac{1}{n_1} + \dfrac{1}{n_2}}}$, the confidence interval for $\mu_1 - \mu_2$ is:

$$P\left(-t_{\alpha/2} \leq T \leq t_{\alpha/2}\right) = P\left(-t_{\alpha/2} \leq \frac{(\bar{y}_1 - \bar{y}_2) - (\mu_1 - \mu_2)}{s_p\sqrt{\dfrac{1}{n_1} + \dfrac{1}{n_2}}} \leq t_{\alpha/2}\right)$$

$$= P\left(-t_{\alpha/2}s_p\sqrt{\frac{1}{n_1} + \frac{1}{n_2}} \leq (\bar{y}_1 - \bar{y}_2) - (\mu_1 - \mu_2) \leq t_{\alpha/2}s_p\sqrt{\frac{1}{n_1} + \frac{1}{n_2}}\right)$$

$$= P\left(-(\bar{y}_1 - \bar{y}_2) - t_{\alpha/2}s_p\sqrt{\frac{1}{n_1} + \frac{1}{n_2}} \leq -(\mu_1 - \mu_2) \leq -(\bar{y}_1 - \bar{y}_2) + t_{\alpha/2}s_p\sqrt{\frac{1}{n_1} + \frac{1}{n_2}}\right)$$

$$= P\left((\bar{y}_1 - \bar{y}_2) + t_{\alpha/2}s_p\sqrt{\frac{1}{n_1} + \frac{1}{n_2}} \geq (\mu_1 - \mu_2) \geq (\bar{y}_1 - \bar{y}_2) - t_{\alpha/2}s_p\sqrt{\frac{1}{n_1} + \frac{1}{n_2}}\right)$$

$$= P\left((\bar{y}_1 - \bar{y}_2) - t_{\alpha/2}s_p\sqrt{\frac{1}{n_1} + \frac{1}{n_2}} \leq (\mu_1 - \mu_2) \leq (\bar{y}_1 - \bar{y}_2) + t_{\alpha/2}s_p\sqrt{\frac{1}{n_1} + \frac{1}{n_2}}\right) = 1 - \alpha$$

Thus, the $100(1 - \alpha)\%$ confidence interval for $\mu_1 - \mu_2$ is $(\bar{y}_1 - \bar{y}_2) \pm t_{\alpha/2}s_p\sqrt{\dfrac{1}{n_1} + \dfrac{1}{n_2}}$

where $t_{\alpha/2}$ is based on $v = n_1 + n_2 - 2$ degrees of freedom.

7.25　For confidence coefficient 0.99, $\alpha = 0.01$ and $\alpha/2 = 0.01/2 = 0.005$. From Table 7, Appendix B, with $v = n - 1 = 4 - 1 = 3$ degrees of freedom, $t_{0.005} = 5.841$. The confidence interval is:

$$\bar{y} \pm t_{0.005}\frac{s}{\sqrt{n}} \Rightarrow 240 \pm 5.841\frac{15}{\sqrt{4}} \Rightarrow 240 \pm 43.81 \Rightarrow (196.19, 283.81)$$

We are 99% confident that the true mean volume of fish layer in the tank is between 196.19 and 28.81 kg. We must assume that the population of volumes of fish layer is normally distributed.

7.27 a. For confidence coefficient 0.90, $\alpha = 0.10$ and $\alpha/2 = 0.10/2 = 0.05$. From Table 5, Appendix B, $z_{0.05} = 1.645$. The confidence interval is:

$$\bar{y} \pm z_{0.05}\frac{s}{\sqrt{n}} \Rightarrow 18 \pm 1.645\frac{20}{\sqrt{500}} \Rightarrow 18 \pm 1.471 \Rightarrow (16.529, 19.471)$$

b. Yes. On average, the absolute deviation is between 16.529% and 19.471%. These values are all below 34%.

7.29   Using MINITAB, the summary statistics are:

One-Sample T: MTBE

Variable	N	Mean	StDev	SE Mean	99% CI
MTBE	12	97.2	113.8	32.8	(-4.8, 199.2)

a. The point estimate for the true mean MTBE level for all well sites located near the New Jersey gasoline station is $\bar{y} = 97.2$.

b. The 99% confidence interval for μ is shown above to be $(-4.8, 199.2)$. We are 99% confident that the true mean MTBE level for all well sites located near the New Jersey gasoline station is between -4.8 and 199.2 parts per billion.

c. Whenever we work with small sample sizes, we need to assume that the population we are sampling from will be approximately normally distributed. In this case, we need to assume that the distribution of MTBE levels in all well sites located near the New Jersey gasoline service station will be approximately normally distributed. The stem-and-leaf plot of the sampled MTBE levels is shown below.

Stem-and-Leaf Display: MTBE

```
Stem-and-leaf of MTBE    N  = 12
Leaf Unit = 10

  6   0   011113
  6   0   6
  5   1   03
  3   1   5
  2   2
  2   2   5
  1   3
  1   3   6
```

It does not appear that the sampled data display an approximate normal distribution. It is doubtful that the assumption will be satisfied for the population of MTBE values.

7.31 a. The population of interest to the researchers is the population of all lichen specimens in all the Alaskan locations of interest.

b. Using MINITAB, the summary statistics are:

One-Sample T: cesium137

Variable	N	Mean	StDev	SE Mean	95% CI
cesium137	9	0.00903	0.00485	0.00162	(0.00530, 0.01276)

The 95% confidence interval for μ is shown to be $(0.00530, 0.01276)$.

    c. We are 95% confident that the true mean amount of cesium-1137 for all lichen specimens falls between 0.00530 and 0.01276 microcuries per milliliter.

    d. Whenever we work with small sample sizes, we need to assume that the population we are sampling from will be approximately normally distributed.

7.33   Using MINITAB, the summary statistics are:

One-Sample T: Wheels

Variable	N	Mean	StDev	SE Mean	99% CI
Wheels	28	3.214	1.371	0.259	(2.497, 3.932)

    a. The 99% confidence interval for μ is shown to be $(2.497, 3.932)$.

    b. We are 99% confident that the true mean number of wheels used on all social robots built with wheels is between 2.497 and 3.932.

    c. In repeated sampling, 99% of all confidence intervals constructed in this manner will contain the true mean, μ.

7.35   Using MINITAB, the calculations are:

One-Sample T: Decay

Variable	N	Mean	StDev	SE Mean	95% CI
Decay	6	1.0733	0.2316	0.0945	(0.8303, 1.3164)

    a. From the printout, the 95% confidence interval is $(0.8303, 1.3164)$. We are 95% confident that true mean decay rate of fine particles produced from oven cooking or toasting is between 0.8303 and 1.3164 µm/hour.

    b. The phrase "95% confident" means that in repeated sampling, 95% of all confidence intervals constructed will contain the true mean.

    c. In order for the inference above to be valid, the distribution of decay rates must be normally distributed.

7.37 a. Using MINITAB, the calculations are:

One-Sample T: Velocity

Variable	N	Mean	StDev	SE Mean	95% CI
Velocity	25	0.26208	0.04669	0.00934	(0.24281, 0.28135)

From the printout, the 95% confidence interval is $(0.24281, 0.28135)$.
We are 95% confident that true mean bubble rising velocity is between 0.24281 and 0.28135.

    b. No. The value of $\mu = 0.338$ does not fall in the 95% confidence interval.

7.39   The 95% confidence interval for the difference in mean drug concentration for tablets produced at the two sites is $(-1.308, 2.338)$. Since 0 is contained in the interval, there is no evidence to indicate a difference in the mean drug concentrations between tablets produced at the two sites.

7.41   Using MINITAB, the calculations are:

Two-Sample T-Test and CI

Sample	N	Mean	StDev	SE Mean
1	406	0.310	0.400	0.020
2	230	0.130	0.300	0.020

Difference = μ (1) - μ (2)
Estimate for difference: 0.1800
95% CI for difference: (0.1205, 0.2395)

```
T-Test of difference = 0 (vs ≠): T-Value = 5.94  P-Value = 0.000
DF = 634
Both use Pooled StDev = 0.367
```

The 95% confidence interval for the difference in mean number of hippo trails between national reserve plots and pastoral ranch plots is $(0.1205, 0.2395)$. Because 0 is not contained in the interval, there is evidence to indicate there is a difference in the mean number of hippo trails between national reserve plots and pastoral ranch plots. Since the interval contains only positive numbers, the mean number of hippo trails is greater in the national reserve plots than in the pastoral ranch plots.

7.43   a.   Let μ_1 = mean yield strength of the RAA alloy and μ_2 = mean yield strength of the current alloy. The small sample confidence interval for $\mu_1 - \mu_2$ is:

$$\left(\bar{y}_1 - \bar{y}_2\right) \pm t_{\alpha/2} s_p \sqrt{\frac{1}{n_1} + \frac{1}{n_2}}$$

where $s_p = \sqrt{\dfrac{(n_1-1)s_1^2 + (n_2-1)s_2^2}{n_1+n_2-2}} = \sqrt{\dfrac{(3-1)(19.3)^2 + (3-1)(12.4)^2}{3+3-2}} = 16.2211$

For confidence coefficient 0.95, $\alpha = 0.05$ and $\alpha/2 = 0.05/2 = 0.025$. From Table 7, Appendix B, $v = n_1 + n_2 - 2 = 3 + 3 - 2 = 4$ degrees of freedom, $t_{0.025} = 2.776$. The 95% confidence interval is:

$$\left(\bar{y}_1 - \bar{y}_2\right) \pm t_{0.025} s_p \sqrt{\frac{1}{n_1} + \frac{1}{n_2}} \Rightarrow (641.0 - 592.7) \pm 2.776 (16.221)\sqrt{\frac{1}{3} + \frac{1}{3}}$$

$$\Rightarrow 48.3 \pm 36.766 \Rightarrow (11.534, 85.066)$$

b.   We agree with the researchers. All of the values in the confidence interval are positive. This indicates that the mean yield strength of the RAA alloy exceeds the mean yield strength of the current alloy.

7.45   Using MINITAB, the output is:

Two-Sample T-Test and CI

```
Sample    N   Mean   StDev   SE Mean
1        431   21.5   33.4      1.6
2        508   22.2   34.9      1.5

Difference = μ (1) - μ (2)

Estimate for difference:   -0.70
95% CI for difference:   (-5.10, 3.70)
T-Test of difference = 0 (vs ≠): T-Value = -0.31  P-Value = 0.755
DF = 937

Both use Pooled StDev = 34.2198
```

From above, the 95% confidence interval is $(-5.10, 3.70)$. No. Because 0 falls in the 95% confidence interval, there is no evidence that the mean amount of surplus hay producers are willing to sell to the biomass market differ for the two areas.

7.47 a. Let μ_1 = mean change in Sv for sintering time of 10 minutes and μ_2 = mean change in Sv for sintering time of 150 minutes.

The large sample confidence interval for $\mu_1 - \mu_2$ is:

$$(\bar{y}_1 - \bar{y}_2) \pm z_{\alpha/2}\sqrt{\frac{\sigma_1^2}{n_1} + \frac{\sigma_2^2}{n_2}}$$

For confidence coefficient 0.95, $\alpha = 0.05$ and $\alpha/2 = 0.05/2 = 0.025$. From Table 5, Appendix B, $z_{0.025} = 1.96$. The 95% confidence interval is:

$$(\bar{y}_1 - \bar{y}_2) \pm z_{\alpha/2}\sqrt{\frac{s_1^2}{n_1} + \frac{s_2^2}{n_2}} \Rightarrow (736.0 - 299.5) \pm 1.96\sqrt{\frac{181.9^2}{100} + \frac{161.0^2}{100}}$$

$$\Rightarrow 436.5 \pm 47.612 \Rightarrow (388.888, 484.112)$$

We are 95% confident that the mean change in Sv for sintering times of 10 minutes exceeds the mean change in Sv for sintering times of 150 minutes.

b. The 95% confidence interval is:

$$(\bar{y}_1 - \bar{y}_2) \pm z_{\alpha/2}\sqrt{\frac{s_1^2}{n_1} + \frac{s_2^2}{n_2}} \Rightarrow (96.73 - 97.82) \pm 1.96\sqrt{\frac{2.1^2}{100} + \frac{1.5^2}{100}}$$

$$\Rightarrow -1.09 \pm 0.506 \Rightarrow (-1.596, -0.584)$$

We are 95% confident that the mean change in Vv for sintering times of 10 minutes is less than the mean change in Vv for sintering times of 150 minutes.

7.49 a. The twin holes at the same location are not independent. Thus, we need to analyze the data as paired differences.

b. The differences are:

Loc	1st Hole	2nd Hole	Diff
1	5.5	5.7	−0.2
2	11.0	11.2	−0.2
3	5.9	6.0	−0.1
4	8.2	5.6	2.6
5	10.0	9.3	0.7
6	7.9	7.0	0.9
7	10.1	8.4	1.7
8	7.4	9.0	−1.6
9	7.0	6.0	1.0
10	9.2	8.1	1.1
11	8.3	10.0	−1.7
12	8.6	8.1	0.5
13	10.5	10.4	0.1
14	5.5	7.0	−1.5
15	10.0	11.2	−1.2

c. $\bar{d} = \dfrac{\sum d_i}{n} = \dfrac{2.1}{15} = 0.14$; $s_d^2 = \dfrac{\sum d_i^2 - \dfrac{\left(\sum d_i\right)^2}{n}}{n-1} = \dfrac{22.65 - \dfrac{(2.1)^2}{15}}{15-1} = 1.5969$;

$s_d = \sqrt{1.5969} = 1.264$

d.   For confidence coefficient 0.90, $\alpha = 0.10$ and $\alpha/2 = 0.10/2 = 0.05$. From Table 7, Appendix B, $v = n-1 = 15-1 = 14$ degrees of freedom, $t_{0.05} = 1.761$. The 90% confidence interval is:

$$\bar{d} \pm t_{\alpha/2}\frac{s_d}{\sqrt{n}} \Rightarrow 0.14 \pm 1.761\frac{1.264}{\sqrt{15}} \Rightarrow 0.14 \pm 0.575 \Rightarrow (-0.435,\ 0.715)$$

e.   We are 90% confident that the true mean difference in THM measurements is between -0.435 and 0.715. Yes, the geologists can conclude that there is no difference in the true THM means of all original holes and their twin holes drilled at the mine because 0 falls in the confidence interval.

7.51   Using MINITAB, the output is:

One-Sample T: Diff

Variable	N	Mean	StDev	SE Mean	95% CI
Diff	3	0.1867	0.1106	0.0639	(-0.0881, 0.4614)

From the output, the 95% confidence interval for $\mu_d = \mu_1 - \mu_2$ is $(-0.0881,\ 0.4614)$. Because 0 is contained in the interval, there is no evidence that the mean initial pH level of mouthwash differs significantly from the mean pH level after 30 days.

7.53   a.   Using MINITAB, the descriptive statistics are:

Descriptive Statistics: Before-S, After-S, Diff-S

Variable	N	Mean	StDev	Minimum	Q1	Median	Q3	Maximum
Before-S	5	112.60	8.3845	105.00	105.50	113.00	119.50	126.00
After-S	5	122.60	8.3845	115.00	116.00	118.00	131.50	134.00
Diff-S	5	-10.00	5.3385	-16.00	-14.50	-11.00	-5.00	-2.00

For confidence coefficient 0.99, $\alpha = 0.01$ and $\alpha/2 = 0.01/2 = 0.005$. From Table 7, Appendix B, $v = n-1 = 5-1 = 4$ degrees of freedom, $t_{0.005} = 4.604$. The 99% confidence interval is:

$$\bar{d} \pm t_{\alpha/2}\frac{s_d}{\sqrt{n}} \Rightarrow -10 \pm 4.604\frac{5.3385}{\sqrt{5}} \Rightarrow -10 \pm 10.992 \Rightarrow (-20.992,\ 0.992)$$

We are 99% confident that the difference between the before and after mean systolic blood pressure readings is between -20.992 and 0.992 mmHg. Since 0 is contained in the interval, there is no evidence of a difference in the mean systolic blood pressure between before and after readings.

b.   Using MINITAB, the descriptive statistics are:

Descriptive Statistics: Before-D, After-D, Diff-D

Variable	N	Mean	StDev	Minimum	Q1	Median	Q3	Maximum
Before-D	5	66.40	9.0167	60.00	60.00	60.00	76.00	79.00
After-D	5	75.40	7.5033	66.00	69.50	73.00	82.50	86.00
Diff-D	5	-9.00	9.8995	-19.00	-16.00	-13.00	0.00	7.00

The 99% confidence interval is:

$$\bar{d} \pm t_{\alpha/2}\frac{s_d}{\sqrt{n}} \Rightarrow -9 \pm 4.604\frac{9.8995}{\sqrt{5}} \Rightarrow -9 \pm 20.383 \Rightarrow (-29.383, 11.383)$$

We are 99% confident that the difference between the before and after mean diastolic blood pressure readings is between -29.383 and 11.383 mmHg. Since 0 is contained in the interval, there is no evidence of a difference in the mean diastolic blood pressure between before and after readings.

c. Using MINITAB, the descriptive statistics are:

```
Descriptive Statistics: Before-HR, After-HR, Diff-HR

Variable    N   Mean    StDev  Minimum     Q1  Median     Q3  Maximum
Before-HR   5  71.20   5.3572    64.00  66.50   70.00  76.50    77.00
After-HR    5  79.20   6.4187    70.00  73.00   80.00  85.00    86.00
Diff-HR     5  -8.00   4.7434   -12.00 -11.50   -9.00  -4.00     0.00
```

The 99% confidence interval is:

$$\bar{d} \pm t_{\alpha/2}\frac{s_d}{\sqrt{n}} \Rightarrow -8 \pm 4.604\frac{4.7434}{\sqrt{5}} \Rightarrow -8 \pm 9.767 \Rightarrow (-17.767, 1.767)$$

We are 99% confident that the difference between the before and after mean heart rate is between -17.767 and 1.767 beats. Since 0 is contained in the interval, there is no evidence of a difference in the mean heart rate between before and after readings.

7.55 Summary information yields the following for the differences: $\bar{d} = 0.0005225$ and $s_d = 0.001291$.

For confidence coefficient 0.95, $\alpha = 0.05$ and $\alpha/2 = 0.05/2 = 0.025$. From Table 5, Appendix B, $z_{0.025} = 1.96$. The 95% confidence interval is:

$$\bar{d} \pm z_{\alpha/2}\frac{s_d}{\sqrt{n}} \Rightarrow 0.0005225 \pm 1.96\frac{0.001291}{\sqrt{40}} \Rightarrow 0.0005225 \pm 0.000401$$

$$\Rightarrow (0.0001224, 0.0009226)$$

Every value in the interval is below the value 0.002. We are 95% confident that the winery should use the alternative method for measuring wine density.

7.57 a. The point estimate of p, the true driver phone cell use rate, is $\hat{p} = \dfrac{Y}{n} = \dfrac{35}{1,165} = 0.030$.

b. To see if the sample size is sufficiently large:

$$n\hat{p} = 1,165(0.03) = 34.95 \geq 4; n\hat{q} = 1,165(0.97) = 1,130.05 \geq 4$$

Since both $n\hat{p} \geq 4$ and $n\hat{q} \geq 4$, we may conclude that the normal approximation is reasonable.

For confidence coefficient 0.95, $\alpha = 0.05$ and $\alpha/2 = 0.05/2 = 0.025$. From Table 5, Appendix B, $z_{0.025} = 1.96$. The 95% confidence interval is:

$$\hat{p} \pm z_{\alpha/2}\sigma_{\hat{p}} \Rightarrow \hat{p} \pm 1.96\sqrt{\frac{\hat{p}\hat{q}}{n}} \Rightarrow 0.030 \pm 1.96\sqrt{\frac{0.03(0.97)}{1,165}}$$

$$\Rightarrow 0.030 \pm 0.010 \Rightarrow (0.020, 0.040)$$

We are 95% confident that the true driver phone cell use rate is between 0.020 and 0.040.

7.59   To see if the sample size is sufficiently large:

$n\hat{p} = 328(0.5427) = 178.0 \geq 4; \ n\hat{q} = 328(0.4573) = 150.0 \geq 4$

Since both $n\hat{p} \geq 4$ and $n\hat{q} \geq 4$, we may conclude that the normal approximation is reasonable.

For confidence coefficient 0.90, $\alpha = 0.10$ and $\alpha / 2 = 0.10 / 2 = 0.05$. From Table 5, Appendix B, $z_{0.025} = 1.645$. The 90% confidence interval is:

$$\hat{p} \pm z_{\alpha/2}\sigma_{\hat{p}} \Rightarrow \hat{p} \pm 1.645\sqrt{\frac{\hat{p}\hat{q}}{n}} \Rightarrow 0.5427 \pm 1.645\sqrt{\frac{0.5427(0.4573)}{328}}$$

$$\Rightarrow 0.5427 \pm 0.0452 \Rightarrow (0.4975, 0.5879)$$

We are 90% confident that the true proportion of all groundwater wells in Bangladesh that have an estimated arsenic level below 50 micro-grams per liter is between 0.4975 and 0.5879.

7.61   The point estimate of p, the true proportion of all social robots designed with legs but no wheels is $\hat{p} = \dfrac{Y}{n} = \dfrac{63}{106} = 0.5943$.

To see if the sample size is sufficiently large:

$$n\hat{p} = 106(0.5943) = 63.0 \geq 4; \ n\hat{q} = 106(0.4057) = 43.0 \geq 4$$

Since both $n\hat{p} \geq 4$ and $n\hat{q} \geq 4$, we may conclude that the normal approximation is reasonable.

a.   For confidence coefficient 0.99, $\alpha = 0.01$ and $\alpha / 2 = 0.01 / 2 = 0.005$. From Table 5, Appendix B, $z_{0.005} = 2.58$. The 99% confidence interval is:

$$\hat{p} \pm z_{\alpha/2}\sigma_{\hat{p}} \Rightarrow \hat{p} \pm 2.58\sqrt{\frac{\hat{p}\hat{q}}{n}} \Rightarrow 0.5943 \pm 2.58\sqrt{\frac{0.5943(0.4057)}{106}}$$

$$\Rightarrow 0.5943 \pm 0.1230 \Rightarrow (0.471, 0.717)$$

We are 99% confident that the true proportion of all social robots designed with legs but no wheels is between 0.471 and 0.717.

b.   No. The 99% confidence interval does not contain 0.40. Therefore, it is not a likely value for the true proportion of all social robots designed with legs but no wheels.

7.63   The point estimate of p, the true proportion of aircraft bird strikes that occur above 100 feet is $\hat{p} = \dfrac{Y}{n} = \dfrac{36}{44} = 0.8182$.

To see if the sample size is sufficiently large:

$$n\hat{p} = 44(0.8182) = 36.0 \geq 4; \ n\hat{q} = 44(0.1818) = 8.0 \geq 4$$

Since both $n\hat{p} \geq 4$ and $n\hat{q} \geq 4$, we may conclude that the normal approximation is reasonable.

For confidence coefficient 0.95, $\alpha = 0.05$ and $\alpha / 2 = 0.05 / 2 = 0.025$. From Table 5, Appendix B, $z_{0.005} = 1.96$. The 95% confidence interval is:

$$\hat{p} \pm z_{\alpha/2}\sigma_{\hat{p}} \Rightarrow \hat{p} \pm 1.96\sqrt{\frac{\hat{p}\hat{q}}{n}} \Rightarrow 0.8182 \pm 1.96\sqrt{\frac{0.8182(0.1818)}{44}}$$

$$\Rightarrow 0.8182 \pm 0.1140 \Rightarrow (0.7042, 0.9322)$$

Because 0.70 does not fall in the interval, there is evidence that the estimate that less than 70% of aircraft bird strikes occur above 100 feet is not accurate.

7.65 a. The point estimate of p, the true proportion of subjects who use the bright color level as a cue to being right-side-up is $\hat{p} = \dfrac{Y}{n} = \dfrac{58}{90} = 0.644$.

To see if the sample size is sufficiently large:

$$n\hat{p} = 90(0.644) = 58.0 \geq 4; \quad n\hat{q} = 90(0.356) = 32.0 \geq 4$$

Since both $n\hat{p} \geq 4$ and $n\hat{q} \geq 4$, we may conclude that thenormal approximation is reasonable.

For confidence coefficient 0.95, $\alpha = 0.05$ and $\alpha / 2 = 0.05 / 2 = 0.025$. From Table 5, Appendix B, $z_{0.005} = 1.96$. The 95% confidence interval is:

$$\hat{p} \pm z_{\alpha/2}\sigma_{\hat{p}} \Rightarrow \hat{p} \pm 1.96\sqrt{\frac{\hat{p}\hat{q}}{n}} \Rightarrow 0.644 \pm 1.96\sqrt{\frac{0.644(0.356)}{90}}$$

$$\Rightarrow 0.644 \pm 0.099 \Rightarrow (0.545, 0.743)$$

We are 95% confident that the true proportion of subjects who use the bright color level as a cue to being right-side-up is between 0.545 and 0.743.

b. Yes. Since both values of the confidence interval are greater than 0.5, we can infer that a majority of subjects would select bright color levels over dark color levels as a cue.

7.67 a. The parameter of interest to the researches is the difference in the proportions of producers who are willing to offer windowing services in Missouri and Illinois, $p_1 - p_2$.

b. From the printout, the 99% confidence interval is $(-0.135179, -0.0031807)$.

c. Because the interval contains only negative numbers, there is evidence to indicate that the proportion of producers who are willing to offer windowing services in Missouri is less than the proportion of producers who are willing to offer windowing services in Illinois.

7.69 a. The point estimated for the true proportion of super experienced bidders who fall prey to the winner's curse is $\hat{p}_1 = \dfrac{Y_1}{n_1} = \dfrac{29}{189} = 0.1534$.

b. The point estimated for the true proportion of less experienced bidders who fall prey to the winner's curse is $\hat{p}_2 = \dfrac{Y_2}{n_2} = \dfrac{32}{149} = 0.2148$.

c. To see if the sample size is sufficiently large:

$$n_1\hat{p}_1 = 189(0.1534) = 29.0 \geq 4; n_1\hat{q}_1 = 189(0.8466) = 160.0 \geq 4$$

$$n_2\hat{p}_2 = 149(0.2148) = 32.0 \geq 4; n_2\hat{q}_2 = 149(0.7852) = 117.0 \geq 4$$

Since $n_1\hat{p}_1 \geq 4$, $n_1\hat{q}_1 \geq 4$, $n_2\hat{p}_2 \geq 4$, and $n_2\hat{q}_2 \geq 4$, we may conclude that the normal approximation is reasonable.

For confidence coefficient 0.90, $\alpha = 0.10$ and $\alpha/2 = 0.10/2 = 0.05$. From Table 5, Appendix B, $z_{0.025} = 1.645$. The 90% confidence interval is:

$$(\hat{p}_1 - \hat{p}_2) \pm z_{\alpha/2}\sqrt{\frac{\hat{p}_1\hat{q}_1}{n_1} + \frac{\hat{p}_2\hat{q}_2}{n_2}} \Rightarrow (0.1534 - 0.2148)$$

$$\pm 1.645\sqrt{\frac{0.1534(0.8466)}{189} + \frac{0.2148(0.7852)}{149}}$$

$$\Rightarrow -0.0614 \pm 0.0702 \Rightarrow (-0.1316, 0.0088)$$

d. We are 90% confident that the difference between the proportion of super experienced and less experienced bidders who fall prey to the winner's curse is between -0.1316 and 0.0088. Since 0 is contained in the interval, there is no evidence to indicate that the bid experience impacts the likelihood of the winner's curse occurring.

7.71 a. Theory 1 states that foragers (Eastern Jomon) with a broad-based economy will have a lower LEH defect prevalence than early agriculturists (Yayoi).

The point estimated for the true proportion of early agriculturists (Yayoi) who have the LEH defect is 0.631.

The point estimated for the true proportion of foragers (Eastern Jomon) with a broad-based economy who have the LEH defect is 0.482.

To see if the sample size is sufficiently large:

$$n_1\hat{p}_1 = 182(0.631) = 114.8 \geq 4; \quad n_1\hat{q}_1 = 182(0.369) = 67.2 \geq 4$$

$$n_2\hat{p}_2 = 164(0.482) = 79.0 \geq 4; \quad n_2\hat{q}_2 = 164(0.518) = 85.0 \geq 4$$

Since $n_1\hat{p}_1 \geq 4$, $n_1\hat{q}_1 \geq 4$, $n_2\hat{p}_2 \geq 4$, and $n_2\hat{q}_2 \geq 4$, we may conclude that the normal approximation is reasonable.

For confidence coefficient 0.99, $\alpha = 0.01$ and $\alpha/2 = 0.01/2 = 0.005$. From Table 5, Appendix B, $z_{0.005} = 2.575$. The 99% confidence interval is:

$$(\hat{p}_1 - \hat{p}_2) \pm z_{\alpha/2}\sqrt{\frac{\hat{p}_1\hat{q}_1}{n_1} + \frac{\hat{p}_2\hat{q}_2}{n_2}} \Rightarrow (0.631 - 0.482) \pm 2.575\sqrt{\frac{0.631(0.369)}{182} + \frac{0.482(0.518)}{164}}$$

$$\Rightarrow 0.149 \pm 0.1363 \Rightarrow (0.0127, 0.2853)$$

Because 0 is not in the interval, there is evidence to support Theory 1.

b. Theory 1 states that foragers (Western Jomon) with a wet rice economy will not differ in LEH defect prevalence from early agriculturists (Yayoi).

The point estimated for the true proportion of foragers (Western Jomon) with a wet rice economy who have the LEH defect is 0.648.

To see if the sample size is sufficiently large:

$$n_1\hat{p}_1 = 182(0.631) = 114.8 \geq 4; \quad n_1\hat{q}_1 = 182(0.369) = 67.2 \geq 4$$

$$n_3\hat{p}_3 = 122(0.648) = 79.1 \geq 4; \quad n_3\hat{q}_3 = 122(0.352) = 42.9 \geq 4$$

Since $n_1\hat{p}_1 \geq 4$, $n_1\hat{q}_1 \geq 4$, $n_3\hat{p}_3 \geq 4$, and $n_3\hat{q}_3 \geq 4$, we may conclude that the normal approximation is reasonable.

$$(\hat{p}_1 - \hat{p}_3) \pm z_{\alpha/2}\sqrt{\frac{\hat{p}_1\hat{q}_1}{n_1} + \frac{\hat{p}_3\hat{q}_3}{n_3}} \Rightarrow (0.631 - 0.648) \pm 2.575\sqrt{\frac{0.631(0.369)}{182} + \frac{0.648(0.352)}{122}}$$

$$\Rightarrow -0.017 \pm 0.144 \Rightarrow (-0.161, 0.127)$$

Because 0 is in the interval, there is evidence to support Theory 2.

7.73 Using Table 8, Appendix B:

a. For $\alpha = 0.05$ and $df = 7$, $\chi^2_{0.05} = 14.0671$

b. For $\alpha = 0.10$ and $df = 16$, $\chi^2_{0.10} = 23.5418$

c. For $\alpha = 0.01$ and $df = 10$, $\chi^2_{0.01} = 23.2093$

d. For $\alpha = 0.025$ and $df = 8$, $\chi^2_{0.025} = 17.5346$

e. For $\alpha = 0.005$ and $df = 5$, $\chi^2_{0.005} = 16.7496$

7.75 The confidence interval for σ^2 is $\dfrac{(n-1)s^2}{\chi^2_{\alpha/2}} \leq \sigma^2 \leq \dfrac{(n-1)s^2}{\chi^2_{1-\alpha/2}}$.

For confidence coefficient 0.95, $\alpha = 0.05$ and $\alpha/2 = 0.05/2 = 0.025$. From Table 8, Appendix B, $v = n - 1 = 6 - 1 = 5$ degrees of freedom, $\chi^2_{0.025} = 12.8325$ and $\chi^2_{0.975} = 0.831211$. The 95% confidence interval is:

$$\frac{(n-1)s^2}{\chi^2_{\alpha/2}} \leq \sigma^2 \leq \frac{(n-1)s^2}{\chi^2_{1-\alpha/2}} \Rightarrow \frac{(6-1)0.011^2}{12.8325} \leq \sigma^2 \leq \frac{(6-1)0.011^2}{0.831211}$$

$$\Rightarrow 0.000047 \leq \sigma^2 \leq 0.000728$$

$$\Rightarrow \sqrt{0.000047} \leq \sigma \leq \sqrt{0.000728} \Rightarrow 0.0069 \leq \sigma \leq 0.0270$$

We are 95% confident that the true standard deviation of the internal oil content distribution for the sweet potato chips is between 0.0069 and 0.0270.

7.77 The confidence interval for σ^2 is $\dfrac{(n-1)s^2}{\chi^2_{\alpha/2}} \leq \sigma^2 \leq \dfrac{(n-1)s^2}{\chi^2_{1-\alpha/2}}$.

For confidence coefficient 0.95, $\alpha = 0.05$ and $\alpha/2 = 0.05/2 = 0.025$. From Table 8, Appendix B, $v = n - 1 = 12 - 1 = 11$ degrees of freedom, $\chi^2_{0.025} = 21.9200$ and $\chi^2_{0.975} = 3.81575$. The 95% confidence interval is:

$$\frac{(n-1)s^2}{\chi^2_{\alpha/2}} \leq \sigma^2 \leq \frac{(n-1)s^2}{\chi^2_{1-\alpha/2}} \Rightarrow \frac{(12-1)4,487^2}{21.9200} \leq \sigma^2 \leq \frac{(12-1)4,487^2}{3.81575}$$

$$\Rightarrow 10,103,323.86 \leq \sigma^2 \leq 58,039,666.91$$

$$\Rightarrow \sqrt{10,103,323.86} \leq \sigma \leq \sqrt{58,039,666.91}$$

$$\Rightarrow 3,178.57 \leq \sigma \leq 7,618.38$$

We are 95% confident that the true standard deviation of radon levels in tombs in the Valley of Kings is between 9.274 and 7,618.38.

7.79   Using MINITAB, the calculations are:

```
Test and CI for One Variance: Y
Method
The chi-square method is only for the normal distribution.
Statistics
Variable   N  StDev  Variance
Y         50   3.18     10.1
99% Confidence Intervals
                          CI for        CI for
Variable  Method          StDev        Variance
Y         Chi-Square  (2.52, 4.27)   (6.3, 18.2)
```

From the printout, the 99% confidence interval is $(6.3, 18.2)$. We are 99% confident that the true variance of drug concentrations for the new method is between 6.3 and 18.2.

7.81   a.   Using a computer package, suppose the random sample of 10 observations selected yields:

100.977, 367.611, 63.369, 185.598, 141.733, 72.648, 64.591, 59.846, 13.587, 2.954.

b.   Using MINITAB, the calculations are:

```
Test and CI for One Variance: Sample
Method
The chi-square method is only for the normal distribution.
Statistics
Variable   N  StDev  Variance
Sample    10   106    11333
95% Confidence Intervals
                          CI for        CI for
Variable  Method          StDev        Variance
Sample    Chi-Square   (73, 194)    (5362, 37770)
```

The 95% confidence interval for the population variance is $(5,362, 37,770)$. We are 95% confident that the true variance of the interarrival times is between 5,362 and 37,770.

c.   Using a computer package, the true variance of all the observations is: $\sigma^2 = 8,348.028$. Yes, the interval in part b contains this value.

7.83   Using Table 10, Appendix B,

a.   For $v_1 = 7$ and $v_2 = 25$, $F_{0.05} = 2.40$

b.   For $v_1 = 10$ and $v_2 = 8$, $F_{0.05} = 3.35$

c.   For $v_1 = 30$ and $v_2 = 60$, $F_{0.05} = 1.65$

d.   For $v_1 = 15$ and $v_2 = 4$, $F_{0.05} = 5.86$

7.85 a. The confidence interval for σ_1^2/σ_2^2 is $\dfrac{s_1^2}{s_2^2} \cdot \dfrac{1}{F_{\alpha/2(v_1,v_2)}} \leq \dfrac{\sigma_1^2}{\sigma_2^2} \leq \dfrac{s_1^2}{s_2^2} F_{\alpha/2(v_1,v_2)}$.

For confidence coefficient 0.90, $\alpha = 0.10$ and $\alpha/2 = 0.10/2 = 0.05$. Using a computer package with $v_1 = n_1 - 1 = 406 - 1 = 405$ and $v_2 = n_2 - 1 = 230 - 1 = 229$ degrees of freedom, $F_{0.05,(405,229)} = 1.2161$ and $F_{0.05,(229,405)} = 1.20886$. The 90% confidence interval is:

$$\frac{0.4^2}{0.3^2} \cdot \frac{1}{1.2161} \leq \frac{\sigma_1^2}{\sigma_2^2} \leq \frac{0.4^2}{0.3^2}(1.20886) \Rightarrow 1.462 \leq \frac{\sigma_1^2}{\sigma_2^2} \leq 2.149$$

   b. Yes. The value 1 is not contained in the 95% confidence interval. Therefore, we can conclude that the variability in the number of hippo trails from a water source in a National Reserve differs from the variability in the number of hippo trails from a water source in a pastoral ranch at $\alpha = 0.10$.

7.87 Summary information reveals the following:

Group 1 – Perturbed Intrinsics, No Perturbed Projections: $s_1^2 = 0.00077$, $n = 5$

Group 2 – No Perturbed Intrinsics, Perturbed Projections: $s_2^2 = 0.02153$, $n = 5$

The confidence interval for σ_1^2/σ_2^2 is $\dfrac{s_1^2}{s_2^2} \cdot \dfrac{1}{F_{\alpha/2(v_1,v_2)}} \leq \dfrac{\sigma_1^2}{\sigma_2^2} \leq \dfrac{s_1^2}{s_2^2} F_{\alpha/2(v_2,v_1)}$.

For confidence coefficient 0.90, $\alpha = 0.10$ and $\alpha/2 = 0.10/2 = 0.05$. From Table 10, Appendix B, with $v_1 = n_1 - 1 = 5 - 1 = 4$ and $v_2 = n_2 - 1 = 5 - 1 = 4$ degrees of freedom, $F_{0.05,(4,4)} = 6.39$. The 90% confidence interval is:

$$\frac{0.00077}{0.02153} \cdot \frac{1}{6.39} \leq \frac{\sigma_1^2}{\sigma_2^2} \leq \frac{0.00077}{0.02153}(6.39) \Rightarrow 0.0056 \leq \frac{\sigma_1^2}{\sigma_2^2} \leq 0.2285$$

7.89 a. The confidence level desired by the researchers is 0.95.
   b. The sampling error desired by the researchers is $H = 0.001$.
   c. For confidence coefficient 0.95, $\alpha = 0.05$ and $\alpha/2 = 0.05/2 = 0.025$. From Table 5, Appendix B, $z_{0.025} = 1.96$.

   The sample size is $n = \dfrac{(z_{0.025})^2 \sigma^2}{H^2} = \dfrac{1.96^2(0.005)^2}{0.001^2} = 96.04 \approx 97$.

7.91 For confidence coefficient 0.95, $\alpha = 0.05$ and $\alpha/2 = 0.05/2 = 0.025$. From Table 5, Appendix B, $z_{0.025} = 1.96$.

   The sample size is $n = \dfrac{(z_{0.025})^2 \sigma^2}{H^2} = \dfrac{1.96^2(15)^2}{5^2} = 34.6 \approx 35$.

7.93 For confidence coefficient 0.99, $\alpha = 0.01$ and $\alpha/2 = 0.01/2 = 0.005$. From Table 5, Appendix B, $z_{0.050} = 2.575$. From Exercise 7.50, $s_1^2 = 69.77$. We will use this to estimate the population variance.

   The sample size is $n = \dfrac{(z_{0.005})^2 \sigma_d^2}{H^2} = \dfrac{2.575^2(69.77)}{2^2} = 115.57 \approx 116$.

   Therefore, we would have to sample an additional $116 - 13 = 103$ structures.

7.95   For confidence coefficient 0.90, $\alpha = 0.10$ and $\alpha / 2 = 0.10 / 2 = 0.05$. From Table 5, Appendix B, $z_{0.05} = 1.645$. From Exercise 7.68, $\hat{p}_1 = 0.768$ and $\hat{p}_2 = 0.613$. We will use these to estimate p_1 and p_2.

The sample sizes are

$$n_1 = n_2 = \frac{(z_{0.05})^2 (p_1 q_1 + p_2 q_2)}{H^2} = \frac{1.645^2 (0.768(0.232) + 0.613(0.387))}{0.05^2} = 449.6 \approx 450.$$

7.97   For confidence coefficient 0.95, $\alpha = 0.05$ and $\alpha / 2 = 0.05 / 2 = 0.025$. From Table 5, Appendix B, $z_{0.025} = 1.96$. From Exercise 7.66, $\hat{p}_1 = 0.4$ and $\hat{p}_2 = 0.2136$. We will use these to estimate p_1 and p_2.

The sample sizes are

$$n_1 = n_2 = \frac{(z_{0.025})^2 (p_1 q_1 + p_2 q_2)}{H^2} = \frac{1.96^2 (0.4(0.6) + 0.2136(0.7864))}{0.06^2} = 435.4 \approx 436.$$

7.99   The formula for determining sample size for estimating a population proportion is:

$$n = \frac{(z_{\alpha/2})^2 (pq)}{H^2} = \frac{(z_{\alpha/2})^2 p(1-p)}{H^2}$$

To show that n is maximized when $p = 0.5$, we take the derivative of n with respect to p, set it equal to 0, and solve:

$$\frac{dn}{dp} = \frac{d\left[\dfrac{(z_{\alpha/2})^2 p(1-p)}{H^2}\right]}{dp} = \frac{(z_{\alpha/2})^2}{H^2}\left[p(-1) + (1-p)(1)\right] = \frac{(z_{\alpha/2})^2}{H^2}(1 - 2p)$$

Now, setting this equal to 0 and solving, we get:

$$\frac{(z_{\alpha/2})^2}{H^2}(1 - 2p) = 0 \Rightarrow 1 - 2p = 0 \Rightarrow 2p = 1 \Rightarrow p = 0.5$$

7.103   From Example 7.20,

$$\hat{p}_B = \left(\frac{n}{n+3}\right)\bar{y} + \frac{1}{3}\left(\frac{3}{n+3}\right) = \left(\frac{n}{n+3}\right)\bar{y} + \left(\frac{1}{n+3}\right) = \left(\frac{1}{n+3}\right)(n\bar{y} + 1).$$

For $\bar{y} = 0.8$, we find $\hat{p}_B = \left(\frac{1}{n+3}\right)(n\bar{y} + 1) = \left(\frac{1}{n+3}\right)(0.8n + 1)$

7.105 $y \sim N(\mu, 1) \Rightarrow \bar{y} \sim N(\mu, 1/n) \Rightarrow f(\bar{y} \mid \mu) \sim \exp\left\{\frac{(\bar{y} - \mu)^2}{-2/n}\right\}$

$\mu \sim N(\theta, 1) \Rightarrow h(\mu) \sim \exp\left\{\frac{(\mu - \theta)^2}{-2}\right\}$

$$g(\mu \mid \bar{y}) \sim f(\bar{y} \mid \mu) \cdot h(\mu) \sim \exp\left\{\frac{(\bar{y}-\mu)^2}{-2/n} + \frac{(\mu-\theta)^2}{2}\right\} \sim \exp\left\{\frac{(\bar{y}-\mu)^2}{-2/n} + \frac{\dfrac{(\mu-\theta)^2}{n}}{2/n}\right\}$$

constant

$$= \exp\left\{\frac{\bar{y}^2 - 2\bar{y}\mu + \mu^2 + \dfrac{\mu^2}{n} - \dfrac{2\theta\mu}{n} + \dfrac{\theta^2}{n}}{-2/n}\right\}$$

$$\sim \exp\left\{\frac{\mu^2\left(\dfrac{n+1}{n}\right) - 2\mu\left(\bar{y}+\dfrac{\theta}{n}\right)}{-2/n}\right\} \sim \exp\left\{\frac{\mu^2 - 2\mu\dfrac{\left(\bar{y}+\dfrac{\theta}{n}\right)}{\left(\dfrac{n}{n+1}\right)}}{-\dfrac{2}{n}\left(\dfrac{n}{n+1}\right)}\right\} \sim \exp\left\{\frac{\mu^2 - 2\mu\dfrac{\left(\bar{y}+\dfrac{\theta}{n}\right)}{\left(\dfrac{n}{n+1}\right)}}{-\dfrac{2}{n}\left(\dfrac{n}{n+1}\right)}\right\}$$

add constant to complete the square

$$\sim \exp\left\{\frac{\mu^2 - 2\mu\dfrac{\left(\bar{y}+\dfrac{\theta}{n}\right)}{\left(\dfrac{n}{n+1}\right)} + \left[\dfrac{\left(\bar{y}+\dfrac{\theta}{n}\right)}{\left(\dfrac{n}{n+1}\right)}\right]^2}{-\dfrac{2}{n}\left(\dfrac{n}{n+1}\right)}\right\} \sim xp\left\{\frac{\left[\mu - \dfrac{\left(\bar{y}+\dfrac{\theta}{n}\right)}{\left(\dfrac{n}{n+1}\right)}\right]^2}{-\dfrac{2}{n}\left(\dfrac{n}{n+1}\right)}\right\} \Rightarrow \text{Normal with mean} = \frac{\left(\bar{y}+\dfrac{\theta}{n}\right)}{\left(\dfrac{n}{n+1}\right)}$$

Weighted average of $\bar{y}+\theta$

7.107   For confidence coefficient 0.95, $\alpha = 0.05$ and $\alpha/2 = 0.05/2 = 0.025$. From Table 5, Appendix B, $z_{0.025} = 1.96$.

The sample size is $n = \dfrac{(z_{0.025})^2\sigma^2}{H^2} = \dfrac{1.96^2(5)^2}{1^2} = 96.04 \approx 97$.

7.109   Using MINITAB the descriptive statistics are:

Descriptive Statistics: A, B

Variable	N	Mean	StDev	Variance	Minimum	Q1	Median	Q3	Maximum
A	5	45.00	5.61	31.50	37.00	40.00	45.00	50.00	52.00
B	5	45.40	3.36	11.30	41.00	42.00	46.00	48.50	49.00

To compare precision, we need to compare the population variances. The confidence interval for σ_1^2/σ_2^2 is $\dfrac{s_1^2}{s_2^2} \cdot \dfrac{1}{F_{\alpha/2(v_1,v_2)}} \le \dfrac{\sigma_1^2}{\sigma_2^2} \le \dfrac{s_1^2}{s_2^2} F_{\alpha/2(v_2,v_1)}$.

For confidence coefficient 0.90, $\alpha = 0.10$ and $\alpha / 2 = 0.10 / 2 = 0.05$. From Table 10, Appendix B, with $v_1 = n_1 - 1 = 5 - 1 = 4$ and $v_2 = n_2 - 1 = 5 - 1 = 4$ degrees of freedom, $F_{0.05,(4,4)} = 6.39$. The 90% confidence interval is:

$$\frac{31.5}{11.3} \cdot \frac{1}{6.39} \leq \frac{\sigma_1^2}{\sigma_2^2} \leq \frac{31.5}{11.3}(6.39) \Rightarrow 0.436 \leq \frac{\sigma_1^2}{\sigma_2^2} \leq 17.813$$

We are 90% confident that the ratio of the variances for the two instruments is between 0.436 and 17.813.

7.111 a.   Let p = proportion of all injuries that are due to falls.

$$\hat{p} = \frac{Y}{n} = 0.23$$

To see if the sample size is sufficiently large:

$$n\hat{p} = 2,514(0.23) = 578.22 \geq 4; n\hat{q} = 2,514(0.77) = 1935.78 \geq 4$$

Since both $n\hat{p} \geq 4$ and $n\hat{q} \geq 4$, we may conclude that the normal approximation is reasonable.

For confidence coefficient 0.95, $\alpha = 0.05$ and $\alpha / 2 = 0.05 / 2 = 0.025$. From Table 5, Appendix B, $z_{0.025} = 1.96$. The 95% confidence interval for p is

$$\hat{p} \pm z_{0.025}\sqrt{\frac{\hat{p}\hat{q}}{n}} \Rightarrow 0.23 \pm 1.96\sqrt{\frac{0.23(0.77)}{2514}} \Rightarrow 0.23 \pm 0.0165 \Rightarrow (0.2135, 0.2465)$$

We are 95% confident that the proportion of all injuries that are due to falls is between 0.2135 and 0.2465.

b.   Let p = proportion of all injuries that are due to burns or scalds.

$$\hat{p} = \frac{Y}{n} = 0.20$$

To see if the sample size is sufficiently large:

$$n\hat{p} = 2,514(0.20) = 502.8 \geq 4; n\hat{q} = 2,514(0.80) = 2011.2 \geq 4$$

Since both $n\hat{p} \geq 4$ and $n\hat{q} \geq 4$, we may conclude that thenormal approximation is reasonable.

For confidence coefficient 0.95, $\alpha = 0.05$ and $\alpha / 2 = 0.05 / 2 = 0.025$. From Table 5, Appendix B, $z_{0.025} = 1.96$. The 95% confidence interval for p is

$$\hat{p} \pm z_{0.025}\sqrt{\frac{\hat{p}\hat{q}}{n}} \Rightarrow 0.20 \pm 1.96\sqrt{\frac{0.20(0.80)}{2514}} \Rightarrow 0.20 \pm 0.0165 \Rightarrow (0.1844, 0.2156)$$

We are 95% confident that the proportion of all injuries that are due to burns or scalds is between 0.1844 and 0.2156.

7.113 a. The confidence interval for σ^2 is $\dfrac{(n-1)s^2}{\chi^2_{\alpha/2}} \le \sigma^2 \le \dfrac{(n-1)s^2}{\chi^2_{1-\alpha/2}}$.

For confidence coefficient 0.95, $\alpha = 0.05$ and $\alpha/2 = 0.05/2 = 0.025$. From Table 8, Appendix B, $v = n-1 = 18-1 = 17$ degrees of freedom, $\chi^2_{0.025} = 30.1910$ and $\chi^2_{0.975} = 7.56418$. The 95% confidence interval is:

$$\frac{(n-1)s^2}{\chi^2_{\alpha/2}} \le \sigma^2 \le \frac{(n-1)s^2}{\chi^2_{1-\alpha/2}} \Rightarrow \frac{(18-1)6.3^2}{30.1910} \le \sigma^2 \le \frac{(18-1)6.3^2}{7.56418} \Rightarrow 22.349 \le \sigma^2 \le 89.201$$

$$\Rightarrow \sqrt{22.349} \le \sigma \le \sqrt{89.201} \Rightarrow 4.727 \le \sigma \le 9.445$$

   b. No. Since 7 is contained in the above interval, there is no evidence that the true standard deviation is less than 7.

7.115   For confidence coefficient 0.95, $\alpha = 0.05$ and $\alpha/2 = 0.05/2 = 0.025$. From Table 5, Appendix B, $z_{0.025} = 1.96$.
The sample sizes are

$$n_1 = n_2 = \frac{(z_{0.025})^2(\sigma_1^2 + \sigma_2^2)}{H^2} = \frac{1.96^2(0.75^2 + 0.75^2)}{0.05^2} = 1{,}728.7 \approx 1{,}729.$$

7.117   The confidence interval for σ^2 is $\dfrac{(n-1)s^2}{\chi^2_{\alpha/2}} \le \sigma^2 \le \dfrac{(n-1)s^2}{\chi^2_{1-\alpha/2}}$.

For confidence coefficient 0.95, $\alpha = 0.05$ and $\alpha/2 = 0.05/2 = 0.025$. From Table 8, Appendix B, $v = n-1 = 7-1 = 6$ degrees of freedom, $\chi^2_{0.025} = 14.4494$ and $\chi^2_{0.975} = 1.237347$. The 95% confidence interval is:

$$\frac{(n-1)s^2}{\chi^2_{\alpha/2}} \le \sigma^2 \le \frac{(n-1)s^2}{\chi^2_{1-\alpha/2}} \Rightarrow \frac{(7-1)9^2}{14.4494} \le \sigma^2 \le \frac{(7-1)9^2}{1.237347} \Rightarrow 33.635 \le \sigma^2 \le 392.776$$

7.119 a. For confidence coefficient 0.90, $\alpha = 0.10$ and $\alpha/2 = 0.10/2 = 0.05$. From Table 5, Appendix B, $z_{0.05} = 1.645$.
The sample size is $n = \dfrac{(z_{0.05})^2 \sigma^2}{H^2} = \dfrac{1.645^2(2)^2}{0.1^2} = 1{,}082.4 \approx 1{,}083$.

   b. In part s, we found $n = 1{,}083$. If we used an n of only 100, the width of the confidence interval for μ would be wider since we would be dividing by a smaller number.

   c. We know $H = \dfrac{z_{\alpha/2}\sigma}{\sqrt{n}} \Rightarrow z_{\alpha/2} = \dfrac{H\sqrt{n}}{\sigma} = \dfrac{0.1\sqrt{100}}{2} = 0.5$.

   $P(-0.5 \le Z \le 0.5) = 0.1915 + 0.1915 = 0.3830$ (using Table 5, Appendix B)
   Thus, the level of confidence is approximately 38.3%.

7.121 a. Let $p_1 =$ proportion of control cells that exhibited altered growth and $p_2 =$ proportion of cells exposed to E2F1 that exhibited altered growth.

$$\hat{p}_1 = \frac{Y_1}{n_1} = \frac{15}{158} = 0.095 \qquad \hat{p}_2 = \frac{Y_2}{n_2} = \frac{41}{92} = 0.446$$

To see if the sample size is sufficiently large:

$$n_1\hat{p}_1 = 158(0.095) = 15.0 \geq 4; \quad n_1\hat{q}_1 = 158(0.905) = 143.0 \geq 4$$

$$n_2\hat{p}_2 = 92(0.446) = 41.0 \geq 4; \quad n_2\hat{q}_2 = 92(0.554) = 51.0 \geq 4$$

Since $n_1\hat{p}_1 \geq 4, n_1\hat{q}_1 \geq 4, n_2\hat{p}_2 \geq 4$, and $n_2\hat{q}_2 \geq 4$, we may conclude that the normal approximation is reasonable.

For confidence coefficient 0.90, $\alpha = 0.10$ and $\alpha / 2 = 0.10 / 2 = 0.05$. From Table 5, Appendix B, $z_{0.05} = 1.645$. The 90% confidence interval is:

$$(\hat{p}_1 - \hat{p}_2) \pm z_{\alpha/2}\sqrt{\frac{\hat{p}_1\hat{q}_1}{n_1} + \frac{\hat{p}_2\hat{q}_2}{n_2}} \Rightarrow (0.095 - 0.446) \pm 1.96\sqrt{\frac{0.095(0.905)}{158} + \frac{0.446(0.554)}{92}}$$

$$\Rightarrow -0.351 \pm 0.093 \Rightarrow (-0.444, -0.258)$$

b. We are 90% confident that the difference in the proportion of cells that exhibited altered growth between control cells and cells exposed to E2F1 is between -0.444 and -0.258. Since the confidence interval contains only negative numbers, there is evidence to indicate the proportion of cells exposed to E2F1 that exhibited altered growth is greater than the proportion of control cells that exhibited altered growth.

7.123 For confidence coefficient 0.99, $\alpha = 0.01$ and $\alpha / 2 = 0.01 / 2 = 0.005$. From Table 5, Appendix B, $z_{0.005} = 2.58$. From the previous estimate, we will use $\hat{p} = 0.333$ to estimate p.

The sample size is $n = \dfrac{(z_{0.005})^2(pq)}{H^2} = \dfrac{2.58^2(0.333)(0.667)}{0.01^2} = 14{,}784.6 \approx 14{,}785$.

7.125 Let $\mu_1 =$ mean protein uptake of fast muscles and $\mu_2 =$ mean protein uptake of slow muscles.

For confidence coefficient 0.95, $\alpha = 0.05$ and $\alpha / 2 = 0.05 / 2 = 0.025$. From Table 7, Appendix B, with $v = n_1 + n_2 - 2 = 12 + 12 - 2 = 22$ degrees of freedom, $t_{0.025} = 2.074$. We must first find the estimate of the common variance:

$$s_p = \sqrt{\frac{(n_1 - 1)s_1^2 + (n_2 - 1)s_2^2}{n_1 + n_1 - 2}} = \sqrt{\frac{(12 - 1)(0.104)^2 + (12 - 1)(0.035)^2}{12 + 12 - 2}} = 0.0814$$

The confidence interval is:

$$(\bar{y}_1 - \bar{y}_1) \pm t_{0.025}\, s_p\sqrt{\frac{1}{n_1} + \frac{1}{n_2}} \Rightarrow (0.57 - 0.37) \pm 2.074(0.0814)\sqrt{\frac{1}{12} + \frac{1}{12}} \Rightarrow 0.20 \pm 0.069$$

$$\Rightarrow (0.131, \ 0.269)$$

We are 95% confident that the difference in mean protein uptake of fast and slow muscles is between 0.131 and 0.269. Since 0 is not in the interval, there is evidence to indicate the mean uptake of protein is greater for fast muscles than for slow muscles.

7.127 a. First, we must find $E(Y)$.

$$E(Y) = \int_{-\infty}^{\infty} yf(y)\,dy = \int_{\theta}^{\theta+1} y\,dy = \frac{y^2}{2}\Bigg|_{\theta}^{\theta+1} = \frac{(\theta+1)^2}{2} - \frac{(\theta)^2}{2}$$

$$= \frac{\theta^2 + 2\theta + 1 - \theta^2}{2} = \frac{2\theta+1}{2} = \theta + \frac{1}{2}$$

Thus, $E(\bar{y}) = E\left(\dfrac{y_1 + y_2 + \cdots + y_n}{n}\right) = \dfrac{1}{n}\left[E(y_1) + E(y_2) + \cdots + E(y_n)\right]$

$$= \frac{1}{n}\left[\left(\theta + \frac{1}{2}\right) + \left(\theta + \frac{1}{2}\right) + \cdots \left(\theta + \frac{1}{2}\right)\right] = \frac{1}{n}n\left(\theta + \frac{1}{2}\right) = \left(\theta + \frac{1}{2}\right)$$

The bias is $\dfrac{1}{2}$.

b. First, we must find $V(Y)$. We know $V(Y) = E(Y^2) - \left[E(Y)\right]^2$

$$E(Y^2) = \int_{-\infty}^{\infty} y^2 f(y)\,dy = \int_{\theta}^{\theta+1} y^2\,dy = \frac{y^3}{3}\Bigg|_{\theta}^{\theta+1} = \frac{(\theta+1)^3}{3} - \frac{(\theta)^3}{3}$$

$$= \frac{\theta^3 + 3\theta^2 + 3\theta + 1 - \theta^3}{3} = \frac{3\theta^2 + 3\theta + 1}{3} = \theta^2 + \theta + \frac{1}{3}$$

$$V(Y) = E(Y^2) - \left[E(Y)\right]^2 = \theta^2 + \theta + \frac{1}{3} - \left(\theta + \frac{1}{2}\right)^2$$

$$= \theta^2 + \theta + \frac{1}{3} - \left(\theta^2 + \theta + \frac{1}{4}\right) = \frac{1}{3} - \frac{1}{4} = \frac{4}{12} - \frac{3}{12}$$

$$V(\bar{y}) = V\left(\frac{y_1 + y_2 + \cdots y_n}{n}\right) = \frac{1}{n^2}\left[V(y_1) + V(y_2) + \cdots V(y_n)\right]$$

$$= \frac{1}{n^2}\left[\frac{1}{12} + \frac{1}{12} + \cdots + \frac{1}{12}\right] = \frac{1}{n^2}\frac{n}{12} = \frac{1}{12n}$$

c. If $E(\bar{y}) = \theta + \dfrac{1}{2}$, then $\bar{y} - \dfrac{1}{2}$ would be an unbiased estimator of θ.

$$E\left(\bar{y} - \frac{1}{2}\right) = \left(\theta + \frac{1}{2}\right) - \frac{1}{2} = \theta$$

7.129 a.   Let $W = \dfrac{2Y}{\beta}$. If Y has a gamma distribution with $\alpha = 1$ and arbitrary β, the density function for Y is:

$$f(y) = \begin{cases} \dfrac{1}{\beta} e^{-y/\beta} & y > 0 \\ 0 & \text{elsewhere} \end{cases}$$

If the range for Y is $y > 0$, then the range for W is $w > 0$.

If $W = \dfrac{2Y}{\beta}$, then $Y = \dfrac{W\beta}{2}$ and $dy = \dfrac{\beta}{2} dw$

Then $\displaystyle\int_0^\infty \dfrac{1}{\beta} e^{-y/\beta} \, dy = \int_0^\infty \dfrac{1}{\beta} e^{-w\beta/2\beta} \dfrac{\beta}{2} \, dw = \int_0^\infty \dfrac{1}{2} e^{-w/2} \, dw.$

Thus, the density function of W is $f(w) = \begin{cases} \dfrac{1}{2} e^{-w/2} & w > 0 \\ 0 & \text{elsewhere} \end{cases}.$

The density function indicates W has a gamma distribution with $\alpha = 1$ and $\beta = 2$.

b.   From Section 5.7, a chi-square random variable has a gamma distribution with $\alpha = v/2$ and $\beta = 2$, where v is the degrees of freedom. Thus, any gamma distribution with $\beta = 2$ can be transformed into a chi-square distribution with $v = 2\alpha$.

From part a, $W = \dfrac{2Y}{\beta}$ then has a chi-square distribution with $v = 2\alpha = 2(1) = 2$ degrees of freedom.

c.   Using the pivotal statistic $W = \dfrac{2Y}{\beta}$, the confidence interval for β is:

$$P\left(\chi_{0.975}^2 \leq W \leq \chi_{0.025}^2 \right) = P\left(\chi_{0.975}^2 \leq \dfrac{2Y}{\beta} \leq \chi_{0.025}^2 \right)$$

$$= P\left(\dfrac{1}{\chi_{0.975}^2} \geq \dfrac{\beta}{2Y} \geq \dfrac{1}{\chi_{0.025}^2} \right) = P\left(\dfrac{1}{\chi_{0.025}^2} \leq \dfrac{\beta}{2Y} \leq \dfrac{1}{\chi_{0.975}^2} \right)$$

$$= P\left(\dfrac{2Y}{\chi_{0.025}^2} \leq \beta \leq \dfrac{2Y}{\chi_{0.975}^2} \right) = 0.95$$

where the critical χ^2 values have 2 df.

7.131 a. The midpoint of the interval $\bar{y} - t_{\alpha/2}\left(\dfrac{s}{\sqrt{n}}\right) \leq \mu \leq \bar{y} + t_{\alpha/2}\left(\dfrac{s}{\sqrt{n}}\right)$ is $\bar{y}$.

If $E(Y_i) = \mu$ for all i, then

$$E(\bar{y}) = E\left(\frac{y_1 + y_2 + \cdots y_n}{n}\right) = \frac{1}{n}\left[E(y_1) + E(y_2) + \cdots E(y_n)\right] = \frac{1}{n}n\mu = \mu$$

Since the expected value of the midpoint of the confidence interval is μ, the confidence interval is unbiased.

b. The midpoint of the interval $\dfrac{(n-1)s^2}{\chi^2_{\alpha/2}} \leq \sigma^2 \leq \dfrac{(n-1)s^2}{\chi^2_{1-\alpha/2}}$ is

$$\frac{\dfrac{(n-1)s^2}{\chi^2_{\alpha/2}} + \dfrac{(n-1)s^2}{\chi^2_{1-\alpha/2}}}{2} = \frac{(n-1)s^2}{2}\left(\frac{1}{\chi^2_{\alpha/2}} + \frac{1}{\chi^2_{1-\alpha/2}}\right)$$

We know $E(S^2) = \sigma^2$.

Therefore, $E\left\{\dfrac{(n-1)s^2}{2}\left(\dfrac{1}{\chi^2_{\alpha/2}} + \dfrac{1}{\chi^2_{1-\alpha/2}}\right)\right\} = \dfrac{(n-1)\sigma^2}{2}\left(\dfrac{1}{\chi^2_{\alpha/2}} + \dfrac{1}{\chi^2_{1-\alpha/2}}\right) \neq \sigma^2$.

Since the expected value of the midpoint of the interval is not σ^2, the interval is biased.

8

Tests of Hypotheses

8.1 a. $\alpha =$ probability of committing a Type I error or probability of rejecting H_0 when H_0 is true.

b. $\beta =$ probability of committing a Type II error or probability of accepting H_0 when H_0 is false.

8.3 a. A false negative would be accepting H_0 when H_0 is false, which is a Type II error.

b. A false positive would be rejecting H_0 when H_0 is true, which is a Type I error.

c. According to Dunnett, a false positive or Type I error would be more serious. Much money and time would be spent with further testing on an ineffective drug.

8.5 a. $\alpha =$ probability of rejecting H_0 when H_0 is true

$$= P(Y \geq 6 \text{ if } p = 0.1) = 1 - P(Y \leq 5) = 1 - \sum_{y=0}^{5} p(y) = 1 - 0.9666 = 0.0334$$

Note: $p(y)$ is found using Table 2, Appendix B, with $n = 25$ and $p = 0.1$

b. $\beta =$ probability of accepting H_0 when H_0 is false

$$= P(Y \leq 5 \text{ if } p = 0.2) = \sum_{y=0}^{5} p(y) = 0.6167$$

Note: $p(y)$ is found using Table 2, Appendix B, with $n = 25$ and $p = 0.2$
The power of the test $= 1 - \beta = 1 - 0.6167 = 0.3833$.

c. $\beta =$ probability of accepting H_0 when H_0 is false

$$= P(Y \leq 5 \text{ if } p = 0.4) = \sum_{y=0}^{5} p(y) = 0.0294$$

Note: $p(y)$ is found using Table 2, Appendix B, with $n = 25$ and $p = 0.4$
The power of the test $= 1 - \beta = 1 - 0.0294 = 0.9706$.

8.7 Answers will vary. Suppose Y is a normal random variable with standard deviation $\sigma = 2$. A random sample of size 100 is drawn from the population and we want to test $H_0 : \mu = 20$ against the alternative $H_a : \mu > 20$. The standard deviation of $\bar{y}$ is $\sigma_{\bar{y}} = \dfrac{\sigma}{\sqrt{n}} = \dfrac{2}{\sqrt{100}} = 0.2$.

If we select $\alpha = 0.01$, then the rejection region is to reject H_0 if $z > 2.33$. In terms of $\bar{y}$, the rejection region would be $\bar{y} > k$. To find k, we solve

$$z = \frac{k - \mu}{\sigma_{\bar{y}}} \Rightarrow 2.33 = \frac{k - 20.5}{0.2} \Rightarrow k = 2.33(0.2) + 20 = 20.466$$

If the true value of $\mu = 20.5$, then when $\alpha = 0.01$, $\beta = P(\bar{y} < 20.466 \mid \mu = 20.5) =$

$$P\left(Z < \frac{20.466 - 20.5}{0.2}\right) = P(Z < -0.17) = 0.5 - 0.0675 = 0.4325$$

If we select $\alpha = 0.025$, then the rejection region is to reject H_0 if $z > 1.96$. In terms of $\bar{y}$, the rejection region would be $\bar{y} > k$. To find k, we solve

$$z = \frac{k - \mu}{\sigma_{\bar{y}}} \Rightarrow 1.96 = \frac{k - 20}{0.2} \Rightarrow k = 1.96(0.2) + 20 = 20.392$$

If the true value of $\mu = 20.5$, then when $\alpha = 0.025$,

$$\beta = P(\bar{y} < 20.392 \mid \mu = 20.5) = P\left(Z < \frac{20.392 - 20.5}{0.2}\right)$$

$$= P(Z < -0.54) = 0.5 - 0.2054 = 0.2946$$

If we select $\alpha = 0.05$, then the rejection region is to reject H_0 if $z > 1.645$. In terms of $\bar{y}$, the rejection region would be $\bar{y} > k$. To find k, we solve

$$z = \frac{k - \mu}{\sigma_{\bar{y}}} \Rightarrow 1.645 = \frac{k - 20}{0.2} \Rightarrow k = 1.645(0.2) + 20 = 20.329$$

If the true value of $\mu = 20.5$, then when $\alpha = 0.025$,

$$\beta = P(\bar{y} < 20.392 \mid \mu = 20.5) = P\left(Z < \frac{20.392 - 20.5}{0.2}\right)$$

$$= P(Z < -0.86) = 0.5 - 0.3051 = 0.1949$$

For this example, when $\alpha = 0.01$, $\beta = 0.4325$.

When $\alpha = 0.025$, $\beta = 0.2946$.

When $\alpha = 0.05$, $\beta = 0.1949$.

as α gets larger, β gets smaller.

8.9   From Exercise 8.8, $L(\mu) = \left(\dfrac{1}{\sqrt{2\pi}}\right)^n e^{-\Sigma(y_i - \mu)^2/2}$

$$\lambda = \frac{L(\mu_0)}{L(\hat{\mu})}$$

Since $\mu_0 = 0$ and $\hat{\mu} = \bar{y}$,

$$\lambda = \frac{L(\mu_0)}{L(\hat{\mu})} = \frac{\left(\dfrac{1}{\sqrt{2\pi}}\right)^n e^{-\Sigma(y_i - 0)^2/2}}{\left(\dfrac{1}{\sqrt{2\pi}}\right)^n e^{-\Sigma(y_i - \bar{y})^2/2}} = \frac{e^{-\Sigma y_i^2/2}}{e^{-\Sigma(y_i - \bar{y})^2/2}}$$

$$= e^{-\Sigma y_i^2/2} e^{\Sigma(y_i - \bar{y})^2/2} = e^{-\Sigma y_i^2/2} e^{\Sigma y_i^2/2 - n\bar{y}^2/2} = e^{-n\bar{y}^2/2}$$

8.11   To determine if the mean tensile strength of this fiber composite exceeds 20 megapascals, we test:

$$H_0 : \mu = 20$$

$$H_a : \mu > 20$$

8.13   To determine if the mean breaking strength is less than 22 pounds, we test:

$$H_0 : \mu = 22$$

$$H_a : \mu < 22$$

8.15   Let μ_1 = mean Datapro rating for the software vendor and let μ_2 = mean Datapro rating for the rival software vendor. To determine if the mean if the software vendor has a higher Datapro rating than its rival, we test:

$$H_0 : \mu_1 - \mu_1 = 0$$

$$H_a : \mu_1 - \mu_1 > 0$$

8.17   Let μ_1 = mean number of items produced by method 1 and let μ_2 = mean number of items produced by method 2. To determine if the mean number of items produced differ for the two methods, we test:

$$H_0 : \mu_1 - \mu_1 = 0$$

$$H_a : \mu_1 - \mu_1 \neq 0$$

8.19   The p-values associated with each test statistic are found in Table 5, Appendix B. Since the alternative hypothesis indicates a two-tailed test, the following p-values are calculated.

   a. $p-\text{value} = P(z < -1.01) + P(z > 1.01) = 2P(z > 1.01) = 2[0.5 - P(0 < z < 1.01)]$

   $$= 2(0.5 - 0.3438) = 2(0.1562) = 0.3124$$

   b. $p-\text{value} = P(z < -2.37) + P(z > 2.37) = 2P(z > 2.37) = 2[0.5 - P(0 < z < 2.37)]$

   $$= 2(0.5 - 0.4911) = 2(0.0089) = 0.0178$$

   c. $p-\text{value} = P(z < -4.66) + P(z > 4.66) = 2P(z > 4.66) = 2[0.5 - P(0 < z < 4.66)]$

   $$= 2(0.5 - 0.5) = 2(0) = 0$$

   d. $p-\text{value} = P(z < -1.45) + P(z > 1.45) = 2P(z > 1.45) = 2[0.5 - P(0 < z < 1.45)]$

   $$= 2(0.5 - 0.4265) = 2(0.0735) = 0.1470$$

8.21 a.   For a one-tailed test with $z = -1.63$, the correct p-value is $0.1032 / 2 = 0.0516$. Since the p-value is not less than $\alpha(p = 0.0516 \nless 0.05)$, H_0 is not rejected.

   b.   For a one-tailed test with $z = 1.63$, the correct p-value is $1 - 0.1032/2 = 1 - 0.0516 = 0.9484$. Since the p-value is not less than $\alpha(p = 0.9484 \nless 0.10)$, H_0 is not rejected.

   c.   For a one-tailed test with $z = -1.63$, the correct p-value is $0.1032 / 2 = 0.0516$. Since the p-value is less than $\alpha(p = 0.0516 < 0.05)$, H_0 is rejected.

   d.   For a one-tailed test with $z = -1.63$, the correct p-value is $0.1032 / 2 = 0.0516$. Since the p-value is not less than $\alpha(p = 0.0516 \nless 0.01)$, H_0 is not rejected.

8.23 a. Let μ = mean fup/fumic ratio. To determine if the mean ratio differs from 1, we test:

$$H_0 : \mu = 1$$

$$H_a : \mu \neq 1$$

b. $\bar{y}$ is a statistic, and thus, a variable. We need to take into account the variability of $\bar{y}$ to see if the observed value of $\bar{y}$ is an unusual value if the true mean is actually 1.

c. From the printout, the test statistic is $t = -47.09$ and the p-value is $p = 0.000$.

d. Suppose we select $\alpha = 0.05$. The probability of rejecting H_0 when it is true is 0.05. The probability of concluding that the mean ratio differs from 1 when, in fact, it is equal to 1 is 0.05.

e. Since the p-value is less than $\alpha(p = 0.000 < 0.05)$, H_0 is rejected. There is sufficient evidence to indicate the mean ratio differs from 1 at $\alpha = 0.05$.

f. We must assume that a random sample was selected from the population. Because the sample size is so large, we do not need to assume that the population is normal because of the Central Limit Theorem.

8.25 a. To determine if the mean daily amount of distilled water collected by the new system is greater than 1.4, we test:

$$H_0 : \mu = 1.4$$

$$H_a : \mu > 1.4$$

b. $\alpha = 0.10$. The probability of concluding the mean daily amount of distilled water collected by the new system is greater than 1.4 when the mean is equal to 1.4 is equal to $\alpha = 0.10$.

c. $\bar{y} = \dfrac{\sum y_i}{n} = \dfrac{5.07 + 5.45 + 5.21}{3} = 5.243$

$$s^2 = \frac{\sum y_i^2 - \dfrac{\left(\sum y_i\right)^2}{n}}{n-1} = \frac{(5.07^2 + 5.45^2 + 5.21^2) - \dfrac{15.73^2}{3}}{3-1} = 0.0369$$

$s = \sqrt{0.0369} = 0.1922$

d. $t = \dfrac{\bar{y} - \mu_0}{s / \sqrt{n}} = \dfrac{5.243 - 1.4}{0.1922 / \sqrt{3}} = 34.63$

e. Using a computer package, with $df = n - 1 = 3 - 1 = 2$, $p = P(t > 34.63) = 0.000$.

f. Since the p-value is less than $\alpha(p = 0.000 < 0.10)$, H_0 is rejected. There is sufficient evidence to indicate the mean daily amount of distilled water collected by the new system is greater than 1.4 at $\alpha = 0.10$.

8.27   To determine if the mean dentary depth of molars is different from 15, we test:

$$H_0 : \mu = 15$$

$$H_a : \mu \neq 15$$

From the printout, the test statistic is $t = 3.229$ and the p-value is $p = 0.005$.

The rejection region requires $\alpha = 0.01$ in the upper tail of the t distribution. From Table 7, Appendix B, with $df = n - 1 = 3 - 1 = 2$, $t_{0.01} = 6.956$. The rejection region is $t > 6.956$.

No p-value is given so we will use $\alpha = 0.05$. Since the p-value is less than $\alpha(p = 0.005 < 0.05)$, H_0 is rejected. There is sufficient evidence to indicate the mean dentary depth of molars is different from 15 at $\alpha = 0.05$. There is evidence to indicate the cheek teeth came from another extinct primate species.

8.29   The descriptive statistics are:

Descriptive Statistics: Heatrate

Variable	N	Mean	StDev	Minimum	Q1	Median	Q3	Maximum
Heatrate	67	11066	1595	8714	9918	10656	11842	16243

Let $\mu =$ mean heat rate of gas turbines augmented with high pressure inlet fogging. To determine if the mean exceeds 10,000kJ/kWh, we test:

$$H_0 : \mu = 10,000$$

$$H_a : \mu > 10,000$$

The test statistic is $Z = \dfrac{\bar{y} - \mu_0}{\sigma / \sqrt{n}} = \dfrac{11,066 - 10,000}{1,595 / \sqrt{67}} = 5.471$.

The rejection region requires $\alpha = 0.05$ in the upper tail of the z distribution. From Table 5, Appendix B, $z_{0.05} = 1.645$. The rejection region is $z > 1.645$.

Since the observed value of the test statistic falls in the rejection region $(z = 5.471 > 1.645)$, H_0 is rejected. There is sufficient evidence to indicate the mean heat rate of gas turbines augmented with high pressure inlet fogging exceeds10,000kJ/kWh at $\alpha = 0.05$.

8.31 a. To determine if the true mean bias differs from 0, we test:

$$H_0 : \mu = 0$$

$$H_a : \mu \neq 0$$

The test statistic is $t = \dfrac{\bar{y} - \mu_0}{s / \sqrt{n}} = \dfrac{0.0853 - 0}{1.6031 / \sqrt{15}} = 0.21$.

The rejection region requires $\alpha / 2 = 0.10 / 2 = 0.05$ in each tail of the t distribution. From Table 7, Appendix B, with $df = n - 1 = 15 - 1 = 14$, $t_{0.05} = 1.761$. The rejection region is $t < -1.761$ or $t > 1.761$.

Since the observed value of the test statistic does not fall in the rejection region $(t = 0.21 \not> -1.761)$, H_0 is not rejected. There is insufficient evidence to indicate the true mean bias differs from 0 at $\alpha = 0.10$.

OR

The p-value is $p = 0.8396$. Since the p-value is not less than $\alpha(p = 0.8396 \not< 0.10)$, H_0 is not rejected.

b. The data in the table is the average bias for each subject over 50 minutes. The numbers do not indicate whether the subjects made circles or not. If a subject made a circle, then the average bias would be 0.

8.35 a. To determine if the students, on average, will overestimate the time it takes to read the report, we test:

$$H_0 : \mu = 48$$

$$H_a : \mu > 48$$

The test statistic is $Z = \dfrac{\bar{y} - \mu_0}{\sigma / \sqrt{n}} = \dfrac{60 - 48}{41 / \sqrt{40}} = 1.85.$

The rejection region requires $\alpha = 0.10$ in the upper tail of the z distribution. From Table 5, Appendix B, $z_{0.10} = 1.28$. The rejection region is $z > 1.28$.

Since the observed value of the test statistic falls in the rejection region $(z = 1.85 > 1.28)$, H_0 is rejected. There is sufficient evidence to indicate the students, on average, will overestimate the time it takes to read the report at $\alpha = 0.10$.

b. To determine if the students, on average, will underestimate the number of report pages that can be read, we test:

$$H_0 : \mu = 32$$

$$H_a : \mu < 32$$

The test statistic is $Z = \dfrac{\bar{y} - \mu_0}{\sigma / \sqrt{n}} = \dfrac{28 - 32}{41 / \sqrt{40}} = 1.85.$

The rejection region requires $\alpha = 0.10$ in the lower tail of the z distribution. From Table 5, Appendix B, $z_{0.10} = 1.28$. The rejection region is $z < -1.28$.

Since the observed value of the test statistic falls in the rejection region $(z = -1.85 < -1.28)$, H_0 is rejected. There is sufficient evidence to indicate the students, on average, will underestimate the number of report pages that can be read at $\alpha = 0.10$.

c. No. The sample sizes are sufficiently large that the Central Limit Theorem will apply.

8.37   Let $\mu_1 =$ mean IBI value of the Muskingum river basin and $\mu_2 =$ mean IBI value of the Hockingriver basin. To compare the mean IBI values of the two river basins, we test:

$$H_0 : \mu_1 - \mu_2 = 0$$

$$H_a : \mu_1 - \mu_2 \neq 0$$

The test statistic is $z = \dfrac{(\bar{y}_1 - \bar{y}_1) - D_0}{\sqrt{\dfrac{s_1^2}{n_1} + \dfrac{s_2^2}{n_2}}} = \dfrac{(0.035 - 0.340) - 0}{\sqrt{\dfrac{1.046^2}{53} + \dfrac{0.960^2}{51}}} = -1.55.$

The rejection region requires $\alpha / 2 = 0.10 / 2 = 0.05$ in each tail of the z distribution. From Table 5, Appendix B, $z_{0.05} = 1.645$. The rejection region is $z < -1.645$ or $z > 1.645$.

Since the observed value of the test statistic does not fall in the rejection region $(z = -1.55 \not< -1.645)$, H_0 is not rejected. There is insufficient evidence to indicate a difference in the mean IBI values for the two river basins at $\alpha = 0.10$.

The two-tailed test of hypothesis and the confidence interval are identical when the reliability level and the choice of alpha match up. In this case, the $\alpha = 0.10$ for the test of hypothesis and the 90% reliability level for the confidence interval insure the results of the two analyses will be identical.

8.39 a. Let μ_1 = mean heat rate of traditional augmented gas turbines and μ_2 = mean heat rate of aeroderivative augmented gas turbines.

Some preliminary calculations are:

$$s_p^2 = \frac{(n_1 - 1)s_1^2 + (n_2 - 1)s_2^2}{n_1 + n_2 - 2} = \frac{(39-1)1,279^2 + (7-1)2,652^2}{39 + 7 - 2} = 2,371,831.409$$

To compare the mean heat rates for the two gas turbines, we test:

$$H_0 : \mu_1 - \mu_2 = 0$$

$$H_a : \mu_1 - \mu_2 \neq 0$$

The test statistic is $t = \dfrac{(\bar{y}_1 - \bar{y}_1) - D_0}{\sqrt{s_p^2 \left(\dfrac{1}{n_1} + \dfrac{1}{n_2} \right)}} = \dfrac{(11,544 - 12,312) - 0}{\sqrt{2,371,831.409 \left(\dfrac{1}{39} + \dfrac{1}{7} \right)}} = -1.215.$

The rejection region requires $\alpha/2 = 0.05/2 = 0.025$ in each tail of the t distribution. Using a computer, with $df = n_1 + n_2 - 2 = 39 + 7 - 2 = 44$, $t_{0.025} = 2.015$. The rejection region is $t < -2.015$ or $t > 2.015$.

Since the observed value of the test statistic does not fall in the rejection region ($t = -1.215 \not< -2.015$), H_0 is not rejected. There is insufficient evidence to indicate a difference in the mean heat rates for the traditional and aeroderivative gas turbines at $\alpha = 0.05$.

b. Let μ_1 = mean heat rate of advanced augmented gas turbines and μ_2 = mean heat rate of aeroderivative augmented gas turbines.

Some preliminary calculations are:

$$s_p^2 = \frac{(n_1 - 1)s_1^2 + (n_2 - 1)s_2^2}{n_1 + n_2 - 2} = \frac{(21-1)639^2 + (7-1)2,652^2}{21 + 7 - 2} = 1,927,117.077$$

To compare the mean heat rates for the two gas turbines, we test:

$$H_0 : \mu_1 - \mu_2 = 0$$

$$H_a : \mu_1 - \mu_2 \neq 0$$

The test statistic is $t = \dfrac{(\bar{y}_1 - \bar{y}_1) - D_0}{\sqrt{s_p^2 \left(\dfrac{1}{n_1} + \dfrac{1}{n_2} \right)}} = \dfrac{(9,764 - 12,312) - 0}{\sqrt{1,937,117.077 \left(\dfrac{1}{21} + \dfrac{1}{7} \right)}} = -4.195.$

The rejection region requires $\alpha/2 = 0.05/2 = 0.025$ in each tail of the t distribution. From Table 7, Appendix B, with $df = n_1 + n_2 - 2 = 21 + 7 - 2 = 26$, $t_{0.025} = 2.056$. The rejection region is $t < -2.056$ or $t > 2.056$.

Since the observed value of the test statistic falls in the rejection region ($t = -4.195 < -2.056$), H_0 is rejected. There is sufficient evidence to indicate a difference in the mean heat rates for the advanced and aeroderivative gas turbines at $\alpha = 0.05$.

8.41 a.   Let μ_1 = mean score of those with flexed arms and μ_2 = mean score of those with extended arms.

Some preliminary calculations are:

$$s_p^2 = \frac{(n_1 - 1)s_1^2 + (n_2 - 1)s_2^2}{n_1 + n_2 - 2} = \frac{(11-1)4^2 + (11-1)2^2}{11+11-2} = 10$$

To see if the mean scores for those with flexed arms is greater than that for those with extended arms, we test:

$$H_0 : \mu_1 - \mu_2 = 0$$

$$H_a : \mu_1 - \mu_2 > 0$$

The test statistic is $t = \dfrac{(\bar{y}_1 - \bar{y}_1) - D_0}{\sqrt{s_p^2 \left(\dfrac{1}{n_1} + \dfrac{1}{n_2} \right)}} = \dfrac{(59-43)-0}{\sqrt{10\left(\dfrac{1}{11} + \dfrac{1}{11} \right)}} = 11.87.$

The rejection region requires $\alpha = 0.05$ in the upper tail of the t distribution. From Table 7, Appendix B, with $df = n_1 + n_2 - 2 = 11 + 11 - 2 = 20$, $t_{0.05} = 1.725$. The rejection region is $t > 1.725$.

Since the observed value of the test statistic falls in the rejection region ($t = 11.87 > 1.725$), H_0 is rejected. There is sufficient evidence to indicate the mean score for those with flexed arms is greater than the mean score for those with extended arms at $\alpha = 0.05$.

b.   Some preliminary calculations are:

$$s_p^2 = \frac{(n_1 - 1)s_1^2 + (n_2 - 1)s_2^2}{n_1 + n_2 - 2} = \frac{(11-1)10^2 + (11-1)15^2}{11+11-2} = 162.5$$

The test statistic is $t = \dfrac{(\bar{y}_1 - \bar{y}_1) - D_0}{\sqrt{s_p^2 \left(\dfrac{1}{n_1} + \dfrac{1}{n_2} \right)}} = \dfrac{(59-43)-0}{\sqrt{162.5\left(\dfrac{1}{11} + \dfrac{1}{11} \right)}} = 2.94.$

Since the observed value of the test statistic falls in the rejection region ($t = 2.94 > 1.725$), H_0 is rejected. There is sufficient evidence to indicate the mean score for those with flexed arms is greater than the mean score for those with extended arms at $\alpha = 0.05$.

8.43   Let μ_1 = mean rate of increase of total phosphorus for the control algal and μ_2 = mean rate of increase of total phosphorus for the water hyacinth.

Some preliminary calculations are:

$$s_p^2 = \frac{(n_1 - 1)s_1^2 + (n_2 - 1)s_2^2}{n_1 + n_2 - 2} = \frac{(8-1)0.008^2 + (8-1)0.006^2}{8+8-2} = 0.00005$$

To compare the mean rates of increase of total phosphorus for the two aquatic plants, we test:

$$H_0 : \mu_1 - \mu_2 = 0$$

$$H_a : \mu_1 - \mu_2 \neq 0$$

The test statistic is $t = \dfrac{(\bar{y}_1 - \bar{y}_1) - D_0}{\sqrt{s_p^2 \left(\dfrac{1}{n_1} + \dfrac{1}{n_2} \right)}} = \dfrac{(0.036 - 0.026) - 0}{\sqrt{0.00005 \left(\dfrac{1}{8} + \dfrac{1}{8} \right)}} = 2.828.$

The rejection region requires $\alpha / 2 = 0.05 / 2 = 0.025$ in each tail of the t distribution. From Table 7, Appendix B, with $df = n_1 + n_2 - 2 = 8 + 8 - 2 = 14$, $t_{0.025} = 2.145$. The rejection region is $t < -2.145$ or $t > 2.145$.

Since the observed value of the test statistic falls in the rejection region ($t = 2.828 > 2.145$), H_0 is rejected. There is sufficient evidence to indicate the mean rates of increase of total phosphorus for the two aquatic plants differ at $\alpha = 0.05$.

8.45 a. Using MINITAB, the descriptive statistics are:

Descriptive Statistics: Excel, EPS, Diff

Variable	N	Mean	StDev	Minimum	Q1	Median	Q3	Maximum
Excel	10	88.3	100.5	9.3	16.7	46.9	161.3	317.4
EPS	10	78.1	95.5	2.0	6.5	41.4	153.7	281.7
Diff	10	10.18	12.03	-10.72	4.86	7.11	15.50	35.66

Let μ_1 = mean skin factor value from Excel spreadsheets and μ_2 = mean skin factor value from EPS software.

To determine if the mean skin factor values differ for the two estimation methods, we test:

$$H_0 : \mu_1 - \mu_2 = 0$$

$$H_a : \mu_1 - \mu_2 \neq 0$$

The test statistic is $t = \dfrac{\bar{d} - D_0}{\dfrac{s_d}{\sqrt{n}}} = \dfrac{10.18 - 0}{\dfrac{12.03}{\sqrt{10}}} = 2.68.$

The rejection region requires $\alpha / 2 = 0.05 / 2 = 0.025$ in each tail of the t distribution. From Table 7, Appendix B, with $df = n - 1 = 10 - 1 = 9$, $t_{0.025} = 2.262$. The rejection region is $t < -2.262$ or $t > 2.262$.

Since the observed value of the test statistic falls in the rejection region ($t = 2.68 > 2.262$), H_0 is rejected. There is sufficient evidence to indicate the mean skin factor values differ for the two estimation methodsat $\alpha = 0.05$.

b. Using MINITAB, the descriptive statistics for the horizontal wells are:

Descriptive Statistics: H-Diff

Variable	N	Mean	StDev	Minimum	Q1	Median	Q3	Maximum
H-Diff	5	7.63	2.69	5.03	5.87	6.93	9.74	12.19

To determine if the mean skin factor values differ for the two estimation methods, we test:

$$H_0 : \mu_1 - \mu_2 = 0$$

$$H_a : \mu_1 - \mu_2 \neq 0$$

The test statistic is $t = \dfrac{\bar{d} - D_0}{\dfrac{s_d}{\sqrt{n}}} = \dfrac{7.63 - 0}{\dfrac{2.69}{\sqrt{5}}} = 6.34.$

The rejection region requires $\alpha / 2 = 0.05 / 2 = 0.025$ in each tail of the t distribution. From Table 7, Appendix B, with $df = n - 1 = 5 - 1 = 4$, $t_{0.025} = 2.776$. The rejection region is $t < -2.776$ or $t > 2.776$.

Since the observed value of the test statistic falls in the rejection region ($t = 6.34 > 2.776$), H_0 is rejected. There is sufficient evidence to indicate the mean skin factor values differ for the two estimation methods for horizontal wells at $\alpha = 0.05$.

c. Using MINITAB, the descriptive statistics for the horizontal wells are:

```
Descriptive Statistics: V-Diff
```

Variable	N	Mean	StDev	Minimum	Q1	Median	Q3	Maximum
V-Diff	5	12.74	17.38	-10.72	-3.18	13.81	28.11	35.66

To determine if the mean skin factor values differ for the two estimation methods, we test:

$$H_0 : \mu_1 - \mu_2 = 0$$

$$H_a : \mu_1 - \mu_2 \neq 0$$

The test statistic is $t = \dfrac{\bar{d} - D_0}{\dfrac{s_d}{\sqrt{n}}} = \dfrac{12.74 - 0}{\dfrac{17.38}{\sqrt{5}}} = 1.64.$

The rejection region requires $\alpha / 2 = 0.05 / 2 = 0.025$ in each tail of the t distribution. From Table 7, Appendix B, with $df = n - 1 = 5 - 1 = 4$, $t_{0.025} = 2.776$. The rejection region is $t < -2.776$ or $t > 2.776$.

Since the observed value of the test statistic does not fall in the rejection region ($t = 1.64 \ngtr 2.776$), H_0 is not rejected. There is insufficient evidence to indicate the mean skin factor values differ for the two estimation methods for vertical wells at $\alpha = 0.05$.

8.47 a. Let μ_1 = mean THM for first hole and μ_3 = mean THM for second hole.

From Exercise 7.49, $\bar{d} = 0.14, s_d^2 = 1.5969$ and $s_d = 1.264$.

To determine if the mean THM differs between the two holes, we test:

$$H_0 : \mu_1 - \mu_2 = 0$$

$$H_a : \mu_1 - \mu_2 \neq 0$$

The test statistic is $t = \dfrac{\bar{d} - D_0}{\dfrac{s_d}{\sqrt{n}}} = \dfrac{0.14 - 0}{\dfrac{1.264}{\sqrt{5}}} = 0.43.$

The rejection region requires $\alpha / 2 = 0.10 / 2 = 0.05$ in each tail of the t distribution. From Table 7, Appendix B, with $df = n - 1 = 15 - 1 = 14$, $t_{0.05} = 1.761$. The rejection region is $t < -1.761$ or $t > 1.761$.

Since the observed value of the test statistic does not fall in the rejection region ($t = 0.43 \ngtr 1.761$), H_0 is not rejected. There is insufficient evidence to indicate the mean THM differs between the two holes at $\alpha = 0.10$.

b. Yes. In Exercise 7.49, the 90% confidence interval was $(-0.435, 0.715)$. Because the interval contains 0, there is no evidence to indicate the two mean THM values differ. This is the same conclusion in part a. Because the $\alpha = 0.10$ level compares to the confidence level (90%), the results should agree.

8.49 a. Let μ_1 = mean standardized growth of genes in the full-dark condition and μ_2 = mean standardized growth of genes in the transient light condition.
Using MINITAB, the descriptive statistics are:

Descriptive Statistics: FD-TL Diff

Variable	N	Mean	StDev	Minimum	Q1	Median	Q3	Maximum
FD-TL Diff	10	-1.212	1.290	-2.676	-2.543	-1.257	-0.342	1.022

To determine if there is a difference in the mean standardized growth of genes between the full-dark condition and the transient light condition, we test:

$$H_0 : \mu_1 - \mu_2 = 0$$

$$H_a : \mu_1 - \mu_2 \neq 0$$

The test statistic is $t = \dfrac{\bar{d} - D_0}{\dfrac{s_d}{\sqrt{n}}} = \dfrac{-1.212 - 0}{\dfrac{1.290}{\sqrt{10}}} = -2.97$.

The rejection region requires $\alpha / 2 = 0.01 / 2 = 0.005$ in each tail of the t distribution. From Table 7, Appendix B, with $df = n - 1 = 10 - 1 = 9$, $t_{0.005} = 3.250$. The rejection region is $t < -3.250$ or $t > 3.250$.

Since the observed value of the test statistic does not fall in the rejection region $(t = -2.97 \nless -3.250)$, H_0 is not rejected. There is insufficient evidence to indicate there is a difference in the mean standardized growth of genes between the full-dark condition and the transient light condition at $\alpha = 0.01$.

b. Using the computer, we find the true difference in means is -0.4197. Based on the conclusion in part a, we did not detect this difference.

c. Let μ_1 = mean standardized growth of genes in the full-dark condition and μ_2 = mean standardized growth of genes in the transient light condition.
Using MINITAB, the descriptive statistics are:

Descriptive Statistics: FD-TD Diff

Variable	N	Mean	StDev	Minimum	Q1	Median	Q3	Maximum
FD-TD Diff	10	0.140	0.774	-1.297	-0.388	0.0730	0.829	1.280

To determine if there is a difference in the mean standardized growth of genes between the full-dark condition and the transient dark condition, we test:

$$H_0 : \mu_1 - \mu_2 = 0$$

$$H_a : \mu_1 - \mu_2 \neq 0$$

The test statistic is $t = \dfrac{\bar{d} - D_0}{\dfrac{s_d}{\sqrt{n}}} = \dfrac{0.140 - 0}{\dfrac{0.774}{\sqrt{10}}} = 0.57$.

The rejection region is $t < -3.250$ or $t > 3.250$.

Since the observed value of the test statistic does not fall in the rejection region $(t = 0.57 \ngtr 3.250)$, H_0 is not rejected. There is insufficient evidence to indicate there is a difference in the mean standardized growth of genes between the full-dark condition and the transient dark condition at $\alpha = 0.01$.

Using the computer, we find the true difference in means is -0.2274. Based on the conclusion above, we did not detect this difference.

c. Let μ_1 = mean standardized growth of genes in the transient light condition
and μ_2 = mean standardized growth of genes in the transient light condition.
Using MINITAB, the descriptive statistics are:

Descriptive Statistics: TL-TD Diff

Variable	N	Mean	StDev	Minimum	Q1	Median	Q3	Maximum
TL-TD Diff	10	1.352	1.322	-0.660	-0.117	1.871	2.485	2.696

To determine if there is a difference in the mean standardized growth of genes
between the transient light condition and the transient dark condition, we test:

$$H_0 : \mu_1 - \mu_2 = 0$$

$$H_a : \mu_1 - \mu_2 \neq 0$$

The test statistic is $t = \dfrac{\bar{d} - D_0}{\dfrac{s_d}{\sqrt{n}}} = \dfrac{1.352 - 0}{\dfrac{1.322}{\sqrt{10}}} = 3.23.$

The rejection region is $t < -3.250$ or $t > 3.250$.

Since the observed value of the test statistic does not fall in the rejection
region ($t = 3.23 \not> 3.250$), H_0 is not rejected. There is insufficient evidence to indicate there is a difference in the mean standardized growth of genes between
the transient light condition and the transient dark condition at $\alpha = 0.01$.

Using the computer, we find the true difference in means is 0.1923. Based on
the conclusion above, we did not detect this difference.

8.51   Let μ_1 = mean daily transverse strain change for the field measurement and
μ_2 = mean daily transverse strain change for the 3D model.
Using MINITAB, the descriptive statistics are:

Descriptive Statistics: F-3D Diff

Variable	N	Mean	StDev	Minimum	Q1	Median	Q3	Maximum
F-3D Diff	6	-2.33	8.02	-12.00	-9.00	-3.50	4.75	10.00

To determine if there is a difference in the mean daily transverse strain change
between the field measurement and the 3D model, we test:

$$H_0 : \mu_1 - \mu_2 = 0$$

$$H_a : \mu_1 - \mu_2 \neq 0$$

The test statistic is $t = \dfrac{\bar{d} - D_0}{\dfrac{s_d}{\sqrt{n}}} = \dfrac{-2.33 - 0}{\dfrac{8.02}{\sqrt{6}}} = -0.71.$

The rejection region requires $\alpha / 2 = 0.05 / 2 = 0.025$ in each tail of the t distribution. From Table 7, Appendix B, with $df = n - 1 = 6 - 1 = 5$, $t_{0.025} = 2.571$. The
rejection region is $t < -2.571$ or $t > 2.571$.

Since the observed value of the test statistic does not fall in the rejection
region ($t = -0.71 \not< -2.571$), H_0 is not rejected. There is insufficient evidence
to indicate there is a difference in the mean daily transverse strain change
between the field measurement and the 3D model at $\alpha = 0.05$.

8.53   Let μ_1 = mean viscosity measurement of the experimental method and μ_2 = mean viscosity measurement of the new method.

   Using MINITAB, the descriptive statistics are:

Descriptive Statistics: Diff

Variable	N	Mean	StDev	Minimum	Q1	Median	Q3	Maximum
Diff	12	-0.00875	0.00958	-0.02400	-0.01775	-0.00700	-0.00300	0.00400

To determine if there is a difference in the mean viscosity measurements between the experimental method and the new method, we test:

$$H_0 : \mu_1 - \mu_2 = 0$$

$$H_a : \mu_1 - \mu_2 \neq 0$$

The test statistic is $t = \dfrac{\bar{d} - D_0}{\dfrac{s_d}{\sqrt{n}}} = \dfrac{-0.00875 - 0}{\dfrac{0.00958}{\sqrt{12}}} = -3.16.$

Since no alpha level was give, we will use $\alpha = 0.05$. The rejection region requires $\alpha / 2 = 0.05 / 2 = 0.025$ in each tail of the t distribution. From Table 7, Appendix B, with $df = n - 1 = 12 - 1 = 11$, $t_{0.025} = 2.201$. The rejection region is $t < -2.201$ or $t > 2.201$.

   Since the observed value of the test statistic falls in the rejection region ($t = -3.16 < -2.201$), H_0 is rejected. There is sufficient evidence to indicate there is a difference in the mean viscosity measurements between the experimental method and the new methodat $\alpha = 0.05$.

   The results indicate that there is not "excellent agreement" between the new calculation and the experiments.

8.55 a.   Let p = true proportion of toxic chemical incidents in Taiwan that occur in a school laboratory. To determine if this percentage is less than 10%, we test:

$$H_0 : p = 0.10$$

$$H_a : p < 0.10$$

   b.   The rejection region requires $\alpha = 0.01$ in the lower tail of the z distribution. From Table 5, Appendix B, $z_{0.01} = 2.33$. The rejection region is $z < -2.33$.

   c.   The point estimate of p is $\hat{p} = \dfrac{Y}{n} = \dfrac{15}{250} = 0.06.$

   The test statistic is $Z = \dfrac{\hat{p} - p_0}{\sqrt{\dfrac{p_0 q_0}{n}}} = \dfrac{0.06 - 0.10}{\sqrt{\dfrac{0.10(0.90)}{250}}} = -2.108.$

   d.   Since the observed value of the test statistic does not fall in the rejection region ($z = -2.108 \not< -2.33$), H_0 is not rejected. There is insufficient evidence to indicate the true proportion of toxic chemical incidents in Taiwan that occur in a school laboratory is less than 10% at $\alpha = 0.01$.

8.57   Let p = true proportion of engineering students who have edited content in wiki-based tools. To determine if more than half of engineering students edit content in wiki-based tools, we test:

$$H_0 : p = 0.50$$

$$H_a : p > 0.50$$

The point estimate of p is $\hat{p} = \dfrac{Y}{n} = \dfrac{72}{136} = 0.5294$.

The test statistic is $Z = \dfrac{\hat{p} - p_0}{\sqrt{\dfrac{p_0 q_0}{n}}} = \dfrac{0.5294 - 0.50}{\sqrt{\dfrac{0.50(0.50)}{136}}} = 0.686$.

The rejection region requires $\alpha = 0.10$ in the upper tail of the z distribution. From Table 5, Appendix B, $z_{0.10} = 1.28$. The rejection region is $z > 1.28$.

Since the observed value of the test statistic does not fall in the rejection region ($z = 0.686 \not> 1.28$), H_0 is not rejected. There is insufficient evidence to indicate that more than half of engineering students edit content in wiki-based tools at $\alpha = 0.10$.

8.59   Let p = true success rate of the feeder. To determine if the true success rate of the feeder exceeds 0.90, we test:

$$H_0 : p = 0.90$$
$$H_a : p > 0.90$$

The point estimate of p is $\hat{p} = \dfrac{Y}{n} = \dfrac{94}{100} = 0.94$.

The test statistic is $Z = \dfrac{\hat{p} - p_0}{\sqrt{\dfrac{p_0 q_0}{n}}} = \dfrac{0.94 - 0.90}{\sqrt{\dfrac{0.90(0.10)}{100}}} = 1.33$.

The rejection region requires $\alpha = 0.10$ in the upper tail of the z distribution. From Table 5, Appendix B, $z_{0.10} = 1.28$. The rejection region is $z > 1.28$.

Since the observed value of the test statistic falls in the rejection region ($z = 1.33 > 1.28$), H_0 is rejected. There is sufficient evidence to indicate the true success rate of the feeder exceeds 0.90 at $\alpha = 0.10$.

8.61   Let p = true proportion of students who prefer the new, computerized method. To determine if the true proportion of students who prefer the new, computerized method is greater than 0.70, we test:

$$H_0 : p = 0.70$$
$$H_a : p > 0.70$$

The point estimate of p is $\hat{p} = \dfrac{Y}{n} = \dfrac{138}{171} = 0.807$.

The test statistic is $Z = \dfrac{\hat{p} - p_0}{\sqrt{\dfrac{p_0 q_0}{n}}} = \dfrac{0.807 - 0.70}{\sqrt{\dfrac{0.70(0.30)}{171}}} = 3.05$.

Since no alpha value was given, we will use $\alpha = 0.05$. The rejection region requires $\alpha = 0.05$ in the upper tail of the z distribution. From Table 5, Appendix B, $z_{0.05} = 1.645$. The rejection region is $z > 1.645$.

Since the observed value of the test statistic falls in the rejection region ($z = 3.05 > 1.645$), H_0 is rejected. There is sufficient evidence to indicate the true proportion of students who prefer the new, computerized method is greater than 0.70 at $\alpha = 0.05$. The study indicates that Confir ID should be added to the curriculum at SRU.

8.63   To determine if the proportion of producers who are willing to offer windrowing services to the biomass market differ for the two areas, we test:

$$H_0 : p_1 - p_2 = 0$$
$$H_a : p_1 - p_2 \neq 0$$

From the printout, the test statistic is $z = -2.67$ and the p-value is $p = 0.008$.

Since the p-value is so small, H_0 is rejected. There is sufficient evidence to indicate the proportion of producers who are willing to offer windrowing services to the biomass market differ for the two areas for any value of $\alpha > 0.008$.

In order for the test of hypothesis and the 99% confidence interval to agree, we need to run the test using $\alpha = 0.01$.

8.65 a.   Let p_1 = the true proportion of corn-strain males trapped by the pheromone in corn fields and p_2 = the true proportion of corn-strain males trapped by the pheromone in grass fields.

Some preliminary calculations are:

$$\hat{p}_1 = \frac{Y_1}{n_1} = \frac{86}{112} = 0.7679 \qquad \hat{p}_2 = \frac{Y_2}{n_2} = \frac{164}{215} = 0.7628$$

$$\hat{p} = \frac{Y_1 + Y_2}{n_1 + n_2} = \frac{86 + 164}{112 + 215} = 0.7645$$

To determine if the proportions of corn-strain males trapped by the pheromone differ for corn and grass fields, we test:

$$H_0 : p_1 - p_2 = 0$$
$$H_a : p_1 - p_2 \neq 0$$

The test statistic is $Z = \dfrac{(\hat{p}_1 - \hat{p}_2) - 0}{\sqrt{\hat{p}\hat{q}\left(\dfrac{1}{n_1} + \dfrac{1}{n_2}\right)}} = \dfrac{(0.7679 - 0.7628) - 0}{\sqrt{0.7645(0.2355)\left(\dfrac{1}{112} + \dfrac{1}{215}\right)}} = 0.10.$

The rejection region requires $\alpha / 2 = 0.10 / 2 = 0.05$ in each tail of the z distribution. From Table 5, Appendix B, $z_{0.05} = 1.645$. The rejection region is $z < -1.645$ or $z > 1.645$.

Since the observed value of the test statistic does not fall in the rejection region ($z = 0.10 \not> 1.645$), H_0 is not rejected. There is insufficient evidence to indicate that the proportions of corn-strain males trapped by the pheromone differ for corn and grass fields at $\alpha = 0.10$.

b. Let p_1 = the true proportion of rice-strain males trapped by the pheromone in corn fields and p_2 = the true proportion of rice-strain males trapped by the pheromone in grass fields.

Some preliminary calculations are:

$$\hat{p}_1 = \frac{Y_1}{n_1} = \frac{92}{150} = 0.6133 \quad \hat{p}_2 = \frac{Y_2}{n_2} = \frac{375}{669} = 0.5605 \quad \hat{p} = \frac{Y_1 + Y_2}{n_1 + n_2} = \frac{92 + 375}{150 + 669} = 0.5702$$

To determine if the proportions of rice-strain males trapped by the pheromone differ for corn and grass fields, we test:

$$H_0 : p_1 - p_2 = 0$$
$$H_a : p_1 - p_2 \neq 0$$

The test statistic is $Z = \dfrac{(\hat{p}_1 - \hat{p}_2) - 0}{\sqrt{\hat{p}\hat{q}\left(\dfrac{1}{n_1} + \dfrac{1}{n_2}\right)}} = \dfrac{(0.6133 - 0.5605) - 0}{\sqrt{0.5702(0.4298)\left(\dfrac{1}{150} + \dfrac{1}{669}\right)}} = 1.18.$

The rejection region requires $\alpha/2 = 0.10/2 = 0.05$ in each tail of the z distribution. From Table 5, Appendix B, $z_{0.05} = 1.645$. The rejection region is $z < -1.645$ or $z > 1.645$.

Since the observed value of the test statistic does not fall in the rejection region $(z = 1.18 \not> 1.645)$, H_0 is not rejected. There is insufficient evidence to indicate that the proportions of rice-strain males trapped by the pheromone differ for corn and grass fields at $\alpha = 0.10$.

8.67   Let p_1 = the true proportion of traffic signs maintained by the NCDOT that fail the minimum FHWA retroreflectivity requirements and p_2 = the true proportion of traffic signs maintained by the county that fail the minimum FHWA retroreflectivity requirements.

Some preliminary calculations are:

$$\hat{p}_1 = \frac{Y_1}{n_1} = \frac{512}{1,000} = 0.512 \quad \hat{p}_2 = \frac{Y_2}{n_2} = \frac{328}{1,000} = 0.328$$

$$\hat{p} = \frac{Y_1 + Y_2}{n_1 + n_2} = \frac{512 + 328}{1,000 + 1,000} = 0.420$$

To determine if the true proportion of traffic signs maintained by the NCDOT that fail the minimum FHWA retroreflectivity requirements differs from the true proportion of traffic signs maintained by the county that fail the minimum FHWA retroreflectivity requirements, we test:

$$H_0 : p_1 - p_2 = 0$$
$$H_a : p_1 - p_2 \neq 0$$

The test statistic is $Z = \dfrac{(\hat{p}_1 - \hat{p}_2) - 0}{\sqrt{\hat{p}\hat{q}\left(\dfrac{1}{n_1} + \dfrac{1}{n_2}\right)}} = \dfrac{(0.512 - 0.328) - 0}{\sqrt{0.420(0.580)\left(\dfrac{1}{1,000} + \dfrac{1}{1,000}\right)}} = 8.34.$

The rejection region requires $\alpha/2 = 0.05/2 = 0.025$ in each tail of the z distribution. From Table 5, Appendix B, $z_{0.025} = 1.96$. The rejection region is $z < -1.96$ or $z > 1.96$.

Since the observed value of the test statistic falls in the rejection region $(z = 8.34 > 1.96)$, H_o is rejected. There is sufficient evidence to indicate that the true proportion of traffic signs maintained by the NCDOT that fail the minimum FHWA retroreflectivity requirements differs from the true proportion of traffic signs maintained by the county that fail the minimum FHWA retroreflectivity requirements at $\alpha = 0.05$.

8.69   Let $p_1 = $ the true proportion of weevils found dead after 4 days and $p_2 = $ the true proportion of weevils found dead after 3.5 days.

Some preliminary calculations are:

$$\hat{p}_1 = \frac{Y_1}{n_1} = \frac{31,386}{31,421} = 0.99889 \quad \hat{p}_2 = \frac{Y_2}{n_2} = \frac{23,516}{23,676} = 0.99324$$

$$\hat{p} = \frac{Y_1 + Y_2}{n_1 + n_2} = \frac{31,386 + 23,516}{31,421 + 23,676} = 0.99646$$

To compare the mortality rates of adult rice weevils exposed to nitrogen at the two exposure times, we test:

$$H_0 : p_1 - p_2 = 0$$
$$H_a : p_1 - p_2 \neq 0$$

The test statistic is

$$Z = \frac{(\hat{p}_1 - \hat{p}_2) - 0}{\sqrt{\hat{p}\hat{q}\left(\frac{1}{n_1} + \frac{1}{n_2}\right)}} = \frac{(0.99889 - 0.99324) - 0}{\sqrt{0.99646(0.00354)\left(\frac{1}{31,421} + \frac{1}{23,676}\right)}} = 11.05.$$

The rejection region requires $\alpha / 2 = 0.10 / 2 = 0.05$ in each tail of the z distribution. From Table 5, Appendix B, $z_{0.05} = 1.645$. The rejection region is $z < -1.645$ or $z > 1.645$.

Since the observed value of the test statistic falls in the rejection region $(z = 11.05 > 1.645)$, H_0 is rejected. There is sufficient evidence to indicate the mortality rates of adult rice weevils exposed to nitrogen differ at the two times at $\alpha = 0.10$.

8.71   a.   Let $p_1 = $ proportion of BE students who withdraw from Engineering Mathematics and let $p_2 = $ proportion of BTech students who withdraw from Engineering Mathematics.

Some preliminary calculations are:

$$\hat{p}_1 = 0.278 \quad \hat{p}_2 = 0.197 \quad \hat{p} = \frac{Y_1 + Y_2}{n_1 + n_2} = \frac{537(0.278) + 117(0.197)}{537 + 117} = 0.2635$$

To determine if the proportion of BE students who withdraw from Engineering Mathematics differs from the proportion of BTech students who withdraw from Engineering Mathematics, we test:

$$H_0 : p_1 - p_2 = 0$$

$$H_a : p_1 - p_2 \neq 0$$

The test statistic is $Z = \dfrac{(\hat{p}_1 - \hat{p}_2) - 0}{\sqrt{\hat{p}\hat{q}\left(\dfrac{1}{n_1} + \dfrac{1}{n_2}\right)}} = \dfrac{(0.278 - 0.197) - 0}{\sqrt{0.2635(0.7365)\left(\dfrac{1}{537} + \dfrac{1}{117}\right)}} = 1.80.$

The rejection region requires $\alpha/2 = 0.05/2 = 0.025$ in each tail of the z distribution. From Table 5, Appendix B, $z_{0.025} = 1.96$. The rejection region is $z < -1.96$ or $z > 1.96$.

Since the observed value of the test statistic does not fall in the rejection region ($z = 1.80 \not> 1.96$), H_o is not rejected. There is insufficient evidence to indicate the proportion of BE students who withdraw from Engineering Mathematics differs from the proportion of BTech students who withdraw from Engineering Mathematics at $\alpha = 0.05$.

b. Let $p_1 =$ proportion of BE students who withdraw from Engineering/Graphics CAD and let $p_2 =$ proportion of BTech students who withdraw from Engineering/Graphics CAD.

Some preliminary calculations are:

$$\hat{p}_1 = 0.395 \quad \hat{p}_2 = 0.521 \quad \hat{p} = \frac{Y_1 + Y_2}{n_1 + n_2} = \frac{727(0.395) + 374(0.521)}{727 + 374} = 0.4378$$

To determine if the proportion of BE students who withdraw from Engineering/Graphics CAD differs from the proportion of BTech students who withdraw from Engineering/Graphics CAD, we test:

$$H_0 : p_1 - p_2 = 0$$
$$H_a : p_1 - p_2 \neq 0$$

The test statistic is $Z = \dfrac{(\hat{p}_1 - \hat{p}_2) - 0}{\sqrt{\hat{p}\hat{q}\left(\dfrac{1}{n_1} + \dfrac{1}{n_2}\right)}} = \dfrac{(0.395 - 0.512) - 0}{\sqrt{0.4378(0.5622)\left(\dfrac{1}{727} + \dfrac{1}{374}\right)}} = -3.99.$

The rejection region requires $\alpha/2 = 0.05/2 = 0.025$ in each tail of the z distribution. From Table 5, Appendix B, $z_{0.025} = 1.96$. The rejection region is $z < -1.96$ or $z > 1.96$.

Since the observed value of the test statistic falls in the rejection region ($z = -3.99 < -1.96$), H_o is rejected. There is sufficient evidence to indicate the proportion of BE students who withdraw from Engineering/Graphics CAD differs from the proportion of BTech students who withdraw from Engineering/Graphics CAD at $\alpha = 0.05$.

8.73 a. Let $\sigma^2 =$ variance of the visible albedo values of all Canadian Arctic ice ponds. To determine if the variance differs from 0.0225, we test:

$$H_0 : \sigma^2 = 0.0225$$
$$H_a : \sigma^2 \neq 0.0225$$

The test statistic is $\chi^2 = \dfrac{(n-1)s^2}{\sigma_0^2} = \dfrac{(504-1)0.01839}{0.0225} = 411.119.$

The rejection region requires $\alpha/2 = 0.10/2 = 0.05$ in each tail of the χ^2 distribution. The rejection region is $\chi^2 < 451.991$ or $\chi^2 > 556.283$.

Since the observed value of the test statistic falls in the rejection region ($\chi^2 = 411.119$ or < 451.991), H_o is rejected. There is sufficient evidence to indicate the variance differs from 0.0225 at $\alpha = 0.10$.

b. From Exercise 7.80, we found the 90% confidence interval to be $(0.0166, 0.0205)$. We are 90% confident that the true value of the variance is between 0.0166 and 0.0205. Because 0.0225 does not fall in the confidence interval, it is not a likely value. Thus, we should reject it, which we did in part a.

8.75 Let $\sigma^2 = $ variance of the population of maximum strand forces. Using MINITAB, the descriptive statistics are:

Descriptive Statistics: Force

Variable	N	Mean	StDev	Variance	Q1	Median	Q3
Force	12	163.22	4.99	24.87	159.95	161.70	165.80

To determine if the true variance is less than $5^2 = 25$, we test:

$$H_0 : \sigma^2 = 25 \quad \text{or} \quad H_0 : \sigma = 5$$
$$H_a : \sigma^2 < 25 \qquad\quad H_a : \sigma < 5$$

The test statistic is $\chi^2 = \dfrac{(n-1)s^2}{\sigma_0^2} = \dfrac{(12-1)24.87}{25} = 10.943$.

The rejection region requires $\alpha = 0.10$ in the lower tail of the χ^2 distribution. From Table 8, Appendix B, with $df = n-1 = 12-1 = 11$, $\chi^2_{0.90} = 5.57779$. The rejection region is $\chi^2 < 5.57779$.

Since the observed value of the test statistic does not fall in the rejection region ($\chi^2 = 10.943 \not< 5.57779$), H_0 is not rejected. There is insufficient evidence to indicate the variance is less than 25 or the standard deviation is less than 5 at $\alpha = 0.10$.

8.77 a. To determine if the SNR variance exceeds 0.54, we test:

$$H_0 : \sigma^2 = 0.54$$
$$H_a : \sigma^2 > 0.54$$

b. Using the normal population assumption, we estimate that
$$\text{Range} \approx 4\sigma \Rightarrow 3.0 - 0.03 \approx 4\sigma \Rightarrow 2.97 \approx 4\sigma \Rightarrow \sigma \approx 0.7425$$

c. The test statistic is $\chi^2 = \dfrac{(n-1)s^2}{\sigma_0^2} = \dfrac{(41-1)0.7425^2}{0.54} = 40.8375$.

The rejection region requires $\alpha = 0.10$ in the upper tail of the χ^2 distribution. From Table 8, Appendix B, with $df = n-1 = 41-1 = 40$, $\chi^2_{0.10} = 51.8050$. The rejection region is $\chi^2 > 51.8050$.

Since the observed value of the test statistic does not fall in the rejection region ($\chi^2 = 40.8375 \not> 51.8050$), H_0 is not rejected. There is insufficient evidence to indicate the SNR variance exceeds 0.54 at $\alpha = 0.10$.

8.79 a. Let $\sigma^2 = $ variance of the amounts of rubber cement dispensed. To determine if the variance is more than 0.3, we test:

$$H_0 : \sigma^2 = 0.3$$
$$H_a : \sigma^2 > 0.3$$

The test statistic is $\chi^2 = \dfrac{(n-1)s^2}{\sigma_0^2} = \dfrac{(10-1)0.48^2}{0.3} = 6.912$.

The rejection region requires $\alpha = 0.05$ in the upper tail of the χ^2 distribution. From Table 8, Appendix B, with $df = n-1 = 10-1 = 9$, $\chi^2_{0.05} = 16.9190$. The rejection region is $\chi^2 > 16.9190$.

Since the observed value of the test statistic does not fall in the rejection region ($\chi^2 = 6.912 \not> 16.9190$), H_0 is not rejected. There is insufficient evidence to indicate the variance is more than 0.3 at $\alpha = 0.05$.

b.   We must assume that the population being sampled from is approximately normal.

8.81   a.   Let $\sigma_1^2 = $ variance of transient surface deflection for mineral subgrade access roads and $\sigma_2^2 = $ variance of transient surface deflection for peat subgrade access roads. To determine if the variances of transient surface deflections differ for the two pavement types, we test:

$$H_0 : \frac{\sigma_1^2}{\sigma_2^2} = 1$$

$$H_a : \frac{\sigma_1^2}{\sigma_2^2} \neq 1$$

The test statistic is $F = \dfrac{\text{Larger sample variance}}{\text{Smaller sample variance}} = \dfrac{14.3^2}{3.39^2} = 17.794$.

The rejection region requires $\alpha/2 = 0.05/2 = 0.025$ in the upper tail of the F distribution. Using a computer, with $v_1 = n_2 - 1 = 40 - 1 = 39$ and $v_2 = n_2 - 1 = 32 - 1 = 31$ degrees of freedom, $F_{0.025} = 1.997$. The rejection region is $F > 1.997$.

Since the observed value of the test statistic falls in the rejection region ($F = 17.794 > 1.997$), H_0 is rejected. There is sufficient evidence to indicate the variances of transient surface deflections differ for the two pavement types at $\alpha = 0.05$.

b.   From Exercise 7.110, the 95% confidence interval is $0.029, 0.112$. Since the interval does not contain 0, there is evidence that the variances for the two pavement types are not equal. This agrees with the teat of hypothesis. As long as the significance level matches the alpha level and the test is a two-tailed test, the results will always agree.

8.83   a.   Some preliminary calculations are:

Descriptive Statistics: Novice, Experienced

Variable	N	Mean	StDev	Variance	Q1	Median	Q3
Novice	12	32.83	8.64	74.70	26.75	32.00	39.00
Experienced	12	20.58	5.74	32.99	17.25	19.50	24.75

Let $\sigma_1^2 = $ variance in inspection errors for novice inspectors and $\sigma_2^2 = $ variance in inspection errors for experienced inspectors. To determine if the variance in inspection errors for experienced inspectors is lower than the variance in inspection errors for novice inspectors, we test:

$$H_0 : \frac{\sigma_1^2}{\sigma_2^2} = 1$$

$$H_a : \frac{\sigma_1^2}{\sigma_2^2} > 1$$

The test statistic is $F = \dfrac{\text{Larger sample variance}}{\text{Smaller sample variance}} = \dfrac{74.70}{32.99} = 2.264.$

The rejection region requires $\alpha = 0.05$ in the upper tail of the F distribution. Using a computer, with $v_1 = n_1 - 1 = 12 - 1 = 11$ and $v_2 = n_2 - 1 = 12 - 1 = 11$ degrees of freedom, $F_{0.05} = 2.818$. The rejection region is $F > 2.818$.

Since the observed value of the test statistic does not fall in the rejection region ($F = 2.264 \not> 2.818$), H_o is not rejected. There is insufficient evidence to indicate the variance in inspection errors for experienced inspectors is lower than the variance in inspection errors for novice inspectors at $\alpha = 0.05$. The sample does not support the manager's belief.

b. The p-value for the test is $P(F > 2.264) = 0.096$, using a computer with $v_1 = n_1 - 1 = 12 - 1 = 11$ and $v_2 = n_2 - 1 = 12 - 1 = 11$.

8.85 Let $\sigma_1^2 =$ variance under foggy conditions and $\sigma_2^2 =$ variance under cloudy/clear conditions.

From Exercise 8.44, $s_1 = 0.1186$ and $s_2 = 0.1865$ with $n_1 = 8$ and $n_2 = 4$.

To determine if the variances differ, we test:

$$H_0 : \frac{\sigma_1^2}{\sigma_2^2} = 1$$

$$H_a : \frac{\sigma_1^2}{\sigma_2^2} \neq 1$$

The test statistic is $F = \dfrac{\text{Larger sample variance}}{\text{Smaller sample variance}} = \dfrac{0.1865^2}{0.1186^2} = 2.473.$

The rejection region requires $\alpha / 2 = 0.05 / 2 = 0.025$ in the upper tail of the F distribution. From Table 11, Appendix B, with $v_1 = n_1 - 1 = 4 - 1 = 3$ and $v_2 = n_2 - 1 = 8 - 1 = 7$ degrees of freedom, $F_{0.025} = 5.89$. The rejection region is $F > 5.89$.

Since the observed value of the test statistic does not fall in the rejection region ($F = 2.473 \not> 5.89$), H_o is not rejected. There is insufficient evidence to indicate the variances differ at $\alpha = 0.05$.

8.87 Using MINITAB, the descriptive statistics are:

Descriptive Statistics: 70-cm, 100-cm

Variable	N	Mean	StDev	Variance	Q1	Median	Q3
70-cm	8	10.250	2.689	7.231	7.700	10.700	12.750
100-cm	8	10.763	2.581	6.663	8.900	10.700	12.850

a. Let $\sigma_1^2 =$ variance in the cracking torsion moments for 70-cm slab width and $\sigma_2^2 =$ variance in the cracking torsion moments for 100-cm slab width. To determine if there is a difference in the variation of the two types of T-beams, we test:

$$H_0 : \frac{\sigma_1^2}{\sigma_2^2} = 1$$

$$H_a : \frac{\sigma_1^2}{\sigma_2^2} \neq 1$$

The test statistic is $F = \dfrac{\text{Larger sample variance}}{\text{Smaller sample variance}} = \dfrac{7.231}{6.663} = 1.085.$

The rejection region requires $\alpha/2 = 0.10/2 = 0.05$ in the upper tail of the F distribution. From Table 10, Appendix B, with $v_1 = n_1 - 1 = 8 - 1 = 7$ and $v_2 = n_2 - 1 = 8 - 1 = 7$ degrees of freedom, $F_{0.05} = 3.79$. The rejection region is $F > 3.79$.

Since the observed value of the test statistic does not fall in the rejection region $(F = 1.085 \not> 3.79)$, H_0 is not rejected. There is insufficient evidence to indicate a difference in the variation in the cracking torsion moments for the two types of T-beams at $\alpha = 0.10$.

b. We must assume that the two populations being sampled from are approximately normal and that the samples are independently and randomly selected.

8.89   We need to show that $\dfrac{s_1^2}{s_2^2} < F_{(1-\alpha/2)}$ and $\dfrac{s_2^2}{s_1^2} > F_{\alpha/2}^*$.

If $\dfrac{s_1^2}{s_2^2} < F_{(1-\alpha/2)}$, then $\dfrac{s_2^2}{s_1^2} > \dfrac{1}{F_{(1-\alpha/2)}}$. However, $F_{(1-\alpha/2)} = \dfrac{1}{F_{\alpha/2}^*} \Rightarrow \dfrac{s_2^2}{s_1^2} > \dfrac{1}{F_{(1-\alpha/2)}} = \dfrac{1}{\frac{1}{F_{\alpha/2}^*}} = F_{\alpha/2}^*$

8.95   We use the results from Exercise 8.62:

$$H_0 : p \geq 0.5 \qquad H_a : p < 0.5$$
$$p : Beta(\alpha = 1, \beta = 2)$$
$$X : Bin(n = 81, p)$$

Posterior distribution: $g(p \mid x) : Beta(\alpha = x+1, \beta = n-x+2)$

$n = 81, x = 29 \Rightarrow \alpha = 29+1 = 30, \beta = 81-29+2 = 54$

$$\left.\begin{array}{l} P(p \geq 0.5 \mid x = 29) = 0.004 \\ P(p < 0.5 \mid x = 29) = 0.996 \end{array}\right\} \text{From MINITAB Beta function}$$

8.97   From Exercise 7.105, $g(\mu \mid \bar{y}) : N\left\{\mu^* = \dfrac{n\bar{y}+5}{n+1}, \sigma^{2^*} = \dfrac{1}{n+1}\right\}$

Reject H_0 if $P(\mu < \mu_0) > P(\mu \geq \mu_0)$ using the normal distribution with mean = $\dfrac{n\bar{y}+5}{n+1}$ and variance = $\dfrac{1}{n+1}$.

8.99 a. Let σ_1^2 = variance of the number of ant species in the Dry Steppe and σ_2^2 = variance of the number of ant species in the Gobi Dessert. To determine if there is a difference in the variation at the two locations, we test:

$$H_0 : \frac{\sigma_1^2}{\sigma_2^2} = 1$$

$$H_a : \frac{\sigma_1^2}{\sigma_2^2} \neq 1$$

b. Using MINITAB, the descriptive statistics are:

Descriptive Statistics: AntSpecies

Variable	Region	N	Mean	StDev	Variance	Q1	Median	Q3

```
AntSpecies  Dry Steppe   5  14.00  21.31    454.00  3.00    5.00  29.50
            Gobi Desert  6  11.83  18.21    331.77  4.00    4.50  16.00
```

The test statistic is $F = \dfrac{\text{Larger sample variance}}{\text{Smaller sample variance}} = \dfrac{454.00}{331.77} = 1.368$.

c. The rejection region requires $\alpha/2 = 0.05/2 = 0.025$ in the upper tail of the F distribution. From Table 11, Appendix B, with $v_1 = n_1 - 1 = 5 - 1 = 4$ and $v_2 = n_2 - 1 = 6 - 1 = 5$ degrees of freedom, $F_{0.025} = 7.39$. The rejection region is $F > 7.39$.

d. Using MINITAB, with $v_1 = n_1 - 1 = 5 - 1 = 4$ and $v_2 = n_2 - 1 = 6 - 1 = 5$,

$P(F > 1.368) = 0.363$

e. Since the observed value of the test statistic does not fall in the rejection region ($F = 1.368 \not> 7.39$), H_0 is not rejected. There is insufficient evidence to indicate a difference in the variation at the two locations at $\alpha = 0.05$.

f. We must assume that both populations being sampled from are approximately normal and that the samples are random and independent.

8.101 a. Using Table 2, Appendix B, with $n = 10$ and $p = 0.5$

$\alpha = P(Y \leq 1) + P(Y \geq 8) = P(Y \leq 1) + 1 - P(Y \leq 7) = 0.0107 + (1 - 0.9453) = 0.0654$

b. Using Table 2, Appendix B, with $n = 10$ and $p = 0.4$

$\beta = P(2 \leq Y \leq 7) = P(Y \leq 7) - P(Y \leq 1) = 0.9877 - 0.0464 = 0.9413$

$Power = 1 - \beta = 1 - 0.9413 = 0.0587$

c. Using Table 2, Appendix B, with $n = 10$ and $p = 0.8$

$\beta = P(2 \leq Y \leq 7) = P(Y \leq 7) - P(Y \leq 1) = 0.3222 - 0.0000 = 0.3222$

$Power = 1 - \beta = 1 - 0.3222 = 0.6778$

8.103 a. Let μ_1 = mean perception of managers at less automated firms and μ_2 = mean perception of managers at highly automated firms.

Some preliminary calculations are:

$$s_p^2 = \frac{(n_1 - 1)s_1^2 + (n_2 - 1)s_2^2}{n_1 + n_2 - 2} = \frac{(17 - 1)0.762^2 + (8 - 1)0.721^2}{17 + 8 - 2} = 0.5621$$

To determine if there is a difference in the mean perception between managers of highly automated and less automated firms, we test:

$$H_0 : \mu_1 - \mu_2 = 0$$

$$H_a : \mu_1 - \mu_2 \neq 0$$

The test statistic is $t = \dfrac{(\bar{y}_1 - \bar{y}_1) - D_0}{\sqrt{s_p^2 \left(\dfrac{1}{n_1} + \dfrac{1}{n_2} \right)}} = \dfrac{(3.274 - 3.280) - 0}{\sqrt{0.5621 \left(\dfrac{1}{17} + \dfrac{1}{8} \right)}} = -0.0187$.

The rejection region requires $\alpha/2 = 0.01/2 = 0.005$ in each tail of the t distribution. From Table 7, Appendix B, with $df = n_1 + n_2 - 2 = 17 + 8 - 2 = 23$, $t_{0.005} = 2.807$. The rejection region is $t < -2.807$ or $t > 2.807$.

Since the observed value of the test statistic does not fall in the rejection region $(t = -0.0187 \not< -2.807)$, H_0 is not rejected. There is insufficient evidence to indicate a difference in the mean perception between managers of highly automated and less automated firms at $\alpha = 0.01$.

b. If the variances are not equal, the test statistic is

$$t = \frac{(\bar{y}_1 - \bar{y}_1) - D_0}{\sqrt{\dfrac{s_1^2}{n_1} + \dfrac{s_2^2}{n_2}}} = \frac{(3.274 - 3.280) - 0}{\sqrt{\dfrac{0.762^2}{17} + \dfrac{0.721^2}{8}}} = -0.019$$

The degrees of freedom is:

$$v = \frac{\left(s_1^2/n_1 + s_2^2/n_2\right)^2}{\dfrac{\left(s_1^2/n_1\right)^2}{n_1 - 1} + \dfrac{\left(s_2^2/n_2\right)^2}{n_2 - 1}} = \frac{(0.762^2/17 + 0.721^2/8)^2}{\dfrac{(0.762^2/17)^2}{17 - 1} + \dfrac{(0.721^2/8)^2}{8 - 1}} = \frac{0.009828}{0.000676114} = 14.54 \approx 14$$

The rejection region requires $\alpha/2 = 0.01/2 = 0.005$ in each tail of the t distribution. From Table 7, Appendix B, with $df = 14$, $t_{0.005} = 2.977$. The rejection region is $t < -2.977$ or $t > 2.977$.

Since the observed value of the test statistic does not fall in the rejection region $(t = -0.019 \not< -2.977)$, H_0 is not rejected. There is insufficient evidence to indicate a difference in the mean perception between managers of highly automated and less automated firms at $\alpha = 0.01$. The conclusion is the same as in part a.

8.105 a. To determine if the mean level of radiation is less than 5 picocuries per liter of water, we test:

$$H_0 : \mu = 5$$

$$H_a : \mu < 5$$

The test statistic is $t = \dfrac{\bar{y} - \mu_0}{s/\sqrt{n}} = \dfrac{4.61 - 5}{0.87/\sqrt{24}} = -2.196.$

The rejection region requires $\alpha = 0.01$ in the lower tail of the t distribution. From Table 7, Appendix B, with $df = n - 1 = 24 - 1 = 23$, $t_{0.01} = 2.500$. The rejection region is $t < -2.500$.

Since the observed value of the test statistic does not fall in the rejection region $(t = -2.196 \not< -2.500)$, H_0 is not rejected. There is insufficient evidence to indicate the mean level of radiation is less than 5 picocuries per liter of water at $\alpha = 0.01$.

b. We want our chance of making a Type I error to be small. That is, we want the probability of saying the mean level of radiation is safe when it really is unsafe to be small.

c. β = probability of accepting H_0 when H_0 is false. H_0 will be accepted if $t > -2.500$. In terms of $\bar{y}$:

$$t = \frac{\overline{y} - \mu_0}{s \big/ \sqrt{n}} = \frac{\overline{y} - 5}{0.87 \big/ \sqrt{24}} > -2.500 \Rightarrow \overline{y} - 5 > -0.44397 \Rightarrow \overline{y} > 4.55603$$

$$\beta = P\left(\overline{y} > 4.55603 \mid \mu_a = 4.5\right) = P\left(t > \frac{4.55603 - 4.5}{0.87 \big/ \sqrt{24}}\right) = P(t > 0.3155)$$

Using a computer with $df = 23$, $P(t > 0.3155) = 0.3776$.

d. Using a computer with $df = 23$, $p = P(t < -2.196) = 0.0192$

8.107 a. Answers will vary. Using a computer, a random sample of 40 observations was selected from the DDT values. The descriptive statistics are:

Descriptive Statistics: Sample

Variable	N	Mean	StDev	Minimum	Q1	Median	Q3	Maximum
Sample	40	21.27	59.17	0.11	2.85	5.85	12.00	360.00

$\overline{y} = 21.27, s = 59.17$

b. To determine if the mean DDT content in individual fish inhabiting the Tennessee River exceeds 5 ppm, we test:

$$H_0 : \mu = 5$$

$$H_a : \mu > 5$$

The test statistic is $t = \dfrac{\overline{y} - \mu_0}{s \big/ \sqrt{n}} = \dfrac{21.27 - 5}{59.17 \big/ \sqrt{40}} = 1.74.$

The rejection region requires $\alpha = 0.01$ in the lower tail of the t distribution. Using a computer with $df = n - 1 = 40 - 1 = 39$, $t_{0.01} = 2.426$. The rejection region is $t > 2.426$.

Since the observed value of the test statistic does not fall in the rejection region $(t = 1.74 \not> 2.426)$, H_0 is not rejected. There is insufficient evidence to indicate the mean DDT content in individual fish inhabiting the Tennessee River exceeds 5 ppm at $\alpha = 0.01$.

c. The disadvantages of using a smaller sample size is that it is much harder to reject H_0 when H_0 is false.

d. Answers will vary. Using a computer, a random sample of 8 observations was selected from the 40 DDT values. The descriptive statistics are:

Descriptive Statistics: Sample 2

Variable	N	Mean	StDev	Minimum	Q1	Median	Q3	Maximum
Sample 2	8	7.68	3.53	2.60	3.65	9.10	10.00	12.00

To determine if the mean DDT content in individual fish inhabiting the Tennessee River exceeds 5 ppm, we test:

$$H_0 : \mu = 5$$
$$H_a : \mu > 5$$

The test statistic is $t = \dfrac{\overline{y} - \mu_0}{s \big/ \sqrt{n}} = \dfrac{7.68 - 5}{3.53 \big/ \sqrt{8}} = 2.15.$

The rejection region requires $\alpha = 0.01$ in the lower tail of the t distribution. From Table 7, Appendix B, with $df = n - 1 = 8 - 1 = 7$, $t_{0.01} = 2.998$. The rejection region is $t > 2.998$.

Since the observed value of the test statistic does not fall in the rejection region ($t = 2.15 \ngtr 2.998$), H_0 is not rejected. There is insufficient evidence to indicate the mean DDT content in individual fish inhabiting the Tennessee River exceeds 5 ppm at $\alpha = 0.01$.

For these two examples, the results are the same.

8.109   Let p_1 = proportion of passive solar-heated homes that required less than 200 gallons of oil in fuel consumption last year and p_2 = proportion of solar-heated homes that required less than 200 gallons of oil in fuel consumption last year.

$$\hat{p}_1 = \frac{Y_1}{n_1} = \frac{37}{50} = 0.74 \quad \hat{p}_2 = \frac{Y_2}{n_2} = \frac{46}{50} = 0.92 \quad \hat{p} = \frac{Y_1 + Y_2}{n_1 + n_2} = \frac{37 + 46}{50 + 50} = 0.83$$

To determine if there is a difference between the proportions, we test:

$$H_0 : p_1 - p_2 = 0$$
$$H_a : p_1 - p_2 \neq 0$$

The test statistic is $Z = \dfrac{(\hat{p}_1 - \hat{p}_2) - 0}{\sqrt{\hat{p}\hat{q}\left(\dfrac{1}{n_1} + \dfrac{1}{n_2}\right)}} = \dfrac{(0.74 - 0.92) - 0}{\sqrt{0.83(.17)\left(\dfrac{1}{50} + \dfrac{1}{50}\right)}} = -2.40.$

The rejection region requires $\alpha / 2 = 0.02 / 2 = 0.01$ in each tail of the z distribution. From Table 5, Appendix B, $z_{0.01} = 2.33$. The rejection region is $z < -2.33$ or $z > 2.33$.

Since the observed value of the test statistic falls in the rejection region ($z = -2.40 < -2.33$), H_0 is rejected. There is sufficient evidence to indicate a difference between the proportions of passive and active solar-heated homes that require less than 200 gallons of oil in fuel consumption last year at $\alpha = 0.02$.

8.111   Let σ_1^2 = population variance of the TOC levels at Bedford and σ_2^2 = population variance of the TOC levels at Foxcote.

To determine if the TOC levels have a greater variation at Foxcote, we test:

$$H_0 : \frac{\sigma_1^2}{\sigma_2^2} = 1$$

$$H_a : \frac{\sigma_1^2}{\sigma_2^2} < 1$$

The test statistic is $F = \dfrac{\text{Larger sample variance}}{\text{Smaller sample variance}} = \dfrac{1.27^2}{0.96^2} = 1.75.$

The rejection region requires $\alpha = 0.05$ in the upper tail of the F distribution. Using a computer, with $v_1 = n_2 - 1 = 52 - 1 = 51$ and $v_2 = n_1 - 1 = 61 - 1 = 60$ degrees of freedom,

$F_{0.05} = 1.556$. The rejection region is $F > 1.556$.

Since the observed value of the test statistic falls in the rejection region ($F = 1.75 > 1.556$), H_0 is rejected. There is sufficient evidence to indicate the TOC levels at Foxcote have greater variation than those at Bedford at $\alpha = 0.05$.

8.113 a. To determine if the mean number of inspections is less than 10 when solder joints are spaced 0.1 inch apart, we test:

$$H_0 : \mu = 10$$
$$H_a : \mu < 10$$

b. A Type I error is rejecting the null hypothesis when it is true. In this problem, we would conclude that the laser-based inspection equipment can inspect an average of less than 10 solder joints per second when, in fact, it can inspect at least 10 solder joints per second.

A Type II error is accepting the null hypothesis when it is false. In this problem, we would conclude that the laser-based inspection equipment can inspect an average of at least 10 solder joints per second when, in fact, it can inspect less than 10 solder joints per second.

c. Using MINITAB, the descriptive statistics are:

Descriptive Statistics: PCB

Variable	N	Mean	StDev	Minimum	Q1	Median	Q3	Maximum
PCB	48	9.292	2.103	0.000	9.000	9.000	10.000	13.000

The test statistic is $Z = \dfrac{\bar{y} - \mu_0}{s / \sqrt{n}} = \dfrac{9.292 - 10}{2.103 / \sqrt{48}} = -2.33$.

The rejection region requires $\alpha = 0.05$ in the lower tail of the z distribution. From Table 5, Appendix B, $z_{0.05} = 1.645$. The rejection region is $z < -1.645$.

Since the observed value of the test statistic falls in the rejection region ($z = -2.33 < -1.645$), H_0 is rejected. There is sufficient evidence to indicate the true mean number of inspections is less than 10 when solder joints are spaced 0.1 inch apart at $\alpha = 0.05$.

8.115 a. Let μ_1 = mean quality performance rating of competitive R&D contracts and μ_2 = mean quality performance rating for sole source R&D contracts.

To determine if the mean rating for the competitive contracts exceeds the mean rating for the sole source contracts, we test:

$$H_0 : \mu_1 - \mu_2 = 0$$
$$H_a : \mu_1 - \mu_2 > 0$$

b. The rejection region requires $\alpha = 0.05$ in the upper tail of the z distribution. From Table 5, Appendix B, $z_{0.05} = 1.645$. The rejection region is $z > 1.645$.

c. Since the p-value is less than $\alpha (p < 0.03 < 0.05)$, we would reject H_0. There is sufficient evidence to indicate the mean rating for competitive contracts exceeds the mean rating for sole source contracts at $\alpha = 0.05$.

9

Categorical Data Analysis

9.1   a.   The qualitative variable is the type of jaw habits patients have. It has four levels: grinding, clenching, both, and neither.

b.   The one-way table for the data is:

	Type of Habit		
Grinding	Clenching	Both	Neither
3	11	30	16

c.   Let p_3 = proportion of patients who admit to both habits.

$$\hat{p}_3 = \frac{n_3}{n} = \frac{30}{60} = 0.5$$

For confidence coefficient 0.95, $\alpha = 0.05$ and $\alpha / 2 = 0.05 / 2 = 0.025$. From Table 5, Appendix

B, $z_{0.025} = 1.96$. The 95% confidence interval is:

$$\hat{p}_3 \pm z_{\alpha/2}\sigma_{\hat{p}_3} \Rightarrow \hat{p}_3 \pm 1.96\sqrt{\frac{\hat{p}_3\hat{q}_3}{n}} \Rightarrow 0.5 \pm 1.96\sqrt{\frac{0.5(0.5)}{60}}$$

$$\Rightarrow 0.5 \pm 0.127 \Rightarrow (0.373, 0.627)$$

We are 95% confident that the true proportion of patients who admit to both habits is between 0.373 and 0.627.

d.   Let p_4 = proportion of patients who admit to neither habit.

$$\hat{p}_4 = \frac{n_4}{n} = \frac{16}{60} = 0.267$$

For confidence coefficient 0.95, $\alpha = 0.05$ and $\alpha / 2 = 0.05 / 2 = 0.025$. From Table 5, Appendix B, $z_{0.025} = 1.96$. The 95% confidence interval is:

$$(\hat{p}_3 - \hat{p}_4) \pm z_{\alpha/2}\sqrt{\frac{\hat{p}_3\hat{q}_3 + \hat{p}_4\hat{q}_4 + 2\hat{p}_3\hat{q}_4}{n}}$$

$$\Rightarrow (0.5 - 0.267) \pm 1.96\sqrt{\frac{0.5(0.5) + 0.267(0.733) + 2(0.5)(0.267)}{60}}$$

$$\Rightarrow 0.233 \pm 0.214 \Rightarrow (0.019, 0.447)$$

We are 95% confident that the difference between the true proportion of dental patients who admit to both habits and the true proportion of dental patients who claim they have neither habit is between 0.019 and 0.447.

9.3 a. Let p_1 = proportion of readers who feel they know enough about CAD.

$\hat{p}_1 = 0.44$

For confidence coefficient 0.95, $\alpha = 0.05$ and $\alpha / 2 = 0.05 / 2 = 0.025$. From Table 5, Appendix B, $z_{0.025} = 1.96$. The 95% confidence interval is:

$$\hat{p}_1 \pm z_{\alpha/2}\sigma_{\hat{p}_1} \Rightarrow \hat{p}_1 \pm 1.96\sqrt{\frac{\hat{p}_1\hat{q}_1}{n}} \Rightarrow 0.44 \pm 1.96\sqrt{\frac{0.44(0.56)}{1,000}}$$

$$\Rightarrow 0.44 \pm 0.031 \Rightarrow (0.409, 0.471)$$

We are 95% confident that the true proportion of readers who feel they know enough about CAD is between 0.409 and 0.471.

b. Let p_2 = proportion of readers who feel they don't know enough about CAD but aren't worried and $\bar{p}_3$ = proportion of readers who feel they don't know enough about CAD but are concerned.

$\hat{p}_2 = 0.12$ $\hat{p}_2 = 0.35$

For confidence coefficient 0.95, $\alpha = 0.05$ and $\alpha / 2 = 0.05 / 2 = 0.025$. From Table 5, Appendix B, $z_{0.025} = 1.96$. The 95% confidence interval is:

$$(\hat{p}_2 - \hat{p}_3) \pm z_{\alpha/2}\sqrt{\frac{\hat{p}_2\hat{q}_2 + \hat{p}_3\hat{q}_3 + 2\hat{p}_2\hat{q}_3}{n}}$$

$$\Rightarrow (0.12 - 0.35) \pm 1.96\sqrt{\frac{0.12(0.88) + 0.35(0.65) + 2(0.12)(0.35)}{1,000}}$$

$$\Rightarrow -0.23 \pm 0.040 \Rightarrow (-0.270, -0.190)$$

We are 95% confident that the difference between the proportions of readers who answered "no, but I'm not worried about it" and of those who answered "no, and it concerns me" is between −0.270 and −0.190.

9.5 a. Let p_1 = the proportion of melt ponds in the Canadian Arctic that have first-year ice.

$\hat{p}_1 = 0.1746$

For confidence coefficient 0.90, $\alpha = 0.10$ and $\alpha / 2 = 0.10 / 2 = 0.05$. From Table 5, Appendix B, $z_{0.05} = 1.645$. The 90% confidence interval is:

$$\hat{p}_1 \pm z_{\alpha/2}\sigma_{\hat{p}_1} \Rightarrow \hat{p}_1 \pm 1.645\sqrt{\frac{\hat{p}_1\hat{q}_1}{n}} \Rightarrow 0.1746 \pm 1.645\sqrt{\frac{0.1746(0.8254)}{504}}$$

$$\Rightarrow 0.1746 \pm 0.0278 \Rightarrow (0.1468, 0.2024)$$

We are 90% confident that the true proportion of melt ponds in the Canadian Arctic that have first-year ice is between 0.1468 and 0.2024.

b. Let p_2 = the proportion of melt ponds in the Canadian Arctic that have multi-year ice.

$\hat{p}_2 = 0.4365$

For confidence coefficient 0.90, $\alpha = 0.10$ and $\alpha / 2 = 0.10 / 2 = 0.05$. From Table 5, Appendix

B, $z_{0.05} = 1.645$. The 90% confidence interval is:

$$(\hat{p}_1 - \hat{p}_2) \pm z_{\alpha/2} \sqrt{\frac{\hat{p}_1 \hat{q}_1 + \hat{p}_2 \hat{q}_2 + 2\hat{p}_1 \hat{q}_2}{n}}$$

$$\Rightarrow (0.1746 - 0.4365) \pm 1.645 \sqrt{\frac{0.1746(0.8254) + 0.4365(0.5635) + 2(0.1746)(0.4365)}{504}}$$

$$\Rightarrow -0.2619 \pm 0.0540 \Rightarrow (-0.3159, -0.2079)$$

We are 90% confident that the difference between the proportion of melt ponds in the Canadian Arctic that have first-year ice and the proportion that have multi-year ice is between −0.3159 and −0.2079.

9.7   a.   Let p_1 = the proportion of all American adults who disagree with the statement.

$$\hat{p}_1 = \frac{311 + 343}{965} = 0.6777$$

For confidence coefficient 0.99, $\alpha = 0.01$ and $\alpha / 2 = 0.01 / 2 = 0.005$. From Table 5, Appendix

B, $z_{0.005} = 2.575$. The 99% confidence interval is:

$$\hat{p}_1 \pm z_{\alpha/2} \sigma_{\hat{p}_1} \Rightarrow \hat{p}_1 \pm 2.575 \sqrt{\frac{\hat{p}_1 \hat{q}_1}{n}} \Rightarrow 0.6777 \pm 2.575 \sqrt{\frac{0.677(0.3223)}{965}}$$

$$\Rightarrow 0.6777 \pm 0.0387 \Rightarrow (0.6390, 0.7164)$$

We are 99% confident that the true proportion of all American adults who disagree with the statement is between 0.6390 and 0.7164.

b.   Let p_2 = the proportion of all American adults who agree with the statement.

$$\hat{p}_2 = \frac{99 + 212}{965} = 0.3223$$

For confidence coefficient 0.99, $\alpha = 0.01$ and $\alpha / 2 = 0.01 / 2 = 0.005$. From Table 5, Appendix B, $z_{0.005} = 2.575$. The 99% confidence interval is:

$$(\hat{p}_1 - \hat{p}_3) \pm z_{\alpha/2} \sqrt{\frac{\hat{p}_1 \hat{q}_1 + \hat{p}_2 \hat{q}_2 + 2\hat{p}_1 \hat{p}_2}{n}}$$

$$\Rightarrow (0.6777 - 0.3223) \pm 2.575 \sqrt{\frac{0.6777(0.3223) + 0.3223(0.6777) + 2(0.6777)(0.3223)}{965}}$$

$$\Rightarrow 0.3554 \pm 0.0775 \Rightarrow (0.2779, 0.4329)$$

We are 99% confident that the difference between the proportions of all American adults who disagree and agree with the statement is between 0.2779 and 0.4329.

9.9   The probability function for the multinomial is:

$$p(n_1, n_2, \cdots, n_k) = \frac{n!}{n_1! n_2! \cdots n_k!} p_1^{n_1} p_2^{n_2} \cdots p_k^{n_k} \text{ where } n = n_1 + n_2 + \cdots + n_k \text{ and}$$

$$p_1 + p_2 + \cdots + p_k = 1$$

Since this is a probability function, we know that if we sum over all possible values of $n_1, n_2, \cdots, n_k$, the result is 1.

$$\sum_{n_1=0}^{n} \sum_{n_2=0}^{n} \cdots \sum_{n_k=0}^{n} \frac{n!}{n_1! n_2! \cdots n_k!} p_1^{n_1} p_2^{n_2} \cdots p_k^{n_k} = 1 \text{ where } n = n_1 + n_2 + \cdots + n_k$$

Without loss of generality, let $i = 1$ and $j = 2$.

$$E(n_1 n_2) = \sum_{n_1} \sum_{n_2} \cdots \sum_{n_k} n_1 n_2 \frac{n!}{n_1! n_2! \cdots n_k!} p_1^{n_1} p_2^{n_2} \cdots p_k^{n_k}$$

$$= n(n-1) p_1 p_2 \sum_{n_1} \sum_{n_2} \cdots \sum_{n_k} \frac{(n-2)!}{(n_1-1)!(n_2-1)! \cdots n_k!} p_1^{n_1-1} p_2^{n_2-1} \cdots p_k^{n_k}$$

$$= n(n-1) p_1 p_2$$

$$Cov(n_i, n_j) = E(n_i n_j) - E(n_i)E(n_j) = n(n-1)p_i p_j - np_i(np_j) = np_i p_j(n-1-n) = -np_i p_j$$

9.11   Some preliminary calculations are:

$$E(n_i) = np_i = 859(1/6) = 143.1667$$

To determine if the proportions of mobile devise users in the six texting categories differ, we test:

$$H_o : p_1 = p_2 = p_3 = p_4 = p_5 = p_6 = 1/6$$

H_a : At least one of the multinomial probabilities differs from its null hypothesized value

The test statistic is

$$\chi^2 = \sum \frac{[n_i - E(n_i)]^2}{E(n_i)} = \frac{(396 - 143.1667)^2}{143.1667} + \frac{(311 - 143.1667)^2}{143.1667} + \frac{(70 - 143.1667)^2}{143.1667}$$

$$+ \frac{(39 - 143.1667)^2}{143.1667} + \frac{(18 - 143.1667)^2}{143.1667} + \frac{(25 - 143.1667)^2}{143.1667} = 963.4002$$

The rejection region requires $\alpha = 0.10$ in the upper tail of the χ^2 distribution with $df = k - 1 = 6 - 1 = 5$. From Table 8, Appendix B, $\chi^2_{0.100} = 9.23635$. The rejection region is $\chi^2 > 9.23635$.

Since the observed value of the test statistic falls in the rejection region ($\chi^2 = 963.4002 > 9.23635$), H_0 is rejected. There is sufficient evidence to indicate the proportions of mobile devise users in the six texting categories differ at $\alpha = 0.10$.

9.13   Some preliminary calculations are:

$$E(n_1) = np_1 = 504(0.15) = 75.6 \qquad E(n_{12}) = np_2 = 504(0.40) = 201.6$$

$$E(n_3) = np_3 = 504(0.45) = 226.8$$

Let p_1 = the proportion of melt ponds in the Canadian Arctic that have first-year ice, p_2 = the proportion of melt ponds in the Canadian Arctic that have landfast ice, and p_3 = the proportion of melt ponds in the Canadian Arctic that have multi-year ice.

To test the engineers' theory, we test:

$H_0 : p_1 = 0.15, p_2 = 0.40, p_3 = 0.45$

H_a : At least one of the multinominal probabilities differes from its null hypothesized value

The test statistic is

$$\chi^2 = \sum \frac{[n_i - E(n_i)]^2}{E(n_i)} = \frac{(88 - 75.6)^2}{75.6} + \frac{(196 - 201.6)^2}{201.6} + \frac{(220 - 226.8)^2}{226.8} = 2.393$$

The rejection region requires $\alpha = 0.01$ in the upper tail of the χ^2 distribution with $df = k - 1 = 3 - 1 = 2$. From Table 8, Appendix B, $\chi^2_{0.01} = 9.21034$. The rejection region is $\chi^2 > 9.21034$.

Since the observed value of the test statistic does not fall in the rejection region ($\chi^2 = 2.393 \not> 9.21034$), H_0 is not rejected. There is insufficient evidence to indicate the engineers' theory is incorrect at $\alpha = 0.01$.

9.15   Some preliminary calculations are:

$$E(n_1) = np_i = 965(0.25) = 241.25$$

To determine if the percentages in the four response categories are different, we test:

$H_0 : p_1 = p_2 = p_3 = p_4 = 0.25$

H_a : At least one of the multinomial probabilities differs from its null hypothesized value

The test statistic is

$$\chi^2 = \sum \frac{[n_i - E(n_i)]^2}{E(n_i)} = \frac{(99 - 241.25)^2}{241.25} + \frac{(212 - 241.25)^2}{241.25}$$

$$+ \frac{(311 - 241.25)^2}{241.25} + \frac{(343 - 241.25)^2}{241.25} = 150.503$$

The rejection region requires $\alpha = 0.01$ in the upper tail of the χ^2 distribution with $df = k - 1 = 4 - 1 = 3$. From Table 8, Appendix B, $\chi^2_{0.01} = 11.3441$. The rejection region is $\chi^2 > 11.3441$.

Since the observed value of the test statistic falls in the rejection region ($\chi^2 = 150.503 > 11.3441$), H_0 is rejected. There is sufficient evidence to indicate the percentages in the four response categories are different at $\alpha = 0.01$.

9.17   Some preliminary calculations:

$E(n_1) = np_1 = 1,000(0.35) = 350$ $E(n_{12}) = np_2 = 1,000(0.45) = 450$

$E(n_3) = np_3 = 1,000(0.10) = 100$ $E(n_4) = np_4 = 1,000(0.10) = 100$

To determine if the distribution of background colors for all road signs maintained by NCDOT match the color distribution of signs in the warehouse, we test:

$H_0 : p_1 = 0.35, p_2 = 0.45, p_3 = 0.10, p_4 = 0.10$

H_a : At least one of the multinomial probabilities differs from its null hypothesized value

The test statistic is

$$\chi^2 = \sum \frac{[n_i - E(n_i)]^2}{E(n_i)} = \frac{(373 - 350)^2}{350} + \frac{(447 - 450)^2}{450} + \frac{(88 - 100)^2}{100}$$

$$+ \frac{(92 - 100)^2}{100} = 3.611$$

The rejection region requires $\alpha = 0.05$ in the upper tail of the χ^2 distribution with $df = k - 1 = 4 - 1 = 3$. From Table 8, Appendix B, $\chi^2_{0.05} = 7.81473$. The rejection region is $\chi^2 > 7.81473$.

Since the observed value of the test statistic does not fall in the rejection region $(\chi^2 > 3.611 \not> 7.81473)$, H_0 is not rejected. There is insufficient evidence to indicate the distribution of background colors for all road signs maintained by NCDOT match the color distribution of signs in the warehouse at $\alpha = 0.05$.

9.19   Some preliminary calculations:

$E(n_1) = np_1 = 2,097(0.02) = 41.94$ $E(n_{12}) = np_2 = 2,094(0.25) = 524.25$

$E(n_3) = np_3 = 2,094(0.73) = 1,530.81$

To determine if the distribution of E4/E4 genotypes for the population of young adults differs from the norm, we test:

$H_0 : p_1 = 0.02, p_2 = 0.25, p_3 = 0.73$

H_a : At least one of the multinomial probabilities differs from its null hypothesized value

The test statistic is

$$\chi^2 = \sum \frac{[n_i - E(n_i)]^2}{E(n_i)} = \frac{(56 - 41.94)^2}{41.94} + \frac{(517 - 524.25)^2}{524.25} + \frac{(1,524 - 1,530.81)^2}{1,530.80} = 4.844$$

The rejection region requires $\alpha = 0.05$ in the upper tail of the χ^2 distribution with $df = k - 1 = 3 - 1 = 2$.. From Table 8, Appendix B, $\chi^2_{0.05} = 5.99147$. The rejection region is $\chi^2 > 5.99147$.

Since the observed value of the test statistic does not fall in the rejection region $(\chi^2 = 4.844 \not> 5.99147)$, H_0 is not rejected. There is insufficient evidence to indicate the distribution of E4/E4 genotypes for the population of young adults differs from the norm at $\alpha = 0.05$.

9.21 a. To determine if the distribution of FIA trends were the same for the Pennsylvania Nappe and Maryland Nappe, we test:

H_0: Nappe and FIA are independent
H_a: Nappe and FIA are dependent

b. Some preliminary calculations are:

$$\hat{E}(n_{ij}) = \frac{n_{ig}n_{gj}}{n}$$

$$\hat{E}(n_{11}) = \frac{26(47)}{70} = 17.457 \qquad \hat{E}(n_{12}) = \frac{26(23)}{70} = 8.543$$

$$\hat{E}(n_{21}) = \frac{27(47)}{70} = 18.129 \qquad \hat{E}(n_{22}) = \frac{27(23)}{70} = 8.871$$

$$\hat{E}(n_{31}) = \frac{17(47)}{70} = 11.414 \qquad \hat{E}(n_{32}) = \frac{17(23)}{70} = 5.586$$

The test statistic is

$$\chi^2 = \sum\sum \frac{[n_{ij} - \hat{E}(n_{ij})]^2}{\hat{E}(n_{ij})} = \frac{(20-17.457)^2}{17.457} + \frac{(6-8.543)^2}{8.543} + \cdots \frac{(7-5.586)^2}{5.586} = 1.874$$

c. The rejection region requires $\alpha = 0.05$ in the upper tail of the χ^2 distribution with $df = (r-1)(c-1) = (3-1)(2-1) = 2$. From Table 8, Appendix B, $\chi^2_{0.05} = 5.99147$. The rejection region is $\chi > 5.99147$.

d. Since the observed value of the test statistic does not fall in the rejection region ($\chi^2 = 1.874 \not> 5.99147$), H_0 is not rejected. There is insufficient evidence to indicate that Nappe and FIA are dependent at $\alpha = 0.05$.

9.23 Some preliminary calculations are:

$$\hat{E}(n_{ij}) = \frac{n_{ig}n_{gj}}{n}$$

$$\hat{E}(n_{11}) = \frac{234(40)}{437} = 21.419 \quad \hat{E}(n_{12}) = \frac{234(397)}{437} = 212.581 \quad \hat{E}(n_{21}) = \frac{203(40)}{437} = 18.581$$

$$E(n_{22}) = \frac{203(394)}{437} = 184.419$$

To determine if the response rate of air traffic controllers to mid-air collision alarms differs for true and false alarms, we test:

H_0: Response rate and type of alert are independent
H_a: Response rate and type of alert are dependent

The test statistic is

$$\chi^2 = \sum\sum \frac{[n_{ij} - \hat{E}(n_{ij})]^2}{\hat{E}(n_{ij})} = \frac{(3-21.419)^2}{21.419} + \frac{(231-212.581)^2}{212.581}$$

$$+ \frac{(37-18.581)^2}{18.581} + \frac{(166-184.419)^2}{184.419} = 37.532$$

The rejection region requires $\alpha = 0.05$ in the upper tail of the χ^2 distribution with $df = (r-1)(c-1) = (2-1)(2-1) = 1$. From Table 8, Appendix B, $\chi^2_{0.05} = 3.84146$. The rejection region is $\chi > 3.84146$.

Since the observed value of the test statistic falls in the rejection region ($\chi^2 = 37.532 > 3.84146$), H_0 is rejected. There is sufficient evidence to indicate that response rate and type of alert are dependent at $\alpha = 0.05$. The response rate differs for true and false alerts. Air traffic controllers tend to respond to true alerts at a higher rate than to false alerts.

9.25 a. The contingency table is:

```
Rows: MTBE    Columns: Wellclass

              Private   Public   All

Below Limit       81       72   153
Detect            22       48    70
All              103      120   223
```

b. To determine if detectable MTBE status depends on well class, we test:

H_0: MTBE and well class are independent
H_a: MTBE and well class are dependent

Using MINITAB, the calculations are:

Chi-Square Test for Association: MTBE, Wellclass
```
       Rows: MTBE    Columns: Wellclass

                  Private   Public   All
       Below Limit     81       72   153
                    70.67    82.33

       Detect          22       48    70
                    32.33    37.67

       All            103      120   223
       Cell Contents:      Count
                           Expected count

       Pearson Chi-Square = 8.943, DF = 1, P-Value = 0.003
       Likelihood Ratio Chi-Square = 9.125, DF = 1, P-Value = 0.003
```

The test statistic is $\chi^2 = 8.943$ and the p-value is $p = 0.003$. Since the p-value is less than $\alpha (p = 0.003 < 0.05)$, H_0 is rejected. There is sufficient evidence to indicate that MTBE status depends on well class at $\alpha = 0.05$.

c. The contingency table is:
```
       Rows: MTBE    Columns: Aquifer

                  Bedrock   Unconsoli   All

       Below Limit    138          15   153
       Detect          63           7    70
       All            201          22   223
```

d. To determine if detectable MTBE status depends on aquifer, we test:

H_0: MTBE and aquifer are independent
H_a: MTBE and aquifer are dependent

Using MINITAB, the calculations are:
Chi-Square Test for Association: MTBE, Aquifer

Rows: MTBE   Columns: Aquifer

	Bedrock	Unconsoli	All
Below Limit	138	15	153
	137.91	15.09	
Detect	63	7	70
	63.09	6.91	
All	201	22	223

Cell Contents: Count
                  Expected count

Pearson Chi-Square = 0.002, DF = 1, P-Value = 0.964
Likelihood Ratio Chi-Square = 0.002, DF = 1, P-Value = 0.964

The test statistic is $\chi^2 = 0.002$ and the p-value is $p = 0.964$. Since the p-value is not less than α ($p = 0.964 \not< 0.05$), H_0 is not rejected. There is insufficient evidence to indicate that MTBE status depends on aquifer at $\alpha = 0.05$.

9.27 a. The expected cell counts are:

$$\hat{E}(n_{ij}) = \frac{n_{ig} n_{gj}}{n}$$

$$\hat{E}(n_{11}) = \frac{5(13)}{73} = 0.890 \qquad \hat{E}(n_{12}) = \frac{5(17)}{73} = 1.164 \qquad \hat{E}(n_{13}) = \frac{5(43)}{73} = 2.945$$

$$\hat{E}(n_{21}) = \frac{51(13)}{73} = 9.082 \qquad \hat{E}(n_{22}) = \frac{51(17)}{73} = 11.877 \qquad \hat{E}(n_{23}) = \frac{51(43)}{73} = 30.041$$

$$\hat{E}(n_{31}) = \frac{17(13)}{73} = 3.027 \qquad \hat{E}(n_{32}) = \frac{17(17)}{73} = 3.959 \qquad \hat{E}(n_{33}) = \frac{17(43)}{73} = 10.014$$

Since some of the expected cell counts are less than 5, the requirements for the chi-square test of independence are not met.

b. The new reformulated table is:

	DwarfShrub	Grasses	Herbs	Total
NS/SR	8	3	11	22
SA	5	14	32	51
Total	13	17	43	73

The expected cell counts are:

$$\hat{E}(n_{11}) = \frac{22(13)}{73} = 3.918 \quad \hat{E}(n_{12}) = \frac{22(17)}{73} = 5.123 \quad \hat{E}(n_{13}) = \frac{22(43)}{73} = 12.959$$

$$\hat{E}(n_{21}) = \frac{51(13)}{73} = 9.082 \quad \hat{E}(n_{22}) = \frac{51(17)}{73} = 11.877 \quad \hat{E}(n_{23}) = \frac{51(43)}{73} = 30.041$$

Since one of the expected cell counts is less than 5, the requirements for the chi-square test of independence are not met.

c. The new reformulated table is:

	DwarfShrub/ Grasses	Herbs	Total
NS/SR	11	11	22
SA	19	32	51
Total	30	43	73

The expected cell counts are:

$$\hat{E}(n_{11}) = \frac{22(30)}{73} = 9.041 \qquad \hat{E}(n_{12}) = \frac{22(43)}{73} = 12.959$$

$$\hat{E}(n_{21}) = \frac{51(30)}{73} = 20.959 \qquad \hat{E}(n_{22}) = \frac{51(43)}{73} = 30.041$$

Since all of the expected cell counts are at least 5, the requirements for the chi-square test of independence are met.

d. To determine if seedling abundance depends on plant type, we test:

H_0: Seedling abundance and plant type are independent
H_a: Seedling abundance and plant type are dependent

The test statistic is

$$\chi^2 = \sum\sum \frac{\left[n_{ij} - \hat{E}(n_{ij})\right]^2}{\hat{E}(n_{ij})} = \frac{(11-9.041)^2}{9.041} + \frac{(11-12.959)^2}{12.959}$$

$$+ \frac{(19-20.959)^2}{20.959} + \frac{(32-30.041)^2}{30.041} = 1.031$$

The rejection region requires $\alpha = 0.10$ in the upper tail of the χ^2 distribution with $df = (r-1)(c-1) = (2-1)(2-1) = 1$. From Table 8, Appendix B, $\chi^2_{0.10} = 2.70554$. The rejection region is $\chi > 2.70554$.

Since the observed value of the test statistic does not fall in the rejection region $(\chi^2 = 1.031 \not> 2.70554)$, H_0 is not rejected. There is insufficient evidence to indicate that seedling abundance and plant type are dependent at $\alpha = 0.10$.

9.29   It is desired to conduct a chi-square test of independence between defect and EVG prediction. The assumption required for the test is that all expected cell counts are at least 5. The expected cell counts for the data are:

$$\hat{E}(n_{ij}) = \frac{n_{ig}n_{gj}}{n}$$

$$\hat{E}(n_{11}) = \frac{449(488)}{498} = 439.98 \qquad \hat{E}(n_{12}) = \frac{449(10)}{498} = 9.02$$

$$\hat{E}(n_{21}) = \frac{49(488)}{498} = 48.02 \qquad \hat{E}(n_{22}) = \frac{49(10)}{498} = 0.98$$

Since the cell counts are not all at least 5, the requirements for the chi-square test of independence fail. The test should not be conducted.

9.31 To determine if one circuit appears to be more difficult to analyze than any other circuit, we test:

H_0: The distributions of observations for the four answers is the same for each circuit

H_a: The distributions of observations for the four answers differ for at least t20 circuits

Using MINITAB, the calculations are:

```
Chi-Square Test for Association: Rows, Worksheet columns

Rows: Rows    Columns: Worksheet columns

                  Circuit 1   Circuit 2   Circuit 3   Circuit 4   All
Both Correct            31          10           5           4    50
                    12.500      12.500      12.500      12.500

Incorrect Volt           0           3          11          12    26
                     6.500       6.500       6.500       6.500

Incorrect graph          5          17          16          14    52
                    13.000      13.000      13.000      13.000

Both Incorrect           4          10           8          10    32
                     8.000       8.000       8.000       8.000

All                     40          40          40          40   160
Cell Contents:       Count
                     Expected count

Pearson Chi-Square = 64.237, DF = 9, P-Value = 0.000
Likelihood Ratio Chi-Square = 66.886, DF = 9, P-Value = 0.000
```

The test statistic is $\chi^2 = 64.237$ and the p-value is $p = 0.000$. Since the p-value is so small, we would reject H_0 for any reasonable value of α. There is sufficient evidence to indicate that at least one circuit appears to be more difficult to analyze than any other circuit.

9.33 a. The row totals for this experiment are fixed since exactly 10 teeth are bonded with each adhesive type. So, the row totals are fixed at the value of 10.

b. To determine if the distributions of ARI scores differ for the two types of bonding adhesives, we test:

H_0: ARI score and adhesive type are independent
H_a: ARI score and adhesive type are dependent

Using MINITAB, the calculations are:

```
Chi-Square Test for Association: Rows, ARI

Rows: Rows    Columns: ARI

                  1       2       3       4   All
Composite         1       5       3       1    10
              1.500   6.500   1.500   0.500

Smartbond         2       8       0       0    10
              1.500   6.500   1.500   0.500

All               3      13       3       1    20
```

```
Cell Contents:        Count
                      Expected count
Pearson Chi-Square = 5.026, DF = 3
Likelihood Ratio Chi-Square = 6.584, DF = 3

* WARNING * 2 cells with expected counts less than 1
* WARNING * Chi-Square approximation probably invalid

* NOTE * 6 cells with expected counts less than 5
```

The test statistic is $\chi^2 = 5.026$ and the p-value is $p = 0.169903$. Since the p-value is not less than $\alpha(p = 0.1699 \not< 0.05)$, H_0 is not rejected. There is insufficient evidence to indicate that the distributions of ARI scores differ for the two types of bonding adhesives at $\alpha = 0.05$.

c. One of the assumptions required for the test to be valid is that all of the expected cell counts must be at least 5. We can see from the table above that most of the expected counts are less than 5. Therefore, this test should be disregarded and any inferences derived from it ignored.

9.35   Some preliminary calculations are:

$$\hat{E}(n_{ij}) = \frac{n_{ig} n_{gj}}{n}$$

$$\hat{E}(n_{11}) = \frac{109(374)}{567} = 71.898 \qquad \hat{E}(n_{12}) = \frac{109(193)}{567} = 37.102$$

$$\hat{E}(n_{21}) = \frac{458(374)}{567} = 302.102 \qquad \hat{E}(n_{22}) = \frac{458(193)}{567} = 155.898$$

To determine if the proportion of patients taking Seldane-D who experience insomnia differs from the corresponding proportion for patients receiving the placebo, we test:

H_0: The proportion of patients experiencing insomnia is the same for those receiving Seldane-D and those receiving the placebo

H_a: The proportion of patients experiencing insomnia is not the same for those receiving Seldane-D and those receiving the placebo

The test statistic is

$$\chi^2 = \sum\sum \frac{\left[n_{ij} - \hat{E}(n_{ij})\right]^2}{\hat{E}(n_{ij})} = \frac{(97 - 71.898)^2}{71.898} + \frac{(12 - 37.102)^2}{37.102}$$

$$+ \frac{(277 - 302.102)^2}{302.102} + \frac{(181 - 155.898)^2}{155.898} = 31.875$$

The rejection region requires $\alpha = 0.10$ in the upper tail of the χ^2 distribution with $df = (r-1)(c-1) = (2-1)(2-1) = 1$. From Table 8, Appendix B, $\chi^2_{0.10} = 2.70554$. The rejection region is $\chi > 2.70554$.

Since the observed value of the test statistic falls in the rejection region ($\chi^2 = 31.875 > 2.70554$), H_0 is rejected. There is sufficient evidence to indicate the proportion of patients taking Seldane-D who experience insomnia differs from the corresponding proportion for patients receiving the placeboat $\alpha = 0.10$.

9.37 a. Fisher's exact test should be used because the sample size requirements for the chi-square test of independence fail.

b. To determine if the distributions of ARI scores differ for the two types of bonding adhesives, we test:

H_0: ARI score and adhesive type are independent
H_a: ARI score and adhesive type are dependent

From the output, the p-value is $p = 0.2616$. Since the p-value is not less than $\alpha (p = 0.2616 \not< 0.05)$, H_0 is not rejected. There is insufficient evidence to indicate that the distributions of ARI scores differ for the two types of bonding adhesives at $\alpha = 0.05$.

9.39 To determine if an NAWIC member's satisfaction with life as an employee and their satisfaction with job challenge are related, we test:

H_0: Life satisfaction and job challenge are independent
H_a: Life satisfaction and job challenge are dependent

Using SAS, the results of Fisher's Exact test are:

Table of JOB_CHALLENGE by LIFE_CHALLENGE

JOB_CHALLENGE	LIFE_CHALLENGE		
Frequency Percent Row Pct Col Pct	SATISFIED	DISSATISFIED	Total
SATISFIED	364	33	397
	81.43	7.38	88.81
	91.69	8.31	
	93.81	55.93	
DISSATISFIED	24	26	50
	5.37	5.82	11.19
	48.00	52.00	
	6.19	44.07	
Total	388	59	447
	86.80	13.20	100.00

Fisher's Exact Test	
Cell (1,1) Frequency (F)	364
Left-sided Pr <= F	1.0000
Right-sided Pr >= F	7.279E-13
Table Probability (P)	6.699E-13
Two-sided Pr <= P	7.279E-13

The p-value is $p = 7.279E-13 \approx 0.0000$. Since the p-value is less than $\alpha (p = 0.000 < 0.05)$, H_0 is rejected. There is sufficient evidence to indicate life satisfaction and job challenge are dependent at $\alpha = 0.05$.

9.41 Some preliminary calculations are:

$$E(n_i) = np_i = 671(0.25) = 167.75$$

To determine if the hourly employees have a preference for one of the work schedules, we test:

$H_0 : p_1 = p_2 = p_3 = p_4 = 0.25$

H_a : At least one of the multinomial probabilities differs from its null hypothesized value

The test statistic is

$$\chi^2 = \sum \frac{\left[n_i - E(n_i) \right]^2}{E(n_i)} = \frac{(389 - 167.75)^2}{167.75} + \frac{(54 - 167.75)^2}{167.75} + \frac{(208 - 167.75)^2}{167.75}$$

$$+ \frac{(20 - 167.75)^2}{167.75} = 508.738$$

The rejection region requires $\alpha = 0.01$ in the upper tail of the χ^2 distribution with $df = k - 1 = 4 - 1 = 3$. From Table 8, Appendix B, $\chi^2_{0.01} = 11.3449$. The rejection region is $\chi^2 > 11.3449$.

Since the observed value of the test statistic falls in the rejection region ($\chi^2 = 508.738 > 11.3449$), H_0 is rejected. There is sufficient evidence to indicate the hourly employees have a preference for one of the work schedules at $\alpha = 0.01$.

9.43   To determine if sex and dose at time of exposure are independent, we test:

H_0: Sex and dose are independent
H_a: Sex and dose are dependent

Using MINITAB, the calculations are:

Chi-Square Test for Association: Rads, Gender

```
Rows: Rads    Columns: Gender
          Male   Female  All

<1             6       13   19
          5.569   13.431

1-10           8       18   26
          7.621   18.379

11+            3       10   13
          3.810    9.190

All           17       41   58
Cell Contents:        Count

                  Expected count
```

```
Pearson Chi-Square = 0.318, DF = 2, P-Value = 0.853
Likelihood Ratio Chi-Square = 0.328, DF = 2, P-Value = 0.849
```

```
* NOTE * 1 cells with expected counts less than 5
```

The test statistic is $\chi^2 = 0.318$ and the p-value is $p = 0.853$. Since the p-value is not less than $\alpha (p = 0.853 \not< 0.01)$, H_0 is not rejected. There is insufficient evidence to indicate that sex and dose at time of exposure are dependent at $\alpha = 0.01$. We note that one of the expected cell counts is less than 5. However, because the p-value is not close to the value of α, the results would be similar if we used Fisher's Exact test.

9.45   Some preliminary calculations are:

$$\hat{E}(n_{ij}) = \frac{n_{ig}n_{gj}}{n}$$

$$\hat{E}(n_{11}) = \frac{194(137)}{363} = 126.661 \qquad \hat{E}(n_{12}) = \frac{194(91)}{363} = 48.63$$

$$\hat{E}(n_{13}) = \frac{194(35)}{363} = 18.705 \qquad \hat{E}(n_{21}) = \frac{169(237)}{363} = 110.339$$

$$\hat{E}(n_{22}) = \frac{169(91)}{363} = 42.366 \qquad \hat{E}(n_{23}) = \frac{169(35)}{363} = 16.295$$

To determine if the percentages of moths caught by the three traps depend on day of the week, we test:

H_0: Percentage of moths caught and day of the week are independent
H_a: Percentage of moths caught and day of the week are dependent

The test statistic is

$$\chi^2 = \sum\sum \frac{\left[n_{ij} - \hat{E}(n_{ij}) \right]^2}{\hat{E}(n_{ij})} = \frac{(136 - 126.661)^2}{126.661} + \frac{(41 - 48.634)^2}{48.634}$$

$$+ \cdots + \frac{(18 - 16.295)^2}{16.295} = 4.387$$

The rejection region requires $\alpha = 0.10$ in the upper tail of the χ^2 distribution with $df = (c-1)(c-1) = (3-1)(2-1) = 2$. From Table 8, Appendix B, $\chi^2_{0.10} = 4.60517$. The rejection region is $\chi > 4.60517$.

Since the observed value of the test statistic does not fall in the rejection region $(\chi^2 = 4.387 \not> 4.60517)$, H_0 is not rejected. There is insufficient evidence to indicate the percentages of moths caught by the three traps depend on day of the week at $\alpha = 0.10$.

9.47   Some preliminary calculations are:

$$E(n_i) = np_i = 100(0.20) = 20$$

To determine if a preference for one or more of the five water management strategies exists, we test:

$H_0 : p_1 = p_2 = p_3 = p_4 = p_5 = 0.20$
$H_a :$ At least one of the multinomial probabilities differs from its null hypothesized value

The test statistic is

$$\chi^2 = \sum \frac{[n_i - E(n_i)]^2}{E(n_i)} = \frac{(17 - 20)^2}{20} + \frac{(27 - 20)^2}{20} + \frac{(22 - 20)^2}{20}$$

$$+ \frac{(15 - 20)^2}{20} + \frac{(19 - 20)^2}{20} = 4.4$$

The rejection region requires $\alpha = 0.05$ in the upper tail of the χ^2 distribution with $df = k - 1 = 5 - 1 = 4$. From Table 8, Appendix B, $\chi^2_{0.05} = 9.48773$. The rejection region is $\chi^2 > 9.48773$.

Since the observed value of the test statistic does not fall in the rejection region $(\chi^2 = 4.4 \not> 9.48773)$, H_0 is not rejected. There is insufficient evidence to indicate if a preference for one or more of the five water management strategies exists at $\alpha = 0.05$.

9.49 a.   Some preliminary calculations are:

$$\hat{E}(n_{ij}) = \frac{n_{ig}n_{gj}}{n}$$

$$\hat{E}(n_{11}) = \frac{57(57)}{200} = 16.245 \qquad \hat{E}(n_{12}) = \frac{57(74)}{200} = 21.09 \qquad \hat{E}(n_{13}) = \frac{57(69)}{200} = 19.665$$

$$\hat{E}(n_{21}) = \frac{143(57)}{200} = 40.755 \qquad \hat{E}(n_{22}) = \frac{143(74)}{200} = 52.91 \qquad \hat{E}(n_{23}) = \frac{143(69)}{200} = 49.335$$

To determine if type of commercial and recall of brand name are dependent, we test:

H_0: Type of commercial and recall of brand name are independent
H_a: Type of commercial and recall of brand name are dependent

The test statistic is

$$\chi^2 = \sum\sum \frac{\left[n_{ij} - \hat{E}(n_{ij})\right]^2}{\hat{E}(n_{ij})} = \frac{(15 - 16.245)^2}{16.245} + \frac{(32 - 21.09)^2}{21.09} + \frac{(10 - 19.665)^2}{21.09}$$

$$+ \frac{(42 - 40.755)^2}{40.755} + \frac{(42 - 52.91)^2}{52.91} + \frac{(59 - 49.335)^2}{49.335} = 14.671$$

The rejection region requires $\alpha = 0.05$ in the upper tail of the χ^2 distribution with $df = (r - 1)(c - 1) = (3 - 1)(2 - 1) = 2$. From Table 8, Appendix B, $\chi^2_{0.05} = 5.99147$. The rejection region is $\chi > 5.99147$.

Since the observed value of the test statistic falls in the rejection region $(\chi^2 = 14.671 > 5.99147)$, H_0 is rejected. There is sufficient evidence to indicate that the type of commercial and recall of brand name are dependent at $\alpha = 0.05$.

b.   Let p_1 = proportion of viewers recalling brand for normal commercials and p_2 = proportion of viewers recalling brand for 24-second time-compressed commercials.

$$\hat{p}_1 = \frac{n_1}{n} = \frac{15}{57} = 0.263 \qquad \hat{p}_2 = \frac{n_2}{n} = \frac{32}{74} = 0.432$$

For confidence coefficient 0.95, $\alpha = 0.05$ and $\alpha/2 = 0.05/2 = 0.025$. From Table 5, Appendix B, $z_{0.025} = 1.96$. The 95% confidence interval is:

$$(\hat{p}_1 - \hat{p}_2) \pm z_{\alpha/2} \sqrt{\frac{\hat{p}_1 \hat{q}_1 + \hat{p}_2 \hat{q}_2}{n}}$$

$$\Rightarrow (0.263 - 0.432) \pm 1.96 \sqrt{\frac{0.263(0.737)}{57} + \frac{0.432(0.568)}{74}}$$

$$\Rightarrow -0.169 \pm 0.161 \Rightarrow (-0.330, -0.008)$$

We are 95% confident that the difference between the proportions recalling brand name for viewers of normal and 24-second time-compressed commercials is between -0.330 and -0.008.

9.51   Some preliminary calculations are:

$E(n_1) = up_i = 151(0.25) = 37.75$

To determine if the proportions of problems are different among the four DSS components, we test:

$H_0 : p_1 = p_2 = p_3 = p_4 = 0.25$
H_a : At least one of the multinomial probabilities differs from its null hypothesized value

The test statistic is

$$\chi^2 = \sum \frac{[n_1 - E(n_i)]^2}{E(n_i)} = \frac{(31 - 37.75)^2}{37.75} + \frac{(28 - 37.75)^2}{37.75} + \frac{(45 - 37.75)^2}{37.75}$$

$$+ \frac{(47 - 37.75)^2}{37.75} = 7.384$$

The rejection region requires $\alpha = 0.05$ in the upper tail of the χ^2 distribution with $df = k - 1 = 4 - 1 = 3$. From Table 8, Appendix B, $\chi^2_{0.05} = 7.81473$. The rejection region is $\chi^2 > 7.81473$.

Since the observed value of the test statistic does not fall in the rejection region $(\chi^2 = 7.384 \not> 7.81473)$, H_0 is not rejected. There is insufficient evidence to indicate the proportions of problems are different among the four DSS components at $\alpha = 0.05$.

9.53   a.   Let p_2 = proportion of gastroenteritis cases who drink 1-2 glasses of water per day.

$$\hat{p}_2 = \frac{n_2}{n} = \frac{11}{40} = 0.275$$

For confidence coefficient 0.99, $\alpha = 0.01$ and $\alpha/2 = 0.01/2 = 0.005$. From Table 5, Appendix B, $z_{0.005} = 2.575$. The 99% confidence interval is:

$$\hat{p}_2 \pm z_{\alpha/2}\sigma_{\hat{p}_2} \Rightarrow \hat{p}_2 \pm 2.575\sqrt{\frac{\hat{p}_2\hat{q}_2}{n}} \Rightarrow 0.275 \pm 2.575\sqrt{\frac{0.275(0.725)}{40}}$$

$$\Rightarrow 0.275 \pm 0.182 \Rightarrow (0.093, 0.457)$$

We are 99% confident that the true proportion of gastroenteritis cases who drink 1-2 glasses of water per day is between 0.093 and 0.457.

   b.   Let p_1 = proportion of gastroenteritis cases who drink 0 glasses of water per day.

$$\hat{p}_1 = \frac{n_1}{n} = \frac{6}{40} = 0.15$$

For confidence coefficient 0.99, $\alpha = 0.01$ and $\alpha/2 = 0.1/2 = 0.005$. From Table 5, Appendix B, $z_{0.005} = 2.575$. The 99% confidence interval is:

$$(\hat{p}_2 - \hat{p}_1) \pm z_{\alpha/2}\sqrt{\frac{\hat{p}_2\hat{q}_2 + \hat{p}_1\hat{q}_1 + 2\hat{p}_1\hat{p}_2}{n}}$$

$$\Rightarrow (0.275 - 0.15) \pm 2.575\sqrt{\frac{0.275(0.725) + 0.15(0.85) + 2(0.15)(0.275)}{40}}$$

$$\Rightarrow 0.125 \pm 0.261 \Rightarrow (-0.136, 0.386)$$

We are 99% confident that the difference between the proportions of gastroenteritis cases who drink 1-2 and 0 glasses of water per day is between -0.136 and 0.386.

   c.   Some preliminary calculations are:

$$E(n_i) = np_i = 40(1/4) = 10$$

To determine if the incidence of gastrointestinal disease during the epidemic is related to water consumption, we test:

$H_0 : p_1 = p_2 = p_3 = p_4 = 0.25$

H_a : At least one of the multinomial probabilities differs from its null hypothesized value

The test statistic is

$$\chi^2 = \sum \frac{[n_i - E(n_i)]^2}{E(n_i)} = \frac{(6-10)^2}{10} + \frac{(11-10)^2}{10} + \frac{(13-10)^2}{10} + \frac{(10-10)^2}{10} = 2.6$$

The rejection region requires $\alpha = 0.01$ in the upper tail of the χ^2 distribution with $df = k - 1 = 4 - 1 = 3$. From Table 8, Appendix B, $\chi^2_{0.01} = 9.34840$. The rejection region is $\chi^2 > 9.34840$.

Since the observed value of the test statistic does not fall in the rejection region ($\chi^2 = 2.6 \not> 9.34840$), H_0 is not rejected. There is insufficient evidence to indicate the incidence of gastrointestinal disease during the epidemic is related to water consumption at $\alpha = 0.01$.

9.55   First, we need to set up a two-way table. We need to find the observed number of manganese nodules for each magnetic age:

Age	Observed
Miocene-recent	$389(0.059) = 23$
Oligocene	$140(0.179) = 25$
Eocene	$214(0.164) = 35$
Paleocene	$84(0.214) = 18$
Lake Cretaceous	$247(0.211) = 52$
Early and Middle Cretaceous	$1120(0.142) = 159$
Jurassic	$99(0.110) = 11$
Total	323

The two-way table is:

Age	Manganese Nodules	No Manganese Nodules	Total
Miocene-recent	23	366	389
Oligocene	25	115	140
Eocene	35	179	214
Paleocene	18	66	84
Lake Cretaceous	52	195	247
Early and Middle Cretaceous	159	691	1120
Jurassic	11	88	99
Total	323	1970	2293

Some preliminary calculations are:

$$\hat{E}(n_{ij}) = \frac{n_{ig}n_{gj}}{n}$$

$$\hat{E}(n_{11}) = \frac{389(323)}{2293} = 54.796 \qquad \hat{E}(n_{12}) = \frac{389(1970)}{2293} = 334.204$$

$$\hat{E}(n_{21}) = \frac{140(323)}{2293} = 19.721 \qquad \hat{E}(n_{22}) = \frac{140(1970)}{2293} = 120.279$$

$$\hat{E}(n_{31}) = \frac{214(323)}{2293} = 30.145 \qquad \hat{E}(n_{32}) = \frac{214(1970)}{2293} = 183.855$$

$$\hat{E}(n_{41}) = \frac{66(323)}{2293} = 11.833 \qquad \hat{E}(n_{42}) = \frac{66(1970)}{2293} = 72.167$$

$$\hat{E}(n_{51}) = \frac{247(323)}{2293} = 34.793 \qquad \hat{E}(n_{52}) = \frac{247(1970)}{2293} = 212.207$$

$$\hat{E}(n_{61}) = \frac{1120(323)}{2293} = 157.767 \qquad \hat{E}(n_{62}) = \frac{1120(1970)}{2293} = 962.233$$

$$\hat{E}(n_{71}) = \frac{99(323)}{2293} = 13.945 \qquad \hat{E}(n_{62}) = \frac{99(1970)}{2293} = 85.055$$

To determine if the probability of finding manganese nodules in the deep-sea Earth's crust is dependent on the magnetic age of the crust, we test:

H_0: Age of crust and manganese nodules are independent
H_a: Age of crust and manganese nodules are dependent

The test statistic is

$$\chi^2 = \sum\sum \frac{\left[n_{ij} - \hat{E}(n_{ij})\right]^2}{\hat{E}(n_{ij})} = \frac{(23 - 54.796)^2}{54.796} + \frac{(366 - 334.204)^2}{334.204} + \frac{(25 - 19.721)^2}{19.721}$$

$$+ \cdots + \frac{(88 - 85.055)^2}{85.055} = 38.411$$

The rejection region requires $\alpha = 0.05$ in the upper tail of the χ^2 distribution with $df = (r-1)(c-1) = (7-1)(2-1) = 6$. From Table 8, Appendix B, $(\chi^2_{0.05} = 12.5916$. The rejection region is $\chi > 12.5916$.

Since the observed value of the test statistic falls in the rejection region $(\chi^2 = 38.411 > 12.5916)$, H_0 is rejected. There is sufficient evidence that the probability of finding manganese nodules in the deep-sea Earth's crust is dependent on the magnetic age of the crust at $\alpha = 0.05$.

10

Simple Linear Regression

10.1   The line passes through the points $(0, 1)$ and $(2, 3)$. Therefore,

$1 = \beta_0 + \beta_1(0) \Rightarrow \beta_0 = 1$ and

$3 = \beta_0 + \beta_1(2)$ Substituting $\beta_0 = 1$ into this equation, we get

$3 = 1 + \beta_1(2) \Rightarrow 2 = \beta_1(2) \Rightarrow \beta_1 = 1$

Thus the line passing through the two points is $y = 1 + x$.

10.3   a.   The y-intercept is 3 and the slope is 2.   b. The y-intercept is 1 and the slope is 1.

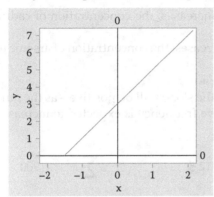
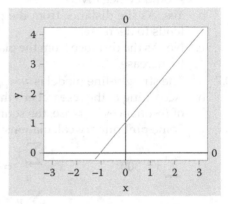

c.   The y-intercept is -2 and the slope is 3.   d. The y-intercept is 0 and the slope is 5.

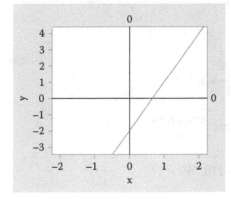
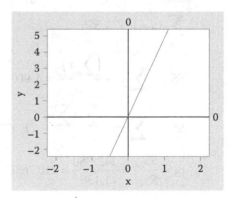

e. The y-intercept is 4 and the slope is -2.

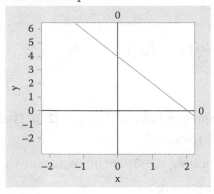

10.5 a. The straight-line model is $y = \beta_0 + \beta_1 x + \varepsilon$. Based on the theory, we would expect the metal level to decrease as the distance increases. Thus, the slope should be negative.

b. Yes. As the distance from the plant increases, the concentration of cadmium tends to decrease.

c. No. As the distance from the plant increases, the concentration of arsenic tends to increase.

10.7 a. The straight-line model is $y = \beta_0 + \beta_1 x + \varepsilon$.

b. According to the researcher's theory, the slope will be positive – as the number of resonances increase, the sound wave frequency is expected to increase.

c. Some preliminary calculations:

$$\sum x = 300 \qquad \bar{x} = \frac{\sum x}{n} = \frac{300}{24} = 12.5 \qquad \sum x^2 = 4,900$$

$$\sum y = 98,494 \qquad \bar{y} = \frac{\sum y}{n} = \frac{98,494}{24} = 4,103.91667 \qquad \sum y^2 = 456,565,950$$

$$\sum xy = 1,473,555$$

$$SS_{xx} = \sum x^2 - \frac{\left(\sum x\right)^2}{n} = 4,900 - \frac{300^2}{24} = 1,150$$

$$SS_{xy} = \sum xy - \frac{\sum x \sum y}{n} = 1,473,555 - \frac{300(98,494)}{24} = 242,380$$

$$\hat{\beta}_1 = \frac{SS_{xy}}{SS_{xx}} = \frac{242,380}{1,150} = 210.7652174 \approx 210.765$$

$$\hat{\beta}_0 = \bar{y} - \hat{\beta}_1 \bar{x} = 4,103.91667 - 210.7652174(12.5) = 1,469.351$$

$\hat{\beta}_0 = 1,469.351$: Since $x = 0$ is not in the observed range, there is no interpretation for $\hat{\beta}_0$. It is simply the y-intercept.

$\hat{\beta}_1 = 210.765$: For each unit increase in resonance, the mean value of frequency is estimated to increase by 210.765.

10.9 a. Some preliminary calculations are:

$$\sum x = 62 \qquad \bar{x} = \frac{\sum x}{n} = \frac{62}{6} = 10.33333 \qquad \sum x^2 = 720.52$$

$$\sum y = 97.8 \qquad \bar{y} = \frac{\sum y}{n} = \frac{97.8}{6} = 16.3 \qquad \sum y^2 = 1,710.2$$

$$\sum xy = 1,087.78$$

$$SS_{xx} = \sum x^2 - \frac{\left(\sum x\right)^2}{n} = 720 - \frac{62^2}{6} = 79.853333$$

$$SS_{xy} = \sum xy - \frac{\sum x \sum y}{n} = 1,087.78 - \frac{62(97.8)}{6} = 77.18$$

$$\hat{\beta}_1 = \frac{SS_{xy}}{SS_{xx}} = \frac{77.18}{79.853333} = 0.966521957 \approx 0.9665$$

$$\hat{\beta}_0 = \bar{y} - \hat{\beta}_1\bar{x} = 16.3 - 0.966521957(10.33333) = 6.3126$$

The least squares line is $\hat{y} = 6.3126 + 0.9665x$.

b. $\hat{\beta}_0 = 6.3126$: Since $x = 0$ is not in the observed range, there is no interpretation for $\hat{\beta}_0$. It is simply the y-intercept.

c. $\hat{\beta}_1 = 0.9665$: For each unit increase in pore diameter, the mean value of porosity is estimated to increase by 0.9665%.

d. When $x = 10$, $\hat{y} = 6.3126 + 0.9665(10) = 15.9776\%$.

10.11 a. The straight-line model is $y = \beta_0 + \beta_1 x + \varepsilon$.

b. Some preliminary calculations are:

$$\sum x = 51.4 \qquad \bar{x} = \frac{\sum x}{n} = \frac{51.4}{15} = 3.426667 \qquad \sum x^2 = 227.5$$

$$\sum y = 45.5 \qquad \bar{y} = \frac{\sum y}{n} = \frac{45.5}{15} = 3.03333 \qquad \sum y^2 = 214.41$$

$$\sum xy = 210.49$$

$$SS_{xx} = \sum x^2 - \frac{\left(\sum x\right)^2}{n} = 227.5 - \frac{51.4^2}{15} = 51.36933$$

$$SS_{xy} = \sum xy - \frac{\sum x \sum y}{n} = 210.49 - \frac{51.4(45.5)}{15} = 54.57667$$

$$\hat{\beta}_1 = \frac{SS_{xy}}{SS_{xx}} = \frac{54.57667}{51.36933} = 1.062436733 \approx 1.0624$$

$$\hat{\beta}_0 = \bar{y} - \hat{\beta}_1\bar{x} = 3.03333 - 1.062436733(3.426667) = -0.6073$$

The least squares line is $\hat{y} = -0.6073 + 1.0624x$.

c. Using MINITAB, the graph is:

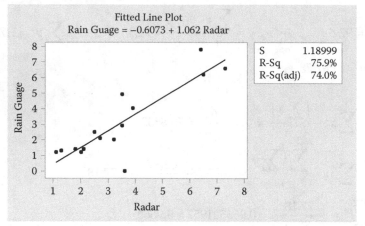

There appears to be a positive linear relationship between the two variables. As the radar rainfall increases, the rain gauge values also increase.

d. $\hat{\beta}_0 = -0.6073$: Since $x = 0$ is not in the observed range, there is no interpretation for $\hat{\beta}_0$. It is simply the y-intercept.

$\hat{\beta}_1 = 1.0624$: For each additional millimeter in radar rainfall, the mean rain gauge rainfall increases by an estimated 1.0624 millimeters.

e. The straight-line model is $y = \beta_0 + \beta_1 x + \varepsilon$.
Some preliminary calculations are:

$$\sum x = 46.7 \qquad \bar{x} = \frac{\sum x}{n} = \frac{46.7}{15} = 3.113333 \qquad \sum x^2 = 210.21$$

$$\sum y = 45.5 \qquad \bar{y} = \frac{\sum y}{n} = \frac{45.5}{15} = 3.03333 \qquad \sum y^2 = 214.41$$

$$\sum xy = 207.89$$

$$SS_{xx} = \sum x^2 - \frac{\left(\sum x\right)^2}{n} = 210.21 - \frac{46.7^2}{15} = 64.817333$$

$$SS_{xy} = \sum xy - \frac{\sum x \sum y}{n} = 207.89 - \frac{46.7(45.5)}{15} = 66.233333$$

$$\hat{\beta}_1 = \frac{SS_{xy}}{SS_{xx}} = \frac{66.233333}{64.812333} = 1.021846008 \approx 1.0218$$

$$\hat{\beta}_0 = \bar{y} - \hat{\beta}_1\bar{x} = 3.03333 - 1.021846008(3.113333) = -0.1480$$

The least squares line is $\hat{y} = -0.1480 + 1.0218x$.
Using MINITAB, the graph is:

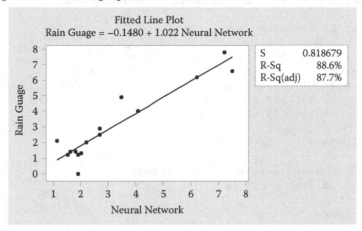

There appears to be a positive linear relationship between the two variables. As the neural rainfall increases, the rain gauge values also increase.

$\hat{\beta}_0 = -0.1480$: Since $x = 0$ is not in the observed range, there is no interpretation for $\hat{\beta}_0$. It is simply the y-intercept.

$\hat{\beta}_1 = 1.0218$: For each additional millimeter in neural rainfall, the mean rain gauge rainfall increases by an estimated 1.0218 millimeters.

10.13   Some preliminary calculations are:

$$\sum x = 25.05 \qquad \bar{x} = \frac{\sum x}{n} = \frac{25.05}{10} = 2.505 \qquad \sum x^2 = 62.7893$$

$$\sum y = 169.1 \qquad \bar{y} = \frac{\sum y}{n} = \frac{169.1}{10} = 16.91 \qquad \sum y^2 = 3,492.49$$

$$\sum xy = 419.613$$

$$SS_{xx} = \sum x^2 - \frac{\left(\sum x\right)^2}{n} = 62.7893 - \frac{25.05^2}{10} = 0.03905$$

$$SS_{xy} = \sum xy - \frac{\sum x \sum y}{n} = 419.613 - \frac{25.05(169.1)}{10} = -3.9825$$

$$\hat{\beta}_1 = \frac{SS_{xy}}{SS_{xx}} = \frac{-3.9825}{0.03905} = -101.9846351 \approx -101.985$$

$$\hat{\beta}_0 = \bar{y} - \hat{\beta}_1 \bar{x} = 16.91 - (-101.9846351)(2.505) = 272.382$$

The least squares line is $\hat{y} = 272.382 - 101.985x$.

For every 1 point increase in a bone tissue's fractal dimension score, the estimated decrease in Young's Modulus is -101.985.

10.15 a. Using MINITAB, the results of the regression analysis are:

Regression Analysis: LnCrackRate versus LnStress

Analysis of Variance

Source	DF	Adj SS	Adj MS	F-Value	P-Value
Regression	1	20.65	20.646	6.90	0.021
LnStress	1	20.65	20.646	6.90	0.021
Error	13	38.88	2.991		
Total	14	59.53			

Model Summary

S	R-sq	R-sq(adj)	R-sq(pred)
1.72943	34.68%	29.66%	12.22%

Coefficients

Term	Coef	SE Coef	T-Value	P-Value	VIF
Constant	-0.145	0.806	-0.18	0.860	
LnStress	1.553	0.591	2.63	0.021	1.00

Regression Equation

LnCrackRate = -0.145 + 1.553 LnStress

The least squares line is $\ln \hat{y} = -0.145 + 1.553 \ln(stress)$

b. As the log of stress intensity increases by 1 unit, we would expect the log of the crack growth rate to increase by 1.553 units.

10.17   Consider the model $E(y) = \mu$.

$$SSE = \sum (y_i - \hat{\mu})^2 \qquad \frac{dSSE}{d\mu} = 2\sum (y_i - \hat{\mu})(-1)$$

Setting this equal to 0 and solving, we get:

$$-2\sum (y_i - \hat{\mu}) = 0 \Rightarrow \sum (y_i - \hat{\mu}) = 0 \Rightarrow \sum y_i - n\hat{\mu} = 0 \Rightarrow \hat{\mu} = \frac{\sum y_i}{n} = \bar{y}$$

10.19   Show $\hat{\beta}_0 = \bar{y} - \hat{\beta}_1 \bar{x}$

Using the hint, $\hat{\beta}_1 = \dfrac{SS_{xy}}{SS_{xx}} = \dfrac{\sum (x_i - \bar{x}) y_i}{SS_{xx}}$

$$\hat{\beta}_0 = \bar{y} - \frac{\sum (x_i - \bar{x}) y_i}{SS_{xx}} \bar{x} = \frac{\sum y_i}{n} - \frac{\sum \bar{x}(x_i - \bar{x}) y_i}{SS_{xx}}$$

$$= \sum \left[\frac{y_i}{n} - \frac{\bar{x}(x_i - \bar{x}) y_i}{SS_{xx}} \right] = \sum \left[\frac{1}{n} - \frac{\bar{x}(x_i - \bar{x})}{SS_{xx}} \right] y_i$$

10.21 Show $V(\hat{\beta}_0) = \dfrac{\sigma^2}{n}\left(\dfrac{\sum x_i^2}{SS_{xx}}\right)$

From Theorem 6.10,

$$V(\hat{\beta}_0) = V\left[\sum\left(\frac{1}{n} - \frac{\bar{x}(x_i - \bar{x})}{SS_{xx}}\right)y_i\right] = \sum\left(\frac{1}{n} - \frac{\bar{x}(x_i - \bar{x})}{SS_{xx}}\right)^2\sigma^2$$

(All y_i's are independent, so all $Cov(y_i, y_j) = 0$)

$$= \sum\left(\frac{1}{n} - \frac{\sum x_i(x_i - \bar{x})}{nSS_{xx}}\right)^2\sigma^2 = \frac{\sigma^2}{n^2}\sum\left(1 - \frac{\sum x_i(x_i - \bar{x})}{SS_{xx}}\right)^2$$

$$= \frac{\sigma^2}{n^2}\sum\left(1 - \frac{\sum x_i(x_i - \bar{x})}{SS_{xx}}\right)\left(1 - \frac{\sum x_i(x_i - \bar{x})}{SS_{xx}}\right)$$

$$= \frac{\sigma^2}{n^2}\sum\left(1 - \frac{2\sum x_i(x_i - \bar{x})}{SS_{xx}} + \frac{\left(\sum x_i\right)^2(x_i - \bar{x})^2}{SS_{xx}^2}\right)$$

$$= \frac{\sigma^2}{n^2}\left(n - \frac{2\sum x_i\sum(x_i - \bar{x})}{SS_{xx}} + \frac{\left(\sum x_i\right)^2\sum(x_i - \bar{x})^2}{SS_{xx}^2}\right)$$

$$= \frac{\sigma^2}{n^2}\left(n - 0 + \frac{\left(\sum x_i\right)^2}{SS_{xx}}\right)$$

because $\sum(x_i - \bar{x}) = 0$ and $\sum(x_i - \bar{x})^2 = SS_{xx}$

$$= \frac{\sigma^2}{n}\left(1 + \frac{\left(\sum x_i\right)^2}{nSS_{xx}}\right) = \frac{\sigma^2}{n}\left(\frac{SS_{xx} + \dfrac{\left(\sum x_i\right)^2}{n}}{SS_{xx}}\right) = \frac{\sigma^2}{n}\left(\frac{\sum x_i^2 - \dfrac{\left(\sum x_i\right)^2}{n} + \dfrac{\left(\sum x_i\right)^2}{n}}{SS_{xx}}\right)$$

$$= \frac{\sigma^2}{n}\left(\frac{\sum x_i^2}{SS_{xx}}\right)$$

10.23 a. From   Exercise   10.7, $\sum y = 98,494$, $\sum y^2 = 456,565,950$, $SS_{xy} = 242,380$,

$$\hat{\beta}_1 = \frac{SS_{xy}}{SS_{xx}} = \frac{242,380}{1,150} = 210.7652174$$

$$SS_{yy} = \sum y^2 - \frac{\left(\sum y\right)^2}{n} = 456,565,950 - \frac{98,494^2}{24} = 52,354,781.83$$

$$SSE = SS_{yy} - \hat{\beta}_1 SS_{xy} = 52,354,781.83 - (210.7652174)(242,380) = 1,269,508.437$$

$$s^2 = \frac{SSE}{n-2} = \frac{1,269,508.437}{24-2} = 57,704.92895 \approx 57,704.929$$

$$s = \sqrt{s^2} = \sqrt{57,704.92895} = 240.2185$$

b. From Exercise 10.8, $\sum y = 106.94$, $\sum y^2 = 891.049$, $SS_{xy} = 2,258.17308$,
$\hat{\beta}_1 = 0.004785591$

$$SS_{yy} = \sum y^2 - \frac{\left(\sum y\right)^2}{n} = 891.049 - \frac{106.94^2}{13} = 11.3441077$$

$$SSE = SS_{yy} - \hat{\beta}_1 SS_{xy} = 11.3441077 - (0.004785591)(2,258.17308) = 0.53741493$$

$$s^2 = \frac{SSE}{n-2} = \frac{0.53741493}{13-2} = 0.048855902 \approx 0.0489$$

$$s = \sqrt{s^2} = \sqrt{0.048855902} = 0.2210$$

c. From Exercise 10.9, $\sum y = 97.8$, $\sum y^2 = 1,710.2$, $SS_{xy} = 77.18$, $\hat{\beta}_1 = 0.966521957$

$$SS_{yy} = \sum y^2 - \frac{\left(\sum y\right)^2}{n} = 1,710.2 - \frac{97.8^2}{6} = 116.06$$

$$SSE = SS_{yy} - \hat{\beta}_1 SS_{xy} = 116.06 - (0.966521957)(77.18) = 41.46383536$$

$$s^2 = \frac{SSE}{n-2} = \frac{41.46383536}{6-2} = 10.36595884 \approx 10.3660$$

$$s = \sqrt{s^2} = \sqrt{10.36595884} = 3.2196$$

d. From Exercise 10.10,

$$\sum x = 526 \quad \sum x^2 = 18,936 \quad \sum y = 60.1 \quad \sum y^2 = 262.2708 \quad \sum xy = 586.95$$

$$SS_{xx} = \sum x^2 - \frac{\left(\sum x\right)^2}{n} = 18,936 - \frac{526^2}{23} = 6,906.608696$$

$$SS_{yy} = \sum y^2 - \frac{\left(\sum y\right)^2}{n} = 262.2708 - \frac{60.1^2}{23} = 105.226887$$

$$SS_{xy} = \sum xy - \frac{\sum x \sum y}{n} = 586.95 - \frac{526(60.1)}{23} = -787.51087$$

$$\hat{\beta}_1 = \frac{SS_{xy}}{SS_{xx}} = \frac{-787.51087}{6,906.608696} = -0.114022801 \approx -0.1140$$

$$SSE = SS_{yy} - \hat{\beta}_1 SS_{xy} = 105.226887 - (-0.114022801)(-787.51087) = 15.43269163$$

$$s^2 = \frac{SSE}{n-2} = \frac{15.43269163}{23-2} = 0.734890077 \approx 0.7349$$

$$s = \sqrt{s^2} = \sqrt{0.734890077} = 0.8573$$

e. From Exercise 10.11b, $\sum y = 45.5$, $\qquad \sum y^2 = 214.41$, $\qquad SS_{xy} = 54.5766667$, $\hat{\beta}_1 = 1.062436733$

$$SS_{yy} = \sum y^2 - \frac{\left(\sum y\right)^2}{n} = 214.41 - \frac{45.5^2}{15} = 76.393333$$

$$SSE = SS_{yy} - \hat{\beta}_1 SS_{xy} = 76.393333 - (1.062436733)(54.5766667) = 18.40907792$$

$$s^2 = \frac{SSE}{n-2} = \frac{18.40907792}{15-2} = 1.416082917 \approx 1.4161$$

$$s = \sqrt{s^2} = \sqrt{1.416082917} = 1.1900$$

From Exercise 10.11e, $\sum y = 45.5$, $\quad \sum y^2 = 214.41$, $SS_{xy} = 66.233333$, $\hat{\beta}_1 = 1.021846008$

$$SS_{yy} = \sum y^2 - \frac{\left(\sum y\right)^2}{n} = 214.41 - \frac{45.5^2}{15} = 76.393333$$

$$SSE = SS_{yy} - \hat{\beta}_1 SS_{xy} = 76.393333 - (1.021846008)(66.2333333) = 8.71306607$$

$$s^2 = \frac{SSE}{n-2} = \frac{8.71306607}{15-2} = 0.670235851 \approx 0.6702$$

$$s = \sqrt{s^2} = \sqrt{0.670235851} = 0.8187$$

f. From Exercise 10.12, $\sum y = 135.8$, $\qquad \sum y^2 = 769.72$, $SS_{xy} = -130.4416667$, $\hat{\beta}_1 = -0.002310626$

$$SS_{yy} = \sum y^2 - \frac{\left(\sum y\right)^2}{n} = 769.72 - \frac{135.8^2}{24} = 1.31833333$$

$$SSE = SS_{yy} - \hat{\beta}_1 SS_{xy} = 1.31833333 - (-0.002310626)(-130.4416667) = 1.016931442$$

$$s^2 = \frac{SSE}{n-2} = \frac{1.016931442}{24-2} = 0.046224156 \approx 0.0462$$

$$s = \sqrt{s^2} = \sqrt{0.046224156} = 0.2150$$

g. From Exercise 10.13, $\sum y = 169.1$, $\sum y^2 = 3,492.49$, $SS_{xy} = -3.9825$, $\hat{\beta}_1 = -101.9846351$

$$SS_{yy} = \sum y^2 - \frac{\left(\sum y\right)^2}{n} = 3,492.49 - \frac{169.1^2}{10} = 633.009$$

$$SSE = SS_{yy} - \hat{\beta}_1 SS_{xy} = 633.009 - (-101.9846351)(-3.9825) = 226.8551908$$

$$s^2 = \frac{SSE}{n-2} = \frac{226.8551908}{10-2} = 28.35689885 \approx 28.3569$$

$$s = \sqrt{s^2} = \sqrt{28.35689885} = 5.3251$$

h. From Exercise 10.14, $\sum y = 114.6$, $\sum y^2 = 575.02$, $SS_{xy} = 9.592$, $\hat{\beta}_1 = 2.426388748$,

$$SS_{yy} = \sum y^2 - \frac{\left(\sum y\right)^2}{n} = 575.02 - \frac{114.6^2}{24} = 27.805$$

$$SSE = SS_{yy} - \hat{\beta}_1 SS_{xy} = 27.805 - (2.426388748)(9.592) = 4.531079126$$

$$s^2 = \frac{SSE}{n-2} = \frac{4.531079126}{24-2} = 0.205958142 \approx 0.20596$$

$$s = \sqrt{s^2} = \sqrt{0.205958142} = 0.4538$$

i. From Exercise 10.15,

$$\sum x = -17.03671272 \qquad \sum x^2 = 27.91071373 \qquad \sum y = -28.62696021$$

$$\sum y^2 = 114.1620057 \qquad \sum xy = 45.80863906$$

$$SS_{xx} = \sum x^2 - \frac{\left(\sum x\right)^2}{n} = 27.91071373 - \frac{(-17.03671272)^2}{15} = 8.560741708$$

$$SS_{yy} = \sum y^2 - \frac{\left(\sum y\right)^2}{n} = 114.1620057 - \frac{(-28.62696021)^2}{15} = 59.5284823$$

$$SS_{xy} = \sum xy - \frac{\sum x \sum y}{n} = 45.80863906$$

$$- \frac{(-17.03671272)(-28.62696021)}{15} = 13.29468591$$

$$\hat{\beta}_1 = \frac{SS_{xy}}{SS_{xx}} = \frac{13.29468591}{8.560741708} = 1.55298295 \approx 1.5530$$

$$SSE = SS_{yy} - \hat{\beta}_1 SS_{xy} = 59.5284823 - (1.55298295)(13.29468591) = 38.88206175$$

$$s^2 = \frac{SSE}{n-2} = \frac{38.88206175}{15-2} = 2.990927827 \approx 2.9909$$

$$s = \sqrt{s^2} = \sqrt{2.990927827} = 1.7294$$

10.25 a. Using MINITAB, the scatterplot of the data is:

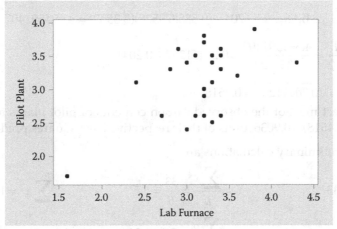

b. Some preliminary calculations are:

$$\sum x = 79.1 \qquad \bar{x} = \frac{\sum x}{n} = \frac{79.1}{25} = 3.183333333 \qquad \sum x^2 = 255.97$$

$$\sum y = 78.2 \qquad \bar{y} = \frac{\sum y}{n} = \frac{78.2}{25} = 3.15 \qquad \sum y^2 = 251.28$$

$$\sum xy = 250.78$$

$$SS_{xx} = \sum x^2 - \frac{\left(\sum x\right)^2}{n} = 255.97 - \frac{79.1^2}{25} = 5.6976$$

$$SS_{xy} = \sum xy - \frac{\sum x \sum y}{n} = 250.78 - \frac{79.1(78.2)}{25} = 3.3552$$

$$\hat{\beta}_1 = \frac{SS_{xy}}{SS_{xx}} = \frac{3.3552}{5.6979} = 0.588879528 \approx 0.5889$$

$$\hat{\beta}_0 = \bar{y} - \hat{\beta}_1 \bar{x} = 3.128 - (0.588879528)(3.2) = 1.2648$$

The least squares line is $\hat{y} = 1.2648 + 0.5889x$

$\hat{\beta}_0 = 1.2648$: Since $x = 0$ is not in the observed range, there is no interpretation for $\hat{\beta}_0$. It is simply the y-intercept.

$\hat{\beta}_1 = 0.5889$: For every one percentage increase in carbon content in a lab furnace, we estimate the mean carbon content in a pilot plant will increase by 0.5889%.

c.   Some additional calculations are:

$$SS_{yy} = \sum y^2 - \frac{\left(\sum y\right)^2}{n} = 251.28 - \frac{(78.2)^2}{25} = 6.6704$$

$$SSE = SS_{yy} - \hat{\beta}_1 SS_{xy} = 6.6704 - (0.588879528)(3.3552) = 4.694591407$$

$$s^2 = \frac{SSE}{n-2} = \frac{4.694591407}{25-2} = 0.20411267 \approx 0.2041$$

d. $s = \sqrt{s^2} = \sqrt{0.20411267} = 0.4518$

We expect most of the observed carbon contents of pilot plants to fall within $2s = 2(0.4518) = 0.9036$ units of their respective least squares predicted values.

10.27 a.   Some preliminary calculations are:

$$\sum x = 33 \qquad \bar{x} = \frac{\sum x}{n} = \frac{33}{6} = 5.5 \qquad \sum x^2 = 199$$

$$\sum y = 3,210.24 \qquad \bar{y} = \frac{\sum y}{n} = \frac{3,210.24}{6} = 535.04 \qquad \sum y^2 = 2,550,024.0014$$

$$\sum xy = 21,368.12$$

$$SS_{xx} = \sum x^2 - \frac{\left(\sum x\right)^2}{n} = 199 - \frac{33^2}{6} = 17.5$$

$$SS_{xy} = \sum xy - \frac{\sum x \sum y}{n} = 21,368.12 - \frac{33(3,210.24)}{6} = 3,711.8$$

$$\hat{\beta}_1 = \frac{SS_{xy}}{SS_{xx}} = \frac{3,711.8}{17.5} = 212.1028571 \approx 212.1029$$

$$\hat{\beta}_0 = \bar{y} - \hat{\beta}_1 \bar{x} = 535.04 - (212.1028571)(5.5) = -631.5257$$

The least squares line is $\hat{y} = -632.5257 + 212.1029x$

b.   Some additional calculations are:

$$SS_{yy} = \sum y^2 - \frac{\left(\sum y\right)^2}{n} = 2,550,024.0014 - \frac{(3,210.24)}{6} = 832,417.1918$$

$$SSE = SS_{yy} - \hat{\beta}_1 SS_{xy} = 832,417.1918 - (212.1028571)(3,711.8) = 45,133.80666$$

$$s^2 = \frac{SSE}{n-2} = \frac{45,133.80666}{6-2} = 11,283.45166 \approx 11,283.452$$

The estimate of σ^2 is $s^2 = 11,283.452$.

$$s = \sqrt{s^2} = \sqrt{11,283.45166} = 106.2236$$

The estimate of σ is $s = 106.2236$.

c. The estimate that can be interpreted is s. We expect most of the VOF values to fall within $2s = 2(106.2236) = 212.4472$ units of their respective least squares predicted values.

10.29 $SSE = \sum (y_i - \hat{y})^2$

$$= \sum \left[y_i - \left(\hat{\beta}_0 + \hat{\beta}_1 x_i \right) \right]^2$$

$$= \sum \left[y_i^2 + \left(\hat{\beta}_0 + \hat{\beta}_1 x_i \right)^2 - 2y_i \left(\hat{\beta}_0 + \hat{\beta}_1 x_i \right) \right]$$

$$= \sum \left[y_i^2 + \hat{\beta}_0^2 + \hat{\beta}_1^2 x_i^2 + 2\hat{\beta}_0 \hat{\beta}_1 x_i - 2y_i \hat{\beta}_0 - 2y_i \hat{\beta}_1 x_i \right]$$

$$= \sum \left[y_i^2 + \left(\bar{y} - \hat{\beta}_1 \bar{x} \right)^2 + \hat{\beta}_1^2 x_i^2 + 2 \left(\bar{y} - \hat{\beta}_1 \bar{x} \right) \hat{\beta}_1 x_i - 2y_i \left(\bar{y} - \hat{\beta}_1 \bar{x} \right) - 2y_i \hat{\beta}_1 x_i \right]$$

$$= \sum \left[y_i^2 + \bar{y}^2 - 2\hat{\beta}_1 \overline{xy} + \hat{\beta}_1^2 \bar{x}^2 + \hat{\beta}_1^2 x_i^2 + 2\hat{\beta}_1 x_i \bar{y} - 2\hat{\beta}_1^2 \bar{x} x_i - 2y_i \bar{y} + 2\hat{\beta}_1 \overline{x} y_i - 2\hat{\beta}_1 y_i x_i \right]$$

$$= \sum \left[\left(y_i^2 - 2y_i \bar{y} + \bar{y}^2 \right) + \hat{\beta}_1 \left(\hat{\beta}_1 \bar{x}^2 - 2\overline{xy} + \hat{\beta}_1 x_i^2 + 2x_i \bar{y} - 2\hat{\beta}_1 \bar{x} x_i + 2\overline{x} y_i - 2y_i x_i \right) \right]$$

$$= \sum (y_i - \bar{y})^2 + \sum \hat{\beta}_1 \left[\hat{\beta}_1 \left(x_i^2 - 2\bar{x} x_i + \bar{x}^2 \right) - 2(x_i - \bar{x})(y_i - \bar{y}) \right]$$

$$= SS_{yy} + \hat{\beta}_1 \left[\sum \hat{\beta}_1 (x_i - \bar{x})^2 - 2\sum (x_i - \bar{x})(y_i - \bar{y}) \right]$$

$$= SS_{yy} + \hat{\beta}_1 \left[\hat{\beta}_1 SS_{xx} - 2SS_{xy} \right]$$

$$= SS_{yy} + \hat{\beta}_1 \left[\frac{SS_{xy}}{SS_{xx}} SS_{xx} - 2SS_{xy} \right]$$

$$= SS_{yy} + \hat{\beta}_1 \left[SS_{xy} - 2SS_{xy} \right]$$

$$= SS_{yy} - \hat{\beta}_1 SS_{xy}$$

10.31 From Exercise 10.8 and Exercise 10.23, $\hat{\beta}_1 = 0.00479$, $s = 0.2210$, and $SS_{xx} = 471,869.2308$.

To determine if the cost ratio increases linearly with pipe diameter, we test:

$$H_0 : \beta_1 = 0$$

$$H_a : \beta_1 > 0$$

The test statistic is $t = \dfrac{\hat{\beta}_1}{\dfrac{s}{\sqrt{SS_{xx}}}} = \dfrac{0.00479}{\dfrac{0.2210}{\sqrt{471,869.2308}}} = 14.89.$

The rejection region requires $\alpha = 0.05$ in the upper tail of the t distribution with $df = n - 2 = 13 - 2 = 11$. From Table 7, Appendix B, $t_{0.05} = 1.796$. The rejection region is $t > 1.796$.

Since the observed value of the test statistic falls in the rejection region $(t = 14.89 > 1.796)$, H_0 is rejected. There is sufficient evidence to indicate that the cost ratio increases linearly with pipe diameter at $\alpha = 0.05$.

For confidence coefficient 0.95, $\alpha = 0.05$ and $\alpha / 2 = 0.05 / 2 = 0.025$. From Table 7, Appendix B, with $df = n - 2 = 13 - 2 = 11$, $t_{0.025} = 2.201$. The 95% confidence interval is:

$$\hat{\beta}_1 \pm t_{\alpha/2} \frac{s}{\sqrt{SS_{xx}}} \Rightarrow 0.00479 \pm 2.201 \frac{0.2210}{\sqrt{471,869.2308}}$$

$$\Rightarrow 0.00479 \pm 0.00071 \Rightarrow (0.00408, 0.005498)$$

10.33 a. From Exercise 10.10 and Exercise 10.23, $\hat{\beta}_1 = -0.1140$, $s = 0.8573$, and $SS_{xx} = 6,906.608696$.

For confidence coefficient 0.90, $\alpha = 0.10$ and $\alpha / 2 = 0.10 / 2 = 0.05$. From Table 7, Appendix B, with $df = n - 2 = 23 - 2 = 21$, $t_{0.05} = 1.721$. The 90% confidence interval is:

$$\hat{\beta}_1 \pm t_{\alpha/2} \frac{s}{\sqrt{SS_{xx}}} \Rightarrow -0.1140 \pm 1.721 \frac{0.8573}{\sqrt{6,906.608696}} \Rightarrow -0.1140 \pm 0.0178$$

$$\Rightarrow (-0.1318, -0.0962)$$

For every one minute increase in elapsed time of spill, we are 95% confident that the mass of the spill will decrease between 0.0962 and 0.1318 pounds.

b. To determine if the true slope of the line differs from 0, we test:

$$H_0 : \beta_1 = 0$$

$$H_a : \beta_1 \neq 0$$

The test statistic is $t = \dfrac{\hat{\beta}_1}{\dfrac{s}{\sqrt{SS_{xx}}}} = \dfrac{-0.1140}{\dfrac{0.8573}{\sqrt{6,906.608696}}} = -11.05.$

The rejection region requires $\alpha / 2 = 0.10 / 2 = 0.05$ in each tail of the t distribution with $df = n - 2 = 23 - 2 = 21$. From Table 7, Appendix B, $t_{0.05} = 1.721$. The rejection region is $t < -1.721$ or $t > 1.721$.

Since the observed value of the test statistic falls in the rejection region $(t = -11.05 < -1.721)$, H_0 is rejected. There is sufficient evidence to indicate that mass and elapsed time of a spill are linearly related at $\alpha = 0.10$.

c. Both the confidence interval and the test of hypothesis conclude that the slope of the regression line is negative. Thus, elapsed time is a useful predictor of mass of the spill.

10.35 From Exercise 10.12 and Exercise 10.23, $\hat{\beta}_1 = -0.0023$, $s = 0.2150$, and $SS_{xx} = 56,452.95833$.

For confidence coefficient 0.90, $\alpha = 0.10$ and $\alpha / 2 = 0.10 / 2 = 0.05$. From Table 7, Appendix B, with $df = n - 2 = 24 - 2 = 22$, $t_{0.05} = 1.717$. The 90% confidence interval is:

$$\hat{\beta}_1 \pm t_{\alpha/2} \frac{s}{\sqrt{SS_{xx}}} \Rightarrow -0.0023 \pm 1.717 \frac{0.2150}{\sqrt{56,452.95833}} \Rightarrow -0.0023 \pm 0.0016$$
$$\Rightarrow (-0.0039, -0.0007)$$

We are 90% confident that the change in the mean sweetness index for each unit change in the pectin is between -0.0039 and -0.0007.

10.37 a. Using MINITAB, the scatterplot is:

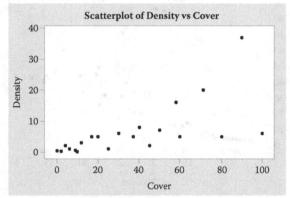

The linear relationship appears to be positive. As the values of cover increase, the values of density tend to increase.

b. Using MINITAB, the regression analysis is:

Regression Analysis: Density versus Cover

Analysis of Variance

Source	DF	Adj SS	Adj MS	F-Value	P-Value
Regression	1	635.1	635.10	14.18	0.001
Cover	1	635.1	635.10	14.18	0.001
Error	19	850.8	44.78		
Total	20	1485.9			

Model Summary

S	R-sq	R-sq(adj)	R-sq(pred)
6.69173	42.74%	39.73%	13.44%

Coefficients

Term	Coef	SE Coef	T-Value	P-Value	VIF
Constant	-0.30	2.31	-0.13	0.898	
Cover	0.1845	0.0490	3.77	0.001	1.00

Regression Equation

Density = -0.30 + 0.1845 Cover

The fitted regression line is $\hat{y} = -0.30 + 0.1845x$.

c.   To determine if bird density increases linearly as percent vegetation coverage
     increases, we test:

$$H_0 : \beta_1 = 0$$

$$H_a : \beta_1 > 0$$

The test statistic is $t = 3.77$ and the p-value is $p = 0.001$.
Since the p-value is less than $\alpha (p = 0.001 < 0.01)$, H_0 is rejected. There is sufficient evidence to indicate bird density increases linearly as percent vegetation
coverage increases at $\alpha = 0.01$.

10.39 a.   Using MINITAB, the scattergram is:

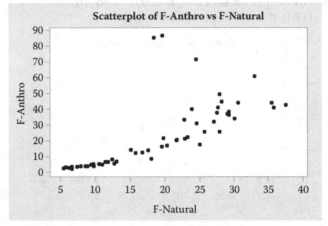

The data appear to have a somewhat linear relationship, thus supporting the
theory.

b.   The three largest anthropogenic indices were removed and the resulting scat-
     tergram is:

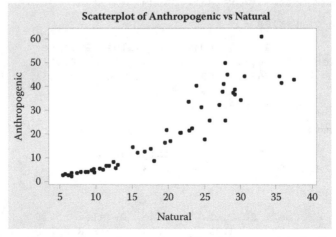

After removing the three observations, the data appear to be linear.

c. Some preliminary calculations are:

$$\sum x = 984.76 \qquad \bar{x} = \frac{\sum x}{n} = \frac{984.76}{51} = 19.30901961 \qquad \sum x^2 = 23,286.2632$$

$$\sum y = 1,039.51 \qquad \bar{y} = \frac{\sum y}{n} = \frac{1,039.41}{51} = 20.38254902 \qquad \sum y^2 = 34,206.1489$$

$$\sum xy = 27,021.6658$$

$$SS_{xx} = \sum x^2 - \frac{\left(\sum x\right)^2}{n} = 23,286.2632 - \frac{984.76^2}{51} = 4,271.51305$$

$$SS_{yy} = \sum y^2 - \frac{\left(\sum y\right)^2}{n} = 34,206.14890 - \frac{1,039.51^2}{51} = 13,018.28537$$

$$SS_{xy} = \sum xy - \frac{\sum x \sum y}{n} = 27,021.6658 - \frac{984.76(1,039.51)}{51} = 6,949.74682$$

$$\hat{\beta}_1 = \frac{SS_{xy}}{SS_{xx}} = \frac{6,949.74682}{4,271.51305} = 1.62699885 \approx 1.6270$$

$$\hat{\beta}_0 = \bar{y} - \hat{\beta}_1 \bar{x} = 20.38254902 - (1.62699885)(19.30901961) = -11.0332$$

$$SSE = SS_{yy} - \hat{\beta}_1 SS_{xy} = 13,018.28537 - (1.62699885)(6,949.74682) = 1,711.055286$$

$$s^2 = \frac{SSE}{n-2} = \frac{1,711.055286}{51-2} = 34.91949563 \approx 34.9195$$

$$s = \sqrt{s^2} = \sqrt{34.91949563} = 5.9093$$

The least squares line is $\hat{y} = -11.0332 + 1.6270x$

d. $\hat{\beta}_0 = -11.0332$: Since $x = 0$ is not in the observed range, there is no interpretation for $\hat{\beta}_0$. It is simply the y-intercept.

$\hat{\beta}_1 = 1.6270$: For each unit increase in natural origin index, the mean anthropogenic index is estimated to increase by 1.6270 units.

e. To determine if the natural origin index and anthropogenic index are positively linearly related, we test:

$$H_0 : \beta_1 = 0$$

$$H_a : \beta_1 > 0$$

The test statistic is $t = \dfrac{\hat{\beta}_1}{\dfrac{s}{\sqrt{SS_{xx}}}} = \dfrac{1.6270}{\dfrac{5.9093}{\sqrt{4,271.51305}}} = 17.99.$

The rejection region requires $\alpha = 0.05$ in the upper tail of the t distribution with $df = n - 2 = 51 - 2 = 49$. Using MINITAB, $t_{0.05} = 1.677$. The rejection region is $t > 1.677$.

Since the observed value of the test statistic falls in the rejection region $(t = 17.99 > 1.677)$, H_0 is rejected. There is sufficient evidence to indicate that the natural origin index and anthropogenic index are positively linearly related at $\alpha = 0.05$.

f. For confidence coefficient 0.95, $\alpha = 0.05$ and $\alpha/2 = 0.05/2 = 0.025$. Using MINITAB, with $df = n - 2 = 51 - 2 = 49$, $t_{0.025} = 2.010$. The 95% confidence interval is:

$$\hat{\beta}_1 \pm t_{\alpha/2} \frac{s}{\sqrt{SS_{xx}}} \Rightarrow 1.6270 \pm 2.010 \frac{5.9093}{\sqrt{4,271.51305}}$$

$$\Rightarrow 1.6270 \pm 0.1817 \Rightarrow (1.4453, 1.8087)$$

We are 95% confident that the change in the mean anthropogenic index for each unit change in the natural origin index is between 1.4453 and 1.8087 units.

10.41   We know that $T = \dfrac{Z}{\sqrt{\dfrac{\chi^2}{df}}}$ has a t distribution with df degrees of freedom.

From Exercise 10.40, $Z = \dfrac{\hat{\beta}_1 - \beta_1}{\sigma/\sqrt{SS_{xx}}}$ has a z distribution.

From Theorem 10.1, $\chi^2 = \dfrac{(n-2)s^2}{\sigma^2}$ has a chi-square distribution with $n - 2$ degrees of freedom.

Then $T = \dfrac{Z}{\sqrt{\dfrac{\chi^2}{df}}} = \dfrac{\dfrac{\hat{\beta}_1 - \beta_1}{\sigma/\sqrt{SS_{xx}}}}{\sqrt{\dfrac{(n-2)s^2}{\sigma^2}}} = \dfrac{\dfrac{\hat{\beta}_1 - \beta_1}{\sigma/\sqrt{SS_{xx}}}}{\dfrac{s}{\sigma}} = \dfrac{\hat{\beta}_1 - \beta_1}{s/\sqrt{SS_{xx}}}$ has a t distribution with

$df = n - 2$.

10.43 a. $r = 0.84$. Since the value is fairly close to 1, there is a moderately strong positive linear relationship between the magnitude of a QSO and the redshift level.

b.   The relationship between r and the estimated slope of the line is that they will both have the same sign. If r is positive, the slope of the line is positive. If r is negative, the slope of the line is negative.

c. $r^2 = 0.84^2 = 0.7056$. 70.56% of the total sample variability around the sample mean magnitude of a QSO is explained by the linear relationship between magnitude of a QSO and the redshift level.

10.45 a.   To determine if the true population correlation coefficient relating NRMSE and bias is positive, we test:

$$H_0 : \rho = 0$$

$$H_a : \rho > 0$$

The test statistic is $t = \dfrac{r\sqrt{n-2}}{\sqrt{1-r^2}} = \dfrac{0.2838\sqrt{3600 - 2}}{\sqrt{1 - 0.2838^2}} = 17.75.$

Since no α was given, we will use $\alpha = 0.05$. The rejection region requires $\alpha = 0.05$ in the upper tail of the t distribution with $df = n - 2 = 3600 - 2 = 3598$. Using MINITAB, $t_{0.05} = 1.645$. The rejection region is $t > 1.645$.

Since the observed value of the test statistic falls in the rejection region $(t = 17.75 > 1.645)$, H_0 is rejected. There is sufficient evidence to indicate the true population correlation coefficient relating NRMSE and bias is positive at $\alpha = 0.05$.

b. No. Even though there is a statistically significant positive relationship between NRMSE scores and bias, the relationship is very weak. With $n = 3600$, the results are almost guaranteed to be statistically significant. However, $r^2 = 0.2838^2 = 0.0805$. Very little of the variation in the bias values is explained by the linear relationship between bias and NRMSE values.

10.47 a. The simple linear regression model is $y = \beta_0 + \beta_1 x + \varepsilon$.

b. $r^2 = 0.92$. 92% of the sum of squares of deviation of the sample metal uptake about their meancan be explained by using the final concentration of metal in the solution as a linear predictor.

10.49 a. We estimate the mean water content to be 0.088 grams per cubic centimeter when the count ratio is equal to 0. (Note: This interpretation is only meaningful if the value of 0 is in the observed range of count ratio values.)

b. For every increase of one unit is the count ratio, we estimate that the mean water content will increase by 0.136 grams per cubic centimeter.

c. Because the p-value is so small $(p = 0.0001)$, H_0 would be rejected for any reasonable value of α. There is sufficient evidence to indicate a linear relationship between water content and count ratio.

d. $r^2 = 0.84$. 84% of the variation in the sample water content values around their mean can be explained by the linear relationship with count ratio.

10.51
$$\hat{\beta}_1 = \frac{SS_{xy}}{SS_{xx}} = \frac{SS_{xy}}{\sqrt{SS_{xx}SS_{yy}}} \cdot \frac{\sqrt{SS_{yy}}}{\sqrt{SS_{xx}}} = r\sqrt{\frac{SS_{yy}}{SS_{xx}}}$$

$$SSE = SS_{yy} - \hat{\beta}_1 SS_{xy} = SS_{yy} - \frac{SS_{xy}}{SS_{xx}} SS_{xy} = SS_{yy} - SS_{yy}\left[\frac{SS_{xy}^2}{SS_{xx}SS_{yy}}\right]$$

$$= SS_{yy} - SS_{yy}\left[r^2\right] = SS_{yy}\left[1 - r^2\right]$$

10.53 a. To determine if the amount of nitrogen removed is linearly related to the amount of ammonium used, we test:

$H_0 : \beta_1 = 0$

$H_a : \beta_1 \neq 0$

The test statistic is $t = 32.80$ and the p-value is $p < 0.0001$. Since the p-value is so small, H_0 is rejected. There is sufficient evidence to indicate the amount of nitrogen removed is linearly related to the amount of ammonium used for any reasonable value of α.

$r^2 = 0.9011$. 90.11% of the variation in the amounts of nitrogen removed around their mean can be explained by the linear relationship with the amounts of ammonium used.

We would recommend using the model for predicting nitrogen amount.

b.   The 95% prediction interval when the amount of ammonium used is 100 milligrams per liter is (41.8558, 77.8634). We are 95% confident that the actual amount of nitrogen removed when 100 milligrams of ammonium is used is between 41.8558 and 77.8634 milligrams per liter.

c.   The confidence interval for the mean nitrogen amount when the amount of ammonium used is 100 milligrams per liter will be narrower than the 95% prediction interval. The prediction interval for the actual value takes into account the variation in locating the mean and the variation in the amounts of nitrogen once the mean has been located. The confidence interval for the mean only takes into account the variation in locating the mean.

10.55 a.   From Exercise 10.9 and Exercise 10.23, $\bar{x} = 10.33333$, $SS_{xx} = 79.853333$,
$SS_{xy} = 77.18$, $\hat{\beta}_1 = 0.9665$, $\hat{\beta}_0 = 6.3126$, $s = 3.2196$

The least squares line is $\hat{y} = 6.3126 + 0.9665x$.

When $x = 10$, $\hat{y} = 6.3126 + 0.9665(10) = 6.3126 + 9.665 = 15.9776$.

For confidence coefficient 0.95, $\alpha = 0.05$ and $\alpha/2 = 0.05/2 = 0.025$. From Table 7, Appendix B, with $df = n - 2 = 6 - 2 = 4$, $t_{0.025} = 2.776$. The 95% confidence interval is:

$$\hat{y} \pm t_{\alpha/2}s\sqrt{1 + \frac{1}{n} + \frac{(x - \bar{x})^2}{SS_{xx}}} \Rightarrow 15.9776 \pm 2.776(3.2196)\sqrt{1 + \frac{1}{6} + \frac{(10 - 10.33333)^2}{79.853333}}$$

$$\Rightarrow 15.9776 \pm 9.6595 \Rightarrow (6.3181, 25.6371)$$

We are 95% confident that the apparent porosity of a brick will fall between 6.3181% and 25.6371% when the mean core diameter of the brick is 10 micrometers.

b.   A 95% confidence interval for the mean porosity percentage when the mean core diameter is 10 will be narrower than the prediction interval. The prediction interval for the actual value takes into account the variation in locating the mean and the variation in the amounts of nitrogen once the mean has been located. The confidence interval for the mean only takes into account the variation in locating the mean.

10.57   From Exercise 10.11 and Exercise 10.23, $\bar{x} = 3.113333$, $SS_{xx} = 64.817333$, $SS_{xy} = 66.233333$,

$\hat{\beta}_1 = 1.0218$, $\hat{\beta}_0 = -0.1480$, $s = 0.8187$.

The least squares line is $\hat{y} = -0.1480 + 1.0218x$.

For $x = 3$, $\hat{y} = -0.1480 + 1.0218(3) = -0.1480 + 3.0654 = 2.9174$

For confidence coefficient 0.99, $\alpha = 0.01$ and $\alpha/2 = 0.01/2 = 0.005$. From Table 7, Appendix B, with $df = n - 2 = 15 - 2 = 13$, $t_{0.005} = 3.012$. The 99% confidence interval is:

$$\hat{y} \pm t_{\alpha/2}s\sqrt{1 + \frac{1}{n} + \frac{(x-\bar{x})^2}{SS_{xx}}} \Rightarrow 2.9174 \pm 3.012(0.8187)\sqrt{1 + \frac{1}{15} + \frac{(3-3.113333)^2}{64.817333}}$$

$$\Rightarrow 2.9174 \pm 2.5470 \Rightarrow (0.3704, 5.4644)$$

We are 99% confident that the rain gauge amount will fall between 0.3704 and 5.4644 millimeters when the neural network estimate is 3 millimeters.

10.59 From Exercise 10.14 and Exercise 10.23, $\bar{x} = 1.88$, $SS_{xx} = 3.9532$, $SS_{xy} = 9.592$, $\hat{\beta}_1 = 2.4264$, $\hat{\beta}_0 = 0.2134$, $s = 0.4538$

The least squares line is $\hat{y} = 0.2134 + 2.4264x$.

For $x = 1.95$, $\hat{y} = 0.2134 + 2.4264(1.95) = 0.2134 + 4.73148 = 4.94488$

For confidence coefficient 0.90, $\alpha = 0.10$ and $\alpha/2 = 0.10/2 = 0.05$. From Table 7, Appendix B, with $df = n - 2 = 24 - 2 = 22$, $t_{0.05} = 1.717$. The 90% confidence interval is:

$$\hat{y} \pm t_{\alpha/2}s\sqrt{\frac{1}{n} + \frac{(x-\bar{x})^2}{SS_{xx}}} \Rightarrow 4.94488 \pm 1.717(0.4538)\sqrt{\frac{1}{24} + \frac{(1.95-1.88)^2}{3.9532}}$$

$$\Rightarrow 4.94488 \pm 0.16140 \Rightarrow (4.7835, 5.1063)$$

We are 90% confident that the mean heat transfer coefficient will fall between 4.7835 and 5.1063 for all fin-tubes that have an unflooded area ratio of 1.95.

10.61 a. Using MINITAB, the results are:

Regression Analysis: Life-A versus Cutting Speed

Analysis of Variance

Source	DF	Adj SS	Adj MS	F-Value	P-Value
Regression	1	15.8413	15.8413	10.81	0.006
Cutting Speed	1	15.8413	15.8413	10.81	0.006
Error	13	19.0560	1.4658		
Lack-of-Fit	3	0.3360	0.1120	0.06	0.980
Pure Error	10	18.7200	1.8720		
Total	14	34.8973			

Model Summary

S	R-sq	R-sq(adj)	R-sq(pred)
1.21072	45.39%	41.19%	29.49%

Coefficients

Term	Coef	SE Coef	T-Value	P-Value	VIF
Constant	6.62	1.15	5.76	0.000	
Cutting Speed	-0.0727	0.0221	-3.29	0.006	1.00

```
Regression Equation

Life-A = 6.62 - 0.0727 Cutting Speed
```

The least squares line is $\hat{y} = 6.62 - 0.0727x$.

b. Using MINITAB, the results are:

Regression Analysis: Life-B versus Cutting Speed

```
Analysis of Variance

Source          DF    Adj SS   Adj MS  F-Value  P-Value
Regression       1   34.7763  34.7763    93.54    0.000
  Cutting Speed  1   34.7763  34.7763    93.54    0.000
Error           13    4.8330   0.3718
  Lack-of-Fit    3    0.6863   0.2288     0.55    0.658
  Pure Error    10    4.1467   0.4147
Total           14   39.6093
```

```
Model Summary

       S    R-sq  R-sq(adj)  R-sq(pred)
0.609729  87.80%     86.86%      83.30%
```

```
Coefficients

Term            Coef  SE Coef  T-Value  P-Value   VIF
Constant       9.310    0.578    16.10    0.000
Cutting Speed -0.1077   0.0111    -9.67    0.000  1.00
```

```
Regression Equation

Life-B = 9.310 - 0.1077 Cutting Speed
```

The least squares line is $\hat{y} = 9.310 - 0.1077x$.

c. Using MINITAB for Brand A, the results are:

Prediction for Life-A
```
Regression Equation

Life-A = 6.62 - 0.0727 Cutting Speed

Variable        Setting
Cutting Speed      45

 Fit    SE Fit        90% CI                90% PI
3.35  0.331570  (2.76281, 3.93719)  (1.12694, 5.57306)
```

The 90% confidence interval for the mean useful life of brand A cutting tool when the cutting speed is 45 meters per minute is (2.76281, 3.93719).

Using MINITAB for Brand B, the results are:

Prediction for Life-B

Regression Equation

Life-B = 9.310 - 0.1077 Cutting Speed

Variable   Setting
Cutting Speed   45

   Fit   SE Fit   90% CI   90% PI
  4.465 0.166981 (4.16929, 4.76071) (3.34545, 5.58455)

The 90% confidence interval for the mean useful life of brand B cutting tool when the cutting speed is 45 meters per minute is (4.16929, 4.76071).

The width of the confidence interval for brand B is narrower.

d. From part c, the 90% prediction interval for the useful life of brand A cutting tool when the cutting speed is 45 meters per minute is (1.12694, 5.57306).

From part c, the 90% prediction interval for the useful life of brand B cutting tool when the cutting speed is 45 meters per minute is (3.34545, 5.58455).

The width of the prediction interval for brand B is narrower. For both brands, the prediction intervals are wider than the confidence intervals in part c.

e. Using MINITAB for Brand A, the results are:

Prediction for Life-A

Regression Equation

Life-A = 6.62 - 0.0727 Cutting Speed

Variable   Setting
Cutting Speed   100

      Fit   SE Fit   95% CI   95% PI
 -0.646667 1.14859 (-3.12805, 1.83471) (-4.25203, 2.95870)   XX

XX denotes an extremely unusual point relative to predictor
levels used to fit the model.

The 95% prediction interval for the useful life of brand A cutting tool when the cutting speed is 100 meters per minute is (−4.25203, 2.95870).

We would have to assume that the relationship between y and x when x is between 30 and 70 meters per minute remains the same when x is increased to 100 meters per minute.

10.63   In Exercise 10.62, it was determined that $y_p - \hat{y}$ is normally distributed with an expected value or mean of 0 and a standard deviation of $\sigma\sqrt{1 + \dfrac{1}{n} + \dfrac{(x_p - \bar{x})^2}{SS_{xx}}}$.

If we take $y_p - \hat{y}$, subtract its mean and divide by its standard deviation, we will form a normally distribution random variable with a mean of 0 and a standard deviation of 1.

10.65 a. There is a U-shaped pattern to this plot. This indicates that the model has been misspecified.

b. The residuals are increasing as the predicted values are increasing. This indicates that the error variances are not constant.

c. The residuals form a football shape. This indicates that the error variances are not constant.

d. The histograms is not mound-shaped. This indicates that the error terms are not normally distributed.

10.67 a. Using MINITAB, the residuals are:

Resonance	Frequency	FITS	RESI
1	979	1680.12	-701.117
2	1572	1890.88	-318.882
3	2113	2101.65	11.353
4	2122	2312.41	-190.412
5	2659	2523.18	135.822
6	2795	2733.94	61.057
7	3181	2944.71	236.292
8	3431	3155.47	275.527
9	3638	3366.24	271.762
10	3694	3577.00	116.996
11	4038	3787.77	250.231
12	4203	3998.53	204.466
13	4334	4209.30	124.701
14	4631	4420.06	210.936
15	4711	4630.83	80.170
16	4993	4841.59	151.405
17	5130	5052.36	77.640
18	5210	5263.13	-53.125
19	5214	5473.89	-259.891
20	5633	5684.66	-51.656
21	5779	5895.42	-116.421
22	5836	6106.19	-270.186
23	6259	6316.95	-57.951
24	6339	6527.72	-188.717

b. Using MINITAB, the plot of the residuals against the resonances is:

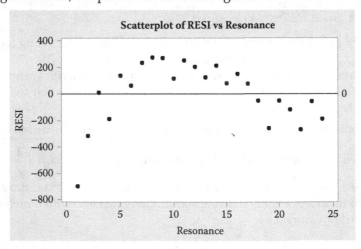

There appears to be an upside down U-shape.

c. It appears that the model has been misspecified.

d. We would suggest that the model should include a term for curvature. We would suggest fitting the model $y = \beta_0 + \beta_1 x + \beta_2 x^2 + \varepsilon$.

10.69 Using MINITAB, the plot of the residuals against the Area ratio values is:

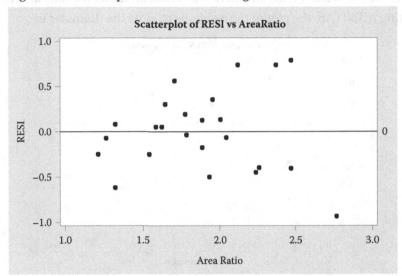

There is no U-shape or football shape. There is no indication that the model has been misspecified.

Additional plots are:

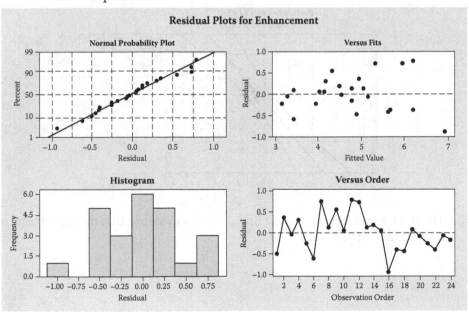

The plot of the residuals against the fitted values does not show a funnel shape or a football shape. There is no indication that the assumption of equal variances is violated. The normal probability plot is almost a straight line and the histogram of the residuals is mound-shaped. There is no indication that the data are not normally distributed. Finally, the plot of the residuals against time does not show a trend. There is no indication that the data are correlated.

10.71   Using MINITAB, the plot of the residuals against the diameter is:

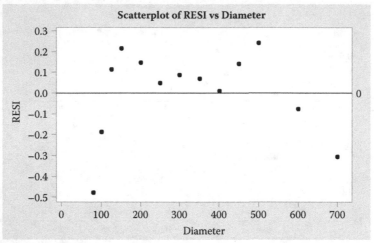

There appears to be an upside down shape to the graph. This indicates that the model has been misspecified.

The normal probability plot is:

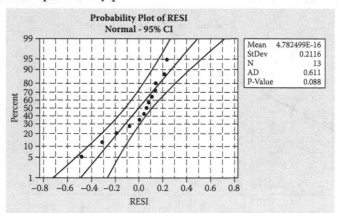

The data do not form a straight line. There is evidence that the error terms are no normally distributed.

10.73 a. Using MINITAB, the scatterplot of the data is:

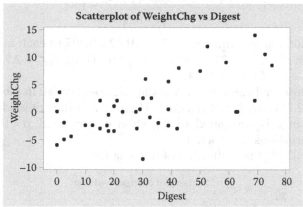

Yes, there appears to be a positive linear trend. As digestion efficiency(%) increases, weight change (%) tends to increase.

b. Using MINITAB, the results are:

Correlation: WeightChg, Digest

```
Pearson correlation of WeightChg and Digest = 0.612
P-Value = 0.000
```

Thus, $r = 0.612$. Since the value is near 0.5, there is a moderate positive linear relationship between weight change (%) and digestions efficiency (%).

c. To determine if weight change is correlated to digestion efficiency, we test:

$$H_0 : \rho = 0$$

$$H_a : \rho \neq 0$$

The test statistic is $t = \dfrac{r\sqrt{n-2}}{\sqrt{1-r^2}} = \dfrac{0.612\sqrt{42-2}}{\sqrt{1-0.612^2}} = 4.89.$

The rejection region requires $\alpha / 2 = 0.01 / 2 = 0.005$ in each tail of the t distribution with $df = n - 2 = 42 - 2 = 40$. Using MINITAB, $t_{0.005} = 2.704$. The rejection region is $t < -2.704$ or $t > 2.704$.

Since the observed value of the test statistic falls in the rejection region $(t = 4.89 > 2.704)$, H_0 is rejected. There is sufficient evidence to indicate weight change is correlated to digestion efficiency at $\alpha = 0.01$.

d. Using MINITAB and excluding observations for duck chow, the results are:

Correlation: WeightChg2, Digest2

```
Pearson correlation of WeightChg2 and Digest2 = 0.309
P-Value = 0.080
```

Thus, $r = 0.309$. Since the value is near 0, there is a weak positive linear relationship between weight change (%) and digestions efficiency (%).

To determine if weight change is correlated to digestion efficiency, we test:

$$H_0 : \rho = 0$$

$$H_a : \rho \neq 0$$

The test statistic is $t = \dfrac{r\sqrt{n-2}}{\sqrt{1-r^2}} = \dfrac{0.309\sqrt{33-2}}{\sqrt{1-0.309^2}} = 1.81$.

The rejection region requires $\alpha/2 = 0.01/2 = 0.005$ in each tail of the t distribution with $df = n - 2 = 33 - 2 = 31$. Using MINITAB, $t_{0.005} = 2.744$. The rejection region is $t < -2.744$ or $t > 2.744$.

Since the observed value of the test statistic does not fall in the rejection region ($t = 1.81 \not> 2.744$), H_0 is not rejected. There is insufficient evidence to indicate weight change is correlated to digestion efficiency at $\alpha = 0.01$ when duck chow observations are deleted.

e.   Using MINITAB, the scatterplot of the data is:

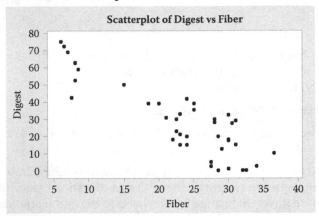

Yes, there appears to be a negative linear trend. As fiber (%) increases, digestion efficiency (%) tends to decrease.

Using MINITAB, the results are:

Correlation: Digest, Fiber

```
Pearson correlation of Digest and Fiber = -0.880
P-Value = 0.000
```

Thus, $r = -0.880$. Since the value is near -1, there is a fairly strong negative linear relationship between fiber (%) and digestions efficiency (%).

To determine if fiber is correlated to digestion efficiency, we test:

$$H_0 : \rho = 0$$
$$H_a : \rho \neq 0$$

The test statistic is $t = \dfrac{r\sqrt{n-2}}{\sqrt{1-r^2}} = \dfrac{-0.880\sqrt{42-2}}{\sqrt{1-(-0.880)^2}} = -11.72$.

The rejection region requires $\alpha/2 = 0.01/2 = 0.005$ in each tail of the t distribution with $df = n - 2 = 42 - 2 = 40$. Using MINITAB, $t_{0.005} = 2.704$. The rejection region is $t < -2.704$ or $t > 2.704$.

Since the observed value of the test statistic falls in the rejection region ($t = -11.72 < -2.704$), H_0 is rejected. There is sufficient evidence to indicate fiber is correlated to digestion efficiency at $\alpha = 0.01$.

Using MINITAB and excluding observations for duck chow, the results are:

Correlation: Digest2, Fiber2

```
Pearson correlation of Digest2 and Fiber2 = -0.646
P-Value = 0.000
```

Thus, $r = -0.646$. Since the value is slightly smaller than -0.5, there is a moderately negative linear relationship between fiber (%) and digestions efficiency (%).

To determine if fiber is correlated to digestion efficiency, we test:

$$H_0 : \rho = 0$$

$$H_a : \rho \neq 0$$

The test statistic is $t = \dfrac{r\sqrt{n-2}}{\sqrt{1-r^2}} = \dfrac{-0.646\sqrt{33-2}}{\sqrt{1-(-0.646)^2}} = -4.71$.

The rejection region requires $\alpha/2 = 0.01/2 = 0.005$ in each tail of the t distribution with $df = n - 2 = 33 - 2 = 31$. Using MINITAB, $t_{0.005} = 2.744$. The rejection region is $t < -2.744$ or $t > 2.744$.

Since the observed value of the test statistic falls in the rejection region ($t < -4.71 < -2.744$), H_0 is rejected. There is sufficient evidence to indicate fiber is correlated to digestion efficiency at $\alpha = 0.01$ when duck chow observations are deleted.

10.75   Using MINITAB, the scatterplot of the data is:

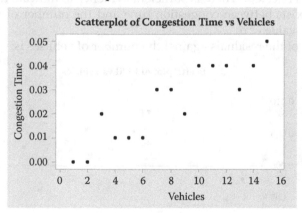

There appears to be a positive linear relationship between congestion time and the number of vehicles.

Using MINITAB, the results are:

```
Regression Analysis: CongestionTime versus Vehicles

Analysis of Variance
```

Source	DF	Adj SS	Adj MS	F-Value	P-Value
Regression	1	0.002893	0.002893	55.27	0.000
Vehicles	1	0.002893	0.002893	55.27	0.000
Error	13	0.000680	0.000052		
Total	14	0.003573			

```
Model Summary

        S    R-sq  R-sq(adj)  R-sq(pred)
0.0072349  80.96%     79.49%      75.32%


Coefficients

Term            Coef  SE Coef  T-Value  P-Value   VIF
Constant    -0.00105  0.00393    -0.27    0.794
Vehicles    0.003214  0.000432     7.43    0.000  1.00


Regression Equation

CongestionTime = -0.00105 + 0.003214 Vehicles
```

To determine if there is a linear relationship between congestion time and the number of vehicles, we test:

$H_0 : \beta_1 = 0$

$H_a : \beta_1 \neq 0$

The test statistic is $t = 7.43$ and the p-value is $p = 0.000$. Since the p-value is so small, H_0 is rejected. There is sufficient evidence to indicate that a linear relationship exists between congestion time and the number of vehicles for any reasonable value of α.

The plot of the residuals against the number of vehicles is:

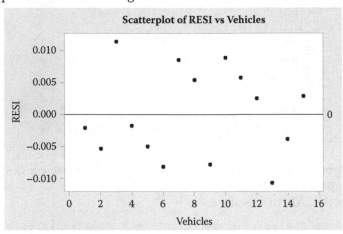

There is no apparent trend in the plot. There is no indication that the model has been misspecified.

Some additional plots are:

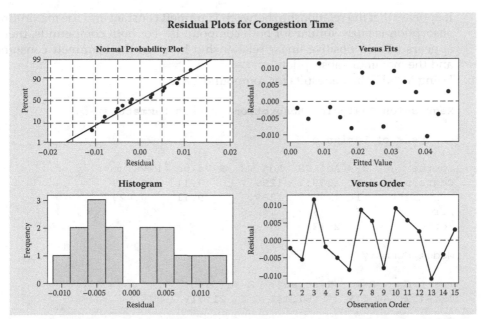

The plot of the residuals against the fitted values does not have a football shape or a funnel shape. There is no indication that the assumption of constant variance is violated. The normal probability plot is close to a straight line and the histogram is somewhat mound-shaped. There is no evidence to indicate the error terms are not normally distributed. Finally, there is no trend in the plot of the residuals against time. There is no evidence that the data are correlated. Thus, the analysis appears to be valid.

10.77 a. Using MINITAB, the scatterplot of the data is:

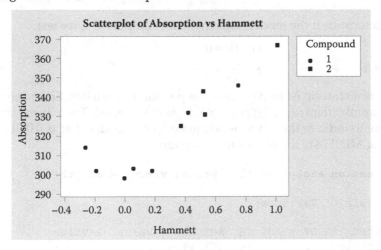

It appears that the relationship between Hammett constant and the maximum absorption is fairly similar for both compounds. For both compounds, there appears to be a positive linear relationship between the Hammett constant and the maximum absorption.

b. Using MINITAB, the results for compound 1 are:

Regression Analysis: Absorption1 versus Hammett1

Analysis of Variance

Source	DF	Adj SS	Adj MS	F-Value	P-Value
Regression	1	1299.7	1299.7	9.11	0.029
Hammett1	1	1299.7	1299.7	9.11	0.029
Error	5	713.2	142.6		
Total	6	2012.9			

Model Summary

S	R-sq	R-sq(adj)	R-sq(pred)
11.9432	64.57%	57.48%	21.71%

Coefficients

Term	Coef	SE Coef	T-Value	P-Value	VIF
Constant	308.14	4.90	62.94	0.000	
Hammett1	41.7	13.8	3.02	0.029	1.00

Regression Equation

Absorption1 = 308.14 + 41.7 Hammett1

c. To determine if the model is adequate for compound 1, we test:

$$H_0 : \beta_1 = 0$$

$$H_a : \beta_1 \neq 0$$

The test statistic is $t = 3.02$ and the p-value is $p = 0.029$. Since the p-value is not smaller than $\alpha(p = 0.029 \not< 0.01)$, H_0 is not rejected. There is insufficient evidence to indicate the model is adequate for compound 1 at $\alpha = 0.01$.

d. Using MINITAB, the results for compound 2 are:

Regression Analysis: Absorption2 versus Hammett2

Analysis of Variance

Source	DF	Adj SS	Adj MS	F-Value	P-Value
Regression	1	952.14	952.14	22.98	0.041
Hammett2	1	952.14	952.14	22.98	0.041
Error	2	82.86	41.43		
Total	3	1035.00			

```
Model Summary

      S    R-sq  R-sq(adj)  R-sq(pred)
6.43656  91.99%     87.99%      81.41%

Coefficients

Term         Coef  SE Coef  T-Value  P-Value   VIF
Constant   302.59     8.73    34.65    0.001
Hammett2    64.1      13.4     4.79    0.041  1.00

Regression Equation

Absorption2 = 302.59 + 64.1 Hammett2
```

To determine if the model is adequate for compound 2, we test:

$$H_0 : \beta_1 = 0$$

$$H_a : \beta_1 \neq 0$$

The test statistic is $t = 4.79$ and the p-value is $p = 0.041$. Since the p-value is not smaller than $\alpha(p = 0.041 \not< 0.01)$, H_0 is not rejected. There is insufficient evidence to indicate the model is adequate for compound 2 at $\alpha = 0.01$.

10.79 a. Some preliminary calculations are:

$$\sum x = 68,074 \qquad \bar{x} = \frac{\sum x}{120} = \frac{68,074}{120} = 567.283333 \qquad \sum x^2 = 60,690,422$$

$$\sum y = 58.206 \qquad \bar{y} = \frac{\sum y}{n} = \frac{58.206}{120} = 0.48505$$

$$\sum y^2 = 41.44977 \qquad \sum xy = 27,388.64$$

$$SS_{xx} = \sum x^2 - \frac{\left(\sum x\right)^2}{n} = 60,690,422 - \frac{(68,074)^2}{120} = 22,073,176.37$$

$$SS_{yy} = \sum y^2 - \frac{\left(\sum y\right)^2}{n} = 41.44977 - \frac{(58.206)^2}{120} = 13.2169497$$

$$SS_{xy} = \sum xy - \frac{\sum x \sum y}{n} = 27,388.64 - \frac{(68,074)(58.206)}{120} = -5,630.6537$$

$$\hat{\beta}_1 = \frac{SS_{xy}}{SS_{xx}} = \frac{-5,630.6537}{22,073,176.37} = -0.00025509 \approx -0.0002551$$

$$\hat{\beta}_0 = \bar{y} - \hat{\beta}_1 \bar{x} = 0.48505 - (-0.00025509)(567.2833333) = 0.62976$$

$$SSE = SS_{yy} - \hat{\beta}_1 SS_{xy} = 13.2169497 - (-0.00025509)(-5,630.6537) = 11.78062625$$

$$s^2 = \frac{SSE}{n-2} = \frac{11.78062625}{120-2} = 0.099835815 \approx 0.09984$$

$$s = \sqrt{s^2} = \sqrt{0.099835815} = 0.31597$$

The fitted regression line is $\hat{y} = 0.62976 - 0.0002551x$

To determine if mercury level decreases linearly as elevation increases, we test:

$$H_0 : \beta_1 = 0$$

$$H_a : \beta_1 < 0$$

The test statistic is $t = \dfrac{\hat{\beta}_1}{\dfrac{s}{\sqrt{SS_{xx}}}} = \dfrac{-0.0002551}{\dfrac{0.31597}{\sqrt{22,073,176.37}}} = -3.79.$

Since no α was given, we will use $\alpha = 0.05$. The rejection region requires $\alpha = 0.05$ in the lower tail of the t distribution with $df = n - 2 = 120 - 2 = 118$. Using MINITAB, $t_{0.05} = 1.658$. The rejection region is $t < -1.658$.

Since the observed value of the test statistic falls in the rejection region ($t < -3.79 < -1.658$), H_0 is rejected. There is sufficient evidence to indicate that the mercury level decreases linearly as elevation increasesat $\alpha = 0.05$.

b. Using MINITAB, the plot of the residuals against elevation is:

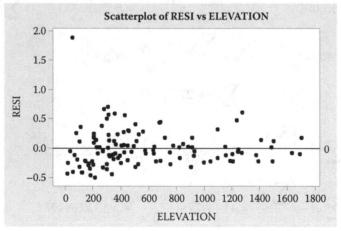

There is no trend to the data. There is no indication that the model has been misspecified.

Additional plots are:

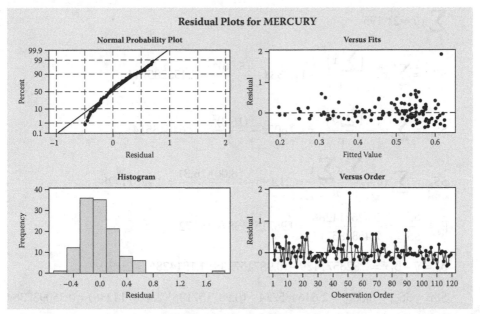

From the normal probability plot, the data do not form a straight line. It appears that the data are not normally distributed. From the plot of the residuals against the fitted values, is appears that the spread of the residual increases as the fitted values increase. It appears that the assumption of constant variance may be violated. There is evidence that the assumptions are not satisfied.

10.81 a. Using MINITAB, the scatterplot of the data is:

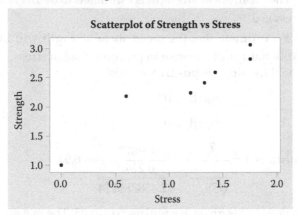

The relationship between shear strength and precompression stress appears to be linear. As precompression stress increases, shear strength tends to increase.

b. Some preliminary calculations are:

$$\sum x = 8.06 \qquad \bar{x} = \frac{\sum x}{n} = \frac{8.06}{7} = 1.151428571 \qquad \sum x^2 = 11.7388$$

$$\sum y = 16.3 \qquad \bar{y} = \frac{\sum y}{n} = \frac{16.3}{7} = 2.328571429 \qquad \sum y^2 = 40.6022$$

$$\sum xy = 21.195$$

$$SS_{xx} = \sum x^2 - \frac{\left(\sum x\right)^2}{n} = 11.7388 - \frac{(8.06)^2}{7} = 2.458285714$$

$$SS_{yy} = \sum y^2 - \frac{\left(\sum y\right)^2}{n} = 40.6022 - \frac{(16.3)^2}{7} = 2.64648514$$

$$SS_{xy} = \sum xy - \frac{\sum x \sum y}{n} = 21.195 - \frac{(8.06)(16.3)}{7} = 2.426714286$$

$$\hat{\beta}_1 = \frac{SS_{xy}}{SS_{xx}} = \frac{2.426714286}{2.458285714} = 0.987157138 \approx 0.9872$$

$$\hat{\beta}_0 = \bar{y} - \hat{\beta}_1 \bar{x} = 2.328571429 - (0.987157138)(1.1514286) = 1.1919$$

$$SSE = SS_{yy} - \hat{\beta}_1 SS_{xy} = 2.646485714 - (0.987157138)(2.426714286) = 0.250937384$$

$$s^2 = \frac{SSE}{n-2} = \frac{0.250937384}{7-2} = 0.05018747 \approx 0.0502$$

$$s = \sqrt{s^2} = \sqrt{0.050187476} = 0.2240$$

The fitted regression line is $\hat{y} = 1.1919 + 0.9872x$

c. $\hat{\beta}_0 = 1.1919$ The mean shear strength is estimated to be 1.1919 when the decompression stress is 0.

$\hat{\beta}_1 = 0.9872$ We estimate that the mean shear strength will increase by 0.9872 for each additional 1 unit increase in precompression stress.

d. To determine if the slope is positive, we test:

$$H_0 : \beta_1 = 0$$

$$H_a : \beta_1 > 0$$

The test statistic is $t = \dfrac{\hat{\beta}_1}{\dfrac{s}{\sqrt{SS_{xx}}}} = \dfrac{0.9872}{\dfrac{0.2240}{\sqrt{2.458285714}}} = 6.91.$

Since no value of α is given, we will use $\alpha = 0.05$. The rejection region requires $\alpha = 0.05$ in the upper tail of the t distribution with $df = n - 2 = 7 - 2 = 5$. From Table 7, Appendix B, $t_{0.05} = 2.015$. The rejection region is $t > 2.015$.

Since the observed value of the test statistic falls in the rejection region ($t = 6.91 > 2.015$), H_0 is rejected. There is sufficient evidence to indicate that the slope is positive at $\alpha = 0.05$.

10.83 a. Some preliminary calculations are:

$$\sum x = 100 \qquad \bar{x} = \frac{\sum x}{n} = \frac{100}{8} = 12.5 \qquad \sum x^2 = 1,332$$

$$\sum y = 8.54 \qquad \bar{y} = \frac{\sum y}{n} = \frac{8.54}{8} = 1.0675 \qquad \sum y^2 = 9.8808$$

$$\sum xy = 114.49$$

$$SS_{xx} = \sum x^2 - \frac{\left(\sum x\right)^2}{n} = 1,332 - \frac{(100)^2}{8} = 82$$

$$SS_{yy} = \sum y^2 - \frac{\left(\sum y\right)^2}{n} = 9.8808 - \frac{(8.54)^2}{8} = 0.76435$$

$$SS_{xy} = \sum xy - \frac{\sum x \sum y}{n} = 114.49 - \frac{(100)(8.54)}{8} = 7.74$$

$$\hat{\beta}_1 = \frac{SS_{xy}}{SS_{xx}} = \frac{7.74}{82} = 0.094390243 \approx 0.09439$$

$$\hat{\beta}_0 = \bar{y} - \hat{\beta}_1 \bar{x} = 1.0675 - (0.094390243)(12.5) = -0.11238$$

$$SSE = SS_{yy} - \hat{\beta}_1 SS_{xy} = 0.76435 - (0.094390243)(7.74) = 0.033769519$$

$$s^2 = \frac{SSE}{n-2} = \frac{0.033769519}{8-2} = 0.005628253 \approx 0.00563$$

$$s = \sqrt{s^2} = \sqrt{0.005628253} = 0.07502$$

The fitted regression line is $\hat{y} = -0.11238 + 0.09439x$.

b. To determine if the model is useful for predicting flow rate, we test:

$$H_0 : \beta_1 = 0$$
$$H_a : \beta_1 \neq 0$$

The test statistic is $t = \dfrac{\hat{\beta}_1}{\dfrac{s}{\sqrt{SS_{xx}}}} = \dfrac{0.09439}{\dfrac{0.07502}{\sqrt{82}}} = 11.39.$

The rejection region requires $\alpha / 2 = 0.05 / 2 = 0.025$ in each tail of the t distribution with $df = n - 2 = 8 - 2 = 6$. From Table 7, Appendix B, $t_{0.025} = 2.447$. The rejection region is $t < -2.447$ or $t > 2.447$.

Since the observed value of the test statistic falls in the rejection region ($t = 11.39 > 2.447$), H_0 is rejected. There is sufficient evidence to indicate that the model is useful for predicting flow rate at $\alpha = 0.05$.

c. For $x = 11$, $\hat{y} = -0.11238 + 0.09439(11) = 0.9259$

For confidence coefficient 0.95, $\alpha = 0.05$ and $\alpha/2 = 0.05/2 = 0.025$. From Table 7, Appendix B, with $df = n - 2 = 8 - 2 = 6$, $t_{0.025} = 2.447$. The 95% confidence interval is:

$$\hat{y} \pm t_{\alpha/2}s\sqrt{1 + \frac{1}{n} + \frac{(x - \bar{x})^2}{SS_{xx}}} \Rightarrow 0.9259 \pm 2.447(0.07502)\sqrt{1 + \frac{1}{8} + \frac{(11 - 12.5)^2}{82}}$$

$$\Rightarrow 0.9259 \pm 0.1971 \Rightarrow (0.7288, 1.1230)$$

We are 95% confident that the flow rate will fall between 0.7288 and 1.1230 when the pressure drop across the filter is 11 inches of water.

10.85   Answers may vary. A possible answer is:

The scaffold-drop survey provides the most accurate estimate of spall rate in a given wall segment. However, the drop areas were not selected at random from the entire complex; rather, drops were made at areas with high spall concentrations. Therefore, if the photo spall rates could be shown to be related to drop spall rates, then the 83 photo spall rates could be used to predict what the drop spall rates would be.

Using MINITAB, a scatterplot of the data is:

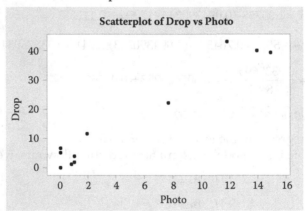

The scatterplot shows a positive linear relationship between the photo spall rate and the drop spall rate.

Using MINITAB, the results are:

Regression Analysis: Drop versus Photo

Analysis of Variance

Source	DF	Adj SS	Adj MS	F-Value	P-Value
Regression	1	2777.51	2777.51	160.23	0.000
Photo	1	2777.51	2777.51	160.23	0.000
Error	9	156.01	17.33		
Lack-of-Fit	6	129.87	21.64	2.48	0.243
Pure Error	3	26.14	8.71		

```
Total           10   2933.53
```

Model Summary

```
     S    R-sq  R-sq(adj)  R-sq(pred)
4.16352  94.68%     94.09%      91.38%
```

Coefficients

```
Term        Coef  SE Coef  T-Value  P-Value   VIF
Constant    2.55     1.64     1.56    0.154
Photo      2.760    0.218    12.66    0.000  1.00
```

Regression Equation

```
Drop = 2.55 + 2.760 Photo
```

The fitted regression line is $\hat{y} = 2.55 + 2.760x$.

To determine if the photo spall rates contribute to the prediction of the drop spall rates, we test:

$$H_0 : \beta_1 = 0$$
$$H_a : \beta_1 \neq 0$$

The test statistic is $t = 12.66$ and the p-value is $p = 0.000$. Since the p-value is so small, H_0 is rejected. There is sufficient evidence to indicate that the photo spall rates contribute to the prediction of the drop spall rates for any reasonable value of α.

$r^2 = 0.9468$ 94.68% of the total variation of the drop spall rates about their means is explained by the linear relationship between drop spall rates and photo spall rates.

One could now use the 83 photo spall rates to predict values for 83 drop spall rates. Then use this information to estimate the true spall rate at a given wall segment and estimate the total spall damage.

11

Multiple Regression Analysis

11.1 a. $\mathbf{Y} = \begin{bmatrix} 18.8 \\ 18.3 \\ 16.3 \\ 6.9 \\ 17.1 \\ 20.4 \end{bmatrix}$ $\mathbf{X} = \begin{bmatrix} 1 & 12.0 \\ 1 & 9.7 \\ 1 & 7.3 \\ 1 & 5.3 \\ 1 & 10.9 \\ 1 & 16.8 \end{bmatrix}$

b. $\mathbf{X'X} = \begin{bmatrix} 1 & 1 & 1 & 1 & 1 & 1 \\ 12.0 & 9.7 & 7.3 & 5.3 & 10.9 & 16.8 \end{bmatrix} \begin{bmatrix} 1 & 12.0 \\ 1 & 9.7 \\ 1 & 7.3 \\ 1 & 5.3 \\ 1 & 10.9 \\ 1 & 16.8 \end{bmatrix} = \begin{bmatrix} 6 & 62 \\ 62 & 720.52 \end{bmatrix}$

$\mathbf{X'Y} = \begin{bmatrix} 1 & 1 & 1 & 1 & 1 & 1 \\ 12.0 & 9.7 & 7.3 & 5.3 & 10.9 & 16.8 \end{bmatrix} \begin{bmatrix} 18.8 \\ 18.3 \\ 16.3 \\ 6.9 \\ 17.1 \\ 20.4 \end{bmatrix} = \begin{bmatrix} 97.8 \\ 1,087.78 \end{bmatrix}$

c. $\mathbf{X'X}^{-1} = \begin{bmatrix} 1.50384 & -0.12940 \\ -0.12940 & 0.012523 \end{bmatrix}$

d. $\hat{\beta} = \mathbf{X'X}^{-1}\mathbf{X'Y} = \begin{bmatrix} 1.5038404 & -0.1294039 \\ -0.1294039 & 0.0125230 \end{bmatrix} \begin{bmatrix} 97.8 \\ 1,087.78 \end{bmatrix} = \begin{bmatrix} 6.312606 \\ 0.966522 \end{bmatrix}$

The least squares line is $\hat{y} = 6.312606 + 0.966522x$.

e. $\mathbf{SSE} = \mathbf{Y'Y} - \hat{\beta}'\mathbf{X'Y}$

$$\mathbf{Y'Y} = \begin{bmatrix} 18.8 & 18.3 & 16.3 & 6.9 & 17.1 & 20.4 \end{bmatrix} \begin{bmatrix} 18.8 \\ 18.3 \\ 16.3 \\ 6.9 \\ 17.1 \\ 20.4 \end{bmatrix} = 1,710.2$$

$$\hat{\boldsymbol{\beta}}'\mathbf{X'Y} = \begin{bmatrix} 6.312606 & 0.966522 \end{bmatrix} \begin{bmatrix} 97.8 \\ 1,087.78 \end{bmatrix} = 1,668.736168$$

$$\mathbf{SSE} = \mathbf{Y'Y} - \hat{\boldsymbol{\beta}}'\mathbf{X'Y} = 1,710.2 - 1,668.736168 = 41.463835$$

$$s^2 = \frac{SSE}{n-2} = \frac{41.463835}{6-2} = 10.365959$$

f. MINITAB is used to find the rest of the calculations and the results are:

Regression Analysis: Y versus X

Analysis of Variance

Source	DF	Adj SS	Adj MS	F-Value	P-Value
Regression	1	74.60	74.60	7.20	0.055
X	1	74.60	74.60	7.20	0.055
Error	4	41.46	10.37		
Total	5	116.06			

Model Summary

S	R-sq	R-sq(adj)	R-sq(pred)
3.21962	64.27%	55.34%	0.00%

Coefficients

Term	Coef	SE Coef	T-Value	P-Value	VIF
Constant	6.31	3.95	1.60	0.185	
X	0.967	0.360	2.68	0.055	1.00

Regression Equation

Y = 6.31 + 0.967 X

Prediction for Y

Regression Equation

Y = 6.31 + 0.967 X

```
Variable   Setting
X               10

    Fit    SE Fit       90% CI              90% PI
15.9778   1.31988   (13.1640, 18.7916)   (8.55972, 23.3959)
```

The standard error of $\hat{\beta}_1$ is $\hat{\sigma}_{\hat{\beta}_1} = s\sqrt{c_{22}} = \sqrt{10.365959}\sqrt{0.012523} = 0.360296.$

g. For confidence coefficient 0.90, $\alpha = 0.10$ and $\alpha/2 = 0.10/2 = 0.05$. From Table 7, Appendix B, with $df = n - 2 = 6 - 2 = 4$, $t_{0.05} = 2.132$. The 90% confidence interval is

$$\hat{\beta}_1 \pm t_{\alpha/2}\hat{\sigma}_{\hat{\beta}_1} \Rightarrow 0.9665 \pm 2.132(0.360296) \Rightarrow 0.9665 \pm 0.7682 \Rightarrow (0.1983, 1.7347)$$

We are 90% confident that the mean apparent porosity percentage will increase between 0.1983 and 1.7347 percent for every one micro-meter increase in mean pore diameter of a brick.

h. From the printout, $R^2 = 0.6427$. 64.27% of the variation of the sampled apparent porosity percentages around their mean can be explained by the linear relationship with mean pore diameter.

i. From the printout, the 90% prediction interval for the apparent porosity percentage when the mean pore diameter is 10 is $(8.55972, 23.3959)$. We are 90% confident that the apparent porosity percentage when the mean pore diameter is 10 is between 8.55973 and 23.3959.

11.3 a. $\mathbf{Y} = \begin{bmatrix} 18.3 \\ 11.6 \\ 32.2 \\ 30.9 \\ 12.5 \\ 9.1 \\ 11.8 \\ 11.0 \\ 19.7 \\ 12.0 \end{bmatrix}$ $\mathbf{X} = \begin{bmatrix} 1 & 2.48 \\ 1 & 2.48 \\ 1 & 2.39 \\ 1 & 2.44 \\ 1 & 2.50 \\ 1 & 2.58 \\ 1 & 2.59 \\ 1 & 2.59 \\ 1 & 2.51 \\ 1 & 2.49 \end{bmatrix}$

b.

$$\mathbf{X'X} = \begin{bmatrix} 1 & 1 & 1 & 1 & 1 & 1 & 1 & 1 & 1 & 1 \\ 2.48 & 2.48 & 2.39 & 2.44 & 2.50 & 2.58 & 2.59 & 2.59 & 2.51 & 2.49 \end{bmatrix} \begin{bmatrix} 1 & 2.48 \\ 1 & 2.48 \\ 1 & 2.39 \\ 1 & 2.44 \\ 1 & 2.50 \\ 1 & 2.58 \\ 1 & 2.59 \\ 1 & 2.59 \\ 1 & 2.51 \\ 1 & 2.49 \end{bmatrix} = \begin{bmatrix} 10 & 25.05 \\ 25.05 & 62.7893 \end{bmatrix}$$

$$\mathbf{X'Y} = \begin{bmatrix} 1 & 1 & 1 & 1 & 1 & 1 & 1 & 1 & 1 & 1 \\ 2.48 & 2.48 & 2.39 & 2.44 & 2.50 & 2.58 & 2.59 & 2.59 & 2.51 & 2.49 \end{bmatrix} \begin{bmatrix} 18.3 \\ 11.6 \\ 32.2 \\ 30.9 \\ 12.5 \\ 9.1 \\ 11.8 \\ 11.0 \\ 19.7 \\ 12.0 \end{bmatrix} = \begin{bmatrix} 169.1 \\ 419.613 \end{bmatrix}$$

c. $\mathbf{X'X}^{-1} = \begin{bmatrix} 160.7920615 & -64.1485275 \\ -64.1485275 & 25.6081946 \end{bmatrix}$

$$\hat{\beta} = \mathbf{X'X}^{-1}\mathbf{X'Y} = \begin{bmatrix} 160.7920615 & -64.1485275 \\ -64.1485275 & 25.6081946 \end{bmatrix} \begin{bmatrix} 169.1 \\ 419.613 \end{bmatrix} = \begin{bmatrix} 272.3815109 \\ -101.9846351 \end{bmatrix}$$

d. $\mathbf{Y'Y} = \begin{bmatrix} 18.3 & 11.6 & 32.2 & 30.9 & 12.5 & 9.1 & 11.8 & 11.0 & 19.7 & 12.0 \end{bmatrix} \begin{bmatrix} 18.3 \\ 11.6 \\ 32.2 \\ 30.9 \\ 12.5 \\ 9.1 \\ 11.8 \\ 11.0 \\ 19.7 \\ 12.0 \end{bmatrix} = 3,492.49$

$$\hat{\beta}'\mathbf{X'Y} = \begin{bmatrix} 272.3815109 & -101.9846351 \end{bmatrix} \begin{bmatrix} 169.1 \\ 419.613 \end{bmatrix} = 3,265.634809$$

$$\mathbf{SSE} = \mathbf{Y'Y} - \hat{\beta}'\mathbf{X'Y} = 3,492.49 - 3,265.634809 = 226.855191$$

$$s^2 = \frac{SSE}{n-2} = \frac{226.855191}{10-2} = 28.35689888$$

e. MINITAB was used to find the rest of the calculations and the results are:

Regression Analysis: y versus x

Analysis of Variance

Source	DF	Adj SS	Adj MS	F-Value	P-Value
Regression	1	406.15	406.15	14.32	0.005
x	1	406.15	406.15	14.32	0.005
Error	8	226.86	28.36		
Lack-of-Fit	6	204.09	34.02	2.99	0.272
Pure Error	2	22.77	11.38		
Total	9	633.01			

Model Summary

S	R-sq	R-sq(adj)	R-sq(pred)
5.32512	64.16%	59.68%	45.91%

Coefficients

Term	Coef	SE Coef	T-Value	P-Value	VIF
Constant	272.4	67.5	4.03	0.004	
x	-102.0	26.9	-3.78	0.005	1.00

Regression Equation

y = 272.4 - 102.0 x

```
Prediction for y

Regression Equation

y = 272.4 - 102.0 x

Variable   Setting

x              2.5

     Fit    SE Fit         95% CI              95% PI
  17.4199  1.68933    (13.5243, 21.3155)   (4.53707, 30.3028)
```

We test:

$$H_0 : \beta_1 = 0$$
$$H_a : \beta_1 < 0$$

The test statistic is $t = -3.78$ and the p-value is $p = 0.005 / 2 = 0.0025$. Since the p-value is less than $\alpha \, (p = 0.0025 < 0.01)$, H_0 is rejected. There is sufficient evidence to indicate that there is a negative linear relationship between Young's Modulus and fractal dimension at $\alpha = 0.01$.

f. From the printout, $R^2 = 0.6416$. 64.16% of the variation of the sampled Young's Modulus values around their mean can be explained by the linear relationship with fractal dimension.

g. From the printout, the 95% prediction interval for the Young's Modulus when the fractal dimension is 2.50 is between 4.53707 and 30.3028. We are 95% confident that the actual Young's Modulus value will be between 4.53707 and 30.3028 when the fractal dimension is 2.50.

11.5 a. $\mathbf{Y} = \begin{bmatrix} 19.29 \\ 17.09 \\ 15.98 \\ 15.17 \\ 14.96 \\ 14.94 \end{bmatrix}$ $\mathbf{X} = \begin{bmatrix} 1 & 5.5452 \\ 1 & 6.2383 \\ 1 & 6.9315 \\ 1 & 8.3178 \\ 1 & 9.9272 \\ 1 & 11.5366 \end{bmatrix}$

$$\mathbf{X'X} = \begin{bmatrix} 1 & 1 & 1 & 1 & 1 & 1 \\ 5.5452 & 6.2383 & 6.9315 & 8.3178 & 9.9272 & 11.5366 \end{bmatrix} \begin{bmatrix} 1 & 5.5452 \\ 1 & 6.2383 \\ 1 & 6.9315 \\ 1 & 8.3178 \\ 1 & 9.9272 \\ 1 & 11.5366 \end{bmatrix} = \begin{bmatrix} 6 & 48.4966 \\ 48.4966 & 418.5395584 \end{bmatrix}$$

$$\mathbf{X'Y} = \begin{bmatrix} 1 & 1 & 1 & 1 & 1 & 1 \\ 5.5452 & 6.2383 & 6.9315 & 8.3178 & 9.9272 & 11.5366 \end{bmatrix} \begin{bmatrix} 19.29 \\ 17.09 \\ 15.98 \\ 15.17 \\ 14.96 \\ 14.94 \end{bmatrix} = \begin{bmatrix} 97.43 \\ 771.3936 \end{bmatrix}$$

$$\mathbf{X'X}^{-1} = \begin{bmatrix} 2.6270843 & -0.3044029 \\ -0.3044029 & 0.0376607 \end{bmatrix}$$

$$\hat{\beta} = \mathbf{X'X}^{-1}\mathbf{X'Y} = \begin{bmatrix} 2.6270843 & -0.3044029 \\ -0.3044029 & 0.0376607 \end{bmatrix} \begin{bmatrix} 97.43 \\ 771.3936 \end{bmatrix} = \begin{bmatrix} 21.14238266 \\ -0.606729048 \end{bmatrix}$$

b. MINITAB was used to find the rest of the calculations and the results are:

Regression Analysis: Y versus LNX

Analysis of Variance

Source	DF	Adj SS	Adj MS	F-Value	P-Value
Regression	1	9.775	9.775	8.16	0.046
LNX	1	9.775	9.775	8.16	0.046
Error	4	4.791	1.198		
Total	5	14.566			

Model Summary

S	R-sq	R-sq(adj)	R-sq(pred)
1.09445	67.11%	58.88%	5.36%

Coefficients

Term	Coef	SE Coef	T-Value	P-Value	VIF
Constant	21.14	1.77	11.92	0.000	
LNX	-0.607	0.212	-2.86	0.046	1.00

Regression Equation

Y = 21.14 - 0.607 LNX

Prediction for Y

Regression Equation

Y = 21.14 - 0.607 LNX

Variable	Setting
LNX	8.51719

Fit	SE Fit	90% CI	90% PI
15.9748	0.456233	(15.0021, 16.9474)	(13.4470, 18.5026)

To determine if the overall model is adequate, we test:

$$H_0 : \beta_1 = 0$$
$$H_a : \beta_1 \neq 0$$

From the printout, the test statistic is $F = 8.16$ and the p-value is $p = 0.046$. Since the p-value is less than $\alpha (p = 0.046 < 0.10)$, H_0 is rejected. There is sufficient evidence to indicate the model is adequate for predicting performance at $\alpha = 0.10$.

c. When $x = 5,000$, $\ln x = 8.51719$. From the printout, the 90% confidence interval for the mean performance when the ln (length) is 8.51719 is between 15.0021 and 16.9474. We are 90% confident that the mean thermal performance value will be between 15.0021 and 16.9474 when the ln (length) is 8.51719.

11.7 a. To determine if the model is useful for prediction rate of conversion, we test:

$$H_0 : \beta_1 = \beta_2 = \beta_3 = 0$$
$$H_a : \text{At least one } \beta_i \neq 0$$

The test statistic is

$$F = \frac{R^2 / k}{(1 - R^2) / [n - (k+1)]} = \frac{0.899 / 3}{(1 - 0.899) / [10 - (3+1)]} = \frac{0.2996666667}{0.016833333} = 17.80$$

The rejection region requires $\alpha = 0.01$ in the upper tail of the F distribution with $v_1 = k = 3$ and $v_2 = n - (k+1) = 10 - (3+1) = 6$. From Table 12, Appendix B, $F_{0.01} = 9.78$. The rejection region is $F > 9.78$.

Since the observed value of the test statistic falls in the rejection region $(F = 17.80 > 9.78)$, H_0 is rejected. There is sufficient evidence to indicate model is useful for prediction rate of conversion at $\alpha = 0.01$.

b. To determine if the atom ratio is a useful predictor rate of conversion, we test:

$$H_0 : \beta_1 = 0$$
$$H_a : \beta_1 \neq 0$$

The test statistic is $t = \dfrac{\hat{\beta}_1}{s_{\hat{\beta}_1}} = \dfrac{-0.808}{0.231} = -3.50$.

The rejection region requires $\alpha / 2 = 0.05 / 2 = 0.025$ in each tail of the t distribution with $df = n - (k+1) = 10 - (3+1) = 6$. From Table 7, Appendix B, $t_{0.025} = 2.447$. The rejection region is $t < -2.447$ or $t > 2.447$.

Since the observed value of the test statistic falls in the rejection region $(t = -3.50 < -2.447)$, H_0 is rejected. There is sufficient evidence to indicate the atom ratio is a useful predictor rate of conversion at $\alpha = 0.01$.

c. For confidence coefficient 0.95, $\alpha = 0.05$ and $\alpha/2 = 0.05/2 = 0.025$. From Table 7, Appendix B, with $df = n-(k+1) = 10-(3+1) = 6$, $t_{0.025} = 2.447$. The 95% confidence interval is:

$$\hat{\beta}_2 \pm t_{\alpha/2}\hat{\sigma}_{\hat{\beta}_2} \Rightarrow -6.38 \pm 2.447(1.93) \Rightarrow -6.38 \pm 4.723 \Rightarrow (-11.103, -1.657)$$

We are 95% confident that the true value of β_2 is between -11.103 and -1.657.

11.9 a. For $x_1 = 10$, $x_2 - 0$, and $x_3 = 1$, $\hat{y} = 52,484 + 2,941(10) + 16,880(0) + 11,108(1) = 93,002$.

    b. For $x_1 = 10$, $x_2 = 1$, and $x_3 = 0$, $\hat{y} = 52,484 + 2,941(10) + 16,880(1) + 11,108(0) = 98,774$.

    c. $R_{adj.}^2 = 0.32$. 32% of the sample variation of the salary values about their means is explained by the model, adjusted for the sample size and the number of parameters in the model.

    d. We are 95% confident that for every additional year of experience, then mean salary increases from \$2,700 to \$3,200, holding PhD status and manager status constant.

    e. We are 95% confident that the difference in mean salary between those with a PhD and those who do not is between \$11,500 to \$22,300, holding years of experience and manager status constant.

    f. We are 95% confident that the difference in mean salary between those who are managers and those who are not is between \$7,600 to \$14,600, holding years of experience and PhD status constant.

11.11   Show $T = \dfrac{\hat{\beta}_i - \beta_i}{s\sqrt{c_{ii}}}$ has a t distribution with $[n-(k+1)]$ degrees of freedom.

From Theorem 11.1, the sampling distribution of $\hat{\beta}_i$ is normal with $E(\hat{\beta}_i) = \beta_i$, $V(\hat{\beta}_i) = c_{ii}\sigma^2$, and $\sigma_{\hat{\beta}_i} = \sigma\sqrt{c_{ii}}$.

Thus, $\dfrac{\hat{\beta}_i - \beta_i}{\sigma\sqrt{c_{ii}}}$ has a standard normal distribution.

From Theorem 11.2, $\chi^2 = \dfrac{SSE}{\sigma^2} = \dfrac{[n-(k+1)]s^2}{\sigma^2}$ has a χ^2 distribution with $v = [n-(k+1)]$ degrees of freedom.

By definition, a t random variable is the ratio of a standard normal random variable and the square root of an independent chi-square random variable divided by its degrees of freedom.

Thus, $T = \dfrac{\dfrac{\hat{\beta}_i - \beta_i}{\sigma\sqrt{c_{ii}}}}{\sqrt{\dfrac{[n-(k+1)]s^2}{\sigma^2}}{[n-(k+1)]}} = \dfrac{\dfrac{\hat{\beta}_i - \beta_i}{\sigma\sqrt{c_{ii}}}}{\sqrt{\dfrac{s^2}{\sigma^2}}} = \dfrac{\hat{\beta}_i - \beta_i}{s\sqrt{c_{ii}}}$ with $[n-(k+1)]$ degrees of freedom.

11.13　Show $T = \dfrac{I - E(I)}{s\sqrt{\mathbf{a'}(\mathbf{X'X})^{-1}\mathbf{a}}}$ has a Student's t distribution with $[n - (k+1)]$ degrees of freedom.

From Theorem 11.3, the sampling distribution of I is normal with

$E(I) = a_0\beta_0 + a_1\beta_1 + \cdots + a_k\beta_k$ and $V(I) = \left[\mathbf{a'}(\mathbf{X'X})^{-1}\mathbf{a}\right]\sigma^2$.

Thus, $T = \dfrac{I - E(I)}{\sqrt{[\mathbf{a'}(\mathbf{X'X})^{-1}\mathbf{a}]\sigma^2}}$ has a standard normal distribution.

From Theorem 11.2, $\chi^2 = \dfrac{SSE}{\sigma^2} = \dfrac{[n-(k+1)]s^2}{\sigma^2}$ has a χ^2 distribution with $\nu = [n-(k+1)]$ degrees of freedom.

By definition, a t random variable is the ratio of a standard normal random variable and the square root of an independent chi-square random variable divided by its degrees of freedom.

Thus, $T = \dfrac{\dfrac{I - E(I)}{\sqrt{[\mathbf{a'}(\mathbf{X'X})^{-1}\mathbf{a}]\sigma^2}}}{\sqrt{\dfrac{\dfrac{[n-(k+1)]s^2}{\sigma^2}}{[n-(k+1)]}}} = \dfrac{\dfrac{I - E(I)}{\sqrt{[\mathbf{a'}(\mathbf{X'X})^{-1}\mathbf{a}]\sigma^2}}}{\sqrt{\dfrac{s^2}{\sigma^2}}} = \dfrac{I - E(I)}{s\sqrt{[\mathbf{a'}(\mathbf{X'X})^{-1}\mathbf{a}]}}$ with $df = [n-(k+1)]$.

11.15 a.　By the assumptions of Multiple Regression Analysis, the probability distribution of ε is normal. Thus, $y = \beta_0 + \beta_1 x_1 + \cdots + \beta_k x_k + \varepsilon$ has a normal distribution for a particular setting of $x_1, x_2, \cdots, x_k$ because ε is the only variable.

$\hat{y} = \hat{\beta}_0 + \hat{\beta}_1 x_1 + \cdots + \hat{\beta}_k x_k$ is a linear combination of normal random variables because each $\hat{\beta}_i$ is normally distributed by Theorem 11.1. Thus, $\hat{y}$ is also normally distributed.

Thus, $(\hat{y} - y)$ is also a linear combination of normal random variables, and therefore, is also normal.

　　b.　$E(\hat{y} - y) = E[(\hat{\beta}_0 + \hat{\beta}_1 x_1 + \cdots + \hat{\beta}_k x_k) - (\beta_0 + \beta_1 x_1 + \cdots + \beta_k x_k + \varepsilon)]$

$= \beta_0 + \beta_1 x_1 + \cdots + \beta_k x_k - (\beta_0 + \beta_1 x_1 + \cdots + \beta_k x_k + 0) = 0$

$V(\hat{y} - y) = V(\hat{y}) + V(y) - 2Cov(\hat{y}, y) = V(\hat{y}) + V(y)$

$= \left[\mathbf{a'}(\mathbf{X'X})^{-1}\mathbf{a}\right]\sigma^2 + \sigma^2$ (by Theorem 11.3)

$= \left[1 + \mathbf{a'}(\mathbf{X'X})^{-1}\mathbf{a}\right]\sigma^2$

11.17 $P\left(-t_{\alpha/2} < \dfrac{(\hat{y}-y)}{s\sqrt{[1+\mathbf{a'(X'X)}^{-1}\mathbf{a}]}} < t_{\alpha/2}\right) = 1-\alpha$

$\Rightarrow P\left(-t_{\alpha/2}s\sqrt{[1+\mathbf{a'(X'X)}^{-1}\mathbf{a}]} < (\hat{y}-y) < t_{\alpha/2}s\sqrt{[1+\mathbf{a'(X'X)}^{-1}\mathbf{a}]}\right) = 1-\alpha$

$\Rightarrow P\left(-\hat{y}-t_{\alpha/2}s\sqrt{[1+\mathbf{a'(X'X)}^{-1}\mathbf{a}]} < -y < -\hat{y}+t_{\alpha/2}s\sqrt{[1+\mathbf{a'(X'X)}^{-1}\mathbf{a}]}\right) = 1-\alpha$

$\Rightarrow P\left(\hat{y}+t_{\alpha/2}s\sqrt{[1+\mathbf{a'(X'X)}^{-1}\mathbf{a}]} > y > \hat{y}-t_{\alpha/2}s\sqrt{[1+\mathbf{a'(X'X)}^{-1}\mathbf{a}]}\right) = 1-\alpha$

$\Rightarrow P\left(\hat{y}-t_{\alpha/2}s\sqrt{[1+\mathbf{a'(X'X)}^{-1}\mathbf{a}]} < y < \hat{y}+t_{\alpha/2}s\sqrt{[1+\mathbf{a'(X'X)}^{-1}\mathbf{a}]}\right) = 1-\alpha$

Thus, the $(1-\alpha)100\%$ prediction interval for y is $\hat{y} \pm t_{\alpha/2}s\sqrt{[1+\mathbf{a'(X'X)}^{-1}\mathbf{a}]}$ with $df = [n-(k+1)]$

11.19 a. To determine if the overall model is adequate, we test:

$$H_0 : \beta_1 = \beta_2 = \beta_3 = \beta_4 = 0$$
$$H_a : \text{At least one } \beta_i \neq 0$$

The test statistic is $F = 4.38$ and the p-value is $p = 0.091$. Since the p-value is less than $\alpha(p < 0.091 < 0.10)$, H_0 is rejected. There is sufficient evidence to indicate the model is adequate for predicting grafting efficiency at $\alpha = 0.10$.

   b. $R_{adj}^2 = 0.629$. 62.9% of the sample variation of the grafting efficiency values about their means can be explained by the fitted model, adjusted for the sample size and the number of parameters in the model.

   c. $s = 11.2206$. We would expect most of the observations to fall within $2s = 2(11.2206) = 22.4412$ units of the least squares prediction line.

   d. For confidence coefficient 0.90, $\alpha = 0.10$ and $\alpha/2 = 0.10/2 = 0.05$. From Table 7, Appendix B, with $df = n-(k+1) = 9-(4+1) = 4$, $t_{0.05} = 2.132$. The 90% confidence interval is:

$$\hat{\beta}_3 \pm t_{\alpha/2}\hat{\sigma}_{\hat{\beta}_3} \Rightarrow 0.4330 \pm 2.132(0.3054) \Rightarrow 0.4330 \pm 0.6511 \Rightarrow (-0.2181,\ 1.0841)$$

We are 90% confident that for each unit increase in reaction temperature, the mean grafting efficiency will increase from -0.2181 and 1.0841 units, holding initiator concentration, cardanol concentration and reaction time constant.

e.   To determine if β_4 differs from 0, we test:

$$H_0 : \beta_4 = 0$$
$$H_a : \beta_4 \neq 0$$

The test statistic is $t = -0.74$ and the p-value is $p = 0.503$. Since the p-value is not less than $\alpha(p = 0.503 \not< 0.10)$, H_0 is not rejected. There is insufficient evidence to indicate there is a linear relationship between grafting efficiency and reaction time, adjusted for initiator concentration, cardanol concentration and reaction temperature at $\alpha = 0.10$.

11.21 a. $R^2 = 0.31$. 31% of the sample variation in the ln(level of CO2 emissions in current year) is explained by the model containing ln(foreign investments 16 years earlier), gross domestic investment 16 years earlier, trade exports 16 years earlier, ln(GNP 16 years earlier), agricultural production 16 years earlier, African country, and ln(level of CO2 emissions 16 years earlier).

b.   To determine if the overall model is adequate, we test:

$$H_0 : \beta_1 = \beta_2 = \cdots = \beta_7 = 0$$

$$H_a : \text{At least one } \beta_i \neq 0$$

The test statistic is $F = \dfrac{R^2 / k}{(1-R^2)/[n-(k+1)]} = \dfrac{0.31/7}{(1-0.31)/[66-(7+1)]} = 3.723.$

The rejection region requires $\alpha = 0.01$ in the upper tail of the F distribution with $v_1 = k = 7$ and $v_2 = n - (k+1) = 66 - (7+1) = 58$. Using MINITAB, $F_{0.01} = 2.965$. The rejection region is $F > 2.965$.

Since the observed value of the test statistic falls in the rejection region $(F = 3.723 > 2.965)$, H_0 is rejected. There is sufficient evidence to indicate that at least one of the 7 independent variables contributes to the prediction of ln(level of CO_2 emissions in current year) at $\alpha = 0.01$.

c.   To determine if agricultural production is a useful predictor of ln(CO2 emissions), we test:

$$H_0 : \beta_5 = 0$$
$$H_a : \beta_5 \neq 0$$

The test statistic is $t = -0.66$ and the p-value is $p > 0.10$.
Since the p-value is not less than $\alpha(p > 0.10 \not< 0.01)$, H_0 is not rejected. There is insufficient evidence to indicate that agricultural production is a useful predictor of ln(CO_2 emissions) at $\alpha = 0.01$.

11.23 a.   The first order model is $y = \beta_0 + \beta_1 x_1 + \beta_2 x_2 + \beta_3 x_3 + \varepsilon$.

    b.   Using MINITAB, the results are:

Regression Analysis: y versus x1, x2, x3

Analysis of Variance

Source	DF	Adj SS	Adj MS	F-Value	P-Value
Regression	3	70.29	23.429	2.66	0.077
x1	1	44.25	44.251	5.03	0.037
x2	1	34.19	34.186	3.89	0.063
x3	1	10.48	10.482	1.19	0.289
Error	19	167.08	8.794		
Total	22	237.37			

Model Summary

S	R-sq	R-sq(adj)	R-sq(pred)
2.96544	29.61%	18.50%	3.79%

Coefficients

Term	Coef	SE Coef	T-Value	P-Value	VIF
Constant	86.90	3.20	27.17	0.000	
x1	-0.2099	0.0936	-2.24	0.037	1.06
x2	0.1515	0.0769	1.97	0.063	1.06
x3	0.0733	0.0671	1.09	0.289	1.09

Regression Equation

```
y = 86.90 - 0.2099 x1 + 0.1515 x2 + 0.0733 x3
```

The least squares prediction equation is $\hat{y} = 86.90 - 0.2099 x_1 + 0.1515 x_2 + 0.0733 x_3$.

    c.   To determine if the overall model is useful in the prediction of the mean project average, we test:

$$H_0 : \beta_1 = \beta_2 = \beta_3 = \beta_4 = 0$$
$$H_a : \text{At least one } \beta_i \neq 0$$

The test statistic is $F = 2.66$ and the p-value is $p = 0.077$.

Since the p-value is not less than $\alpha (p = 0.077 \not< 0.05)$, H_0 is not rejected. There is insufficient evidence to indicate that the overall model is useful in the prediction of the mean project average at $\alpha = 0.05$.

    d. $R_{adj}^2 = 0.1850$. 18.50% of the sample variation of the project average values about their means can be explained by the fitted model including interpersonal scores, range of stress management scores, and range of mood scores, adjusted for the sample size and the number of parameters in the model.

$2s = 2(2.96544) = 5.93088$. We would expect most of the project average scores to fall within 5.93088 units of their least squares predicted values.

e. For $x_1 = 20$, $x_2 = 30$, and $x_3 = 25$,
$\hat{y} = 86.90 - 0.2099(20) + 0.1515(30) + 0.0733(25) = 89.0795$. Using MINITAB, the results are:

Prediction for y

Regression Equation

y = 86.90 - 0.2099 x1 + 0.1515 x2 + 0.0733 x3
Variable Setting
x1   20
x2   30
x3   25

```
     Fit     SE Fit         95% CI              95% PI
  89.0837   0.892843   (87.2149, 90.9524)   (82.6017, 95.5656)
```

The 95% prediction interval is $(82.6017, 95.5656)$. We are 95% confident that the actual project average score is between 82.6017 and 95.5656 when the interpersonal score is 20, the range of stress management score is 30, and range of mood score is 25.

11.25 a. The first order model is $y = \beta_0 + \beta_1 x_1 + \beta_2 x_2 + \beta_3 x_3 + \beta_4 x_4 + \beta_5 x_5 + \varepsilon$.
       b. Using MINITAB, the results are:

Regression Analysis: HEATRATE versus RPM, INLET-TEMP, EXH-TEMP, CPRATIO, ...

Analysis of Variance

Source	DF	Adj SS	Adj MS	F-Value	P-Value
Regression	5	155055273	31011055	147.30	0.000
RPM	1	8574188	8574188	40.73	0.000
INLET-TEMP	1	7929432	7929432	37.67	0.000
EXH-TEMP	1	3641364	3641364	17.30	0.000
CPRATIO	1	30	30	0.00	0.991
AIRFLOW	1	774427	774427	3.68	0.060
Error	61	12841935	210524		
Total	66	167897208			

Model Summary

S	R-sq	R-sq(adj)	R-sq(pred)
458.828	92.35%	91.72%	90.35%

Coefficients

Term	Coef	SE Coef	T-Value	P-Value	VIF
Constant	13614	870	15.65	0.000	
RPM	0.0888	0.0139	6.38	0.000	2.99
INLET-TEMP	-9.20	1.50	-6.14	0.000	13.31
EXH-TEMP	14.39	3.46	4.16	0.000	7.32
CPRATIO	0.4	29.6	0.01	0.991	4.93
AIRFLOW	-0.848	0.442	-1.92	0.060	3.15

Regression Equation

HEATRATE = 13614 + 0.0888 RPM - 9.20 INLET-TEMP + 14.39 EXH-TEMP
+ 0.4 CPRATIO - 0.848 AIRFLOW

The least squares prediction equation is
$$\hat{y} = 13,614 + 0.0888x_1 - 9.20x_2 + 14.39x_3 + 0.4x_4 - 0.848x_5.$$

c. $\hat{\beta}_0 = 13,614$: Since $x_1 = 0$, $x_2 = 0$, $x_3 = 0$, $x_4 = 0$, and $x_5 = 0$ is not in the observed range, $\hat{\beta}_0$ has no practical meaning. It is simply the y-intercept.

$\hat{\beta}_1 = 0.0888$: For every one unit increase in RPM, we estimate the heat rate to increase by 0.0888 kilojoules per kilowatt hour, holding all other variables constant.

$\hat{\beta}_2 = -9.20$: For every one degree increase in Celsius inlet temperature, we estimate the heat rate to decrease by 9.20 kilojoules per kilowatt hour, holding all other variables constant.

$\hat{\beta}_3 = 14.39$: For every one degree increase in Celsius exhaust temperature, we estimate the heat rate to increase by 14.39 kilojoules per kilowatt hour, holding all other variables constant.

$\hat{\beta}_4 = 0.4$: For every one unit increase in cycle pressure ratio, we estimate the heat rate to increase by 0.40 kilojoules per kilowatt hour, holding all other variables constant.

$\hat{\beta}_5 = -0.848$: For every one kilogram per second increase in air mass flow rate, we estimate the heat rate to decrease by 0.848 kilojoules per kilowatt hour, holding all other variables constant.

d. $s = 458.828$: We expect most of the sampled heat rate values to fall within $2s = 2(458.828) = 917.656$ kilojoules per kilowatt hour of their least squares predicted values.

11.27 a. The first order model is $y = \beta_0 + \beta_1 x_1 + \beta_2 x_2 + \beta_3 x_3 + \beta_4 x_4 + \beta_5 x_5 + \beta_6 x_6 + \beta_7 x_7 + \varepsilon$.

b. Using MINITAB, the results are:

Regression Analysis: Voltage versus Volume, Salinity, Temp, Delay, Surfactant, Triton, Solid

Analysis of Variance

Source	DF	Adj SS	Adj MS	F-Value	P-Value
Regression	7	7.9578	1.13682	5.29	0.007
Volume	1	4.2563	4.25625	19.81	0.001
Salinity	1	0.9438	0.94381	4.39	0.060
Temp	1	0.4512	0.45119	2.10	0.175
Delay	1	0.2107	0.21072	0.98	0.343
Surfactant	1	3.7565	3.75649	17.49	0.002
Triton	1	0.1951	0.19511	0.91	0.361
Solid	1	0.2345	0.23454	1.09	0.319
Error	11	2.3632	0.21484		
Lack-of-Fit	9	2.3622	0.26246	492.12	0.002
Pure Error	2	0.0011	0.00053		
Total	18	10.3210			

Model Summary

S	R-sq	R-sq(adj)	R-sq(pred)
0.463508	77.10%	62.53%	18.97%

```
Coefficients

Term             Coef   SE Coef   T-Value   P-Value    VIF
Constant         0.998   0.248      4.03     0.002
Volume        -0.02243  0.00504    -4.45     0.001    1.83
Salinity        0.1557   0.0743     2.10     0.060    1.33
Temp          -0.0172   0.0119     -1.45     0.175    1.25
Delay         -0.00953  0.00962    -0.99     0.343    1.13
Surfactant      0.421    0.101      4.18     0.002    1.83
Triton          0.417    0.438      0.95     0.361    1.45
Solid          -0.155    0.149     -1.04     0.319    1.33
```

Regression Equation

```
Voltage = 0.998 - 0.02243 Volume + 0.1557 Salinity - 0.0172 Temp
- 0.00953 Delay + 0.421 Surfactant + 0.417 Triton - 0.155 Solid
```

The least squares prediction equation is

$\hat{y} = 0.998 - 0.02243x_1 + 0.1557x_2 - 0.0172x_3 - 0.00953x_4 + 0.421x_5 + 0.417x_6 - 0.155x_7.$

c. $\hat{\beta}_0 = 0.998$: We estimate the mean voltage to be 0.998 kw/cm when all the independent variables are 0.

$\hat{\beta}_1 = -0.02243$: For each 1% increase in disperse phase volume, we estimate the mean voltage to decrease by 0.02243 kw/cm, holding all other variables constant.

$\hat{\beta}_2 = 0.1557$: For each 1% increase in salinity, we estimate the mean voltage to increase by 0.1557 kw/cm, holding all other variables constant.

$\hat{\beta}_3 = -0.0172$: For every one degree increase in Celsius temperature, we estimate the mean voltage to decrease by 0.0172 kw/cm, holding all other variables constant.

$\hat{\beta}_4 = -0.00953$: For every one hour increase in time delay, we estimate the mean voltage to decrease by 0.00953 kw/cm, holding all other variables constant.

$\hat{\beta}_5 = 0.421$: For each 1% increase in surfactant concentration, we estimate the mean voltage to increase by 0.421 kw/cm, holding all other variables constant.

$\hat{\beta}_6 = 0.417$: For each unit increase in span triton, we estimate the mean voltage to increase by 0.417 kw/cm, holding all other variables constant.

$\hat{\beta}_7 = -0.155$: For each 1% increase in solid particles, we estimate the mean voltage to decrease by 0.155 kw/cm, holding all other variables constant.

d. To determine if the model is adequate for predicting voltage, we test:

$$H_0 : \beta_1 = \beta_2 = \cdots = \beta_7 = 0$$
$$H_a : \text{At least one } \beta_i \neq 0$$

The test statistic is $F = 5.29$ and the p-value is $p = 0.007$.

Since no α value was given, we will use $\alpha = 0.05$. Since the p-value is less than $\alpha(p = 0.007 < 0.05)$, H_0 is rejected. There is sufficient evidence to indicate that the overall model is useful in the prediction of the voltage at $\alpha = 0.05$.

$R_{adj}^2 = 0.6253$: 62.53% of the sample variation of the voltage values around their mean can be explained by the multiple regression model including the 7 independent variables, adjusted for the sample size and the number of parameters in the model.

$s = 0.463508$: We expect most of the sampled voltage measurements to fall within $2s = 2(0.463508) = 0.927016$ kw/cm of their least squares predicted values.

e. Using MINITAB to fit the model $y = \beta_0 + \beta_1 x_1 + \beta_2 x_2 + \beta_5 x_5 + \varepsilon$, the results are:

Regression Analysis: Voltage versus Volume, Salinity, Surfactant

Analysis of Variance

Source	DF	Adj SS	Adj MS	F-Value	P-Value
Regression	3	6.8701	2.29004	9.95	0.001
Volume	1	5.6439	5.64386	24.53	0.000
Salinity	1	0.8095	0.80949	3.52	0.080
Surfactant	1	3.5422	3.54219	15.40	0.001
Error	15	3.4509	0.23006		
Lack-of-Fit	5	2.4642	0.49285	5.00	0.015
Pure Error	10	0.9867	0.09867		
Total	18	10.3210			

Model Summary

S	R-sq	R-sq(adj)	R-sq(pred)
0.479646	66.56%	59.88%	47.10%

Coefficients

Term	Coef	SE Coef	T-Value	P-Value	VIF
Constant	0.933	0.248	3.76	0.002	
Volume	-0.02427	0.00490	-4.95	0.000	1.62
Salinity	0.1421	0.0757	1.88	0.080	1.29
Surfactant	0.3846	0.0980	3.92	0.001	1.62

Regression Equation

Voltage = 0.933 - 0.02427 Volume + 0.1421 Salinity + 0.3846 Surfactant

Prediction for Voltage

Regression Equation

Voltage = 0.933 - 0.02427 Volume + 0.1421 Salinity + 0.3846
Surfactant

Variable	Setting
Volume	80
Salinity	1
Surfactant	2

Fit	SE Fit	95% CI	95% PI
-0.0979508	0.231826	(-0.592077, 0.396176)	(-1.23344, 1.03754)

For $x_1 = 80$, $x_2 = 1$, and $x_3 = 2$, the 95% prediction interval is $(-1.23344, 1.03754)$. We are 95% confident that when the disperse phase value is 80%, the salinity is 1% and the surfactant concentration is 2%, the voltage will be between -1.23344 and 1.03754 kw/cm.

11.29 a. The interaction model would be $y = \beta_0 + \beta_1 x_1 + \beta_2 x_2 + \beta_3 x_1 x_2 + \varepsilon$.

b. Using MINITAB, the results are:

Regression Analysis: y versus x1, x2, x1x2

Analysis of Variance

Source	DF	Adj SS	Adj MS	F-Value	P-Value
Regression	3	1.96608E+11	65535959189	110.44	0.000
x1	1	165532568	165532568	0.28	0.606
x2	1	47892440959	47892440959	80.71	0.000
x1x2	1	3623223340	3623223340	6.11	0.027
Error	14	8307933599	593423829		
Total	17	2.04916E+11			

Model Summary

S	R-sq	R-sq(adj)	R-sq(pred)
24360.3	95.95%	95.08%	93.76%

Coefficients

Term	Coef	SE Coef	T-Value	P-Value	VIF
Constant	-63238	31150	-2.03	0.062	
x1	18.8	35.5	0.53	0.606	4.20
x2	445486	49589	8.98	0.000	7.00
x1x2	-139.8	56.6	-2.47	0.027	10.21

Regression Equation

y = -63238 + 18.8 x1 + 445486 x2 - 139.8 x1x2

The least squares prediction equation is
$\hat{y} = -63,238 + 18.8 x_1 + 445,486 x_2 - 139.8 x_1 x_2$.

c. To determine if the overall model is adequate for predicting bubble density, we test:

$$H_0 : \beta_1 = \beta_2 = \beta_3 = 0$$
$$H_a : \text{At least one } \beta_i \neq 0$$

The test statistic is $F = 110.44$ and the p-value is $p = 0.000$.

Since the p-value is less than $\alpha (p = 0.000 < 0.05)$, H_0 is rejected. There is sufficient evidence to indicate that the overall model is useful in the prediction of bubble density at $\alpha = 0.05$.

$R_{adj}^2 = 0.9508$: 95.058% of the sample variation of the bubble density values around their mean can be explained by the multiple regression model including the 2 independent variables and their interaction, adjusted for the sample size and the number of parameters in the model.

$s = 24,360.3$: We expect most of the sampled bubble density values to fall within $2s = 2(24,360.3) = 48,720.6$ liters/m^2 of their least squares predicted values.

d. To determine if mass flux and heat flux interact, we test:

$$H_0 : \beta_3 = 0$$
$$H_a : \beta_3 \neq 0$$

The test statistic is $t = -2.47$ and the p-value is $p = 0.027$. Since the p-value is less than $\alpha (p = 0.027 < 0.05)$, H_0 is rejected. There is sufficient evidence to indicate that mass flux and heat flux interact to affect bubble density at $\alpha = 0.05$.

e. When heat flux is 0.5, then

$$\hat{y} = -63,238 + 18.8x_1 + 445,486(0.5) - 139.8x_1(0.5) = 159,505 - 51.1x_1$$

Thus, when heat flux is 0.5 megawatts/m^2, we would expect bubble density to decrease by 51.1 liters/m^2 squared for each 1 kg/m^2 –sec increase in mass flux.

11.31 a. The interaction model is $y = \beta_0 + \beta_1 x_1 + \beta_2 x_2 + \beta_3 x_1 x_2 + \varepsilon$ where y = number of black streaks in the manufactured ring, x_1 = turntable speed, and x_2 = cutting blade position.

b. The researchers found that the linear relationship between y and x_1 depends on the value of x_2. This indicates that the interaction term is important in the model. If the linear relationship between y and x_1 is much steeper at lower values of x_2, then the interaction term will be negative.

11.33 a. The interaction model would be

$$y = \beta_0 + \beta_1 x_1 + \beta_2 x_2 + \beta_3 x_3 + \beta_4 x_4 + \beta_5 x_5 + \beta_6 x_2 x_5 + \beta_7 x_3 x_5 + \varepsilon$$

b. Using MINITAB, the results are:

Regression Analysis: HEATRATE versus RPM, INLET-TEMP, EXH-TEMP, CPRATIO, ...

Analysis of Variance

Source	DF	Adj SS	Adj MS	F-Value	P-Value
Regression	7	158234406	22604915	138.02	0.000
RPM	1	1350402	1350402	8.25	0.006
INLET-TEMP	1	11072478	11072478	67.61	0.000
EXH-TEMP	1	6104233	6104233	37.27	0.000
CPRATIO	1	2144	2144	0.01	0.909
AIRFLOW	1	21411	21411	0.13	0.719
IN-TEMPxAIRFLOW	1	3166357	3166357	19.33	0.000
EX-TEMPxAIRFLOW	1	2329786	2329786	14.23	0.000
Error	59	9662802	163776		
Total	66	167897208			

Model Summary

S	R-sq	R-sq(adj)	R-sq(pred)
404.693	94.24%	93.56%	91.85%

Coefficients

Term	Coef	SE Coef	T-Value	P-Value	VIF
Constant	13646	1068	12.77	0.000	
RPM	0.0460	0.0160	2.87	0.006	5.10
INLET-TEMP	-12.68	1.54	-8.22	0.000	18.09
EXH-TEMP	23.00	3.77	6.11	0.000	11.15
CPRATIO	-3.0	26.4	-0.11	0.909	5.06
AIRFLOW	1.29	3.56	0.36	0.719	262.90
IN-TEMPxAIRFLOW	0.01615	0.00367	4.40	0.000	504.93
EX-TEMPxAIRFLOW	-0.0414	0.0110	-3.77	0.000	850.66

Regression Equation

HEATRATE = 13646 + 0.0460 RPM - 12.68 INLET-TEMP + 23.00 EXH-TEMP
- 3.0 CPRATIO + 1.29 AIRFLOW + 0.01615 IN-TEMPxAIRFLOW
- 0.0414 EX-TEMPxAIRFLOW

The fitted regression line is

$\hat{y} = 13,646 + 0.0460x_1 - 12.68x_2 + 23.00x_3 - 3.0x_4 + 1.29x_5 + 0.01615x_2x_5 - 0.0414x_3x_5$

c. To determine if inlet temperature and airflow rate interact, we test:

$$H_0 : \beta_6 = 0$$
$$H_a : \beta_6 \neq 0$$

The test statistic is $t = 4.40$ and the p-value is $p = 0.000$. Since the p-value is less than $\alpha (p = 0.000 < 0.05)$, H_0 is rejected. There is sufficient evidence to indicate that inlet temperature and airflow rate interact to affect heat rate at $\alpha = 0.05$.

d. To determine if exhaust temperature and airflow rate interact, we test:

$$H_0 : \beta_7 = 0$$
$$H_a : \beta_7 \neq 0$$

The test statistic is $t = -3.77$ and the p-value is $p = 0.000$. Since the p-value is less than $\alpha(p = 0.000 < 0.05)$, H_0 is rejected. There is sufficient evidence to indicate that exhaust temperature and airflow rate interact to affect heat rate at $\alpha = 0.05$.

e. Based on the results of the tests in parts c and d, we would use the interaction model to predict heat rate.

11.35 a. The first order model is $y = \beta_0 + \beta_1 x_1 + \beta_2 x_2 + \beta_3 x_1 x_2 + \varepsilon$.

b. Using MINITAB, the results are:

Regression Analysis: Voltage versus Volume, Salinity, VxS

Analysis of Variance

Source	DF	Adj SS	Adj MS	F-Value	P-Value
Regression	3	4.5580	1.51934	3.95	0.029
Volume	1	0.4135	0.41349	1.08	0.316
Salinity	1	2.7070	2.70698	7.05	0.018
VxS	1	1.2301	1.23010	3.20	0.094
Error	15	5.7630	0.38420		
Lack-of-Fit	1	0.0843	0.08433	0.21	0.655
Pure Error	14	5.6787	0.40562		
Total	18	10.3210			

Model Summary

S	R-sq	R-sq(adj)	R-sq(pred)
0.619838	44.16%	33.00%	11.27%

Coefficients

Term	Coef	SE Coef	T-Value	P-Value	VIF
Constant	1.008	0.335	3.01	0.009	
Volume	-0.00718	0.00692	-1.04	0.316	1.93
Salinity	0.517	0.195	2.65	0.018	5.12
VxS	-0.00599	0.00335	-1.79	0.094	7.03

Regression Equation

Voltage = 1.008 - 0.00718 Volume + 0.517 Salinity - 0.00599 VxS

The least squares prediction equation is
$$\hat{y} = 1.008 - 0.00718 x_1 + 0.517 x_2 - 0.00599 x_1 x_2.$$

c. To determine if volume and salinity interact, we test:

$$H_0 : \beta_3 = 0$$
$$H_a : \beta_3 \neq 0$$

The test statistic is $t = -1.79$ and the p-value is $p = 0.094$. Since the p-value is less than $\alpha(p = 0.094 < 0.10)$, H_0 is rejected. There is sufficient evidence to indicate that volume and salinity interact to affect voltage level at $\alpha = 0.10$.

d. When $x_2 = 4\%$, $\hat{y} = 1.008 - 0.00718 x_1 + 0.517(4) - 0.00599 x_1(4) = 3.076 - 0.03114 x_1$. Thus, the estimated change in voltage for every 1% increase in volume when salinity is 4% is -0.03114.

11.37 a. The least squares prediction equation is $\hat{y} = 6.266 + 0.0079145x - 0.00000426x^2$.

  b. To determine if the overall model is adequate, we test:

$$H_0 : \beta_1 = \beta_2 = 0$$
$$H_a : \text{At least one } \beta_i \neq 0$$

The test statistic is $F = 210.56$ and the p-value is $p = 0.000$.

Since the p-value is less than $\alpha(p = 0.000 < 0.01)$, H_0 is rejected. There is sufficient evidence to indicate that the overall model is useful in the prediction of the ratio of repair to replacement cost of commercial pipe at $\alpha = 0.01$.

  c. $R_{adj}^2 = 0.972 : 97.2\%$ of the sample variation of the ratio of repair to replacement cost of commercial pipe values around their mean can be explained by the second order regression model including the independent variable pipe diameter, adjusted for the sample size and number of parameters in the model.

  d. To determine if the rate of increase of ratio with diameter is slower for larger pipe sizes, we test:

$$H_0 : \beta_2 = 0$$
$$H_a : \beta_2 < 0$$

  e. The test statistic is $t = -3.23$ and the p-value is $p = 0.009/2 = 0.0045$.

Since the p-value is less than $\alpha(p = 0.0045 < 0.0$, H_0 is rejected. There is sufficient evidence to indicate that the rate of increase of ratio with diameter is slower for larger pipe sizes at $\alpha = 0.01$.

  f. The 95% prediction interval is $(7.5947, 8.3624)$. We are 95% confident that the ratio of repair to replacement cost of commercial pipe is between 7.5947 and 8.3624 when the pipe diameter is 240 mm.

11.39 a. The curve will look something like:

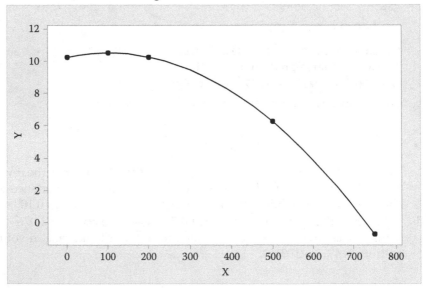

As the dosage increases, the rate of change in the weight change decreases.

b. For $x = 500$, $\hat{y} = 10.25 + 0.0053(500) - 0.0000266(500)^2 = 6.25$

c. For $x = 0$, $\hat{y} = 10.25 + 0.0053(0) - 0.0000266(0)^2 = 10.25$

d. From the graph, the value of x that yields an estimate below 10.25 is $x = 200$ millimeters per kilogram of body weight. This will be the change-point dosage.

11.41　Using MINITAB, the results are:

Regression Analysis: y versus x, x-sq

Analysis of Variance

Source	DF	Adj SS	Adj MS	F-Value	P-Value
Regression	2	635.42	317.709	6.72	0.007
x	1	45.41	45.409	0.96	0.340
x-sq	1	0.32	0.321	0.01	0.935
Error	18	850.48	47.249		
Total	20	1485.90			

Model Summary

S	R-sq	R-sq(adj)	R-sq(pred)
6.87380	42.76%	36.40%	0.00%

Coefficients

Term	Coef	SE Coef	T-Value	P-Value	VIF
Constant	-0.13	3.12	-0.04	0.966	
x	0.171	0.174	0.98	0.340	11.98
x-sq	0.00015	0.00183	0.08	0.935	11.98

Regression Equation

y = -0.13 + 0.171 x + 0.00015 x-sq

To determine if the rate of increase of bird density with percent vegetation coverage is steeper for greener habitats, we test:

$$H_0 : \beta_2 = 0$$
$$H_a : \beta_2 > 0$$

The test statistic is $t = 0.08$ and the p-value is $p = 0.935/2 = 0.4675$.

Since the p-value is not small, H_0 is not rejected. There is insufficient evidence to indicate the rate of increase of bird density with percent vegetation coverage is steeper for greener habitats for any reasonable level of α.

11.43 a. Due to the curvilinear nature of the data, the hypothesized model would be $y = \beta_0 + \beta_1 x + \beta_2 x^2 + \varepsilon$.

b. Using MINITAB, the results are:

Regression Analysis: broadband-alb versus Depth, Depth-sq

Analysis of Variance

Source	DF	Adj SS	Adj MS	F-Value	P-Value
Regression	2	0.9609	0.480460	62.17	0.000
Depth	1	0.8051	0.805146	104.19	0.000

```
Depth-sq          1   0.5400   0.540049    69.88    0.000
Error           501   3.8716   0.007728
  Lack-of-Fit    66   1.1320   0.017151     2.72    0.000
  Pure Error    435   2.7397   0.006298
Total           503   4.8326
```

Model Summary

```
        S     R-sq   R-sq(adj)   R-sq(pred)
0.0879081   19.88%      19.56%       18.62%
```

Coefficients

```
Term          Coef   SE Coef   T-Value   P-Value    VIF
Constant    0.3338    0.0118     28.24     0.000
Depth      -0.8100    0.0794    -10.21     0.000   10.82
Depth-sq    0.941     0.113       8.36     0.000   10.82
```

Regression Equation

```
broadband-alb = 0.3338 - 0.8100 Depth + 0.941 Depth-sq
```

The least-squares prediction equation is $\hat{y} = 0.3338 - 0.8100x + 0.941x^2$.

c. To determine if the overall model is adequate for prediction broadband surface albedo, we test:

$$H_0 : \beta_1 = \beta_2 = 0$$
$$H_a : \text{At least one} \beta_i \neq 0$$

The test statistic is $F = 62.17$ and the p-value is $p = 0.000$.
Since the p-value is less than $\alpha (p = 0.000 < 0.01)$, H_0 is rejected. There is sufficient evidence to indicate that the overall model is adequate for prediction broadband surface albedo at $\alpha = 0.01$.

d. To determine if the quadratic term is useful in predicting broadband surface albedo, we test:

$$H_0 : \beta_2 = 0$$
$$H_a : \beta_2 \neq 0$$

The test statistic is $t = 8.36$ and the p-value is $p = 0.000$.
Since the p-value is less than $\alpha (p = 0.000 < 0.01)$, H_0 is rejected. There is sufficient evidence to indicate the quadratic term is useful in predicting broadband surface albedo at $\alpha = 0.01$.

e. $R^2_{adj} = 0.1956$. 19.56% of the variation of the sampled broadband surface albedo level values around their means can be explained by the curvilinear relationship with pond depth, adjusted for the sample size and the number of parameters in the model.

$s = 0.087081$. We expect most of the sampled broadband surface albedo level values to fall within $2s = 2(0.0879081) = 0.1758162$ units of their least squares predicted values.

11.45 a. Using MINITAB, the scattergram is:

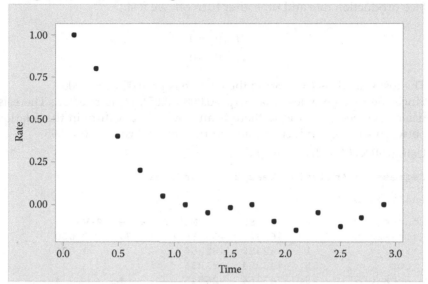

There is a curvilinear relationship between rate and time. The quadratic model should be used.

b. Using MINITAB, the results are:

Regression Analysis: Rate versus Time, Time-Sq

```
Analysis of Variance

Source        DF   Adj SS   Adj MS   F-Value   P-Value
Regression     2   1.5478   0.77391    75.65     0.000
  Time         1   0.9376   0.93761    91.66     0.000
  Time-Sq      1   0.5542   0.55416    54.17     0.000
Error         12   0.1228   0.01023
Total         14   1.6706

Model Summary

      S     R-sq   R-sq(adj)   R-sq(pred)
0.101142   92.65%     91.43%       87.73%

Coefficients

Term        Coef   SE Coef   T-Value   P-Value    VIF
Constant   1.0070   0.0790     12.75     0.000
Time       -1.167    0.122     -9.57     0.000   16.27
Time-Sq    0.2898   0.0394      7.36     0.000   16.27

Regression Equation

Rate = 1.0070 - 1.167 Time + 0.2898 Time-Sq
```

The least squares prediction equation is $\hat{y} = 1.007 - 1.167x + 0.2898x^2$.

c. To determine if there is an upward curvature in the relationship between sur-
    face production rate and time after turn off, we test:

    $$H_0 : \beta_2 = 0$$
    $$H_a : \beta_2 > 0$$

    The test statistic is $t = 7.36$ and the p-value is $p = 0.000/2 = 0.000$.
    Since the p-value is less than $\alpha (p = 0.000 < 0.05)$, H_0 is rejected. There is suf-
    ficient evidence to indicate there is an upward curvature in the relationship
    between surface production rate and time after turn of at $\alpha = 0.05$.

11.47 a. Using MINITAB, the results are:

Regression Analysis: Year_2 versus Year_1

Analysis of Variance

Source	DF	Adj SS	Adj MS	F-Value	P-Value
Regression	1	462350	462350	21.77	0.000
Year_1	1	462350	462350	21.77	0.000
Error	35	743254	21236		
Lack-of-Fit	28	743254	26545	*	*
Pure Error	7	0	0		
Total	36	1205604			

Model Summary

S	R-sq	R-sq(adj)	R-sq(pred)
145.725	38.35%	36.59%	0.00%

Coefficients

Term	Coef	SE Coef	T-Value	P-Value	VIF
Constant	85.0	24.4	3.48	0.001	
Year_1	0.04045	0.00867	4.67	0.000	1.00

Regression Equation

Year_2 = 85.0 + 0.04045 Year_1

The least squares prediction equation is $\hat{y} = 85.0 + 0.04045x$.

b. To determine if the model is adequate for predicting Year 2 PCB concentration,
    we test:

    $$H_0 : \beta_1 = 0$$
    $$H_a : \beta_1 \neq 0$$

    The test statistic is $t = 4.67$ and the p-value is $p = 0.000$.
    Since the p-value is so small, H_0 is rejected. There is sufficient evidence to indi-
    cate the model is adequate for predicting Year 2 PCB concentration for any
    reasonable value of α.

c. Using MINITAB, the plot of the residuals is:

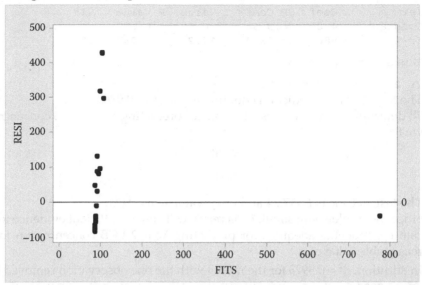

In order for an observation to be considered an outlier, it needs to be more than 3 standard deviations from 0. From the printout, $s = 145.725$. Three standard deviations is $3(145.725) = 437.175$. The largest residual has a value of 429.581 which is less than 437.175. Thus, there is no evidence of any outliers. However, observation #4 looks to be quite unusual with its extremely large value of PCB for Year 1.

d. After removing the extreme value, the results are:

Regression Analysis: Yr_2 versus Yr_1

Analysis of Variance

Source	DF	Adj SS	Adj MS	F-Value	P-Value
Regression	1	717107	717107	297.20	0.000
Yr_1	1	717107	717107	297.20	0.000
Error	34	82039	2413		
Lack-of-Fit	27	82039	3038	*	*
Pure Error	7	0	0		
Total	35	799146			

Model Summary

S	R-sq	R-sq(adj)	R-sq(pred)
49.1213	89.73%	89.43%	85.48%

```
Coefficients

Term         Coef   SE Coef   T-Value   P-Value    VIF
Constant     4.01      9.58      0.42     0.678
Yr_1       0.9863    0.0572     17.24     0.000    1.00

Regression Equation

Yr_2 = 4.01 + 0.9863 Yr_1
```

The least squares prediction equation is $\hat{y} = 4.01 + 0.9863x$.

To determine if the model is adequate for predicting Year 2 PCB concentration, we test:

$$H_0 : \beta_1 = 0$$
$$H_a : \beta_1 \neq 0$$

The test statistic is $t = 17.24$ and the p-value is $p = 0.000$.

Since the p-value is so small, H_0 is rejected. There is sufficient evidence to indicate the model is adequate for predicting Year 2 PCB concentration for any reasonable value of α.

In addition, $R^2 = 0.8973$ for the model with the one observation removed while $R^2 = 0.3835$ for the model with all the observations. This implies a better fit with the one observation removed. Also, Root MSE for the model with the one observation removed is 49.1213 compared to Root MSE for the model with all the observations of 145.725. Again, this implies a better fit for the model with the one observation removed.

e. Using MINITAB, the results of fitting the model to the transformed data is:

Regression Analysis: ln(Yr_2+1) versus ln(Yr_1+1)

```
Analysis of Variance

Source          DF    Adj SS    Adj MS   F-Value   P-Value
Regression       1   145.582   145.582    251.17     0.000
  ln(Yr_1+1)     1   145.582   145.582    251.17     0.000
Error           35    20.286     0.580
  Lack-of-Fit   28    20.286     0.725        *         *
  Pure Error     7     0.000     0.000
Total           36   165.868

Model Summary

      S    R-sq   R-sq(adj)   R-sq(pred)
0.761321  87.77%     87.42%       84.63%

Coefficients

Term            Coef   SE Coef   T-Value   P-Value    VIF
Constant       0.425     0.202      2.10     0.043
ln(Yr_1+1)    0.8508    0.0537     15.85     0.000    1.00

Regression Equation

ln(Yr_2+1) = 0.425 + 0.8508 ln(Yr_1+1)
```

The least squares prediction equation is $\hat{y}^* = 0.425 + 0.8508x^*$.

To determine if the model is adequate for predicting the transformed Year 2 PCB concentration, we test:

$$H_0 : \beta_1 = 0$$
$$H_a : \beta_1 \neq 0$$

The test statistic is $t = 15.85$ and the p-value is $p = 0.000$.

Since the p-value is so small, H_0 is rejected. There is sufficient evidence to indicate the model is adequate for predicting the transformed Year 2 PCB concentration for any reasonable value of α.

$R^2 = 0.8777$. This value is almost as large as the R^2 value for the data when one observation was removed.

The residual plot is:

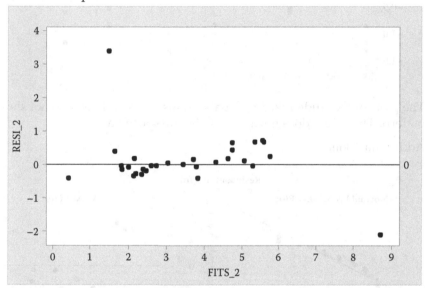

The value of s is $s = 0.761321$. In order for an observation to be considered an outlier, it needs to be more than 3 standard deviations from 0. Three standard deviations is $3(0.761321) = 2.283963$. The largest residual is 3.38931 which is larger than 2.283963. Thus, it appears that there is one outlier.

Also, the residual for Boston Harbor is -2.11567. The z-score for this residual is $z = \dfrac{-2.11567}{0.761321} = -2.779$. Although this value is less than 3 in magnitude, it is still fairly close to 3. This point may be a suspect outlier.

11.49   Observation #11 appears to be a very influential observation as both the _RSTUDENT and _DFFITS values for this observation are very large.

Observations #32, #36, and #47 also appear to be influential as they have large values for either _RSTUDENT or _DFFITS.

11.51   Using MINITAB, plots of the studentized residuals versus the independent variables are:

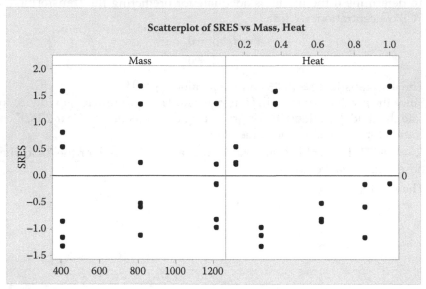

The plots of the studentized residuals versus Mass and Heat do not show any pattern. The model does not appear to be misspecified.

Additional plots are:

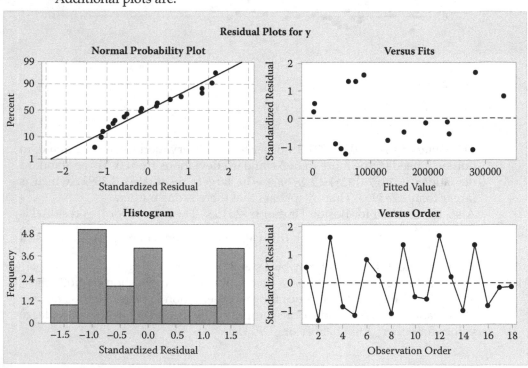

The normal probability plot shows that the plot is not exactly straight. This could indicate that the data may not be normally distributed. Similarly, the histogram does not have a mound-shape. This also indicates the data may not be normally distributed.

The plot of the studentized residuals versus the fitted values does not show any trend. There is no indication that the assumption of constant variance is not valid.

The fitted values (FITS), studentized residuals (SRES), studentized deleted residuals (TRES), and the difference between fits (DFITTS) are shown in the table.

Bubble	FITS	SRES	TRES	DFIT
1	2,691.58666	0.53991	0.52577	0.40584
2	57,111.70061	-1.31603	-1.35471	-0.75600
3	88,208.90858	1.58376	1.68450	0.80477
4	185,387.68349	-0.86281	-0.85445	-0.35388
5	278,679.30740	-1.15829	-1.17382	-0.73637
6	333,099.42135	0.81508	0.80475	0.71387
7	1,800.50681	0.24385	0.23548	0.09870
8	48,292.24715	-1.11014	-1.12020	-0.36301
9	74,858.95591	1.34058	1.38367	0.39220
10	157,879.92079	-0.52010	-0.50609	-0.12626
11	237,580.04707	-0.58303	-0.56876	-0.20302
12	284,071.78741	1.67108	1.79976	0.83245
13	907.22677	0.21728	0.20973	0.16202
14	39,451.01726	-0.96503	-0.96248	-0.53746
15	61,476.04039	1.34297	1.38650	0.66280
16	130,304.23769	-0.80883	-0.79828	-0.33080
17	196,379.30709	-0.16849	-0.16253	-0.10203
18	234,923.09758	-0.14888	-0.14358	-0.12748

There do not appear to be any outliers as no studentized residuals are greater than 3 in magnitude.

There do not appear to be any unusually influential observations. No observations have unusually large values for the studentized deleted residuals (TRES) or the difference between fits (DFITTS).

11.53 a. There appears to be a problem with the assumption of equal variances. As the values of $\hat{y}$ increase, the spread of the residuals increases. This indicates that as $\hat{y}$ increases, the variance increases.

b. Since the data appear to be Poisson in nature, we would suggest the variance stabilizing transformation:

$$y^* = \sqrt{y}$$

11.55   Using MINITAB, pairwise correlations are:

Correlation: y, x1, x2, x3

```
            Y        x1        x2
x1    -0.368
       0.084

x2     0.310     0.083
       0.150     0.705

x3     0.054     0.201    -0.198
       0.808     0.359     0.366
```

```
Cell Contents: Pearson correlation
               P-Value
```

No pairwise correlation among the independent variables is large. Therefore, there is no indication of multicollinearity. Also, the estimate of the beta coefficients in the fitted regression model should be negative for $\hat{\beta}_1$ and positive for $\hat{\beta}_2$ and $\hat{\beta}_3$. From Exercise 11.23, the signs of the estimates of the beta parameters are what we would expect. This indicates that no multicollinearity exists.

11.57   One possible reason why this phenomenon occurred is that x_1 and x_2 could be highly correlated.

11.59   To determine if the model is adequate, we test:

$$H_0 : \beta_1 = \beta_2 = 0$$

$$H_a : \text{At least one } \beta_i \neq 0$$

We are unable to test for model adequacy since there are no degrees of freedom for estimating σ^2, $(df = n - (k+1) = 3 - (2+1) = 0)$.

11.61 a.   Using MINITAB, the results are:

Regression Analysis: Y versus X1

```
Analysis of Variance

Source        DF   Adj SS   Adj MS   F-Value   P-Value
Regression     1   494.28   494.281   253.37     0.000
  X1           1   494.28   494.281   253.37     0.000
Error         23    44.87     1.951
Total         24   539.15
```

```
Model Summary

      S     R-sq   R-sq(adj)   R-sq(pred)
1.39672   91.68%      91.32%       86.54%
```

```
Coefficients

Term        Coef   SE Coef   T-Value   P-Value    VIF
Constant   2.743     0.675      4.06     0.000
X1        0.8010    0.0503     15.92     0.000   1.00
```

```
Regression Equation

Y = 2.743 + 0.8010 X1
```

The least squares prediction equation is $\hat{y} = 2.743 + 0.801x_1$.

To determine if tar content is useful for predicting carbon monoxide content, we test:

$$H_0 : \beta_1 = 0$$
$$H_a : \beta_1 \neq 0$$

The test statistic is $t = 15.92$ and the p-value is $p = 0.000$.

Since the p-value is so small, H_0 is rejected. There is sufficient evidence to indicate tar content is useful for predicting carbon monoxide for any reasonable value of α.

b. Using MINITAB, the results are:

Regression Analysis: Y versus X2

Analysis of Variance

Source	DF	Adj SS	Adj MS	F-Value	P-Value
Regression	1	462.256	462.256	138.27	0.000
X2	1	462.256	462.256	138.27	0.000
Error	23	76.894	3.343		
Lack-of-Fit	21	71.564	3.408	1.28	0.530
Pure Error	2	5.330	2.665		
Total	24	539.150			

Model Summary

S	R-sq	R-sq(adj)	R-sq(pred)
1.82845	85.74%	85.12%	78.40%

Coefficients

Term	Coef	SE Coef	T-Value	P-Value	VIF
Constant	1.665	0.994	1.68	0.107	
X2	12.40	1.05	11.76	0.000	1.00

Regression Equation

Y = 1.665 + 12.40 X2

The least squares prediction equation is $\hat{y} = 1.665 + 12.40x_2$.

To determine if nicotine content is useful for predicting carbon monoxide content, we test:

$$H_0 : \beta_2 = 0$$
$$H_a : \beta_2 \neq 0$$

The test statistic is $t = 11.76$ and the p-value is $p = 0.000$.

Since the p-value is so small, H_0 is rejected. There is sufficient evidence to indicate nicotine content is useful for predicting carbon monoxide for any reasonable value of α.

c. Using MINITAB, the results are:

Regression Analysis: Y versus X3

Analysis of Variance

Source	DF	Adj SS	Adj MS	F-Value	P-Value
Regression	1	116.1	116.06	6.31	0.019
X3	1	116.1	116.06	6.31	0.019

```
Error        23    423.1    18.40
Total        24    539.2
```

Model Summary

```
      S    R-sq   R-sq(adj)   R-sq(pred)
4.28898   21.53%     18.11%        1.33%
```

Coefficients

```
Term          Coef   SE Coef   T-Value   P-Value    VIF
Constant    -11.80      9.72     -1.21     0.237
X3           25.07      9.98      2.51     0.019   1.00
```

Regression Equation

```
Y = -11.80 + 25.07 X3
```

The least squares prediction equation is $\hat{y} = -11.80 + 25.07 x_3$.

To determine if weight is useful for predicting carbon monoxide content, we test:

$$H_0 : \beta_3 = 0$$
$$H_a : \beta_3 \neq 0$$

The test statistic is $t = 2.51$ and the p-value is $p = 0.019$.

Since the p-value is so small, H_0 is rejected. There is sufficient evidence to indicate weight is useful for predicting carbon monoxide for any value of $\alpha > 0.019$.

d. In Example 11.14, $\hat{\beta}_1 = 0.96257$, $\hat{\beta}_2 = -2.63166$, and $\hat{\beta}_3 = -0.13048$. In parts a, b, and c above, the parameter estimates of $\hat{\beta}_1$, $\hat{\beta}_2$ and $\hat{\beta}_3$ are all positive.

11.63 a. $\mathbf{Y} = \begin{bmatrix} 0.231 \\ 0.107 \\ 0.053 \\ 0.129 \\ 0.069 \\ 0.030 \\ 1.005 \\ 0.559 \\ 0.321 \\ 2.948 \\ 1.633 \\ 0.934 \end{bmatrix}$ $\mathbf{X} = \begin{bmatrix} 1 & 740 & 1.10 \\ 1 & 740 & 0.62 \\ 1 & 740 & 0.31 \\ 1 & 805 & 1.10 \\ 1 & 805 & 0.62 \\ 1 & 805 & 0.31 \\ 1 & 980 & 1.10 \\ 1 & 980 & 0.62 \\ 1 & 980 & 0.31 \\ 1 & 1235 & 1.10 \\ 1 & 1235 & 0.62 \\ 1 & 1235 & 0.31 \end{bmatrix}$

b.

$$\mathbf{X'X} = \begin{bmatrix} 1 & 1 & 1 & \cdots & 1 \\ 740 & 740 & 740 & \cdots & 1235 \\ 1.10 & 0.62 & 0.31 & \cdots & 0.31 \end{bmatrix} \begin{bmatrix} 1 & 740 & 1.10 \\ 1 & 740 & 0.62 \\ 1 & 740 & 0.31 \\ \vdots & \vdots & \vdots \\ 1 & 1235 & 0.31 \end{bmatrix}$$

$$= \begin{bmatrix} 12 & 11280 & 8.12 \\ 11280 & 11043750 & 7632.8 \\ 8.12 & 7632.8 & 6.762 \end{bmatrix}$$

$$\mathbf{X'Y} = \begin{bmatrix} 1 & 1 & 1 & \cdots & 1 \\ 740 & 740 & 740 & \cdots & 1235 \\ 1.10 & 0.62 & 0.31 & \cdots & 0.31 \end{bmatrix} \begin{bmatrix} 0.231 \\ 0.107 \\ 0.053 \\ \vdots \\ 0.934 \end{bmatrix} = \begin{bmatrix} 8.019 \\ 9131.205 \\ 6.62724 \end{bmatrix}$$

c. $\mathbf{X'X^{-1}} = \begin{bmatrix} 2.4502646 & -0.0021340 & -0.5338730 \\ -0.0021340 & 0.00000227 & 0.0000000 \\ -0.5338730 & 0.0000000 & 0.7889754 \end{bmatrix}$

d. $\hat{\beta} = \mathbf{X'X^{-1}X'Y} = \begin{bmatrix} 2.4502646 & -0.0021340 & -0.5338730 \\ -0.0021340 & 0.00000227 & 0.0000000 \\ -0.5338730 & 0.0000000 & 0.7889754 \end{bmatrix} \begin{bmatrix} 8.019 \\ 9131.205 \\ 6.62724 \end{bmatrix} = \begin{bmatrix} -3.372673 \\ 0.0036167 \\ 0.9475989 \end{bmatrix}$

The least squares prediction equation is $\hat{y} = -3.3727 + 0.00362x_1 + 0.9476x_2$.

e. $\mathbf{YY} = \begin{bmatrix} 0.231 & 0.107 & 0.053 & \cdots & 0.934 \end{bmatrix} \begin{bmatrix} 0.231 \\ 0.107 \\ 0.053 \\ \vdots \\ 0.934 \end{bmatrix} = 13.745217$

$$\hat{\beta}'\mathbf{X'Y} = \begin{bmatrix} -3.372673 & 0.0036167 & 0.9475989 \end{bmatrix} \begin{bmatrix} 8.019 \\ 9131.205 \\ 6.62724 \end{bmatrix} = 12.25932967$$

$\mathbf{SSE} = \mathbf{Y'Y} - \hat{\beta}'\mathbf{X'Y} = 13.745217 - 12.25932967 = 1.4857275$

$s^2 = \dfrac{SSE}{n-3} = \dfrac{1.4857275}{12-3} = 0.165080833 \qquad s = \sqrt{0.165080833} = 0.4063$

Most of the observed values of optical density will fall within $2s = 2(0.4063) = 0.8126$ units of their least squares predicted values.

f. MINITAB was used to find the rest of the calculations and the results are:

```
Regression Analysis: Y versus X1, X2

Analysis of Variance

Source        DF   Adj SS   Adj MS   F-Value   P-Value
Regression     2    6.901   3.4504     20.90     0.000
   X1          1    5.763   5.7627     34.91     0.000
   X2          1    1.138   1.1381      6.89     0.028
Error          9    1.486   0.1651
Total         11    8.387

Model Summary

       S    R-sq   R-sq(adj)   R-sq(pred)
0.406301   82.28%     78.35%       56.75%

Coefficients

Term           Coef    SE Coef   T-Value   P-Value    VIF
Constant     -3.373      0.636     -5.30     0.000
X1         0.003617   0.000612      5.91     0.000   1.00
X2            0.948      0.361      2.63     0.028   1.00

Regression Equation

Y = -3.373 + 0.003617 X1 + 0.948 X2
```

$R_{adj}^2 = 0.7835$. 78.35% of the variation in the sampled optical density values around their means can be explained by the relationship with band frequency and film thickness, adjusted for the sample size and the number of parameters in the model.

g. To determine if the overall model is adequate, we test:

$$H_0 : \beta_1 = \beta_2 = 0$$
$$H_a : \text{At least one } \beta_i \neq 0$$

The test statistic is $F = 20.90$ and the p-value is $p = 0.000$. Since the p-value is less than $\alpha(p = 0.000 < 0.10)$, H_0 is rejected. There is sufficient evidence to indicate the model is adequate for predicting optical density at $\alpha = 0.10$.

h. For confidence coefficient 0.90, $\alpha = 0.10$ and $\alpha/2 = 0.10/2 = 0.05$. From Table 7, Appendix B, with $df = n - 3 = 12 - 3 = 9$, $t_{0.05} = 1.833$. The 90% confidence interval is

$$\hat{\beta}_1 \pm t_{\alpha/2} \hat{\sigma}_{\hat{\beta}_1} \Rightarrow 0.00362 \pm 1.833(0.000612) \Rightarrow 0.00362 \pm 0.00112 \Rightarrow (0.00250, 0.00474)$$

We are 90% confident that the mean optical density will increase between 0.00250 and 0.00474 units for every one cm^{-1} increase in band frequency, holding film thickness constant.

i. For confidence coefficient 0.90, $\alpha = 0.10$ and $\alpha/2 = 0.10/2 = 0.05$. From Table 7, Appendix B, with $df = n - 3 = 12 - 3 = 9$, $t_{0.05} = 1.833$. The 90% confidence interval is

$$\hat{\beta}_2 \pm t_{\alpha/2} \hat{\sigma}_{\hat{\beta}_2} \Rightarrow 0.948 \pm 1.833(0.361) \Rightarrow 0.948 \pm 0.6617 \Rightarrow (0.2863, 1.6097)$$

We are 90% confident that the mean optical density will increase between 0.2863 and 1.6097 units for every one millimeter increase in film thickness, holding band frequency constant.

j. Using MINITAB, the results are:

Prediction for Y

Regression Equation

Y = -3.373 + 0.003617 X1 + 0.948 X2

Variable	Setting
X1	950
X2	0.62

Fit	SE Fit	90% CI	90% PI
0.650720	0.119216	(0.432184, 0.869256)	(-0.125476, 1.42692)

The 90% confidence interval is (−0.1255, 1.4269). We are 90% confident that the actual optical density will be between -0.1255 and 1.4269 when the band frequency is 950 cm^{-1} and the film thickness is 0.62 millimeters. Sine optical density cannot be negative, the interval is (0, 1.4269).

11.65 a. Using MINITAB, the results are:

Regression Analysis: Yield versus Temp, Pressure, TxP

Analysis of Variance

Source	DF	Adj SS	Adj MS	F-Value	P-Value
Regression	3	167.504	55.8346	74.57	0.001
Temp	1	4.325	4.3245	5.78	0.074
Pressure	1	31.460	31.4603	42.02	0.003
TxP	1	35.701	35.7013	47.68	0.002
Error	4	2.995	0.7487		
Total	7	170.499			

Model Summary

S	R-sq	R-sq(adj)	R-sq(pred)
0.865303	98.24%	96.93%	92.97%

Coefficients

Term	Coef	SE Coef	T-Value	P-Value	VIF
Constant	10.62	5.88	1.81	0.145	
Temp	0.2325	0.0967	2.40	0.074	10.00
Pressure	2.413	0.372	6.48	0.003	37.00
TxP	-0.04225	0.00612	-6.91	0.002	46.00

Regression Equation

Yield = 10.62 + 0.2325 Temp + 2.413 Pressure - 0.04225 TxP

The least squares prediction equation is
$\hat{y} = 10.62 + 0.2325x_1 + 2.413x_2 - 0.04225x_1x_2$.

b. To determine if the overall model is adequate, we test:

$$H_0 : \beta_1 = \beta_2 = \beta_3 = 0$$
$$H_a : \text{At least one } \beta_i \neq 0$$

The test statistic is $F = 74.57$ and the p-value is $p = 0.001$. Since the p-value is less than $\alpha(p = 0.000 < 0.01)$, H_0 is rejected. There is sufficient evidence to indicate the model is adequate for predicting yield at $\alpha = 0.01$.

c. To determine if the interaction between temperature and pressure is significant, we test:

$$H_0 : \beta_3 = 0$$
$$H_a : \beta_3 \neq 0$$

The test statistic is $t = -6.9$ and the p-value is $p = 0.002$. Since the p-value is less than $\alpha(p = 0.002 < 0.01)$, H_0 is rejected. There is sufficient evidence to indicate temperature and pressure interact at $\alpha = 0.01$.

11.67 a. Using MINITAB, the results are:

Regression Analysis: Permeability versus Porosity, Slope

Analysis of Variance

Source	DF	Adj SS	Adj MS	F-Value	P-Value
Regression	2	1.65932	0.82966	35.84	0.000
Porosity	1	0.07840	0.07840	3.39	0.115
Slope	1	1.65932	1.65932	71.69	0.000
Error	6	0.13888	0.02315		
Total	8	1.79820			

Model Summary

S	R-sq	R-sq(adj)	R-sq(pred)
0.152141	92.28%	89.70%	80.28%

Coefficients

Term	Coef	SE Coef	T-Value	P-Value	VIF
Constant	0.132	0.190	0.69	0.513	
Porosity	-9.31	5.06	-1.84	0.115	1.05
Slope	1.558	0.184	8.47	0.000	1.05

Regression Equation

Permeability = 0.132 - 9.31 Porosity + 1.558 Slope

The prediction equation is $\hat{y} = 0.132 - 9.31x_1 + 1.558x_2$.

$\hat{\beta}_0 = 0.132$: Since $x = 0$ and $x_2 = 0$ is not in the observed range, $\hat{\beta}_0$ has no meaningful interpretation. It is simply the y-intercept.

$\hat{\beta}_1 = -9.31$: For each unit increase in porosity, the coefficient of permeability is estimated to decrease by 9.31, holding slope constant.

$\hat{\beta}_2 = 1.558$: For each unit increase in slope, the coefficient of permeability is estimated to increase by 1.558, holding porosity constant.

b. To determine if the overall model is useful, we test:

$$H_0 : \beta_1 = \beta_2 = 0$$
$$H_a : \text{At least one} \, \beta_i \neq 0$$

The test statistic is $F = 35.84$ and the p-value is $p = 0.001$. Since the p-value is so small, H_0 is rejected. There is sufficient evidence to indicate the model is adequate for predicting the coefficient of permeability for any reasonable value of α.

c. To determine if concrete porosity is a useful predictor of permeability, we test:

$$H_0 : \beta_1 = 0$$
$$H_a : \beta_i \neq 0$$

The test statistic is $t = -1.84$ and the p-value is $p = 0.115$. Since the p-value is not less than $\alpha(p = 0.115 \nless 0.05)$, H_0 is not rejected. There is insufficient evidence to indicate concrete porosity is a useful predictor of permeability, adjusted for the slope at $\alpha = 0.05$.

d. To determine if the estimated water outflow-time slope is a useful predictor of permeability, we test:

$$H_0 : \beta_2 = 0$$

$$H_a : \beta_2 \neq 0$$

The test statistic is $t = 8.47$ and the p-value is $p = 0.000$. Since the p-value is less than $\alpha(p = 0.000 < 0.05)$, H_0 is rejected. There is sufficient evidence to indicate the estimated water outflow-time slope is a useful predictor of permeability, adjusted for porosity at $\alpha = 0.05$.

e. $R^2 = 0.9228$: 92.28% of the variation in the sampled permeability values around their means is explained by the model containing concrete porosity and the estimated water outflow-time slope.

f. The estimate of σ is $s = 0.152141$. Most of the observed values of the permeability values will fall within $2s = 2(0.152141) = 0.304282$ units of their least squares predicted values.

g. Using MINITAB, the results are:

Prediction for Permeability

```
Regression Equation

Permeability = 0.132 - 9.31 Porosity + 1.558 Slope

Variable   Setting
Porosity      0.05
Slope          0.3

     Fit      SE Fit          95% CI                    95% PI
0.133934   0.0876404   (-0.0805147, 0.348382)   (-0.295691, 0.563558)
```
The 95% prediction interval is (−0.2957, 0.5636). we are 95% confident that the actual permeability value will fall between -0.2957 and 0.5636 when the concrete porosity is 0.05 and the slope is 0.30. Since the coefficient of permeability cannot be negative, the interval will be (0, 0.5636).

11.69 a.   The hypothesized model would be $y = \beta_0 + \beta_1 x + \beta_2 x^2 + \varepsilon$.

   b.   Using MINITAB, the scatterplot of the data is:

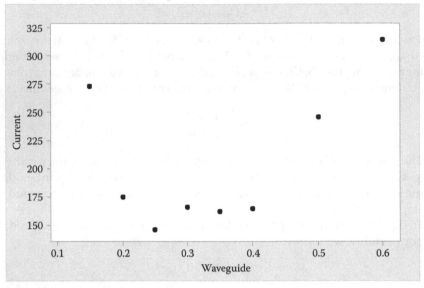

It appears that the researchers' theory is correct.

   c.   Using MINITAB, the results are:

Regression Analysis: Current versus Waveguide, Wave-sq

Analysis of Variance

Source	DF	Adj SS	Adj MS	F-Value	P-Value
Regression	2	24163	12081.3	18.37	0.005
Waveguide	1	14617	14617.3	22.22	0.005
Wave-sq	1	18600	18599.6	28.28	0.003
Error	5	3289	657.8		
Total	7	27452			

Model Summary

S	R-sq	R-sq(adj)	R-sq(pred)
25.6472	88.02%	83.23%	31.38%

Coefficients

Term	Coef	SE Coef	T-Value	P-Value	VIF
Constant	438.3	60.5	7.24	0.001	
Waveguide	-1684	357	-4.71	0.005	31.48
Wave-sq	2502	471	5.32	0.003	31.48

Regression Equation

Current = 438.3 - 1684 Waveguide + 2502 Wave-sq

The fitted regression model is $\hat{y} = 438.3 - 1684x + 2502x^2$.

d. To determine if there is a curvilinear relationship between threshold current and waveguide al mole fraction, we test:

$$H_0 : \beta_2 = 0$$
$$H_a : \beta_2 \neq 0$$

The test statistic is $t = 5.32$ band the p-value is $p = 0.003$. Since the p-value is less than $\alpha(p = 0.003 < 0.10)$, H_0 is rejected. There is sufficient evidence to indicate there is a curvilinear relationship between threshold current and waveguide al mole fraction at $\alpha = 0.10$.

11.71   Using MINITAB, the results are:

Regression Analysis: Unrooted versus Length, Len-sq

Analysis of Variance

Source	DF	Adj SS	Adj MS	F-Value	P-Value
Regression	2	115583	57792	26.66	0.002
Length	1	17101	17101	7.89	0.038
Len-sq	1	38312	38312	17.68	0.008
Error	5	10837	2167		
Total	7	126420			

Model Summary

S	R-sq	R-sq(adj)	R-sq(pred)
46.5546	91.43%	88.00%	53.16%

Coefficients

Term	Coef	SE Coef	T-Value	P-Value	VIF
Constant	112.1	65.0	1.73	0.145	
Length	-93.0	33.1	-2.81	0.038	21.25
Len-sq	15.10	3.59	4.20	0.008	21.25

Regression Equation

Unrooted = 112.1 - 93.0 Length + 15.10 Len-sq

To determine if there is an upward concave curvilinear relationship, we test:

$$H_0 : \beta_2 = 0$$
$$H_a : \beta_2 > 0$$

The test statistic is $t = 4.20$ and the p-value is $p = 0.008/2 = 0.004$. Since the p-value is less than $\alpha(p = 0.004 < 0.10)$, H_0 is rejected. There is sufficient evidence to indicate there is an upward concave curvilinear relationship between unrooted walks and walk length at $\alpha = 0.10$.

11.73   Using MINITAB, the results are:

Regression Analysis: Acid versus Oxidant

Analysis of Variance

Source	DF	Adj SS	Adj MS	F-Value	P-Value
Regression	1	2.414	2.4136	17.08	0.001
Oxidant	1	2.414	2.4136	17.08	0.001
Error	17	2.402	0.1413		

```
    Lack-of-Fit  13   1.382  0.1063     0.42     0.897
    Pure Error    4   1.020  0.2551
Total            18   4.816
```

Model Summary

```
        S    R-sq  R-sq(adj)  R-sq(pred)
0.375918  50.12%     47.18%      36.48%
```

Coefficients

```
Term         Coef  SE Coef  T-Value  P-Value   VIF
Constant   -0.024    0.246    -0.10    0.924
Oxidant   0.01958  0.00474     4.13    0.001  1.00
```

Regression Equation

```
Acid = -0.024 + 0.01958 Oxidant
```

The residual plots are:

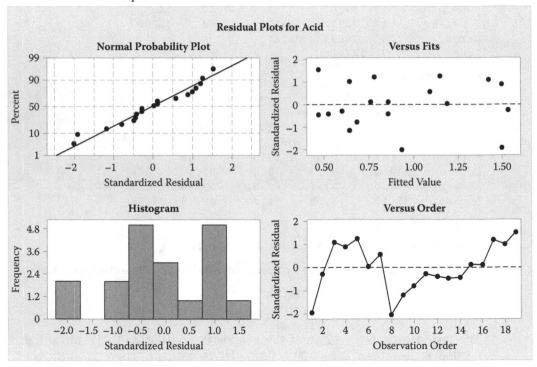

First, we will look at the standardized residuals versus the fitted values. There is no U shape or upside down U shape. This indicates that the model has not been misspecified. Also, there is no funnel shape or football shape. This indicates that the assumption of equal variances is not violated.

Next, we look at the normal probability plot. The pot forms a fairly straight line. This indicates that the assumption of normality is not violates. This is also confirmed in the histogram. The histogram is fairly mound-shaped.

The data were not collected sequentially, so we cannot check for independence. There are no apparent outliers because no standardized residuals are greater than 3 in magnitude.

From the above analyses, it appears that all the assumptions are met.

11.75 a. Using MINITAB, the scatterplot of the data is:

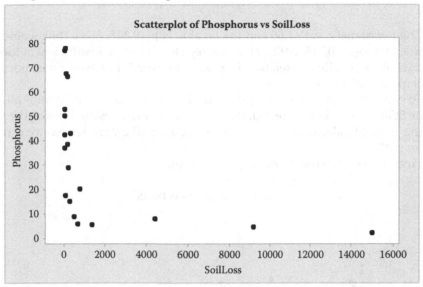

There appears to be a curvilinear trend to the data.

b. Using MINITAB, the results are:

Regression Analysis: Phosphorus versus SoilLoss, Soil-sq

Analysis of Variance

Source	DF	Adj SS	Adj MS	F-Value	P-Value
Regression	2	4325	2162.7	4.55	0.026
SoilLoss	1	2423	2422.7	5.09	0.037
Soil-sq	1	1312	1312.1	2.76	0.115
Error	17	8085	475.6		
Total	19	12411			

Model Summary

S	R-sq	R-sq(adj)	R-sq(pred)
21.8086	34.85%	27.19%	0.00%

Coefficients

Term	Coef	SE Coef	T-Value	P-Value	VIF
Constant	42.25	5.71	7.40	0.000	
SoilLoss	-0.01140	0.00505	-2.26	0.037	14.87
Soil-sq	0.000001	0.000000	1.66	0.115	14.87

Regression Equation

Phosphorus = 42.25 - 0.01140 SoilLoss + 0.000001 Soil-sq

The least squares prediction equation is $\hat{y} = 42.25 - 0.0114x + 0.000001x^2$.

c.   To determine if a curvilinear relationship exists between dissolved phosphorus percentage and soil loss, we test:

$$H_0 : \beta_2 = 0$$
$$H_a : \beta_2 \neq 0$$

The test statistic is $t = 1.66$ and the p-value is $p = 0.115$. Since the p-value is not less than $\alpha(p = 0.115 \not< 0.05)$, H_0 is not rejected. There is insufficient evidence to indicate a curvilinear relationship exists between dissolved phosphorus percentage and soil loss at $\alpha = 0.05$.

11.77   To conduct a residual analysis, the model including depth and depth^2 must first be fit and the residuals created. We then analyze these residuals using the following plots. Standardized residuals were used for all graphs so that we can check for outliers.

First, we analyze the residuals plotted depth.

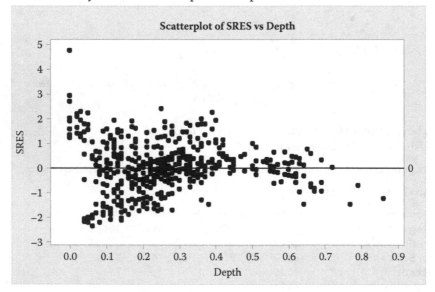

There still appears to be some curvature to the plot. It may be helpful to add depth^3 to the model.

Next, we will look at the normal probability plot, the histograms of the residuals and the residuals versus the fitted values.

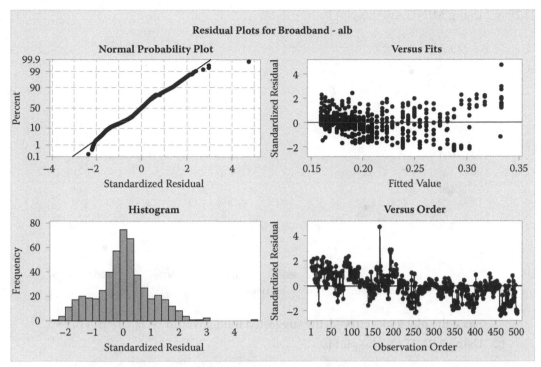

Looking at the normal probability plot, the plot forms a fairly straight line. However, there is one point that is way off the line. This indicates it may be an outlier. Other than this one point, the data look fairly normal. This same conclusion can be drawn from the histogram of the residuals. The histogram indicates that the data are fairly mound shaped except for the one data point.

Looking at the plot of the residuals versus the fitted values, we see a somewhat funnel shape. As the fitted values increase, the spread of the residuals tends to increases. This indicates that the data do not have constant variance.

There is one standardized residual that is greater than 3 in magnitude. This indicates that this point is an outlier. There are also several suspect outliers. We need to go back to the original data set to see if there are any errors in recording for these data points. These outliers could also indicate that other important independent variables have been left out of the model.

The model may be misspecified. We may be able to improve the model if we add depth^3 to the model. Also, because the assumption of equal variance appears to be violated, we could also try transforming the data.

11.79 a.   Using MINITAB, the plot is:

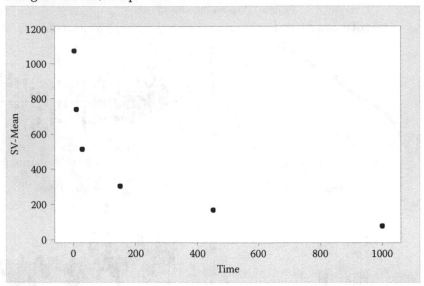

Since the plot looks curved, the suggested model would be $E(s_v) = \beta_0 + \beta_1 x + \beta_2 x^2$.

   b.   Using MINITAB, the plot is:

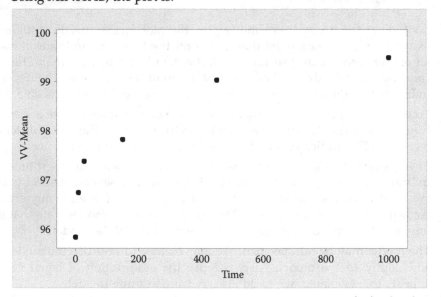

Since the plot looks curved, the suggested model would be $E(V_v) = \beta_0 + \beta_1 x + \beta_2 x^2$.

   c.   Using MINITAB, the results are:

Regression Analysis: SV-Mean versus Time

Analysis of Variance

Source	DF	Adj SS	Adj MS	F-Value	P-Value
Regression	1	412210	412210	5.36	0.082
Time	1	412210	412210	5.36	0.082
Error	4	307467	76867		

```
Total        5   719677
```

Model Summary

```
     S     R-sq  R-sq(adj)  R-sq(pred)
277.248  57.28%    46.60%       0.00%
```

Coefficients

```
Term          Coef  SE Coef  T-Value  P-Value   VIF
Constant       675      142     4.75    0.009
Time        -0.728    0.314    -2.32    0.082  1.00
```

Regression Equation

```
SV-Mean = 675 - 0.728 Time
```

Plots of the residuals are:

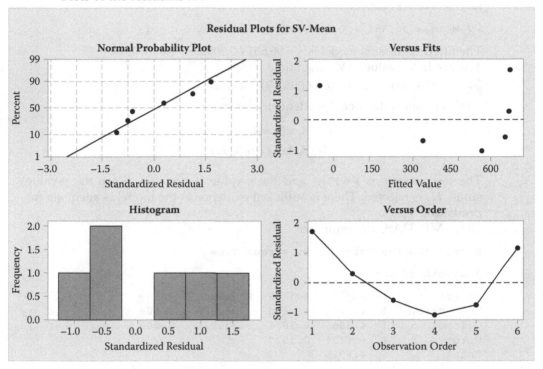

From the plot of the residuals versus the fitted values, we see a U shaped trend. This indicates that the model is misspecified. A second-order model is recommended. From the normal probability plot, there is no indication that the data are not normal. From the data, the standard deviation decreases as the time increases. This indicates that the assumption of constant variance is not met. A variance stabilizing transformation is recommended.

d. Using MINITAB, the results are:

Regression Analysis: VV-Mean versus Time, t-sq

Analysis of Variance

```
Source       DF  Adj SS  Adj MS  F-Value  P-Value
Regression    2  8.5805  4.2903    13.61    0.031
```

```
Time         1   3.2921   3.2921    10.44     0.048
t-sq         1   1.3969   1.3969     4.43     0.126
Error        3   0.9460   0.3153
Total        5   9.5265
```

Model Summary

```
       S    R-sq   R-sq(adj)   R-sq(pred)
0.561552  90.07%      83.45%        0.00%
```

Coefficients

```
Term          Coef   SE Coef   T-Value   P-Value    VIF
Constant    96.551     0.328    294.17     0.000
Time       0.00823   0.00255      3.23     0.048  16.00
t-sq      -0.000005  0.000003     -2.10     0.126  16.00
```

Regression Equation

VV-Mean = 96.551 + 0.00823 Time - 0.000005 t-sq

The fitted regression model is $\hat{y} = 96.551 + 0.00823x - 0.000005x^2$.
The predicted value of V_v when time = 150 is
$\hat{y} = 96.551 + 0.00823(150) - 0.000005(150)^2 = 97.67$.

To determine if the model is adequate, we test:

$$H_0 : \beta_1 = \beta_2 = 0$$
$$H_a : \text{At least one } \beta_i \neq 0$$

The test statistic is $F = 13.61$ and the p-value is $p = 0.031$. Since the p-value small, H_0 is rejected. There is sufficient evidence to the model is adequate for predicting V_v.

e. Using MINITAB, the results are:

Regression Analysis: ln-Sv versus Time

Analysis of Variance

```
Source        DF   Adj SS   Adj MS   F-Value   P-Value
Regression     1   4.3866   4.3866     28.72     0.006
  Time         1   4.3866   4.3866     28.72     0.006
Error          4   0.6109   0.1527
Total          5   4.9975
```

Model Summary

```
       S    R-sq   R-sq(adj)   R-sq(pred)
0.390811  87.78%      84.72%        48.30%
```

Coefficients

```
Term          Coef   SE Coef   T-Value   P-Value    VIF
Constant     6.468     0.200     32.29     0.000
Time     -0.002375  0.000443     -5.36     0.006   1.00
```

```
Regression Equation

ln-Sv = 6.468 - 0.002375 Time
```

The fitted regression model is $\widehat{\log(y)} = 6.468 - 0.002375x$.

f. To determine if the model is adequate, we test:

$$H_0 : \beta_1 = 0$$
$$H_a : \beta_1 \neq 0$$

The test statistic is $t = -5.36$ and the p-value is $p = 0.006$. Since the p-value is less than $\alpha(p = 0.006 < 0.05)$, H_0 is rejected. There is sufficient evidence to the model is adequate for predicting $\ln(S_v)$ at $\alpha = 0.05$.

g. Using MINITAB, the results are:

Prediction for ln-Sv

```
Regression Equation

ln-Sv = 6.468 - 0.002375 Time

Variable  Setting
Time          150

    Fit    SE Fit         95% CI               95% PI
6.11152  0.168636  (5.64331, 6.57973)  (4.92975, 7.29329)
```

With the transformed data, the 95% prediction interval for $\ln(S_v)$ is (4.92975, 7.29329). Taking the antilogs of both ends of the interval, we get (138.34, 1,470.40). We are 95% confident that the true value of S_v is between 138.34 and 1,470.40 when time is equal to 150 minutes.

12

Model Building

12.1 a. The dependent variable, y, is the nitrate concentration of a water sample.
b. The independent variable is water source. It is a qualitative variable with three levels-groundwater, subsurface flow, and over ground flow.

12.3 a. Type of sheathing is qualitative. It has 3 levels.
b. Limit state observed at peak is qualitative. It has 3 levels.
c. Peak load of single stud is measured in kilo-Newtons. This is a quantitative variable.
d. Linear position transducer displacement is measured in millimeters. This is a quantitative variable.

12.5 a. Number of preincident psychological symptoms is quantitative. This number could range from 0 to about 20.
b. Years of experience is quantitative. Years of experience could range from 0 to about 50.
c. Cigarette smoking behavior is qualitative. Levels of this variable might be "smokes" or "does not smoke".
d. Level of social support is qualitative. Levels of this variable might be "gets no social support", "gets some social support", or "gets much social support".
e. Marital status is qualitative. Levels of this variable might be "single", "married", "divorced", "widowed", or "other".
f. Age is quantitative. Age could range from 18 to about 70.
g. Ethnic status is qualitative. Levels of this variable might be "White", "American Indian", "Black", "Asian", "Hispanic" or "other".
h. Exposure to chemical fire is qualitative. Levels of this variable might be "exposed to chemical fire" or "not exposed to chemical fire".
i. Educational level is qualitative. Levels might include "Junior High School", "High School graduate", "some college", "College graduate", or "advanced degree".
j. Distance lived from site of incident is quantitative. The distance could range from 0 to about 100 miles.
k. Gender is qualitative. Levels of gender could be "male" or "female".

12.7 a. Temperature is quantitative as the values will be 50°,, 60°,, 75°, or 90°,.
b. Relative humidity is quantitative as the values will be 30%, 50%, or 70%.
c. Organic compound is qualitative as the values will fall into one of the following categories: benzene, toluene, chloroform, methanol, or anisole.

12.9 a. First-order: b. Second-order

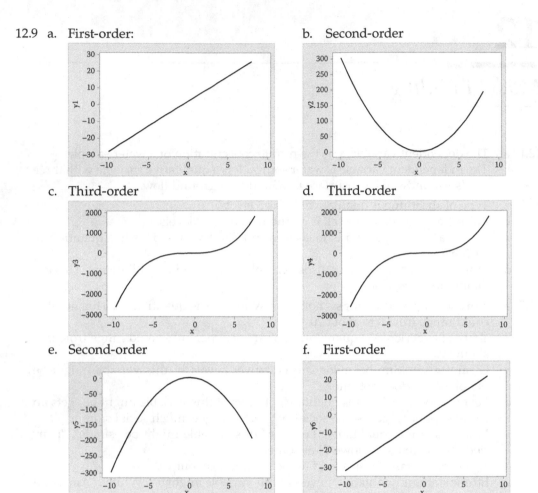

c. Third-order   d. Third-order

e. Second-order   f. First-order

12.11 a. For a curvilinear relationship, the model would be $E(y) = \beta_0 + \beta_1 x_2 + \beta_2 x_2^2$. A sketch of what this relationship might look like is:

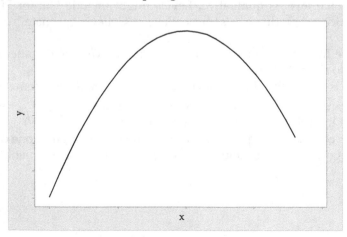

b. For a third-order relationship, the model would be $E(y) = \beta_0 + \beta_1 x_1 + \beta_2 x_1^2 + \beta_3 x_1^3$. A sketch of what this relationship might look like is:

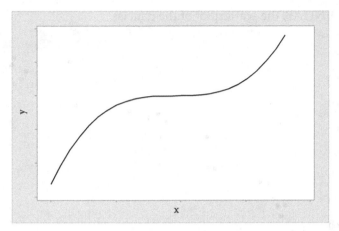

12.13 a. $R^2 = 0.73$. 73% of the total sample variation of the lane utilization values about their means is explained by the second-order model with total traffic flow.

b. To determine if the overall model is useful, we test:

$$H_0 : \beta_1 = \beta_2 = 0$$

$$H_a : \text{At least one } \beta_i \neq 0$$

The test statistic is $F = \dfrac{R^2 / k}{\left(1 - R^2\right)\big/\left[n - \left(k+1\right)\right]} = \dfrac{0.73/2}{(1 - 0.73)/[2000 - (2+1)]} = 2,699.65.$

The rejection region requires $\alpha = 0.01$ in the upper tail of the F distribution with $v_1 = k = 2$ and $v_2 = n - (k+1) = 2,000 - (2+1) = 1,997$. Using MINITAB, $F_{0.01} = 4.616$. The rejection region is $F > 4.616$.

Since the observed value of the test statistic falls in the rejection region ($F = 2,699.65 > 4.616$), H_0 is rejected. There is sufficient evidence to indicate the overall model is adequate at $\alpha = 0.01$.

c. A possible sketch of the model is:

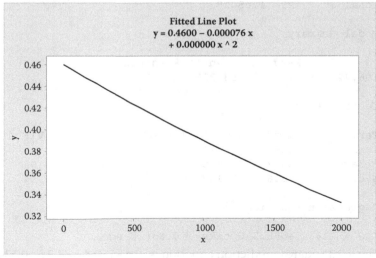

d. A polynomial that could fit the data is $E(y) = \beta_0 + \beta_1 x + \beta_2 x^2$.

12.15 a.   Using MINITAB, the scatterplot of the data is:

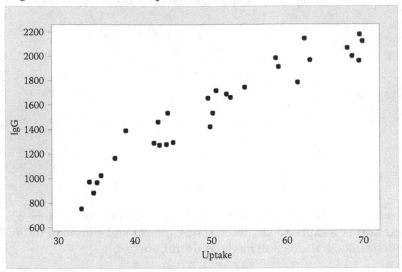

b.   There appears to be a slight curve to the scattergram. Thus, the model would be $E(y) = \beta_0 + \beta_1 x + \beta_2 x^2$.

c.   Using MINITAB, the results are:

Regression Analysis: IgG versus Uptake, U-sq

Analysis of Variance

Source	DF	Adj SS	Adj MS	F-Value	P-Value
Regression	2	4602211	2301105	203.16	0.000
Uptake	1	325476	325476	28.74	0.000
U-sq	1	130164	130164	11.49	0.002
Error	27	305818	11327		
Total	29	4908029			

Model Summary

S	R-sq	R-sq(adj)	R-sq(pred)
106.427	93.77%	93.31%	92.48%

Coefficients

Term	Coef	SE Coef	T-Value	P-Value	VIF
Constant	-1464	411	-3.56	0.001	
Uptake	88.3	16.5	5.36	0.000	99.94
U-sq	-0.536	0.158	-3.39	0.002	99.94

Regression Equation

IgG = -1464 + 88.3 Uptake - 0.536 U-sq

The least squares prediction equation is $\hat{y} = -1,464 + 88.3x - 0.536x^2$.

d. To determine if the overall model is adequate, we test:

$$H_0 : \beta_1 = \beta_2 = 0$$

$$H_a : \text{At least one } \beta_i \neq 0$$

The test statistic is $F = 203.16$ and the p-value is $p = 0.000$. Since the p-value is so small, H_0 is rejected. There is sufficient evidence to indicate the model is useful in predicting the IgG level for any reasonable value of α.

12.17 a. Using MINITAB, the results are:

Regression Analysis: Force versus Mesh

Analysis of Variance

Source	DF	Adj SS	Adj MS	F-Value	P-Value
Regression	1	0.044808	0.044808	14.71	0.062
Mesh	1	0.044808	0.044808	14.71	0.062
Error	2	0.006092	0.003046		
Total	3	0.050900			

Model Summary

S	R-sq	R-sq(adj)	R-sq(pred)
0.0551914	88.03%	82.05%	0.00%

Coefficients

Term	Coef	SE Coef	T-Value	P-Value	VIF
Constant	0.0670	0.0474	1.41	0.294	
Mesh	0.3158	0.0823	3.84	0.062	1.00

Regression Equation

Force = 0.0670 + 0.3158 Mesh

The fitted regression equation is $\hat{y} = 0.067 + 0.3158x$.

b. From MINITAB, the influence diagnostics are:

Force	Mesh	FITS	SRES	TRES	DFIT
0.14	0.125	0.106435	0.871512	0.782493	0.803179
0.15	0.25	0.145913	0.092313	0.065415	0.048691
0.16	0.5	0.22487	−1.35916	−3.47824	−2.01981
0.41	1	0.382783	1.413384	29.1874	78.39567

The fitted values (FITS), studentized residuals (SRES), studentized deleted residuals (TRES), and the difference between fits (DFITTS) are shown in the table. There do not appear to be any outliers as no studentized residuals are greater than 3 in magnitude.

There might be one unusually influential observation. Observation #4 has very large values for the studentized deleted residuals (TRES) and the difference between fits (DFITTS).

c. Using MINITAB, the scatterplot is:

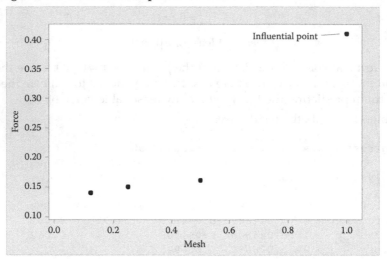

This point suggests that the relationship between force and mesh size may not be linear but curvilinear.

d. Using MINITAB, the results are:

Regression Analysis: Force versus Mesh, M-sq

Analysis of Variance

Source	DF	Adj SS	Adj MS	F-Value	P-Value
Regression	2	0.050665	0.025332	107.72	0.068
Mesh	1	0.001148	0.001148	4.88	0.271
M-sq	1	0.005857	0.005857	24.91	0.126
Error	1	0.000235	0.000235		
Total	3	0.050900			

Model Summary

S	R-sq	R-sq(adj)	R-sq(pred)
0.0153350	99.54%	98.61%	0.00%

Coefficients

Term	Coef	SE Coef	T-Value	P-Value	VIF
Constant	0.1717	0.0248	6.93	0.091	
Mesh	-0.260	0.118	-2.21	0.271	26.41
M-sq	0.4972	0.0996	4.99	0.126	26.41

Regression Equation

Force = 0.1717 - 0.260 Mesh + 0.4972 M-sq

To determine if the model is adequate for predicting force, we test:

$$H_0 : \beta_2 = 0$$

$$H_a : \beta_2 \neq 0$$

The test statistic is $F = 4.99$ and the p-value is $p = 0.126$. If we use $\alpha = 0.10$, the p-value is not less than $\alpha (p = 0.126 \nless 0.10)$, H_0 is not rejected. There is insufficient evidence to indicate that adding the squared term improves the prediction of force at $\alpha = 0.10$.

12.19 a. For $E(y) = 4 - x_1 + 2x_2 + x_1 x_2$, the surface will be three-dimensional. It would appear as a twisted plane similar to the one in Figure 12.10.

 b. For $x_1 = 2, E(y) = 4 - 2 + 2x_2 + 2x_2 = 2 + 4x_2$

 $x_1 = 3, E(y) = 4 - 3 + 2x_2 + 3x_2 = 1 + 5x_2$

 $x_1 = 4, E(y) = 4 - 4 + 2x_2 + 4x_2 = 6x_2$

Using MINITAB, the plot is:

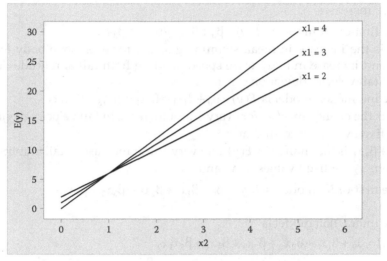

 c. For $x_2 = 2, E(y) = 4 - x_1 + 2(2) + 2x_1 = 8 + x_1$

 $x_2 = 3, E(y) = 4 - x_1 + 2(3) + 3x_1 = 10 + 2x_1$

 $x_2 = 4, E(y) = 4 - x_1 + 2(4) + 4x_1 = 12 + 3x_1$

Using MINITAB, the plot is:

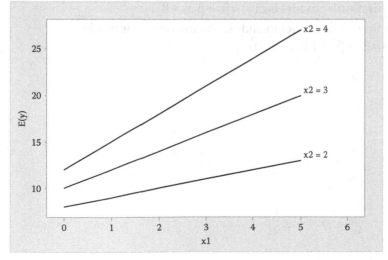

d. From the plot in part b, as x_2 increases from 0 to 5, $E(y)$ also increases. However, the rate of increase depends on the value of x_1. The rate of increase in $E(y)$ is smallest when $x_1 = 2$ and largest when $x_1 = 4$.

   From the plot in part c, as x_1 increases from 0 to 5, $E(y)$ also increases. However, the rate of increase depends on the value of x_2. The rate of increase in $E(y)$ is smallest when $x_2 = 2$ and largest when $x_2 = 4$.

e. From the graph in part b, when $x_1 = 4$ and $x_2 = 1$, $E(y) = 4 - 4 + 2(1) + 4(1) = 6$ and when $x_1 = 2$ and $x_2 = 2$, $E(y) = 4 - 2 + 2(2) + 2(2) = 10$. Thus, $E(y)$ increases by 4 units.

   In Exercise 12.18, when no interaction term was present, $E(y)$ increased as x_2 increased, but decreased as x_1 increased. With the interaction term present, within the given ranges of x_1 and x_2, $E(y)$ increases as x_1 and also increases as x_2 increases.

12.21 a. The first-order model is $E(y) = \beta_0 + \beta_1 x_1 + \beta_2 x_2 + \beta_3 x_3$.

b. β_1 is the increase in mean swimming speed for every one body length per second increase in body wave speed, holding both tail amplitude deviations and tail velocity deviation constant.

c. The interaction model is $E(y) = \beta_0 + \beta_1 x_1 + \beta_2 x_2 + \beta_3 x_3 + \beta_4 x_1 x_2$

d. β_1 is the change in $E(y)$ for every 1-unit increase in tail velocity deviation, x_3, for fixed values of x_1 and x_2.

e. $\beta_2 + \beta_4 x_1$ is the change in $E(y)$ for every 1-unit increase in tail amplitude deviation, x_2, for fixed values of x_1 and x_3.

12.23 a. The first-order model is $E(y) = \beta_0 + \beta_1 x_1 + \beta_2 x_2 + \beta_3 x_3 + \beta_4 x_4$.

b. β_3

c. The interaction model is

$$E(y) = \beta_0 + \beta_1 x_1 + \beta_2 x_2 + \beta_3 x_3 + \beta_4 x_4 + \beta_5 x_1 x_2$$

$$+ \beta_6 x_1 x_3 + \beta_7 x_1 x_4 + \beta_8 x_2 x_3 + \beta_9 x_2 x_4 + \beta_{10} x_3 x_4.$$

d. $\left(\beta_3 + 50\beta_6 + 30\beta_8 + 2\beta_{10}\right)$

12.25 a. Both independent variables frequency and amplitude of wavelet are quantitative.

b. The first-order model is $E(y) = \beta_0 + \beta_1 x_1 + \beta_2 x_2$.

c. The first-order model including the interaction terms is
$E(y) = \beta_0 + \beta_1 x_1 + \beta_2 x_2 + \beta_3 x_1 x_2$.

Answers will vary. A possible plot is:

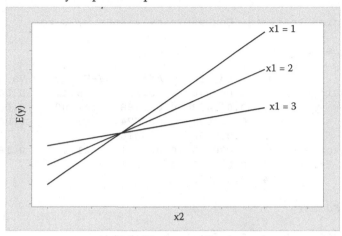

d. The complete second-order model is $E(y) = \beta_0 + \beta_1 x_1 + \beta_2 x_2 + \beta_3 x_1 x_2 + \beta_4 x_1^2 + \beta_5 x_2^2$.

12.27 a. The complete second-order model is $E(y) = \beta_0 + \beta_1 x_1 + \beta_2 x_2 + \beta_3 x_1 x_2 + \beta_4 x_1^2 + \beta_5 x_2^2$.

 b. The straight-line model is $E(y) = \beta_0 + \beta_1 x_1 + \beta_2 x_2$.

 c. The straight-line model with interaction is $E(y) = \beta_0 + \beta_1 x_1 + \beta_2 x_2 + \beta_3 x_1 x_2$.

 d. The slope is $\beta_1 + \beta_3 x_2$.

 e. The slope is $\beta_2 + \beta_3 x_1$.

12.29 a. Using MINITAB, the correlation is:

Correlation: X, X-sq

```
Pearson correlation of X and X-sq = 0.994
P-Value = 0.000
```

The correlation between x and x^2 is $r = 0.994$.

Since x and x^2 are highly correlated, multicollinearity will exist. $\mathbf{X'X}^{-1}$ is difficult to compute when variables are highly correlated, leading to large round off errors.

 b. Let $u_i = \dfrac{x_i - \bar{x}}{s_x}$. Then the regression equation becomes $E(y) = \beta_0^* + \beta_1^* u + \beta_2^* u^2$.

 c. From the data, $\bar{x} = 83.36$ and $s = 24.0518$. Thus, we will subtract 83.36 from each x value and then divide the result by 24.0518 to obtain the u values. Using MINITAB, the correlation is:

Correlation: U, U-sq

```
Pearson correlation of U and U-sq = 0.170
P-Value = 0.417
```

The correlation between u and u^2 is $r = 0.170$.

d. Using MINITAB, the results are:

Regression Analysis: Conductivity versus u, u-sq

Analysis of Variance

Source	DF	Adj SS	Adj MS	F-Value	P-Value
Regression	2	0.002576	0.001288	125.42	0.000
u	1	0.001488	0.001488	144.89	0.000
u-sq	1	0.000674	0.000674	65.60	0.000
Error	22	0.000226	0.000010		
Lack-of-Fit	21	0.000225	0.000011	21.47	0.169
Pure Error	1	0.000001	0.000001		
Total	24	0.002802			

Model Summary

S	R-sq	R-sq(adj)	R-sq(pred)
0.0032047	91.94%	91.20%	89.25%

Coefficients

Term	Coef	SE Coef	T-Value	P-Value	VIF
Constant	0.04873	0.00102	47.57	0.000	
u	0.007990	0.000664	12.04	0.000	1.03
u-sq	0.006741	0.000832	8.10	0.000	1.03

Regression Equation

Conductivity = 0.04873 + 0.007990 u + 0.006741 u-sq

The fitted regression line is $\hat{y} = 0.04873 + 0.007990u + 0.006741u^2$.
The model is adequate for predicting thermal conductivity ($F = 125.42$, $p = 0.000$).
In addition, $R^2_{adj} = 0.9120$ which means that the model fits the data well.

12.31 a. Using MINITAB, the correlation is:

Correlation: Volume, V-sq

Pearson correlation of Volume and V-sq = 0.974
P-Value = 0.000

The correlation between volume and volume^2 is $r = 0.974$. Since the value of r
is so close to 1, we recommend coding the variable, x.

b. Some preliminary calculations:

$$\bar{x} = \frac{302}{20} = 15.1 \qquad s_x = 8.13634$$

$$u = \frac{x - \bar{x}}{s_x} = \frac{x - 15.1}{8.13634}$$

c. Using MINITAB, the correlation is:

Correlation: u, u-sq

Pearson correlation of u and u-sq = -0.046
P-Value = 0.848

The sample correlation coefficient between u and u^2 is $r = -0.046$. Yes, the correlation between u and u^2 is very close to 0, while the correlation between x and x^2 is close to 1.

d. Using MINITAB, the results are:

Regression Analysis: Distance versus u, u-sq

Analysis of Variance

Source	DF	Adj SS	Adj MS	F-Value	P-Value
Regression	2	0.70785	0.353924	65.39	0.000
u	1	0.51020	0.510201	94.27	0.000
u-sq	1	0.16929	0.169285	31.28	0.000
Error	17	0.09201	0.005412		
Total	19	0.79986			

Model Summary

S	R-sq	R-sq(adj)	R-sq(pred)
0.0735672	88.50%	87.14%	79.81%

Coefficients

Term	Coef	SE Coef	T-Value	P-Value	VIF
Constant	0.0983	0.0250	3.93	0.001	
u	-0.1640	0.0169	-9.71	0.000	1.00
u-sq	0.1107	0.0198	5.59	0.000	1.00

Regression Equation

Distance = 0.0983 - 0.1640 u + 0.1107 u-sq

The fitted regression line is $\hat{y} = 0.0983 - 0.164u + 0.1107u^2$.
The model is adequate for predicting mileage $(F = 65.39, p = 0.000)$. In addition, $R_{adj}^2 = 0.8714$ which means that the model fits the data well.

12.33 a. The model is $E(y) = \beta_0 + \beta_1 x_1 + \beta_2 x_2$.

$$\text{where } x_1 = \begin{cases} 1 & \text{if groundwater} \\ 0 & \text{if not} \end{cases} \quad \text{and } x_2 = \begin{cases} 1 & \text{if sub-surface flow} \\ 0 & \text{if not} \end{cases}$$

b. $\beta_0 = \mu_3$ is the mean nitrate concentration of all overground flow water samples.
$\beta_1 = \mu_1 - \mu_3$ is the difference in mean nitrate concentration between the groundwater and the overground flow water samples.
$\beta_2 = \mu_2 - \mu_3$ is the difference in mean nitrate concentration between the subsurface flow and the overground flow water samples.

12.35 a.   The model is $E(y) = \beta_0 + \beta_1 x_1 + \beta_2 x_2 + \beta_3 x_3 + \beta_4 x_4$.

where $x_1 = \begin{cases} 1 & \text{if benzene} \\ 0 & \text{if not} \end{cases}$ $x_2 = \begin{cases} 1 & \text{if toluene} \\ 0 & \text{if not} \end{cases}$ $x_3 = \begin{cases} 1 & \text{if chloroform} \\ 0 & \text{if not} \end{cases}$

$x_4 = \begin{cases} 1 & \text{if methanol} \\ 0 & \text{if not} \end{cases}$

    b. $\beta_0 = \mu_A$ is the mean retention coefficient for anisole.

         $\beta_1 = \mu_B - \mu_A$ is the difference in mean retention coefficient between benzene and anisole.

         $\beta_2 = \mu_T - \mu_A$ is the difference in mean retention coefficient between toluene and anisole.

         $\beta_3 = \mu_C - \mu_A$ is the difference in mean retention coefficient between chloroform and anisole.

         $\beta_4 = \mu_M - \mu_A$ is the difference in mean retention coefficient between methanol and anisole.

    c.   To test for differences among the mean retention coefficients for the five organic compounds, we test:

$$H_0 : \beta_1 = \beta_2 = \beta_3 = \beta_4 = 0$$

$$H_a : \text{At least one } \beta_i \neq 0$$

12.37 a.   The mean body length for whales entangled in gill nets is β_0.

    b. $\beta_1 = \mu_S - \mu_G$ is the difference in mean body length between whales entangled in set nets and whales entangled in gill nets.

    c.   To determine if the mean body length of entangled whales differ among the three types of fishing gear, we test:

$$H_0 : \beta_1 = \beta_2 = 0$$

$$H_a : \text{At least one } \beta_i \neq 0$$

12.39 a.   The independent variable in the experiment is the type of shade.

    b.   The model is $E(y) = \beta_0 + \beta_1 x_1 + \beta_2 x_2$.

where $x_1 = \begin{cases} 1 & \text{if tree shade} \\ 0 & \text{if not} \end{cases}$ $x_2 = \begin{cases} 1 & \text{if no shade} \\ 0 & \text{if not} \end{cases}$

    c. $\beta_0 = \mu_1$ is the mean milk production for cows in artificial shade.

         $\beta_1 = \mu_2 - \mu_1$ is the difference in mean milk production between cows in tree shade and artificial shade.

         $\beta_2 = \mu_3 - \mu_1$ is the difference in mean milk production between cows in no shade and artificial shade.

12.41 a.   Let y = mean body mass. The model is $E(y) = \beta_0 + \beta_1 x_1$ where $x_1 = \begin{cases} 1 & \text{if flightless} \\ 0 & \text{if not} \end{cases}$

b. Let y = mean body mass. The model is $E(y) = \beta_0 + \beta_1 x_1 + \beta_2 x_2 + \beta_3 x_3$

where $x_1 = \begin{cases} 1 & \text{if vertebrates} \\ 0 & \text{if not} \end{cases}$ $x_2 = \begin{cases} 1 & \text{if vegetables} \\ 0 & \text{if not} \end{cases}$ $x_3 = \begin{cases} 1 & \text{if invertebrates} \\ 0 & \text{if not} \end{cases}$

c. Let y = mean egg length. The model is $E(y) = \beta_0 + \beta_1 x_1 + \beta_2 x_2 + \beta_3 x_3$

where $x_1 = \begin{cases} 1 & \text{if cavity within ground} \\ 0 & \text{if not} \end{cases}$ $x_2 = \begin{cases} 1 & \text{if tree} \\ 0 & \text{if not} \end{cases}$ $x_3 = \begin{cases} 1 & \text{if cavity above gound} \\ 0 & \text{if not} \end{cases}$

d. Using MINITAB, the results are:

Regression Analysis: Body Mass versus x1

Analysis of Variance

Source	DF	Adj SS	Adj MS	F-Value	P-Value
Regression	1	23246186220	23246186220	33.05	0.000
x1	1	23246186220	23246186220	33.05	0.000
Error	130	91439576421	703381357		
Total	131	1.14686E+11			

Model Summary

S	R-sq	R-sq(adj)	R-sq(pred)
26521.3	20.27%	19.66%	15.22%

Coefficients

Term	Coef	SE Coef	T-Value	P-Value	VIF
Constant	641	2665	0.24	0.810	
x1	30647	5331	5.75	0.000	1.00

Regression Equation

Body Mass = 641 + 30647 x1

The fitted regression line is $\hat{y} = 641 + 30{,}647 x_1$.

$\hat{\beta}_0 = 641$. The mean body mass for Volant birds is estimated to be 641 grams.

$\hat{\beta}_1 = 30{,}647$. The difference in mean body mass between flightless birds and Volant birds is estimated to be 30,647 grams.

e. To determine if the model is useful for estimating mean body mass, we test:

$$H_0 : \beta_1 = 0$$

$$H_a : \beta_1 \neq 0$$

The test statistic is $F = 33.05$ and the p-value is $p = 0.000$. Since the p-value is less than $\alpha(p = 0.000 < 0.01)$, H_0 is rejected. There is sufficient evidence to indicate a difference in mean body mass between Volant and flightless birds at $\alpha = 0.01$.

f. Using MINITAB, the results are:

```
Regression Analysis: Body Mass versus x1, x2, x3

Analysis of Variance
```

Source	DF	Adj SS	Adj MS	F-Value	P-Value
Regression	3	18924196564	6308065521	8.43	0.000
x1	1	32872371	32872371	0.04	0.834
x2	1	13853412371	13853412371	18.52	0.000
x3	1	9777622	9777622	0.01	0.909
Error	128	95761566077	748137235		
Total	131	1.14686E+11			

```
Model Summary
```

S	R-sq	R-sq(adj)	R-sq(pred)
27352.1	16.50%	14.54%	11.87%

```
Coefficients
```

Term	Coef	SE Coef	T-Value	P-Value	VIF
Constant	903	4171	0.22	0.829	
x1	2997	14298	0.21	0.834	1.06
x2	26206	6090	4.30	0.000	1.34
x3	-660	5772	-0.11	0.909	1.35

```
Regression Equation

Body Mass = 903 + 2997 x1 + 26206 x2 - 660 x3
```

The fitted regression line is $\hat{y} = 903 + 2,997x_1 + 26,206x_2 - 660x_3$.

$\hat{\beta}_0 = 903$. The mean body mass for birds on a fish diet is estimated to be 903 grams.

$\hat{\beta}_1 = 2,997$. The difference in mean body mass between birds on a vertebrate diet and birds on a fish diet is estimated to be 2,997 grams.

$\hat{\beta}_2 = 26,206$. The difference in mean body mass between birds on a vegetable diet and birds on a fish diet is estimated to be 26,206 grams.

$\hat{\beta}_3 = -660$. The difference in mean body mass between birds on an invertebrate diet and birds on a fish diet is estimated to be -660 grams.

g. To determine if the model is useful for estimating mean body mass, we test:

$$H_0 : \beta_1 = \beta_2 = \beta_3 = 0$$

$$H_a : \text{At least one } \beta_i \neq 0$$

The test statistic is $F = 8.43$ and the p-value is $p = 0.000$. Since the p-value is less than $\alpha\left(p = 0.000 < 0.01\right)$, H_0 is rejected. There is sufficient evidence to indicate a difference in mean body mass among birds on the four types of diets at $\alpha = 0.01$.

h. Using MINITAB, the results are:

Regression Analysis: Egg Length versus x1, x2, x3

Method

Rows unused   2

Analysis of Variance

Source	DF	Adj SS	Adj MS	F-Value	P-Value
Regression	3	40230	13410	8.07	0.000
x1	1	1530	1530	0.92	0.339
x2	1	30293	30293	18.23	0.000
x3	1	16129	16129	9.71	0.002
Error	126	209356	1662		
Total	129	249586			

Model Summary

S	R-sq	R-sq(adj)	R-sq(pred)
40.7622	16.12%	14.12%	13.07%

Coefficients

Term	Coef	SE Coef	T-Value	P-Value	VIF
Constant	73.73	4.91	15.03	0.000	
x1	-9.13	9.52	-0.96	0.339	1.10
x2	-39.51	9.25	-4.27	0.000	1.10
x3	-45.0	14.4	-3.12	0.002	1.05

Regression Equation

Egg Length = 73.73 - 9.13 x1 - 39.51 x2 - 45.0 x3

The fitted regression line is $\hat{y} = 73.73 - 9.13x_1 - 39.51x_2 - 45.0x_3$.

$\hat{\beta}_0 = 903$. The egg length for birds nesting on the ground is estimated to be 73.73 millimeters.

$\hat{\beta}_1 = -9.13$. The difference in mean egg length between birds nesting in cavity within ground and birds nesting on the ground is estimated to be −9.13 millimeters.

$\hat{\beta}_2 = -39.51$. The difference in mean egg length between birds nesting in trees and birds nesting on the ground is estimated to be −39.51 millimeters.

$\hat{\beta}_3 = -45.01$. The difference in mean egg length between birds nesting in cavities above ground and birds nesting on the ground is estimated to be −45.01 millimeters.

i. To determine if the model is useful for estimating mean egg length, we test:

$$H_0 : \beta_1 = \beta_2 = \beta_3 = 0$$

$$H_a : \text{At least one } \beta_i \neq 0$$

The test statistic is $F = 8.07$ and the p-value is $p = 0.000$. Since the p-value is less than $\alpha \left(p = 0.000 < 0.01 \right)$, H_0 is rejected. There is sufficient evidence to indicate a difference in mean egg length among birds nesting in the four locations at $\alpha = 0.01$.

12.43 a. Let $x_1 = \begin{cases} 1 & \text{if large/public} \\ 0 & \text{if not} \end{cases}$ $x_1 = \begin{cases} 1 & \text{if large/private} \\ 0 & \text{if not} \end{cases}$ $x_3 = \begin{cases} 1 & \text{if small/public} \\ 0 & \text{if not} \end{cases}$

b. The model is $E(y) = \beta_0 + \beta_1 x_1 + \beta_2 x_2 + \beta_3 x_3$.

$\beta_0 = \mu_4$ is the mean likelihood of reporting sustainability policies for small/ private firms.

$\beta_1 = \mu_1 - \mu_4$ is the difference in mean likelihood of reporting sustainability policies between large/public firms and small/private firms.

$\beta_2 = \mu_2 - \mu_4$ is the difference in mean likelihood of reporting sustainability policies between large/private firms and small/private firms.

$\beta_3 = \mu_3 - \mu_4$ is the difference in mean likelihood of reporting sustainability policies between small/public firms and small/private firms.

c. Reject H_0. There is sufficient evidence to indicate a difference in mean likelihood of reporting sustainability policies among the four types of firms.

d. Let $x_1 = \begin{cases} 1 & \text{if large} \\ 0 & \text{if not} \end{cases}$ $x_2 = \begin{cases} 1 & \text{if public} \\ 0 & \text{if not} \end{cases}$

e. The main effects model is $E(y) = \beta_0 + \beta_1 x_1 + \beta_2 x_2$.

f. For large/public firms, the model is $E(y) = \beta_0 + \beta_1 (1) + \beta_2 (1) = \beta_0 + \beta_1 + \beta_2$.

For large/private firms, the model is $E(y) = \beta_0 + \beta_1 (1) + \beta_2 (0) = \beta_0 + \beta_1$.

For small/public firms, the model is $E(y) = \beta_0 + \beta_1 (0) + \beta_2 (1) = \beta_0 + \beta_2$.

For small/private firms, the model is $E(y) = \beta_0 + \beta_1 (0) + \beta_2 (0) = \beta_0$.

g. For public firms, the difference between large and small firms is:

$$\beta_0 + \beta_1 + \beta_2 - (\beta_0 + \beta_2) = \beta_1.$$

For private firms, the difference between large and small firms is:

$$\beta_0 + \beta_1 - (\beta_0) = \beta_1.$$

h. The interaction model is $E(y) = \beta_0 + \beta_1 x_1 + \beta_2 x_2 + \beta_3 x_1 x_2$.

i. For large/public firms, the model is

$$E(y) = \beta_0 + \beta_1 (1) + \beta_2 (1) + \beta_3 (1)(1) = \beta_0 + \beta_1 + \beta_2 + \beta_3.$$

For large/private firms, the model is $E(y) = \beta_0 + \beta_1 (1) + \beta_2 (0) + \beta_3 (1)(0) = \beta_0 + \beta_1$.

For small/public firms, the model is $E(y) = \beta_0 + \beta_1(0) + \beta_2(1) + \beta_3(0)(1) = \beta_0 + \beta_2$.
For small/private firms, the model is $E(y) = \beta_0 + \beta_1(0) + \beta_2(0) + \beta_3(0)(0) = \beta_0$.

j. For public firms, the difference between large and small firms is:

$$\beta_0 + \beta_1 + \beta_2 + \beta_3 - (\beta_0 + \beta_2) = \beta_1 + \beta_3.$$

For private firms, the difference between large and small firms is:

$$\beta_0 + \beta_1 - \beta_0 = \beta_1.$$

12.45 a. The model is $E(y) = \beta_0 + \beta_1 x_1 + \beta_2 x_3 + \beta_3 x_4 + \beta_4 x_5 + \beta_5 x_6$.

where $x_1 =$ temperature, $x_3 = \begin{cases} 1 & \text{if benzene} \\ 0 & \text{if not} \end{cases}$ $x_4 = \begin{cases} 1 & \text{if toluene} \\ 0 & \text{if not} \end{cases}$

$x_5 = \begin{cases} 1 & \text{if chloroform} \\ 0 & \text{if not} \end{cases}$ $x_6 = \begin{cases} 1 & \text{if methanol} \\ 0 & \text{if not} \end{cases}$

A possible sketch of the model is

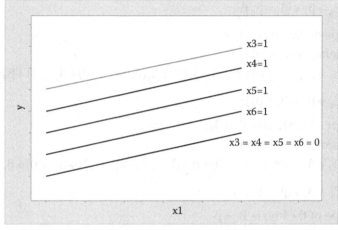

b. $\beta_0 =$ mean retention coefficient when temperature is 0 and the organic compound is anisole.

$\beta_1 =$ is the change in the mean retention coefficient for each degree increase in temperature, holding organic compound constant.

$\beta_2 =$ is the difference in the mean retention coefficient between benzene and anisole, holding temperature constant.

$\beta_3 =$ is the difference in the mean retention coefficient between toluene and anisole, holding temperature constant.

$\beta_4 =$ is the difference in the mean retention coefficient between chloroform and anisole, holding temperature constant.

$\beta_5 =$ is the difference in the mean retention coefficient between methanol and anisole, holding temperature constant.

c. The model would be

$$E(y) = \beta_0 + \beta_1 x_2 + \beta_2 x_3 + \beta_3 x_4 + \beta_4 x_5 + \beta_5 x_6 + \beta_6 x_2 x_3 + \beta_7 x_2 x_4 + \beta_8 x_2 x_5 + \beta_9 x_2 x_6$$

A possible sketch is:

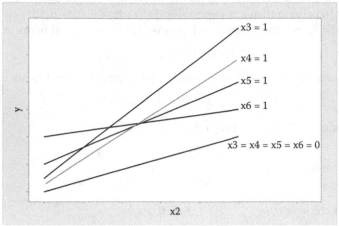

d. For benzene, the model is

$$E(y) = \beta_0 + \beta_1 x_2 + \beta_2(1) + \beta_3(0) + \beta_4(0) + \beta_5(0) + \beta_6 x_2(1) + \beta_7 x_2(0) + \beta_8 x_2(0) + \beta_9 x_2(0)$$

$$= \beta_0 + \beta_2 + (\beta_1 + \beta_6) x_2$$

The slope of the line is $\beta_1 + \beta_6$.

For toluene, the model is

$$E(y) = \beta_0 + \beta_1 x_2 + \beta_2(0) + \beta_3(1) + \beta_4(0) + \beta_5(0) + \beta_6 x_2(0) + \beta_7 x_2(1) + \beta_8 x_2(0) + \beta_9 x_2(0)$$

$$= \beta_0 + \beta_3 + (\beta_1 + \beta_7) x_2$$

The slope of the line is $\beta_1 + \beta_7$.

For chloroform, the model is

$$E(y) = \beta_0 + \beta_1 x_2 + \beta_2(0) + \beta_3(0) + \beta_4(1) + \beta_5(0) + \beta_6 x_2(0) + \beta_7 x_2(0) + \beta_8 x_2(1) + \beta_9 x_2(0)$$

$$= \beta_0 + \beta_4 + (\beta_1 + \beta_8) x_2$$

The slope of the line is $\beta_1 + \beta_8$.

For methanol, the model is

$$E(y) = \beta_0 + \beta_1 x_2 + \beta_2(0) + \beta_3(0) + \beta_4(0) + \beta_5(1) + \beta_6 x_2(0) + \beta_7 x_2(0) + \beta_8 x_2(0) + \beta_9 x_2(1)$$

$$= \beta_0 + \beta_5 + (\beta_1 + \beta_9) x_2$$

The slope of the line is $\beta_1 + \beta_9$.

For anisole, the model is

$$E(y) = \beta_0 + \beta_1 x_2 + \beta_2(0) + \beta_3(0) + \beta_4(0) + \beta_5(0) + \beta_6 x_2(0) + \beta_7 x_2(0) + \beta_8 x_2(0) + \beta_9 x_2(0)$$

$$= \beta_0 + \beta_1 x_2$$

The slope of the line is β_1.

12.47 a. The first-order model is $E(y) = \beta_0 + \beta_1 x_3 + \beta_2 x_7$.

   b. $\beta_2 =$ difference in mean SULMA between Timberjack vehicle and Valmet vehicle, holding dominant hand power level constant.

   c. The interaction model is $E(y) = \beta_0 + \beta_1 x_3 + \beta_2 x_7 + \beta_3 x_3 x_7$.

d. β_1

e. $\beta_2 + 75\beta_3$

f. The complete second-order model would be

$$E(y) = \beta_0 + \beta_1 x_3 + \beta_2 x_7 + \beta_3 x_3 x_7 + \beta_4 x_3^2 + \beta_5 x_3^2 x_7.$$

A possible sketch would be:

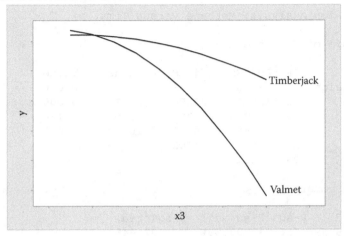

g. β_4

12.49 a. The first-order model is $E(y) = \beta_0 + \beta_1 x_1 + \beta_2 x_2 + \beta_3 x_3$

where $x_2 = \begin{cases} 1 & \text{if method G} \\ 0 & \text{if not} \end{cases}$ and $x_3 = \begin{cases} 1 & \text{if method R}_1 \\ 0 & \text{if not} \end{cases}$

b. $\beta_0 =$ mean percent of shelf life for method R_2 when the potency of the drug is 0.

$\beta_1 =$ change is mean percent of shelf life for each unit increase in drug potency, with method held constant.

$\beta_2 =$ difference in mean percent of shelf life between method G and method R_2, holding drug potency constant.

$\beta_3 =$ difference in mean percent of shelf life between method R_1 and method R_2, holding drug potency constant.

c. The interaction model is $E(y) = \beta_0 + \beta_1 x_1 + \beta_2 x_2 + \beta_3 x_3 + \beta_4 x_1 x_2 + \beta_5 x_1 x_3.$

d. For method G, the model is

$$E(y) = \beta_0 + \beta_1 x_1 + \beta_2(1) + \beta_3(0) + \beta_4 x_1(1) + \beta_5 x_1(0) = \beta_0 + \beta_2 + (\beta_1 + \beta_4) x_1$$

The slope is $\beta_1 + \beta_4$.

For method R_1, the model is

$$E(y) = \beta_0 + \beta_1 x_1 + \beta_2(0) + \beta_3(1) + \beta_4 x_1(0) + \beta_5 x_1(1) = \beta_0 + \beta_3 + (\beta_1 + \beta_5) x_1$$

The slope is $\beta_1 + \beta_5$.

For method R_2, the model is

$$E(y) = \beta_0 + \beta_1 x_1 + \beta_2(0) + \beta_3(0) + \beta_4 x_1(0) + \beta_5 x_1(0) = \beta_0 + \beta_1 x_1$$

The slope is β_1.

12.51   The model would be $E(y) = \beta_0 + \beta_1 x_1 + \beta_2 x_2 + \beta_3 x_1 x_2$ where $x_1 = $ length of time
and $x_2 = \begin{cases} 1 & \text{if no nutrients} \\ 0 & \text{if not} \end{cases}$.

Using MINITAB, the results are:

```
Regression Analysis: Methane versus Time, x2, T_x2

Analysis of Variance

Source       DF  Adj SS  Adj MS  F-Value  P-Value
Regression    3  238658   79553   132.10    0.000
  Time        1  133274  133274   221.30    0.000
  x2          1     395     395     0.66     0.426
  T_x2        1    6457    6457    10.72     0.003
Error        23   13851     602
Total        26  252510
```

```
Model Summary

      S   R-sq  R-sq(adj)  R-sq(pred)
24.5405  94.51%     93.80%      92.44%
```

```
Coefficients

Term        Coef  SE Coef  T-Value  P-Value    VIF
Constant  -164.5     24.0    -6.86    0.000
Time       9.758    0.656    14.88    0.000   2.00
x2          27.5     34.0     0.81    0.426  12.91
T_x2      -3.038    0.928    -3.27    0.003  13.91
```

```
Regression Equation

Methane = -164.5 + 9.758 Time + 27.5 x2 - 3.038 T_x2
```

The least squares prediction equation is $\hat{y} = -164.5 + 9.758 x_1 + 27.5 x_2 - 3.038 x_1 x_2$.
To determine if the emission rates differ for the two types of sludge, we test:

$$H_0 : \beta_3 = 0$$

$$H_a : \beta_3 \neq 0$$

The test statistic is $t = -3.27$ and the p-value is $p = 0.003$. Since the p-value is so small, H_0 is rejected. There is sufficient evidence to indicate the emission rates differ for the two types of sludge at $\alpha = 0.05$.
For no nutrients, the fitted equation is
$\hat{y} = -164.5 + 9.758 x_1 + 27.5(1) - 3.08 x_1 (1) = -137 + 6.72 x_1$

The rate of change is estimated to be 6.72.

For yes nutrients, the fitted equation is

$$\hat{y} = -164.5 + 9.758x_1 + 27.5(0) - 3.08x_1(0) = -164.5 + 9.758x_1$$

The rate of change is estimated to be 9.758.

12.53 a. The model would be $E(y) = \beta_0 + \beta_1 x_1 + \beta_2 x_2 + \beta_3 x_3$.

b. The interaction model would be $E(y) = \beta_0 + \beta_1 x_1 + \beta_2 x_2 + \beta_3 x_3 + \beta_4 x_1 x_2 + \beta_5 x_1 x_3$.

c. For waste type TDS-3A, the model is

$$E(y) = \beta_0 + \beta_1 x_1 + \beta_2(1) + \beta_3(0) + \beta_4 x_1(1) + \beta_5 x_1(0) = \beta_0 + \beta_2 + (\beta_1 + \beta_4)x_1.$$

The slope is $\beta_1 + \beta_4$.

For waste type FE, the model is

$$E(y) = \beta_0 + \beta_1 x_1 + \beta_2(0) + \beta_3(1) + \beta_4 x_1(0) + \beta_5 x_1(1) = \beta_0 + \beta_3 + (\beta_1 + \beta_5)x_1.$$

The slope is $\beta_1 + \beta_5$.

For waste type AL, the model is

$$E(y) = \beta_0 + \beta_1 x_1 + \beta_2(0) + \beta_3(0) + \beta_4 x_1(0) + \beta_5 x_1(0) = \beta_0 + \beta_1 x_1.$$

The slope is β_1.

d. To test for the presence of temperature-waste type interaction, test if the interaction terms are needed in the model.

$$H_0 : \beta_4 = \beta_5 = 0$$

$$H_a : \text{At least one } \beta_i \neq 0$$

To perform this test, we would first fit the complete model (model given in **b**) and then fit the reduced model (model given in part **a**). By comparing the two models, we could decide whether to reject H_0 or not.

12.55 a. To determine if there is a difference in the mean lengths of entangles whales for the three gear types, we test:

$$H_0 : \beta_2 = \beta_3 = \beta_4 = \beta_5 = 0$$

$$H_a : \text{At least one } \beta_i \neq 0$$

b. The reduced model would be $E(y) = \beta_0 + \beta_1 x_1$.

c. We would conclude that there is sufficient evidence to indicate that there were differences in the mean lengths of entangled whales for the three gear types.

d. To determine if the rate of change of whale length with water depth is the same for all three types of fishing gear, we test:

$$H_0 : \beta_4 = \beta_5 = 0$$

$$H_a : \text{At least one } \beta_i \neq 0$$

e. The reduced model would be $E(y) = \beta_0 + \beta_1 x_1 + \beta_2 x_2 + \beta_3 x_3$.

f. If we fail to reject the null hypothesis, we would conclude that there is insufficient evidence to indicate that the rate of change of whale length with water depth differs among the three types of fishing gear.

12.57 a. The main effects model is
$$E(y) = \beta_0 + \beta_1 x_1 + \beta_2 x_2 + \beta_3 x_3 + \beta_4 x_4 + \beta_5 x_5 + \beta_6 x_6 + \beta_7 x_7$$
$$+ \beta_8 x_8 + \beta_9 x_9 + \beta_{10} x_{10} + \beta_{11} x_{11}.$$

b. The interaction model is
$$E(y) = \beta_0 + \beta_1 x_1 + \beta_2 x_2 + \beta_3 x_3 + \beta_4 x_4 + \beta_5 x_5 + \beta_6 x_6 + \beta_7 x_7 + \beta_8 x_8 + \beta_9 x_9 + \beta_{10} x_{10} + \beta_{11} x_{11}$$
$$+ \beta_{12} x_1 x_9 + \beta_{13} x_1 x_{10} + \beta_{14} x_1 x_{11} + \beta_{15} x_2 x_9 + \beta_{16} x_2 x_{10} + \beta_{17} x_2 x_{11}$$
$$+ \beta_{18} x_3 x_9 + \beta_{19} x_3 x_{10} + \beta_{20} x_3 x_{11} + \beta_{21} x_4 x_9 + \beta_{22} x_4 x_{10} + \beta_{23} x_4 x_{11}$$
$$+ \beta_{24} x_5 x_9 + \beta_{25} x_5 x_{10} + \beta_{26} x_5 x_{11} + \beta_{27} x_6 x_9 + \beta_{28} x_6 x_{10} + \beta_{29} x_6 x_{11}$$
$$+ \beta_{30} x_7 x_9 + \beta_{31} x_7 x_{10} + \beta_{32} x_7 x_{11} + \beta_{33} x_8 x_9 + \beta_{34} x_8 x_{10} + \beta_{35} x_8 x_{11}$$

c. To test for interaction, we test:
$$H_0 : \beta_{12} = \beta_{13} = \cdots = \beta_{35} = 0$$
$$H_a : \text{At least one } \beta_i \neq 0$$

We would compare the complete model in part b to the reduced model in part a.

12.59   To determine if the second-order model contributes more information than the first-order model, we test:
$$H_0 : \beta_3 = \beta_4 = \beta_5 = 0$$
$$H_a : \text{At least one } \beta_i \neq 0$$

The test statistic is
$$F = \frac{(SSE_R - SSE_C)/(k-g)}{SSE_C/[n-(k+1)]} = \frac{(SSE_R - SSE_C)/(k-g)}{MSE_C}$$
$$= \frac{(2094.4 - 159.94)/(5-2)}{26.66} = 24.19.$$

The rejection region requires $\alpha = 0.05$ in the upper tail of the F distribution with $v_1 = k - g = 5 - 2 = 3$ and $v_2 = n - (k+1) = 12 - (5+1) = 6$. From Table 10, Appendix B, $F_{0.05} = 4.76$. The rejection region is $F > 4.76$.

Since the observed value of the test statistic falls in the rejection region ($F = 24.19 > 4.76$), H_0 is rejected. There is sufficient evidence to indicate the second-order model contributes more information for the prediction of the signal-to-noise ratio than the first-order model at $\alpha = 0.05$.

12.61 a. To determine whether the rate of increase of emotional distress with experience is different for the two groups of firefighters, we test:
$$H_0 : \beta_4 = \beta_5 = 0$$
$$H_a : \text{At least one } \beta_i \neq 0$$

b. To determine if there are differences in mean emotional distress levels that are attributable to exposure group, we test:
$$H_0 : \beta_3 = \beta_4 = \beta_5 = 0$$
$$H_a : \text{At least one } \beta_i \neq 0$$

c. To test the hypothesis in part b, the test statistic is

$$F = \frac{(SSE_R - SSE_C)/(k-g)}{SSE_C/[n-(k+1)]} = \frac{(795.23 - 783.90)/(5-2)}{783.90/[200-(5+1)]} = 0.935$$

The rejection region requires $\alpha = 0.05$ in the upper tail of the F distribution with $v_1 = k - g = 5 - 2 = 3$ and $v_2 = n - (k+1) = 200 - (5+1) = 194$. Using MINITAB, $F_{0.05} = 2.65$. The rejection region is $F > 2.65$.

Since the observed value of the test statistic does not fall in the rejection region $(F = 0.935 \not> 2.65)$, H_0 is not rejected. There is insufficient evidence to indicate there are differences in mean emotional distress levels that are attributable to exposure group at $\alpha = 0.05$.

12.63 a. To determine if the mean compressive strength differs for the three cement mixes, we test:

$$H_0 : \beta_2 = \beta_3 = 0$$

$$H_a : \text{At least one } \beta_i \neq 0$$

b. The test statistic is $F = \dfrac{(SSE_R - SSE_C)/(k-g)}{SSE_C/[n-(k+1)]} = \dfrac{(183.2 - 140.5)/(3-1)}{140.5/[50-(3+1)]} = 6.99$.

The rejection region requires $\alpha = 0.05$ in the upper tail of the F distribution with $v_1 = k - g = 3 - 1 = 2$ and $v_2 = n - (k+1) = 50 - (3+1) = 46$. Using MINITAB, $F_{0.05} = 3.20$. The rejection region is $F > 3.20$.

Since the observed value of the test statistic falls in the rejection region $(F = 6.99 > 3.20)$, H_0 is rejected. There is sufficient evidence to indicate the mean compressive strength differs for the three cement mixes at x_1 $\alpha = 0.05$.

c. To test if the slope of the linear relationship between mean compressive strength and hardening time varies according to type of cement, first fit the reduced model stated in the text $E(y) = \beta_0 + \beta_1 x_1 + \beta_2 x_2 + \beta_3 x_3$. Then fit the model with the interaction terms $E(y) = \beta_0 + \beta_1 x_1 + \beta_2 x_2 + \beta_3 x_3 + \beta_4 x_1 x_2 + \beta_5 x_1 x_3$. To determine if the interaction terms are significant, we test:

$$H_0 : \beta_4 = \beta_5 = 0$$

$$H_a : \text{At least one } \beta_i \neq 0$$

d. The second-order model with interactions is

$$E(y) = \beta_0 + \beta_1 x_1 + \beta_2 x_1^2 + \beta_3 x_2 + \beta_4 x_3 + \beta_5 x_1 x_2 + \beta_6 x_1 x_3 + \beta_7 x_1^2 x_2 + \beta_8 x_1^2 x_3$$

e. To determine if the three response curves have the same shape but different y-intercepts, we test:

$$H_0 : \beta_5 = \beta_6 = \beta_7 = \beta_8 = 0$$

$$H_a : \text{At least one } \beta_i \neq 0$$

12.65 a. There are a total of six models fit in the first step of the stepwise regression, one for each of the independent variables considered in the model.

b. The best one-variable predictor will be the one with the largest $t = \left| \dfrac{\hat{\beta}_i}{s_{\hat{\beta}_i}} \right|$ value. The values are computed for each of the parameters:

Independent variable	t
x_1	3.81
x_2	−90.00
x_3	2.98
x_4	1.21
x_5	−6.03
x_6	0.86

Thus, x_2 is the variable that will be the best one-variable predictor.

c. There will be five models fit in the second step of the stepwise regression. They will be of the form $E(y) = \beta_0 + \beta_1 x_2 + \beta_2 x_i$, where $i = 1, 3, 4, 5, 6$.

d. The stepwise regression procedure continues to add independent variables to the model until the significance level of the next best independent variable exceeds the α at which we are testing.

e. Two drawbacks to stepwise regression are:

1. The stepwise procedure conducts an extremely large number of t-tests which result in high probabilities of making either Type I or Type II errors.

2. The stepwise procedure does not include any higher-order or interaction terms in the modeling process and may not yield the best predicting model for y.

12.67 a. There will be a total of 11 t-tests performed in the first step.

b. There will be a total of 10 t-tests performed in the second step.

c. To determine the overall utility of the model, we test:

$$H_0 : \beta_1 = \beta_2 = 0$$

$$H_a : \text{At least one } \beta_i \neq 0$$

Since the p-value is so small $p = 0.001$, H_0 is rejected. There is sufficient evidence to indicate the overall model is useful for $\alpha > 0.001$.

$R^2 = 0.988$. 98.81% of the sample variation of TME values about their means is explained by the regression model containing AMAP and NDF.

d. The two major drawbacks to stepwise regression are:

1. The stepwise procedure conducts an extremely large number of t-tests which result in high probabilities of making either Type I or Type II errors.

2. The stepwise procedure does not include any higher-order or interaction terms in the modeling process and may not yield the best predicting model for y.

e. The complete second-order model is

$$E(y) = \beta_0 + \beta_1 AMAP + \beta_2 NDF + \beta_3 AMAP^2 + \beta_4 NDF^2 + \beta_5 AMAPxNDF.$$

f. We would first fit the complete model stated in part e. Then we would fit the reduced model $E(y) = \beta_0 + \beta_1 AMAP + \beta_2 NDF + \beta_3 AMAPxNDF$. We would then compare the complete model to the reduced model to see if the quadratic terms are statistically significant.

12.69 a. In the first step of stepwise regression, 11 models are fit. These would be all the one variable models.

b. In step 2, all two variable models that include the variable selected in step one are fit. There will be a total of 10 models fit in this step.

c. In step 11, there is one model fit – the model containing all the independent variables.

d. The model would be
$$E(y) = \beta_0 + \beta_1 x_{11} + \beta_2 x_4 + \beta_3 x_2 + \beta_4 x_7 + \beta_5 x_{10} + \beta_6 x_1 + \beta_7 x_9 + \beta_8 x_3.$$

e. $R^2 = 0.677$. 67.7% of the sample variation in the overall satisfaction with BRT scores is explained by the model containing the 8 independent variables.

f. In any stepwise regression, many t-tests are performed. This inflates the probability of committing a Type I error. In addition, there are certain combinations of variables that may never be reached because of the first few variables put in the model. Finally, you should consider including interaction and higher order terms in the model.

12.71 a. In step 1, 8 one-variable models will be fit.

b. Of all the one-variable models, the model containing x_1 had the smallest p-value and the largest R^2 value.

c. In step 2, there would be 7 two-variable model fit where x_1 would be one of the variables in the model.

d. $\hat{\beta}_1 = -0.28$. The difference in the mean relative error in estimating effort between the developer and the project leader is estimated to be -0.28, holding the other variables constant.

$\hat{\beta}_2 = 0.27$. The difference in the mean relative error in estimating effort between those whose previous accuracy was greater than 20% and those whose previous accuracy was less than 20% is estimated to be 0.27, holding the other variables constant.

e. In any stepwise regression, many t-tests are performed. This inflates the probability of committing a Type I error. In addition, there are certain combinations of variables that may never be reached because of the first few variables put in the model. Finally, you should consider including interaction and higher order terms in the model.

12.73 Since the maximum amount of pollution that can be emitted increases as the plant's output increases, the model would be $E(y) = \beta_0 + \beta_1 x$, where y = maximum amount of pollution permitted (in parts per million) and x = plant's output (in megawatts).

12.75 To determine if the rate of increase in output per unit increase of input decreases as the input increases, we test:
$$H_0 : \beta_2 = 0$$
$$H_a : \beta_2 < 0$$

The test statistic is $t = -6.60$ and the p-value is $p = 0.000 / 2 = 0.000$.

Since the p-value is less than $\alpha \left(p = 0.000 < 0.05 \right)$, H_0 is rejected. There is sufficient evidence to indicate the rate of increase in output per unit increase of input decreases as the input increases at $\alpha = 0.05$.

12.77 If only 3 data points are used to fit a second-order model, $SSE = 0$ and $df = n - (k = 1) = 3 - (2 + 1) = 0$. Thus, there is no estimate for σ^2. With no estimate for σ^2, no tests can be performed.

12.79 a. The estimated difference between the predicted regulated price and deregulated price is $-0.782 + 0.0399 x_1 - 0.021 x_2 - 0.0033 x_1 x_2$.

b. For $x_2 = 10$ and $x_4 = 1$ and deregulation $x_3 = 1$,

$\hat{y} = 12.192 - 0.598x_1 - 0.00598(10) - 0.01078x_1(10) + 0.086x_1^2 + 0.00014(10)^2$

$\qquad + 0.677(1) - 0.275x_1(1) - 0.026(10)(1) + 0.013x_1(10)(1) - 0.782(1) + 0.0399x_1(1)$

$\qquad - 0.021(10)(1) - 0.0033x_1(10)(1)$

$\qquad = 11.5712 - 0.8439x_1 + 0.086x_1^2$

For $x_2 = 10$ and $x_4 = 1$ and regulation $x_3 = 0$,

$\hat{y} = 12.192 - 0.598x_1 - 0.00598(10) - 0.01078x_1(10) + 0.086x_1^2 + 0.00014(10)^2$

$\qquad + 0.677(1) - 0.275x_1(1) - 0.026(10)(1) + 0.013x_1(10)(1) - 0.782(0) + 0.0399x_1(0)$

$\qquad - 0.021(10)(0) - 0.0033x_1(10)(0)$

$\qquad = 12.5632 - 0.8508x_1 + 0.086x_1^2$

The y-intercept for the regulated prices (12.5632) is larger than the y-intercept for the deregulated prices (11.5712). Also, although the equations have the same rate of curvature, the estimated shift parameters differ.

12.81 a. The complete second-order model is

$E(y) = \beta_0 + \beta_1 x_1 + \beta_2 x_1^2 + \beta_3 x_2 + \beta_4 x_1 x_2 + \beta_5 x_1^2 x_2.$

b. A possible graph is:

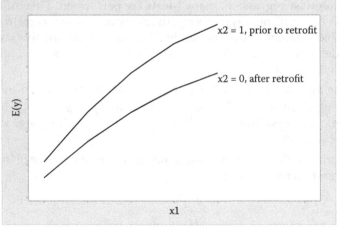

If $x_2 = 1, E(y) = \beta_0 + \beta_1 x_1 + \beta_2 x_1^2 + \beta_3(1) + \beta_4 x_1(1) + \beta_5 x_1^2(1)$

$\qquad = \beta_0 + \beta_3 + (\beta_1 + \beta_4)x_1 + (\beta_2 + \beta_5)x_1^2$

If $x_2 = 0$, $E(y) = \beta_0 + \beta_1 x_1 + \beta_2 x_1^2 + \beta_3(0) + \beta_4 x_1(0) + \beta_5 x_1^2(0) = \beta_0 + \beta_1 x_1 + \beta_2 x_1^2$

The two response curves have different shapes and different y-intercepts.

c. The first-order model is $E(y) = \beta_0 + \beta_1 x_1 + \beta_2 x_2.$

d. A possible graph is:

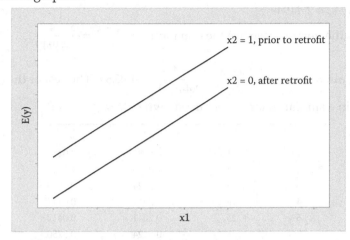

If $x_2 = 1$, $E(y) = \beta_0 + \beta_1 x_1 + \beta_2(1) = \beta_0 + \beta_2 + \beta_1 x_1$

$= \beta_0 + \beta_3 + (\beta_1 + \beta_4) x_1 + (\beta_2 + \beta_5) x_1^2$

If $x_2 = 0$, $E(y) = \beta_0 + \beta_1 x_1 + \beta_2(0) = \beta_0 + \beta_1 x_1$

The two response curves have the same slope, but they have different y-intercepts.

12.83 a. The model would be $E(y) = \beta_0 + \beta_1 x_1 + \beta_2 x_2 + \beta_3 x_3$

Where $x_2 = \begin{cases} 1 & \text{if program B} \\ 0 & \text{if not} \end{cases}$ and $x_3 = \begin{cases} 1 & \text{if program C} \\ 0 & \text{if not} \end{cases}$

b. To determine if the mean work-hours differ for the three safety programs, we test:

$$H_0 : \beta_2 = \beta_3 = 0$$

$$H_a : \text{At least one } \beta_i \neq 0$$

c. The test statistic is

$$F = \frac{(SSE_R - SSE_C)/(k-g)}{SSE_C /[n-(k+1)]} = \frac{(3,113.14 - 1,527.27)/(3-1)}{1,527.27/[9-(3+1)]} = 2.596.$$

The rejection region requires $\alpha = 0.05$ in the upper tail of the F distribution with $v_1 = k - g = 3 - 1 = 2$ and $v_2 = n - (k+1) = 9 - (3+1) = 5$. From Table 10, Appendix B, $F_{0.05} = 5.79$. The rejection region is $F > 5.79$.

Since the observed value of the test statistic does not fall in the rejection region $(F = 2.596 \not> 5.79)$, H_0 is not rejected. There is insufficient evidence to indicate the mean work-hours differ for the three safety programs at x_1 $\alpha = 0.05$.

12.85 a. Some preliminary calculations are:
$\bar{x} = 4.5 \quad s_x = 2.44949$

To find the u-values, we use the equation $u = \dfrac{x - \bar{x}}{s_x} = \dfrac{x - 4.5}{2.44949}$.

b. For a walk of 1 step, $u = \dfrac{x - \bar{x}}{s_x} = \dfrac{1 - 4.5}{2.44949} = -1.4289$. The rest of the u-values are found in a similar manner and are shown below.

x	x^2	u	u_2
1	1	−1.4289	2.0417
2	4	−1.0206	1.0417
3	9	−0.6124	0.3750
4	16	−0.2041	0.0417
5	25	0.2041	0.0417
6	36	0.6124	0.3750
7	49	1.0206	1.0417
8	64	1.4289	2.0417

c. Using MINITAB, the correlation between x and x^2 is:

Correlation: x, x-sq

```
Pearson correlation of x and x-sq = 0.976
P-Value = 0.000
```

The correlation coefficient is 0.976.

d. Using MINITAB, the correlation between u and u^2 is:

Correlation: u, u-sq

```
Pearson correlation of u and u-sq = 0.000
P-Value = 1.00
```

The correlation coefficient is 0.000. This is much smaller than the correlatin coefficient between x and x^2.

e. Using MINITAB, the results are:

Regression Analysis: Unrooted versus u, u-sq

```
Analysis of Variance
```

Source	DF	Adj SS	Adj MS	F-Value	P-Value
Regression	2	115583	57792	26.66	0.002
u	1	77271	77271	35.65	0.002
u-sq	1	38312	38312	17.68	0.008
Error	5	10837	2167		
Total	7	126420			

```
Model Summary

       S    R-sq  R-sq(adj)  R-sq(pred)
 46.5546  91.43%     88.00%      53.16%


Coefficients

Term        Coef  SE Coef  T-Value  P-Value   VIF
Constant    -0.7     25.0    -0.03    0.980
u          105.1     17.6     5.97    0.002  1.00
u-sq        90.6     21.6     4.20    0.008  1.00


Regression Equation

Unrooted = -0.7 + 105.1 u + 90.6 u-sq
```

The least squares prediction equation is $\hat{y} = -0.7 + 105.1u + 90.6u^2$.
To determine if the relationship between the number of unrooted walks and number of steps is curvilinear, we test:

$$H_0 : \beta_2 = 0$$

$$H_a : \beta_2 \neq 0$$

The test statistic is $t = 4.20$ and the p-value is $p = 0.008$. Since the p-value is so small, there is evidence to reject H_0. There is sufficient evidence to indicate that the relationship between the number of unrooted walks and number of steps is curvilinear for $\alpha > 0.008$.

12.87   The complete second-order model is $E(y) = \beta_0 + \beta_1 x_1 + \beta_2 x_2 + \beta_3 x_1^2 + \beta_4 x_2^2 + \beta_5 x_1 x_2$.

12.89   A stepwise regression was run using SAS and the following terms:

$$x_1 = \begin{cases} 1 & \text{if bulb surface D} \\ 0 & \text{if not} \end{cases} \qquad x_2 = \text{length of operation}$$

and the higher order terms $x_1 x_2$, x_1^2, and x_2^2

```
                    The REG Procedure
                     Model: MODEL1
                 Dependent Variable: Y

             Number of Observations Read      14
             Number of Observations Used      14

               Stepwise Selection: Step 1

      Variable X1 Entered: R-Square = 0.4740 and C(p) = 229.1691

                    Analysis of Variance

                             Sum of        Mean
    Source          DF      Squares      Square   F Value    Pr > F
```

Source	DF	Sum of Squares	Mean Square	F Value	Pr > F
Model	1	1045.78571	1045.78571	10.81	0.0065
Error	12	1160.57143	96.71429		
Corrected Total	13	2206.35714			

Variable	Parameter Estimate	Standard Error	Type II SS	F Value	Pr > F
Intercept	24.42857	3.71703	4177.28571	43.19	<.0001
X1	-17.28571	5.25668	1045.78571	10.81	0.0065

Bounds on condition number: 1, 1

Stepwise Selection: Step 2

Variable X2 Entered: R-Square = 0.8551 and C(p) = 57.8826

Analysis of Variance

Source	DF	Sum of Squares	Mean Square	F Value	Pr > F
Model	2	1886.66071	943.33036	32.46	<.0001
Error	11	319.69643	29.06331		
Corrected Total	13	2206.35714			

Variable	Parameter Estimate	Standard Error	Type II SS	F Value	Pr > F
Intercept	12.80357	2.97032	540.00945	18.58	0.0012
X1	-17.28571	2.88163	1045.78571	35.98	<.0001
X2	0.00969	0.00180	840.87500	28.93	0.0002

Bounds on condition number: 1, 4

Stepwise Selection: Step 3

Variable X1X2 Entered: R-Square = 0.9548 and C(p) = 14.5416

Analysis of Variance

Source	DF	Sum of Squares	Mean Square	F Value	Pr > F
Model	3	2106.67857	702.22619	70.45	<.0001
Error	10	99.67857	9.96786		
Corrected Total	13	2206.35714			

Variable	Parameter Estimate	Standard Error	Type II SS	F Value	Pr > F
Intercept	6.85714	2.15126	101.27473	10.16	0.0097
X1	-5.39286	3.04235	31.32005	3.14	0.1067
X2	0.01464	0.00149	960.57143	96.37	<.0001
X1X2	-0.00991	0.00211	220.01786	22.07	0.0008

Bounds on condition number: 4.25, 28.5

Stepwise Selection: Step 4

Variable X2SQ Entered: R-Square = 0.9802 and C(p) = 5.0000

The SAS System   21:38 Sunday, January 3, 2016   3

Analysis of Variance

Sum of   Mean

Source	DF	Squares	Square	F Value	Pr > F
Model	4	2162.68452	540.67113	111.42	<.0001
Error	9	43.67262	4.85251		
Corrected Total	13	2206.35714			

Variable	Parameter Estimate	Standard Error	Type II SS	F Value	Pr > F
Intercept	3.97024	1.72483	25.71018	5.30	0.0469
X1	-5.39286	2.12271	31.32005	6.45	0.0317
X2	0.02330	0.00275	347.55612	71.62	<.0001
X1X2	-0.00991	0.00147	220.01786	45.34	<.0001
X2SQ	-0.00000361	0.00000106	56.00595	11.54	0.0079

Bounds on condition number: 14, 138

All variables left in the model are significant at the 0.1500 level.

No other variable met the 0.1500 significance level for entry into the model.

Summary of Stepwise Selection

Step	Variable Entered	Variable Removed	Number Vars In	Partial R-Square	Model R-Square	C(p)	F Value	Pr > F
1	X1		1	0.4740	0.4740	229.169	10.81	0.0065
2	X2		2	0.3811	0.8551	57.8826	28.93	0.0002
3	X1X2		3	0.0997	0.9548	14.5416	22.07	0.0008
4	X2SQ		4	0.0254	0.9802	5.0000	11.54	0.0079

All the terms left in the model are significant at the 0.05 level of significance. The fitted model is $\hat{y} = 3.97024 - 5.39286x_1 + 0.02330x_2 - 0.00991x_1x_2 - 0.00000361x_2^2$.

13

Principles of Experimental Design

13.1 a. The two factors that affect the quantity of information in an experiment are noise (variability) and volume (n).

   b. The randomized block design increases the quantity of information in an experiment by utilizing a variable that causes variation in the response variable. By using this variable as a block, it is easier to detect the difference between treatment means.

13.3 a. The experimental units are the buried steel pipe locations.

   b. This design is a randomized block design. The factor in this experiment is the type of test used and it has two levels: instant-off and instant-on. Because there is only one factor, the treatments are the same as the factor levels. The blocks are the 19 different pipe locations. Each treatment was used at each location.

   c. The response variable is the corrosion prediction.

   d. The model for this design is $E(y) = \beta_0 + \beta_1 x_1 + \beta_2 x_2 + \beta_3 x_3 + \cdots + \beta_{19} x_{19}$

   where $x_1 = \begin{cases} 1 \text{ if instant-off} \\ 0 \text{ if not} \end{cases}$ $x_2 = \begin{cases} 1 \text{ if location 1} \\ 0 \text{ if not} \end{cases}$ $x_3 = \begin{cases} 1 \text{ if location 2} \\ 0 \text{ if not} \end{cases}$

   $x_{19} = \begin{cases} 1 \text{ if location 18} \\ 0 \text{ if not} \end{cases}$

13.5 a. The experimental units are the cockatiels.

   b. This experiment is a designed experiment. The birds are randomly divided into 3 groups and each group received a different treatment.

   c. There is one factor in this study. The factor is group.

   d. There are three levels of the group variable – Group 1 received purified water, Group 2 received purified water and liquid sucrose, and Group 3 received purified water and liquid sodium chloride.

   e. There are three treatments in the study. Because there is only one factor, the treatments are the same as the factor levels.

   f. The response variable is the water consumption.

   g. The model is $E(y) = \beta_0 + \beta_1 x_1 + \beta_2 x_2$ where $x_1 = \begin{cases} 1 \text{ if Group 1} \\ 0 \text{ if not} \end{cases}$ and $x_2 = \begin{cases} 1 \text{ if Group 2} \\ 0 \text{ if not} \end{cases}$

13.7 a. The month-to-month variation can be accounted for in a randomized block design. This variation would not be accounted for in a completely randomized design and detecting differences between state means would be more difficult.

   b. The treatments in this experiment are the three states, California, Utah, and Alaska.

   c. The blocks in this experiment are the month/years that were sampled, November 2000, October 2001, and November 2001.

13.9 a. The model for the first observation for engineer B ($x_1 = 0$, $x_2 = 1$, $x_3 = 0$, $x_4 = 1, x_5 = \cdots = x_{12} = 0$) is:

$$y_{B1} = \beta_0 + \beta_2 + \beta_4 \varepsilon_{B1}$$

The models for the rest of the observations for engineer B are:

$$y_{B2} = \beta_0 + \beta_2 + \beta_5 + \varepsilon_{B2} \qquad\qquad y_{B3} = \beta_0 + \beta_2 + \beta_6 + \varepsilon_{B3}$$
$$y_{B4} = \beta_0 + \beta_2 + \beta_7 + \varepsilon_{B4} \qquad\qquad y_{B5} = \beta_0 + \beta_2 + \beta_8 + \varepsilon_{B5}$$
$$y_{B6} = \beta_0 + \beta_2 + \beta_9 + \varepsilon_{B6} \qquad\qquad y_{B7} = \beta_0 + \beta_2 + \beta_{10} + \varepsilon_{B7}$$
$$y_{B8} = \beta_0 + \beta_2 + \beta_{11} + \varepsilon_{B8} \qquad\qquad y_{B9} = \beta_0 + \beta_2 + \beta_{12} + \varepsilon_{B9}$$
$$y_{B10} = \beta_0 + \beta_2 + \varepsilon_{B10}$$

The average of the 10 observations for engineer B is:

$$\bar{y}_B = \beta_0 + \beta_2 + \frac{\beta_4 + \beta_5 + \cdots \beta_{12}}{10} + \bar{\varepsilon}_B$$

   b. The model for the first observation for engineer D ($x_1 = 0$, $x_2 = 0$, $x_3 = 0$, $x_4 = 1, x_5 = \cdots = x_{12} = 0$) is:

$$y_{D1} = \beta_0 + \beta_4 + \varepsilon_{D1}$$

The models for the rest of the observations for engineer D are:

$$y_{D2} = \beta_0 + \beta_5 + \varepsilon_{D2} \qquad\qquad y_{D3} = \beta_0 + \beta_6 + \varepsilon_{D3}$$
$$y_{D4} = \beta_0 + \beta_1 + \varepsilon_{D4} \qquad\qquad y_{D5} = \beta_0 + \beta_8 + \varepsilon_{D5}$$
$$y_{D6} = \beta_0 + \beta_9 + \varepsilon_{D6} \qquad\qquad y_{D7} = \beta_0 + \beta_{10} + \varepsilon_{D7}$$
$$y_{D8} = \beta_0 + \beta_{11} + \varepsilon_{D8} \qquad\qquad y_{D9} = \beta_0 + \beta_{12} + \varepsilon_{D9}$$
$$y_{D10} = \beta_0 + \varepsilon_{D10}$$

The average of the 10 observations for engineer D is:

$$\bar{y}_D = \beta_0 + \frac{\beta_4 + \beta_5 + \cdots \beta_{12}}{10} + \bar{\varepsilon}_D$$

   c. $(\bar{y}_B - \bar{y}_D) = \left(\beta_0 + \beta_2 + \dfrac{\beta_4 + \beta_5 + \cdots \beta_{12}}{10} + \bar{\varepsilon}_B \right) - \left(\beta_0 + \dfrac{\beta_4 + \beta_5 + \cdots \beta_{12}}{10} + \bar{\varepsilon}_D \right) = \beta_2 + (\bar{\varepsilon}_B - \bar{\varepsilon}_D)$

13.11 a. A factorial design was employed in this study.

   b. There are two factors in the experiment, each at 6 levels.
   Factor 1: Level of coagulant – Levels: 5, 10, 20, 50, 100, and 200 milligrams per liter
   Factor 2: pH level – Levels: 4.0, 5.0, 6.0, 7.0, 8.0, and 9.0
   The treatments in this experiment are the $6 \times 6 = 36$ combinations of the factor levels. They are: $(5, 4.0), (5, 5.0), (5, 6.0), \ldots, (200, 9.0)$

13.13 a. The response variable is the quality of the steel ingot.

b. There are two factors: temperature and pressure. These are both quantitative variables since they are numeric.

c. The treatments are the $3 \times 5 = 15$ factor-level combinations of temperature and pressure.

d. The experimental units are the steel ingots.

13.15 a. This is not a complete factorial experiment since the treatments do not include all $3 \times 3 = 9$ factor-level combinations.

b. Interaction between two factors implies the effect of one factor on the dependent variable depends in the level of the second factor. There are three levels of factor A at B_1. However, there is only a level of factor A at B_2 and only one level of factor A at B_3. Thus, we cannot measure the effect of factor A at levels B_2 and B_3. Therefore, we cannot determine if the effect of factor A differs at different levels of factor B.

13.17 Replication allows for an estimate of the interaction effect in a factorial design. With no replication, there would be 0 degrees of freedom for error.

13.19 To solve for the number of blocks, b, we want to solve the equation:

$$t_{\alpha/2} s \sqrt{2/b} = B \Rightarrow b = \frac{2(t_{\alpha/2})^2 (s)^2}{B^2}$$

Since the degrees of freedom for $t_{\alpha/2}$ is $(a-1)(b-1) = (2-1)(b-1) = (b-1)$. To be conservative, we will start with $df = 10$. For $\alpha = 0.05$, $\alpha/2 = 0.05/2 = 0.025$. Thus, $t_{0.025} = 2.228$.

$$b = \frac{2(t_{\alpha/2})^2 (s)^2}{B^2} = \frac{2(2.228)^2 (15)^2}{10^2} = 22.34 \approx 23$$

Now, suppose that we use 20 for degrees of freedom. Thus, $t_{0.025} = 2.086$.

$$b = \frac{2(t_{\alpha/2})^2 (s)^2}{B^2} = \frac{2(2.086)^2 (15)^2}{10^2} = 19.58 \approx 20$$

Thus, we should use 20 blocks.

13.21 The steps in the design of an experiment that affect the volume of the signal are:

Step 1 Selecting the factors
Step 2 Choosing the treatments
Step 3 Determining the sample size

13.23 In both designs, there is one factor of primary interest to the study. In the randomized block design, a second factor is suspected of contribution information to the response variable. The block design is advantageous when such a contribution actually exists. The randomized block design allows for "blocking" out the effect of this second factor.

13.25 The complete factorial model is:

$$E(y) = \beta_0 + \beta_1 x_1 + \beta_2 x_2 + \beta_3 x_3 + \beta_4 x_4 + \beta_5 x_5 + \beta_6 x_1 x_2 + \beta_7 x_1 x_3 + \beta_8 x_1 x_4$$

$$+ \beta_9 x_1 x_5 + \beta_{10} x_2 x_3 + \beta_{11} x_2 x_4 + \beta_{12} x_2 x_5 + \beta_{13} x_1 x_2 x_3 + \beta_{14} x_1 x_2 x_4 + \beta_{15} x_1 x_2 x_5$$

where $x_1 = \begin{cases} 1 \text{ if Level 1 of factor 1} \\ 0 \text{ if not} \end{cases}$ $\qquad$ $x_2 = \begin{cases} 1 \text{ if Level 1 of factor 2} \\ 0 \text{ if not} \end{cases}$

$$x_3 = \begin{cases} 1 \text{ if Level 1 of factor 3} \\ 0 \text{ if not} \end{cases} \qquad x_4 = \begin{cases} 1 \text{ if Level 2 of factor 3} \\ 0 \text{ if not} \end{cases}$$

$$x_5 = \begin{cases} 1 \text{ if Level 3 of factor 3} \\ 0 \text{ if not} \end{cases}$$

If we have one replication, then we would have a total of $2 \times 2 \times 4 = 16$ observations. If we fit the complete model, the number of degrees of freedom for estimating σ^2 is $n - (k+1) = 16 - (15+1) = 0$.

13.27 a. This experiment involves a single factor, work scheduling, at three levels: flextime, staggered starting hours, and fixed hours. Since work scheduling is the only factor, these levels represent the treatments.

 b. To assign treatments in a completely random manner, with equal numbers of workers in each treatment, number the 60 workers from 1 to 60. Use Table 1 in Appendix B to select two-digit numbers, discarding those that are larger than 60 or are identical, until there is a total of 40 two-digit numbers. The workers who have been assigned the first 20 numbers in the sequence are assigned to flextime, the second group of 20 workers is assigned to staggered starting times, and the remaining workers are assigned to fixed hours.

 c. The linear model is $E(y) = \beta_0 + \beta_1 x_1 + \beta_2 x_2$

$$\text{where } x_1 = \begin{cases} 1 \text{ if flextime} \\ 0 \text{ if not} \end{cases} \qquad x_2 = \begin{cases} 1 \text{ if staggered} \\ 0 \text{ if not} \end{cases}$$

13.29 a. Since y will be observed for all factor-level combinations, this is a complete 3×3 factorial experiment.

 b. The factors are pay rate (quantitative) and length of workday (quantitative).

 c. The treatments will include all $3 \times 3 = 9$ factor-level combinations.

$P_1L_1, P_1L_2, P_1L_3, P_2L_1, P_2L_2, P_2L_3, P_3L_1, P_3L_2, P_3L_3$

13.31 a. The two factors are gender and weight.

 b. Gender is a qualitative variable and weight is a quantitative variable.

 c. There are four treatments in the study:

$$(M, L), (M, H), (F, L), (F, H)$$

14

The Analysis of Variance for Designed Experiments

14.1 a. Twenty boxes were randomly sampled for each of the box types. This results in a completely randomized design.

    b. Yes. For box types B and D, there is no overlap between the graphed means. It appears, therefore, that the mea compressive strengths of these two box types are significantly different.

    c. It does not appear that the mean compressive strengths of all five box types differ. It appears that A and D differ as well as B and D.

14.3 a. The experimental units are the teeth. The treatments are the 3 different bonding times: 1 hour, 24 hours, and 48 hours. The response variable is breaking strength (in Mpa).

    b. To determine if there is a difference in the mean breaking strength among the 3 bonding times, we test:

$$H_0 : \mu_1 = \mu_2 = \mu_3$$

$$H_a : \text{At least two treatment means differ}$$

    c. The rejection region requires $\alpha = 0.01$ in the upper tail of the F distribution with $v_1 = k - 1 = 3 - 1 = 2$ and $v_2 = n - k = 30 - 3 = 27$. Using Table 12, Appendix B, $F_{0.01} = 5.49$. The rejection region is $F > 5.49$.

    d. Since the observed value of the test statistic falls in the rejection region $(F = 61.62 > 5.49)$, H_0 is rejected. There is sufficient evidence to indicate a difference in mean breaking strength among the three bonding times at $\alpha = 0.01$.

    e. The conditions required for a valid ANOVA F-test in a completely randomized design are:

        1. The samples are randomly selected in an independent manner from k treatment populations.

        2. All k sampled populations have distributions that are approximately normal.

        3. The k population variances are equal $(i.e.\ \sigma_1^2 = \sigma_2^2 = \cdots = \sigma_k^2)$

14.5 a. To determine if the mean body length of entangled whales differs for the three types of fishing gear, we test:

$$H_0 : \mu_1 = \mu_2 = \mu_3$$

$$H_a : \text{At least two treatment means differ}$$

    b. The test statistic is $F = 34.81$. The p-value is $p < 0.0001$. Since the p-value is less than $\alpha (p < 0.0001 < 0.05)$. H_0 is rejected. There is sufficient evidence to indicate the mean body length of entangled whales differs for the three types of fishing gear at $\alpha = 0.05$.

14.7 a.   To find MST, divide SST by its degrees of freedom: $MST = \dfrac{SST}{k-1} = \dfrac{0.010}{3} = 0.003333$.

To find MSE, divide SET by its degrees of freedom:

$$MSE = \frac{SSE}{n-k} = \frac{0.029}{140} = 0.000207.$$

$$F = \frac{MST}{MSE} = \frac{0.00333}{0.000207} = 16.08$$

The ANOVA table is:

Source	df	SS	MS	F	p
Exposure	3	0.010	0.003333	16.08	<0.001
Error	140	0.029	0.000207		
Total	143	0.039			

  b.   To determine if the mean dot area differs depending on the exposure time, we test:

$$H_0 : \mu_1 = \mu_2 = \mu_3 = \mu_4$$
$$H_a : \text{At least two treatment means differ}$$

The test statistic is $F = 16.08$ and the p-value is $p < 0.001$. Since the p-value is less than $\alpha(p < 0.001 < 0.05)$, H_0 is rejected. There is sufficient evidence to indicate the mean dot area differs depending on the exposure time at $\alpha = 0.05$.

14.9 a.   The linear model is $E(y) = \beta_0 + \beta_1 x$ where $x = \begin{cases} 1 \text{ if current treatment} \\ 0 \text{ if not} \end{cases}$
  b.   Some preliminary calculations are:

$$\sum x = 3 \qquad \bar{x} = \frac{\sum x}{n} = \frac{3}{6} = 0.5 \qquad \sum x^2 = 3$$

$$\sum y = 3{,}702 \qquad \bar{y} = \frac{\sum y}{n} = \frac{3{,}702}{6} = 617 \qquad \sum y^2 = 2{,}288{,}630$$

$$\sum xy = 1{,}779$$

$$SS_{xx} = \sum x^2 - \frac{\left(\sum x\right)^2}{n} = 3 - \frac{(3)^2}{6} = 1.5$$

$$SS_{yy} = \sum y^2 - \frac{\left(\sum y\right)^2}{n} = 2{,}288{,}630 - \frac{(3{,}702)^2}{6} = 4{,}496$$

$$SS_{xy} = \sum xy - \frac{\sum x \sum y}{n} = 1{,}779 - \frac{(3)(3{,}702)}{6} = -72$$

$$\hat{\beta}_1 = \frac{SS_{xy}}{SS_{xx}} = \frac{-72}{1.5} = -48$$

$$\hat{\beta}_0 = \bar{y} - \hat{\beta}_1 \bar{x} = 617 - (-48)(0.5) = 641$$

The fitted model is $\hat{y} = 641 - 48x$.

$$SSE = SS_{yy} - \hat{\beta}_1 SS_{xy} = 4,496 - (-48)(-72) = 1,040$$

$$SST = SS_{yy} - SSE = 4,496 - 1,040 = 3,456$$

The resulting ANOVA table is:

Source	df	SS	MS	F
Treatment	1	3,456	3,456	13.29
Error	4	1,040	260	
Total	5	4,496		

c. $CM = \dfrac{\left(\sum y\right)^2}{n} = \dfrac{3,702^2}{6} = 2,284,134$

   $SST = \sum \dfrac{T_i^2}{n_i} - CM = \left(\dfrac{1,779^2}{3} + \dfrac{1,923^2}{3}\right) - 2,284,134 = 3,456$

   $MST = \dfrac{SST}{p-1} = \dfrac{3,456}{2-1} = 3,456$

   This agrees with MST from part b. This measures the variation within the treatment groups.

d. $MSE = \dfrac{SSE}{n-p} = \dfrac{1,040}{6-2} = 260$

   This agrees with MSE from part b. This measures the variation that is between the treatment groups.

e. The degrees of freedom associated with MST is $df = p - 1 = 2 - 1 = 1$.

f. The degrees of freedom associated with MSE is $df = n - p = 6 - 2 = 4$.

g. The test statistic is $F = \dfrac{MST}{MSE} = \dfrac{3,456}{260} = 13.29$. This is the same as the test statistic found in part b.

h. The rejection region requires $\alpha = 0.05$ in the upper tail of the F distribution with $v_1 = p - 1 = 2 - 1 = 1$ and $v_2 = n - p = 6 - 2 = 4$. Using Table 10, Appendix B, $F_{0.05} = 7.71$. The rejection region is $F > 7.71$.

i. Since the observed value of the test statistic falls in the rejection region ($F = 13.9 > 7.71$), H_0 is rejected. There is sufficient evidence to indicate a difference in mean Mpa between the two types of alloys at $\alpha = 0.05$.

j. Some preliminary calculations are:

   $$s_1^2 = 147 \qquad s_2^2 = 373$$

   $$s_p^2 = \dfrac{(n_1 - 1)s_1^2 + (n_2 - 1)s_2^2}{n_1 + n_2 - 2} = \dfrac{(3-1)147 + (3-1)373}{3+3-2} = 260$$

   $$T = \dfrac{y_1 - y_2}{\sqrt{s_p^2\left(\dfrac{1}{n_1} + \dfrac{1}{n_2}\right)}} = \dfrac{593 - 641}{\sqrt{260\left(\dfrac{1}{3} + \dfrac{1}{3}\right)}} = -3.646$$

The rejection region requires $\alpha/2 = 0.05/2 = 0.025$ in each tail of the t distribution with $df = n_1 + n_2 - 2 = 3 + 3 - 2 = 4$ and $v_2 = n - p = 6 - 2 = 4$. Using Table 7, Appendix B, $t_{0.025} = 2.776$. The rejection region is $t < -2.776$ or $t > 2.776$.

Since the observed value of the test statistic falls in the rejection region ($t = -3.646 < -2.776$), H_0 is rejected. There is sufficient evidence to indicate a difference in mean Mpa between the two types of alloys at $\alpha = 0.05$.

k. $T^2 = (-3.646)^2 = 13.29 = F$

l.   The F test for comparing two population means is a two tailed test. We are testing for differences in population means, but not specifying any directions.

14.11 a.   Using MINITAB, the results are:

One-way ANOVA: UMRB-1, UMRB-2, UMRB-3, SWRA, SD

Analysis of Variance

Source	DF	Adj SS	Adj MS	F-Value	P-Value
Factor	4	5.836	1.4589	7.25	0.001
Error	21	4.225	0.2012		
Total	25	10.061			

Model Summary

S	R-sq	R-sq(adj)	R-sq(pred)
0.448565	58.00%	50.00%	38.11%

Means

Factor	N	Mean	StDev	95% CI
UMRB-1	7	3.497	0.364	(3.145, 3.850)
UMRB-2	6	4.017	0.573	(3.636, 4.397)
UMRB-3	7	3.761	0.483	(3.409, 4.114)
SWRA	3	2.643	0.358	(2.105, 3.182)
SD	3	2.780	0.259	(2.241, 3.319)

Pooled StDev = 0.448565

To determine if there are differences in the mean Al/Be ratios among the 5 different boreholes, we test:

$$H_0 : \mu_{UMRB-1} = \mu_{UMRB-2} = \mu_{UMRB-3} = \mu_{SWRA} = \mu_{SD}$$

H_a : At least two treatment means differ

The test statistic is $F = 7.25$ and the p-value is $p = 0.001$.

Since the p-value is less than $\alpha(p = 0.001 < 0.10)$, H_0 is rejected. There is sufficient evidence to indicate there are differences in the mean Al/Be ratios among the 5 different boreholes at $\alpha = 0.10$.

14.13 a.   The subjects were randomly divided into 3 treatment groups with one group receiving an injection of scopolamine, another group receiving an injection of glycopyrrolate, and the third group receiving nothing. Thus, this is a completely randomized design.

b.   There are three treatments: injection of scopolamine, injection of glycopyrrolate, and nothing. The response variable is the number of pairs recalled.

c. Using MINITAB, the descriptive statistics are:

Descriptive Statistics: Scopolamine, Placebo, None

```
Variable       N   Mean StDev Minimum    Q1 Median      Q3 Maximum
Scopolamine  12  6.167 1.267   4.000 5.250  6.000   7.500   8.000
Placebo       8  9.375 1.506   7.000 8.250  9.500  10.000  12.000
None          8 10.625 1.506   8.000 9.250 11.000  12.000  12.000
```

The sample means for the three groups are: 6.167, 9.375, and 10.625. There is not sufficient information to support the researcher's theory. We need to take into account the variability within each group.

d. Using MINITAB, the results are:

One-way ANOVA: Scopolamine, Placebo, None

Analysis of Variance

```
Source  DF  Adj SS  Adj MS  F-Value  P-Value
Factor   2  107.01  53.506    27.07    0.000
Error   25   49.42   1.977
Total   27  156.43
```

Model Summary

```
      S   R-sq  R-sq(adj)  R-sq(pred)
1.40594  68.41%     65.88%      60.05%
```

Means

```
Factor          N    Mean   StDev        95% CI
Scopolamine    12   6.167   1.267  (5.331,   7.003)
Placebo         8   9.375   1.506  (8.351,  10.399)
None            8  10.625   1.506  (9.601,  11.649)
```

Pooled StDev = 1.40594

To determine if the mean number of words recalled differed among the three groups, we test:

$$H_0 : \mu_1 = \mu_2 = \mu_3$$

H_a : At least two treatment means differ

The test statistic is $F = 27.07$ and the p-value is $p = 0.000$.

Since the p-value is less than $\alpha(p = 0.000 < 0.05)$, H_0 is rejected. There is sufficient evidence to indicate the mean number of words recalled differed among the three groups at $\alpha = 0.05$.

Prior to the experiment, the researchers theorized that the mean number of word pairs recalled for the scopolamine group would be less than the corresponding means for the other 2 groups. From the printout, the sample mean for the scopolamine group is 6.167. The means for the placebo group and the nothing group are 9.375 and 10.625.

Since the sample mean for the scopolamine group is much smaller than the other two means, it appears that the researchers' theory was correct.

14.15 a.   Some preliminary calculations are:

$$\text{Since } \bar{y}_i = \frac{\sum y_i}{n_i} \Rightarrow \sum y_i = n_i \bar{y}_i.$$

Temperature	n_i	$\bar{y}_i$	$n_i \bar{y}_i = T_i$
100	16	52	832
200	16	112	1,792
300	16	143	2,288
400	16	186	2,976
500	14	257	3,598
	$\sum n_i = 78$		$\sum y = 11,486$

$$CM = \frac{\left(\sum y\right)^2}{n} = \frac{11,486^2}{78} = 1,691,387.128$$

$$SST = \sum \frac{T_i^2}{n_i} - CM = \left(\frac{832^2}{16} + \frac{1,792^2}{16} + \frac{2,288^2}{16} + \frac{2,976^2}{16} + \frac{3,598^2}{15}\right) - 1,691,387.128$$

$$= 357,986.872$$

b. $SSE = (n_1 - 1)s_1^2 + (n_2 - 1)s_2^2 + (n_3 - 1)s_3^2 + (n_4 - 1)s_4^2 + (n_5 - 1)s_5^2$

$$= 15(55)^2 + 15(108)^2 + 15(127)^2 + 15(136)^2 + 13(178)^2 = 1,151,602$$

c. $SS(\text{Total}) = SST + SSE = 357,986.872 + 1,151,602 = 1,509,588.872$

d. $MST = \dfrac{SST}{p-1} = \dfrac{357,986.872}{5-1} = 89,496.718 \quad MSE = \dfrac{SSE}{n-p} = \dfrac{1,151,602}{78-5} = 15,775.37$

$$F = \frac{MST}{MSE} = \frac{89,496.718}{15,775.37} = 5.67$$

Source	df	SS	MS	F
Exposure	4	357,986.872	89,496.718	5.67
Error	140	1,151,602	15,775.37	
Total	143	1,509,588.872		

e.   To determine if the heating temperature affects the mean total thermal strain of concrete, we test:

$$H_0 : \mu_1 = \mu_2 = \mu_3 = \mu_4 = \mu_5$$
$$H_a : \text{At least two treatment means differ}$$

The test statistic is $F = 5.67$.

The rejection region requires $\alpha = 0.01$ in the upper tail of the F distribution with $v_1 = p - 1 = 5 - 1 = 4$ and $v_2 = n - p = 78 - 5 = 73$. Using MINITAB, $F_{0.01} = 3.59$. The rejection region is $F > 3.59$.

Since the observed value of the test statistic falls in the rejection region ($F = 5.67 > 3.59$), H_0 is rejected. There is sufficient evidence to indicate the heating temperature affects the mean total thermal strain of concrete at $\alpha = 0.01$.

14.17 a. The experimental design is a randomized block design with the months as the blocks.

b. $MST = \dfrac{SST}{p-1} \Rightarrow SST = MST(p-1) = 0.195(2) = 0.39$

$MSB = \dfrac{SSB}{b-1} \Rightarrow SSB = MSB(b-1) = 10.78(3) = 32.34$

$F_B = \dfrac{MSB}{MSE} = \dfrac{10.78}{0.069} = 156.23$

Source	df	SS	MS	F	p-value
Forecast Method	2	0.39	0.195	2.83	0.08
Month	3	32.34	10.780	156.23	<0.01
Error	6	0.414	0.069		
Total	11	33.144			

c. To determine if the mean electrical consumption values differ for the three types of forecasts, we test:

$$H_0 : \mu_1 = \mu_2 = \mu_3$$
$$H_a : \text{At least two treatment means differ}$$

The test statistic is $F = 2.83$ and the p-value is $p = 0.08$.

Since the p-value is not less than $\alpha(p = 0.08 \not< 0.05)$, H_0 is not rejected. There is insufficient evidence to indicate the mean electrical consumption values differ for the three types of forecasts at $\alpha = 0.05$.

14.19 To determine if the mean crack widths differ for the four time periods, we test:

$$H_0 : \mu_1 = \mu_2 = \mu_3 = \mu_4$$
$$H_a : \text{At least two treatment means differ}$$

The test statistic is $F = 57.99$ and the p-value is $p < 0.0001$.

Since the p-value is less than $\alpha(p < 0.0001 < 0.05)$, H_0 is rejected. There is sufficient evidence to indicate the mean crack widths differ for the four time periods at $\alpha = 0.05$.

14.21 a. The data were collected as a randomized block design because the skin factor values were collected by each software product each week.

b. The dependent variable is the skin factor value. The treatments are the four software products. The blocks are the 10 weeks the data were collected.

c. Using MINITAB, the results are:

```
Two-way ANOVA: Y versus Software, Week

Source     DF       SS        MS       F      P
Software    3      911     303.8    6.45  0.002
Week        9   340469   37829.8  803.79  0.000
Error      27     1271      47.1
Total      39   342651

S = 6.860   R-Sq = 99.63%   R-Sq(adj) = 99.46%
```

The ANOVA table is:

Source	df	SS	MS	F	p-value
Software	3	911	303.8	6.45	0.002
Week	9	340,469	37,829.8	803.79	0.000
Error	27	1,271	47.1		
Total	39	342,651			

d. To determine if the mean skin factors differ among the four software products, we test:

$$H_0 : \mu_1 = \mu_2 = \mu_3 = \mu_4$$

$$H_a : \text{At least two treatment means differ}$$

The test statistic is $F = 6.45$ and the p-value is $p = 0.002$.

Since the p-value is less than $\alpha(p = 0.002 < 0.01)$, H_0 is rejected. There is suffi-cient evidence to indicate the mean skin factors differ among the four software products at $\alpha = 0.01$.

14.23   The regression model is $E(y) = \beta_0 + \beta_1 x_1 + \beta_1 x_2 + \beta_1 x_3 + \beta_1 x_4$

Where $x_1 = \begin{cases} 1 \text{ if standard} \\ 0 \text{ if not} \end{cases}$ $x_2 = \begin{cases} 1 \text{ if supervent} \\ 0 \text{ if not} \end{cases}$ $x_3 = \begin{cases} 1 \text{ if row 1} \\ 0 \text{ if not} \end{cases}$ $x_4 = \begin{cases} 1 \text{ if row 2} \\ 0 \text{ if not} \end{cases}$

We will fit the complete model and then fit the reduced model without the variables x_1 and x_2.

The results of the complete model are:

Regression Analysis: Time versus x1, x2, x3, x4

```
The regression equation is
Time = 219 + 15.4 x1 - 29.9 x2 - 106 x3 - 52.6 x4

Predictor      Coef   SE Coef      T       P
Constant     218.75     15.79   13.85   0.001
x1            15.42     13.52    1.14   0.337
x2           -29.92     13.52   -2.21   0.114
x3          -105.92     13.52   -7.83   0.004
x4           -52.58     13.52   -3.89   0.030

S = 14.1235    R-Sq = 96.5%    R-Sq(adj) = 91.7%

Analysis of Variance

Source            DF        SS       MS       F       P
Regression         4   16274.5   4068.6   20.40   0.016
Residual Error     3     598.4    199.5
Total              7   16872.9

Source   DF   Seq SS
x1        1   3619.0
x2        1     14.7
x3        1   9624.4
x4        1   3016.4
```

$SSE_c = 598.4$

The results of fitting the reduce model is:

Regression Analysis: Time versus x3, x4

```
The regression equation is
Time = 212 - 103 x3 - 50.2 x4

Predictor        Coef   SE Coef        T        P
Constant       211.50     19.37    10.92    0.000
x3            -103.50     25.00    -4.14    0.009
x4             -50.17     25.00    -2.01    0.101

S = 27.3904    R-Sq = 77.8%    R-Sq(adj) = 68.9%

Analysis of Variance

Source            DF        SS       MS      F        P
Regression         2   13121.7   6560.9   8.75    0.023
Residual Error     5    3751.2    750.2
Total              7   16872.9

Source   DF    Seq SS
x3        1   10101.7
x4        1    3020.0
```

$SSE_R = 3,751.2$

To determine if the mean half-cooling times differ among the three designs, we test:

$$H_0 : \mu_1 = \mu_2 = \mu_3$$

H_a : At least two treatment means differ

The test statistic is $F = \dfrac{(SSE_R - SSE_c)/(k-g)}{SSE_c/[n-(k=1)]} = \dfrac{(3,751.2 - 598.4)/(4-2)}{598.4/[8-(4+1)]} = 7.90.$

The rejection region requires $\alpha = 0.10$ in the upper tail of the F distribution with $v_1 = k - g = 4 - 2 = 2$ and $v_2 = n - (k+1) = 8 - (4+1) = 3$. From Table 9, Appendix B, $F_{0.10} = 5.46$. The rejection region is $F > 5.46$.

Since the observed value of the test statistic falls in the rejection region ($F = 7.90 > 5.46$), H_0 is rejected. There is sufficient evidence to indicate the mean half-cooling times differ among the three designs at $\alpha = 0.10$.

14.25 a. The model is $E(y) = \beta_0 + \beta_1 x_1 + \beta_2 x_2 + \beta_3 x_3 + \beta_4 x_4 + \cdots + \beta_{104} x_{104}$

where $x_1 = \begin{cases} 1 \text{ if Full-dard} \\ 0 \text{ if not} \end{cases}$ $\qquad x_2 = \begin{cases} 1 \text{ if TR-light} \\ 0 \text{ if not} \end{cases}$ $\qquad x_3 = \begin{cases} 1 \text{ if Gene 1} \\ 0 \text{ if not} \end{cases}$

$x_4 = \begin{cases} 1 \text{ if Gene 2} \\ 0 \text{ if not} \end{cases}$ $\quad \cdots \quad x_{104} = \begin{cases} 1 \text{ if Gene 102} \\ 0 \text{ if not} \end{cases}$

b. To determine if the mean standardized growth measurements for the three light/dark conditions differ, we test:

$$H_0 : \beta_1 = \beta_2 = 0$$

H_a : At least one $\beta_i \neq 0$

c. Using MINITAB, the results are:

```
ANOVA: Growth versus LightCond, GeneID
Analysis of Variance for Growth
Source         DF        SS       MS      F      P
LightCond       2     9.093   4.54660   5.33   0.006
GeneID       1022     0.715   0.49720   0.58   0.999
Error         204   174.138   0.85362
Total         308   233.946

S = 0.92392    R-Sq = 25.56%
```

The test statistic is $F = 5.33$ and the p-value is $p = 0.006$.

Since the p-value is less than $\alpha(p = 0.006 < 0.05)$, H_0 is rejected. There is sufficient evidence to indicate the mean standardized growth measurements for the three light/dark conditions differ at $\alpha = 0.05$.

14.27 a. The type of design is a 2×2 factorial design.

b. There are two factors: Diet and Age of hen.
Diet has two levels – fine limestone and course limestone
Age of hen has two levels – younger and older
There are four treatments: (FL,Y), (FL,O), (CL,Y), (CL,O)

c. The experimental units are the eggs.

d. The dependent variable is the shell thickness.

e. There is no diet-age interaction. This means that the effect of diet on the shell thickness does not depend on the age of the hen.

f. There is no effect due to hen age. This means that the mean shell thickness is no different for young and older hens.

g. There is an effect doe to diet. The mean shell thickness for hens fed course limestone is significantly greater than the mean shell thickness for hen fed fine limestone.

14.29 a. If temperature and type of yeast extract interact, then the relationship between autolysis yield and temperature may depends on the type of yeast extract. A possible graph would be:

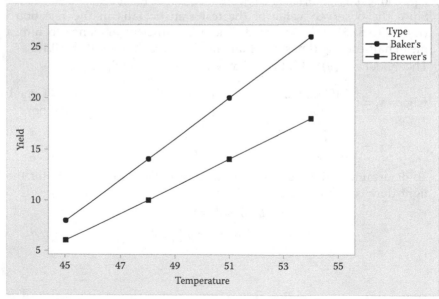

b. If no interaction exists, then the slopes of the two lines will be the same. A possible graph is:

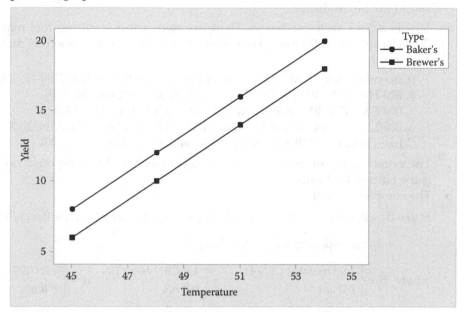

c. The complete model would be

$$E(y) = \beta_0 + \beta_1 x_1 + \beta_2 x_2 + \beta_3 x_3 + \beta_4 x_4 + \beta_5 x_1 x_2 + \beta_6 x_1 x_3 + \beta_7 x_1 x_4$$

where $x_1 = \begin{cases} 1 \text{ if Baker's yeast} \\ 0 \text{ if not} \end{cases}$ $x_2 = \begin{cases} 1 \text{ if } 45°C \\ 0 \text{ if not} \end{cases}$ $x_3 = \begin{cases} 1 \text{ if } 45°C \\ 0 \text{ if not} \end{cases}$ $x_4 = \begin{cases} 1 \text{ if } 45°C \\ 0 \text{ if not} \end{cases}$

d. To determine if type of yeast and temperature interact, we test:

$$H_0 : \beta_5 = \beta_6 = \beta_7 = 0$$

$$H_a : \text{At least one } \beta_i \neq 0$$

We need to conduct a partial F-test to test for interaction. We will fit the complete model listed in part a to compute SSE_C. We will then fit the reduced model that does not include the interaction terms to compute SSE_R.

e. Since the p-value is less than $\alpha(p = 0.0027 < 0.01)$, H_0 is rejected. There is sufficient evidence to indicate that type of yeast and temperature interact to affect autolysis at $\alpha = 0.01$.

f. To conduct the main effect test for type of yeast, we test:

$$H_0 : \beta_1 = 0$$

$$H_a : \beta_1 \neq 0$$

We could use a t-test for this test.

To conduct the main effect test for temperature, we test:

$$H_0 : \beta_2 = \beta_3 = \beta_4 = 0$$

$$H_a : \text{At least one } \beta_1 \neq 0$$

For this test, we would use a partial F-test.

g. Once the interaction term was found to be significant, we do not test for the main effects of the individual variables. The tests in part f for this exercise would not be conducted.

14.31 a. This is a 5×3 factorial design. There are two factors – cutting tool material (5 levels) and speed (3 levels). The treatments are the 15 combinations of material and speed:

(Uncoated CBN-H, 100), (Uncoated CBN-H, 140), (Uncoated CBN-H, 200),
(CBN-H w/TiN, 100), (CBN-H w/TiN, 140), (CBN-H w/TiN, 200),
(CBN-L w/TiN, 100), (CBN-L w/TiN, 140), (CBN-L w/TiN, 200),
(CBN-L w/TiAIN, 100), (CBN-L w/TiAIN, 140), (CBN-L w/TiAIN, 200),
(Mixed ceramic, 100), (Mixed ceramic, 140), (Mixed ceramic, 200)

The experimental units are the coated cutting tools and the response variable is the cutting feed force.

b. The complete model is

$$E(y) = \beta_0 + \beta_1 x_1 + \beta_2 x_2 + \beta_3 x_3 + \beta_4 x_4 + \beta_5 x_5 + \beta_6 x_6 + \beta_7 x_1 x_5 + \beta_8 x_1 x_6 + \beta_9 x_2 x_5 + \beta_{10} x_2 x_6$$

$$+ \beta_{11} x_3 x_5 + \beta_{12} x_3 x_6 + \beta_{13} x_4 x_5 + \beta_{14} x_4 x_6$$

where $x_1 = \begin{cases} 1 \text{ if Uncoated CBN-H} \\ 0 \text{ if not} \end{cases}$ $x_2 = \begin{cases} 1 \text{ if CBN-H w/TiN} \\ 0 \text{ if not} \end{cases}$ $x_3 = \begin{cases} 1 \text{ if CBN-L w/TiN} \\ 0 \text{ if not} \end{cases}$

$x_4 = \begin{cases} 1 \text{ if CBN-L w/TiAIN} \\ 0 \text{ if not} \end{cases}$ $x_5 = \begin{cases} 1 \text{ if 100 m/min} \\ 0 \text{ if not} \end{cases}$ $x_6 = \begin{cases} 1 \text{ if 140 m/min} \\ 0 \text{ if not} \end{cases}$

c. To determine if cutting tool and speed interact, we test:

$$H_0 : \beta_7 = \beta_8 = \cdots = \beta_{14} = 0$$

$$H_a : \text{At least one } \beta_i \neq 0$$

d. Using MINITAB, the results of fitting the complete model is:

Regression Analysis: FEED versus x1, x2, x3, x4, x5, x6, x1x5, x1x6, ...

Analysis of Variance

Source	DF	Adj SS	Adj MS	F-Value	P-Value
Regression	14	28813.9	2058.13	91.34	0.000
x1	1	1892.2	1892.25	83.98	0.000
x2	1	4900.0	4900.00	217.46	0.000
x3	1	506.2	506.25	22.47	0.000
x4	1	676.0	676.00	30.00	0.000
x5	1	420.2	420.25	18.65	0.001
x6	1	20.3	20.25	0.90	0.358
x1x5	1	741.1	741.12	32.89	0.000
x1x6	1	1740.5	1740.50	77.24	0.000
x2x5	1	1081.1	1081.13	47.98	0.000
x2x6	1	3.1	3.13	0.14	0.715
x3x5	1	210.1	210.12	9.33	0.008
x3x6	1	50.0	50.00	2.22	0.157
x4x5	1	105.1	105.13	4.67	0.047
x4x6	1	72.0	72.00	3.20	0.094
Error	15	338.0	22.53		
Total	29	29151.9			

Model Summary

```
       S    R-sq  R-sq(adj)  R-sq(pred)
 4.74693  98.84%     97.76%      95.36%
```

Coefficients

Term	Coef	SE Coef	T-Value	P-Value	VIF
Constant	56.50	3.36	16.83	0.000	
x1	43.50	4.75	9.16	0.000	4.80
x2	70.00	4.75	14.75	0.000	4.80
x3	22.50	4.75	4.74	0.000	4.80
x4	-26.00	4.75	-5.48	0.000	4.80
x5	20.50	4.75	4.32	0.001	6.67
x6	4.50	4.75	0.95	0.358	6.67
x1x5	-38.50	6.71	-5.73	0.000	3.73
x1x6	-59.00	6.71	-8.79	0.000	3.73
x2x5	-46.50	6.71	-6.93	0.000	3.73
x2x6	-2.50	6.71	-0.37	0.715	3.73
x3x5	-20.50	6.71	-3.05	0.008	3.73
x3x6	-10.00	6.71	-1.49	0.157	3.73
x4x5	-14.50	6.71	-2.16	0.047	3.73
x4x6	-12.00	6.71	-1.79	0.094	3.73

Regression Equation

FEED = 56.50 + 43.50 x1 + 70.00 x2 + 22.50 x3 - 26.00 x4 + 20.50 x5
       + 4.50 x6 - 38.50 x1x5 - 59.00 x1x6 - 46.50 x2x5 - 2.50 x2x6
       - 20.50 x3x5 - 10.00 x3x6 - 14.50 x4x5 - 12.00 x4x6

$SSE_c = 338.00$

The results from fitting the reduced model without the interaction terms are:
Regression Analysis: FEED versus x1, x2, x3, x4, x5, x6

Analysis of Variance

Source	DF	Adj SS	Ad jMS	F-Value	P-Value
Regression	6	24855.1	4142.52	22.17	0.000
x1	1	363.0	363.00	1.94	0.177
x2	1	8640.3	8640.33	46.25	0.000
x3	1	456.3	456.33	2.44	0.132
x4	1	3640.1	3640.08	19.49	0.000
x5	1	61.3	61.25	0.33	0.572
x6	1	744.2	744.20	3.98	0.058
Error	23	4296.7	186.81		
Lack-of-Fit	8	3958.7	494.84	21.96	0.000
Pure Error	15	338.0	22.53		
Total	29	29151.9			

Model Summary

```
       S    R-sq  R-sq(adj)  R-sq(pred)
13.6680  85.26%     81.42%      74.92%
```

```
Coefficients

Term        Coef   SE Coef  T-Value  P-Value   VIF
Constant    70.07     6.60    10.61    0.000
x1          11.00     7.89     1.39    0.177    1.60
x2          53.67     7.89     6.80    0.000    1.60
x3          12.33     7.89     1.56    0.132    1.60
x4         -34.83     7.89    -4.41    0.000    1.60
x5          -3.50     6.11    -0.57    0.572    1.33
x6         -12.20     6.11    -2.00    0.058    1.33
```

Regression Equation

```
FEED = 70.07 + 11.00 x1 + 53.67 x2 + 12.33 x3 - 34.83 x4 - 3.50 x5
       - 12.20 x6
```

$SSE_R = 4296.7$

The test statistic is $F = \dfrac{(SSE_R - SSE_C)/(k - g)}{SSE_C/[n - (k = 1)]} = \dfrac{(4,296.7 - 338.00)/(14 - 6)}{338.00/[30 - (14 + 1)]} = 21.96.$

Since no α was given, we will use $\alpha = 0.05$. The rejection region requires $\alpha = 0.05$. in the upper tail of the F distribution with $v_1 = k - g = 14 - 6 = 8$ and $v_2 = n - (k + 1) = 30 - (14 + 1) = 15$. From Table 10, Appendix B, $F_{0.05} = 2.64$. The rejection region is $F > 2.64$.

Since the observed value of the test statistic falls in the rejection region ($F = 21.96 > 2.64$), H_0 is rejected. There is sufficient evidence to indicate cutting tool and speed interact to affect cutting feed force at $\alpha = 0.05$.

e. Yes. From the printout, the test statistic is $F > 21.96$ and the p-value is $p = 0.000$. Since the p-value is less than $\alpha (p = 0.000 < 0.05)$, H_0 is rejected.

f. No. Since the interaction is significant, the main effect tests should not be run.

14.33   Using MINITAB, the results are:

ANOVA: BEETLES versus COLOR, BAIT

```
Analysis of Variance for BEETLES

Source       DF       SS       MS       F       P
COLOR         1  1106.29  1106.29   54.86   0.000
BAIT          1    46.29    46.29    2.30   0.143
COLOR*BAIT    1     5.14     5.14    0.26   0.618
Error        24   484.00    20.17
Total        27  1641.71

S = 4.49073    R-Sq = 70.52%    R-Sq(adj) = 66.83%
```

To determine if color and bait interact, we test:

H_0: Color and bait do not interact
H_a: Color and bait interact

The test statistic is $F = 0.26$ and the p-value is $p = 0.618$. Since no α was given, we will use $\alpha = 0.05$. Since the p-value is not less than, $\alpha (p = 0.618 \not< 0.05)$, H_0 is not rejected. There is insufficient evidence to indicate color and bait interact to affect the mean number of beetles captured in the traps at $\alpha = 0.05$.

Since there is no evidence of interaction, we can test for the main effects.

Color:

To determine if differences exist in the mean number of beetles captured due to color, we test:

$$H_0 : \mu_Y = \mu_G$$
$$H_a : \mu_Y \neq \mu_G$$

The test statistic is $F = 58.86$ and the p-value is $p = 0.000$. Since the p-value is less than $\alpha(p = 0.000 < 0.05)$, H_0 is rejected. There is sufficient evidence to indicate differences exist in the mean number of beetles captured due to color at $\alpha = 0.05$.

Bait:

To determine if differences exist in the mean number of beetles captured due to type of bait, we test:

$$H_0 : \mu_L = \mu_N$$
$$H_a : \mu_L \neq \mu_N$$

The test statistic is $F = 2.30$ and the p-value is $p = 0.143$. Since the p-value is not less than $\alpha(p = 0.143 \nless 0.05)$, H_0 is not rejected. There is insufficient evidence to indicate differences exist in the mean number of beetles captured due to type of bait at $\alpha = 0.05$.

14.35   Using MINITAB, the results are:

ANOVA: V-Height versus M-Height, M-Freq

```
Factor     Type    Levels   Values
M-Height   fixed        3   5, 10, 20
M-Freq     fixed        3   1, 2, 3
```

Analysis of Variance for V-Height

Source	DF	SS	MS	F	P
M-Height	2	77.902	38.951	17.57	0.000
M-Freq	2	387.977	193.989	87.52	0.000
M-Height*M-Freq	4	90.228	22.557	10.18	0.000
Error	27	59.842	2.216		
Total	35	615.950			

```
S = 1.48875    R-Sq = 90.28%    R-Sq(adj) = 87.41%
```

To determine if mowing height and mowing frequency interact, we test:

H_0: Mowing height and mowing frequency do not interact
H_a: Mowing height and mowing frequency interact

The test statistic is $F = 10.18$ and the p-value is $p = 0.000$. Since the p-value so small, H_0 is rejected. There is sufficient evidence to indicate mowing height and mowing frequency interact to affect the mean vegetation height for any reasonable value of α.

Since there is evidence of interaction, we should not test for the main effects. The next step is to find which treatment combination leads to the minimum mean vegetation height.

14.37 a. To use the traditional analysis of variance, we need replications for each factor-level combination. In this problem, we have only one observation per factor level combination.

b. Using MINITAB, the plot of the data is:

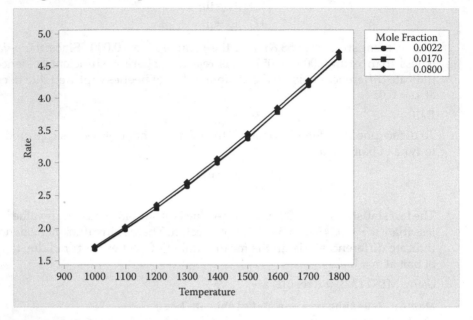

The data appear to be fairly linear.

c. The interaction model is $E(y) = \beta_0 + \beta_1 x_1 + \beta_2 x_2 + \beta_3 x_1 x_2$ where x_1 = temperature and x_2 = mole fraction.

d. When no interaction is present, the relationship between the rate of combustion and temperature is independent of the level of the mole fraction or the effect of temperature on the rate of combustion does not depend on the value of mole fraction.

e. To determine if mole fraction and temperature interact, we test:

$$H_0 : \beta_3 = 0$$
$$H_a : \beta_3 \neq 0$$

The test statistic is $t = 0.55$ and the p-value is $p = 0.591$. Since the p-value is not less than α ($p = 0.591 \not< 0.05$), H_0 is not rejected. There is insufficient evidence to indicate that mole fraction and temperature interact to affect the rate of combustion at $\alpha = 0.05$.

f. From the least squares prediction equation is
$\hat{y} = -2.09528 + 0.003684 x_1 - 0.238 x_2 + 0.000733 x_1 x_2$.

g. For $x_1 = 1{,}300$ and $x_2 = 0.017$,
$\hat{y} = -2.09528 + 0.003684(1{,}300) - 0.238(0.017) + 0.000733(1{,}300)(0.017) = 2.7061$

h. We are 95% confident that the mean diffusivity when the temperature is 1,300 K and the mole fraction of water is 0.017 is between 2.6774 and 2.7350.

14.39 a. The complete model is:

$$E(y) = \beta_0 + \beta_1 x_1 + \beta_2 x_2 + \beta_3 x_3 + \beta_4 x_4 + \beta_5 x_1 x_2 + \beta_6 x_1 x_3 + \beta_7 x_1 x_4 + \beta_8 x_2 x_3 + \beta_9 x_2 x_4$$
$$+ \beta_{10} x_3 x_4 + \beta_{11} x_1 x_2 x_3 + \beta_{12} x_1 x_2 x_4 + \beta_{13} x_1 x_3 x_4 + \beta_{14} x_2 x_3 x_4 + \beta_{15} x_1 x_2 x_3 x_4$$

Where $x_1 = \begin{cases} 1 \text{ if Agent-to-Mineral is L} \\ 0 \text{ if not} \end{cases}$ $x_2 = \begin{cases} 1 \text{ if Collector-to-Mineral is L} \\ 0 \text{ if not} \end{cases}$

$x_1 = \begin{cases} 1 \text{ if Liquid-to-solid is L} \\ 0 \text{ if not} \end{cases}$ $x_4 = \begin{cases} 1 \text{ if Foaming Agent is SABO} \\ 0 \text{ if not} \end{cases}$

b. We will not be able to fit the complete model without replication because $df(\text{error}) = 0$.

c. The new model is

$E(y) = \beta_0 + \beta_1 x_1 + \beta_2 x_2 + \beta_3 x_3 + \beta_4 x_4 + \beta_5 x_1 x_2 + \beta_6 x_1 x_3 + \beta_7 x_1 x_4 + \beta_8 x_2 x_3 + \beta_9 x_2 x_4 + \beta_{10} x_3 x_4$

d. Using MINITAB, the results are:

ANOVA: Copper versus A-M, C-M, L-S, Foam

Analysis of Variance for Copper

Source	DF	SS	MS	F	P
A-M	1	0.47956	0.47956	20.62	0.006
C-M	1	1.18266	1.18266	50.85	0.001
L-S	1	0.49351	0.49351	21.22	0.006
Foam	1	1.64481	1.64481	70.73	0.000
A-M*C-M	1	0.00141	0.00141	0.06	0.816
A-M*L-S	1	0.01891	0.01891	0.81	0.409
A-M*Foam	1	0.03331	0.03331	1.43	0.285
C-M*L-S	1	0.00181	0.00181	0.08	0.792
C-M*Foam	1	0.18276	0.18276	7.86	0.038
L-S*Foam	1	0.07981	0.07981	3.43	0.123
Error	5	0.11628	0.02326		
Total	15	4.23479			

S = 0.1525   R-Sq = 97.25%   R-Sq(adj) = 91.76%

Source	DF	Adj SS	Ad jMS	F-Value	P-Value
Regression	10	4.11851	0.411851	17.71	0.003
Error	5	0.11628	0.023256		
Total	15	4.23479			

Coefficients

Term	Coef	SE Coef	T-Value	P-Value	VIF
Constant	7.828	0.126	61.91	0.000	
x1	-0.305	0.153	-2.00	0.102	4.00
x2	-0.332	0.153	-2.18	0.081	4.00
x3	-0.300	0.153	-1.97	0.106	4.00
x4	-0.195	0.153	-1.28	0.257	4.00
x1x2	-0.038	0.153	-0.25	0.816	3.00
x1x3	0.138	0.153	0.90	0.409	3.00
x1x4	-0.183	0.153	-1.20	0.285	3.00
x2x3	0.042	0.153	0.28	0.792	3.00
x2x4	-0.428	0.153	-2.80	0.038	3.00
x3x4	-0.282	0.153	-1.85	0.123	3.00

Regression Equation

Copper = 7.828 - 0.305 x1 - 0.332 x2 - 0.300 x3 - 0.195 x4 - 0.038 x1x2
         + 0.138 x1x3- 0.183 x1x4 + 0.042 x2x3 - 0.428 x2x4 - 0.282 x3x4

The least squares prediction equation is

$$\hat{y} = 7.828 - 0.305x_1 - 0.332x_2 - 0.300x_3 - 0.195x_4 - 0.038x_1x_2 + 0.138x_1x_3 - 0.183x_1x_4$$

$$+ 0.042x_2x_3 - 0.428x_2x_4 - 0.282x_3x_4$$

e. Since all the terms in the factorial design are independent, we can test each of the interaction terms.

Looking at the p-values associated with the interaction terms, there is only one p-value that is less than 0.05. This p-value is associated with the interaction term between collector-to-mineral mass ratio and foam agent.

The test statistic is $F = 7.86$ and the p-value is $p = 0.038$. There is sufficient evidence that collector-to-mineral mass ratio and foam agent interact to affect percentage copper. No other interaction terms are significant.

f. Since the interaction term between collector-to-mineral mass ratio and foam agent is significant, tests for the main effects for collector-to-mineral mass ratio and foam agent are not performed. However, the tests for the main effects of agent-to-mineral mass ratio and liquid-to-solid ratio can be performed.

To determine if the mean percentage of copper differs for the two levels of agent-to-mineral mass ratio, we test:

$$H_0 : \mu_L = \mu_H$$
$$H_a : \mu_L \neq \mu_H$$

The test statistic is $F = 20.62$ and the p-value is $p = 0.006$. Since the p-value is less than $\alpha(p = 0.006 < 0.05)$, H_0 is rejected. There is sufficient evidence to indicate that the mean percentage of copper differs for the two levels of agent-to-mineral mass ratio at $\alpha = 0.05$.

To determine if the mean percentage of copper differs for the two levels of liquid-to-solid ratio, we test:

$$H_0 : \mu_L = \mu_H$$
$$H_a : \mu_L \neq \mu_H$$

The test statistic is $F = 21.22$ and the p-value is $p = 0.006$. Since the p-value is less than $\alpha(p = 0.006 < 0.05)$, H_0 is rejected. There is sufficient evidence to indicate that the mean percentage of copper differs for the two levels of liquid-to-solid ratio at $\alpha = 0.05$.

14.41 a. There are $3 \times 3 \times 3 \times 3 = 81$ treatments in this factorial design. To write down all treatments, we will use the order IC, CC, R-Temp, and R-Time. The 81 treatments are:

$(1,5,35,6), (1,5,35,8), (1,5,35,10), (1,5,50,6), (1,5,50,8), (1,5,50,10),$
$(1,5,65,6), (1,5,65,8), (1,5,65,10), (1,10,35,6), (1,10,35,8), (1,10,35,10),$
$(1,10,50,6), (1,10,50,8), (1,10,50,10), (1,10,65,6), (1,10,65,8), (1,10,65,10),$
$(1,15,35,6), (1,15,35,8), (1,15,35,10), (1,15,50,6), (1,15,50,8), (1,15,50,10),$
$(1,15,65,6), (1,15,65,8), (1,15,65,10),$
$(2,5,35,6), (2,5,35,8), (2,5,35,10), (2,5,50,6), (2,5,50,8), (2,5,50,10),$
$(2,5,65,6), (2,5,65,8), (2,5,65,10), (2,10,35,6), (2,10,35,8), (2,10,35,10)$
$(2,10,50,6), (2,10,50,8), (2,10,50,10), (2,10,65,6), (2,10,65,8), (2,10,65,10)$
$(2,15,35,6)(2,15,35,8), (2,15,35,10), (2,15,50,6), (2,15,50,8), (2,15,50,10),$
$(2,15,65,6), (2,15,65,8), (2,15,65,10),$
$(3,5,35,6), (3,5,35,8), (3,5,35,10), (3,5,50,6), (3,5,50,8), (3,5,50,10),$

$(3,5,65,6), (3,5,65,8), (3,5,65,10), (3,10,35,6), (3,10,35,8), (3,10,35,10),$
$(3,10,50,6), (3,10,50,8), (3,10,50,10), (3,10,65,6), (3,10,65,8), (3,10,65,10),$
$(3,15,35,6), (3,15,35,8), (3,15,35,10), (3,15,50,6), (3,15,50,8), (3,15,50,10),$
$(3,15,65,6), (3,15,65,8), (3,15,65,10)$

b. The complete model is

$$
\begin{aligned}
E(y) = {}&\beta_0 + \beta_1 x_1 + \beta_2 x_2 + \beta_3 x_3 + \beta_4 x_4 + \beta_5 x_5 + \beta_6 x_6 + \beta_7 x_7 + \beta_8 x_8 \\
&+ \beta_9 x_1 x_3 + \beta_{10} x_1 x_4 + \beta_{11} + \beta_{12} x_1 x_6 + \beta_{13} x_1 x_7 + \beta_{14} x_1 x_8 + \beta_{15} x_2 x_3 \\
&+ \beta_{16} x_2 x_4 + \beta_{17} x_2 x_5 + \beta_{18} x_2 x_6 + \beta_{19} x_2 x_7 + \beta_{20} x_2 x_8 + \beta_{21} x_3 x_5 + \beta_{22} x_3 x_6 \\
&+ \beta_{23} x_3 x_7 + \beta_{24} x_3 x_8 + \beta_{25} x_4 x_5 + \beta_{26} x_4 x_6 + \beta_{27} x_4 x_7 + \beta_{28} x_4 x_8 \\
&+ \beta_{29} x_5 x_7 + \beta_{30} x_5 x_8 + \beta_{31} x_6 x_7 + \beta_{32} x_6 x_8 + \beta_{33} x_1 x_3 x_5 + \beta_{34} x_1 x_3 x_6 \\
&+ \beta_{35} x_1 x_3 x_7 + \beta_{36} x_1 x_3 x_8 + \beta_{37} x_1 x_4 x_5 + \beta_{38} x_1 x_4 x_6 + \beta_{39} x_1 x_4 x_7 + \beta_{40} x_1 x_4 x_8 \\
&+ \beta_{41} x_1 x_5 x_7 + \beta_{42} x_1 x_5 x_8 + \beta_{43} x_1 x_6 x_7 + \beta_{44} x_1 x_6 x_8 + \beta_{45} x_2 x_3 x_5 + \beta_{46} x_2 x_3 x_6 \\
&+ \beta_{47} x_2 x_3 x_7 + \beta_{48} x_2 x_3 x_8 + \beta_{49} x_2 x_4 x_5 + \beta_{50} x_2 x_4 x_6 + \beta_{51} x_2 x_4 x_7 \\
&+ \beta_{52} x_2 x_4 x_8 + \beta_{53} x_2 x_5 x_7 + \beta_{54} x_2 x_5 x_8 + \beta_{55} x_2 x_6 x_7 + \beta_{56} x_2 x_6 x_8 \\
&+ \beta_{57} x_3 x_5 x_7 + \beta_{58} x_3 x_5 x_8 + \beta_{59} x_3 x_6 x_7 + \beta_{60} x_3 x_6 x_8 + \beta_{61} x_4 x_5 x_7 \\
&+ \beta_{62} x_4 x_5 x_8 + \beta_{63} x_4 x_6 x_7 + \beta_{64} x_4 x_6 x_8 + \beta_{65} x_1 x_3 x_5 x_7 + \beta_{66} x_1 x_3 x_5 x_8 \\
&+ \beta_{67} x_1 x_3 x_6 x_7 + \beta_{68} x_1 x_3 x_6 x_8 + \beta_{69} x_1 x_4 x_5 x_7 + \beta_{70} x_1 x_4 x_5 x_8 \\
&+ \beta_{71} x_1 x_4 x_6 x_7 + \beta_{72} x_1 x_4 x_6 x_8 + \beta_{73} x_2 x_3 x_5 x_7 + \beta_{74} x_2 x_3 x_5 x_8 \\
&+ \beta_{75} x_2 x_3 x_6 x_7 + \beta_{76} x_2 x_3 x_6 x_8 + \beta_{77} x_2 x_4 x_5 x_7 + \beta_{78} x_2 x_4 x_5 x_8 \\
&+ \beta_{79} x_2 x_4 x_6 x_7 + \beta_{80} x_2 x_4 x_6 x_8
\end{aligned}
$$

c. The sources of variation and their corresponding degrees of freedom are shown in the ANOVA table below:

Source	df
IC	2
CC	2
R-Temp	2
R-Time	2
IC*CC	4
IC*R-Temp	4
IC*R-Time	4
CC*R-Temp	4
CC*R-Time	4
R-Temp*R-Time	4
IC*CC*R-Temp	8
IC*CC*R-Time	8
IC*R-Temp*R-Time	8
CC*R-Temp*R-Time	8
IC*CC*R-Temp*R-Time	16
Error	81
Total	161

d. No. Since there are 81 different treatments, we need to have more than 81 observations to test all the effects.

e. Because all of the factors are quantitative, we will fit a regression model with the factors as independent quantitative variables. Using MINITAB, the results are:

Regression Analysis: Eff versus x1, x2, x3, x4, x5, x6, x7, x8

Regression Analysis: Eff versus IC, CC, R-Temp, R-Time

Analysis of Variance

Source	DF	Adj SS	Ad jMS	F-Value	P-Value
Regression	4	2208.21	552.05	4.38	0.091
IC	1	196.54	196.54	1.56	0.280
CC	1	1690.42	1690.42	13.43	0.022
R-Temp	1	253.11	253.11	2.01	0.229
R-Time	1	68.14	68.14	0.54	0.503
Error	4	503.61	125.90		
Total	8	2711.82			

Model Summary

S	R-sq	R-sq(adj)	R-sq(pred)
11.2206	81.43%	62.86%	30.08%

Coefficients

Term	Coef	SE Coef	T-Value	P-Value	VIF
Constant	97.3	27.4	3.55	0.024	
IC	-5.72	4.58	-1.25	0.280	1.00
CC	-3.357	0.916	-3.66	0.022	1.00
R-Temp	0.433	0.305	1.42	0.229	1.00
R-Time	-1.68	2.29	-0.74	0.503	1.00

Regression Equation

Eff = 97.3 - 5.72 IC - 3.357 CC + 0.433 R-Temp - 1.68 R-Time

For testing the main effect of IC, the test statistic is $F = -1.25$ and the p-value is $p = 0.280$.

Since the p-value is not less than $\alpha(p = 0.280 \not< 0.05)$, H_0 is not rejected. There is insufficient evidence to indicate that there is an effect on mean efficiency due to IC at $\alpha = 0.05$.

For testing the main effect of CC, the test statistic is $F = -3.66$ and the p-value is $p = 0.022$.

Since the p-value is less than $\alpha(p = 0.22 < 0.05)$, H_0 is rejected. There is sufficient evidence to indicate that there is an effect on mean efficiency due to CC at $\alpha = 0.05$.

For testing the main effect of R-Temp, the test statistic is $F = 1.42$ and the p-value is $p = 0.229$. Since the p-value is not less than $\alpha(p = 0.229 \not< 0.05)$, H_0 is not rejected. There is insufficient evidence to indicate that there is an effect on mean efficiency due to R-Temp at $\alpha = 0.05$.

For testing the main effect of R-Time, the test statistic is $F = -0.74$ and the p-value is $p = 0.503$. Since the p-value is not less than $\alpha(p = 0.503 \not< 0.05)$, H_0 is not rejected. There is insufficient evidence to indicate that there is an effect on mean efficiency due to R-Time at $\alpha = 0.05$.

14.43 a. To calculate the sum of squares for the interaction terms, we sum the ANOVA SS for the 4 interaction terms:

$$SS(INT) = 339.75375 + 7.9858333 + 4.1725 + 0.4375 = 352.3495833.$$

The total degrees of freedom for the interaction terms is 7. Thus, the mean square for interaction is $MS(INT) = \dfrac{SS(INT)}{df} = \dfrac{352.3495833}{7} = 50.335655$.

To determine if any of the interaction terms are useful for predicting yield, we test:

$$H_0 : \beta_3 = \beta_5 = \beta_6 = \beta_7 = 0$$

$$H_a : \text{At least one } \beta_i \neq 0$$

The test statistic is $F = \dfrac{MS(INT)}{MSE} = \dfrac{50.335655}{0.67875} = 74.16$.

The rejection region requires $\alpha = 0.05$ in the upper tail of the F distribution with $v_1 = 7$ and $v_2 = 12$. Using Table 10, Appendix B, $F_{0.05} = 2.91$. The rejection region is $F > 2.91$.

Since the observed value of the test statistic falls in the rejection region ($F = 74.16 > 2.91$), H_0 is rejected. There is sufficient evidence to indicate at least one of the interaction terms is useful for predicting yield at $\alpha = 0.05$.

b. For testing the three-way interaction of alloy, material condition and time, the test statistic is $F = 0.32$ and the p-value is $p = 0.7306$. Since the p-value is not less than $\alpha (p = 0.7306 \not< 0.05)$, H_0 is not rejected. There is insufficient evidence to indicate that the three-way interaction of alloy, material condition and time is useful for predicting yield at $\alpha = 0.05$.

For testing the two-way interaction of alloy and material condition, the test statistic is $F = 500.56$ and the p-value is $p < 0.0001$. Since the p-value is less than $\alpha (p < 0.0001 < 0.05)$, H_0 is rejected. There is sufficient evidence to indicate that the two-way interaction of alloy and material condition is useful for predicting yield at $\alpha = 0.05$.

For testing the two-way interaction of alloy and time, the test statistic is $F = 5.88$ and the p-value is $p = 0.0166$. Since the p-value is less than $\alpha (p = 0.0166 < 0.05)$, H_0 is rejected. There is sufficient evidence to indicate that the two-way interaction of alloy and time is useful for predicting yield at $\alpha = 0.05$.

For testing the two-way interaction of material condition and time, the test statistic is $F = 3.07$ and the p-value is $P = 0.0836$. Since the p-value is not less than $\alpha (p = 0.0836 \not< 0.05)$, H_0 is not rejected. There is insufficient evidence to indicate that the two-way interaction of material condition and time is useful for predicting yield at $\alpha = 0.05$.

14.45 Using MINITAB, the results are:

Regression Analysis: YIELD versus x1, x1sq

Analysis of Variance

Source	DF	Adj SS	Adj MS	F-Value	P-Value
Regression	2	71.04	35.52	0.40	0.676
x1	1	36.22	36.22	0.41	0.530
x1sq	1	55.04	55.04	0.62	0.440
Error	21	1868.84	88.99		
Total	23	1939.88			

```
Model Summary

      S    R-sq   R-sq(adj)   R-sq(pred)
9.43357   3.66%       0.00%        0.00%

Coefficients

Term         Coef   SE Coef   T-Value   P-Value     VIF
Constant    45.65      3.34     13.69     0.000
x1          0.217     0.340      0.64     0.530   13.00
x1sq     -0.00514   0.00654     -0.79     0.440   13.00

Regression Equation

YIELD = 45.65 + 0.217 x1 - 0.00514 x1sq
```

To determine if differences among the second-order models relation $E(y)$ to x_1 for the four categories of alloy type and material condition, we test:

$$H_0 : \beta_3 = \beta_4 = \cdots = \beta_{11} = 0$$
$$H_a : \text{At least one } \beta_i \neq 0$$

The test statistic is $F = \dfrac{(SSE_R - SSE_C)/(k-g)}{SSE_c/[n-(k=1)]} = \dfrac{(1868.84 - 8.1452)/(11-2)}{8.1452/[24-(11+1)]} = 304.59.$

The rejection region requires $\alpha = 0.05$ in the upper tail of the F distribution with $v_1 = k - g = 11 - 2 = 9$ and $v_2 = n - (k+1) = 24 - (11+1) = 12$. Using Table 10, Appendix B, $F_{0.05} = 2.80$. The rejection region is $F > 2.80$.

Since the observed value of the test statistic falls in the rejection region ($F = 304.59 > 2.80$), H_0 is rejected. There is sufficient evidence to indicate differences among the second-order models relation $E(y)$ to x_1 for the four categories of alloy type and material condition at $\alpha = 0.05$.

14.47 a. There are 3 first-stage observations in this sample: Site 1, Site 2, and Site 3.

   b. In each first-stage unit, there are 5 second-stage units selected.

   c. The total number of observations is $3 \times 5 = 15$.

   d. The probabilistic model is $y_{ij} = \mu + \alpha_i + \varepsilon_{ij}$. $(i = 1,2,3; \; j = 1,2,3,4,5)$

   e. From the printout, $\hat{\sigma}^2 = 0.0558$.

   From Table 14.10, $E(MS(A)) = \sigma^2 + n_2 \sigma_\alpha^2$.

   Thus, $\hat{\sigma}_\alpha^2 = \dfrac{MS(A) - \hat{\sigma}^2}{n_2} = \dfrac{0.001287 - 0.0558}{5} = -0.0109$

   Since this number is less than 0, our estimate of σ_α^2 is 0.

   f. To determine if the variation in specimen densities between sites exceeds the variation within sites, we test:

   $$H_0 : \sigma_\alpha^2 = 0$$
   $$H_a : \sigma_\alpha^2 > 0$$

   The test statistic is $F = \dfrac{MS(A)}{MS(B \text{ in } A)} = \dfrac{0.001287}{0.0558} = 0.023.$

   The rejection region requires $\alpha = 0.10$ in the upper tail of the F distribution with $v_1 = n_1 - 1 = 3 - 1 = 2$ and $v_2 = n_1(n_2 - 1) = 3(5-1) = 12$. Using Table 9, Appendix B, $F_{0.10} = 2.81$. The rejection region is $F > 2.81$.

Since the observed value of the test statistic does not fall in the rejection region ($F = 0.023 \ngtr 2.81$), H_0 is not rejected. There is insufficient evidence to indicate the variation in specimen densities between sites exceeds the variation within sites at $\alpha = 0.10$.

14.49 The sources of variation and their corresponding degrees of freedom are shown in the ANOVA table below:

Source	df
Lot (A)	$n_1 - 1 = 10 - 1 = 9$
Batch within Lot (B in A)	$n_1(n_2 - 1) = 10(5 - 1) = 40$
Shipping Lot within Batch (C in B)	$n_1 n_2(n_3 - 1) = 10(5)(20 - 1) = 950$
Total	999

14.51 a. Using MINITAB, the results are:

```
Analysis of Variance

Source   DF   Adj SS   Adj MS   F-Value   P-Value
Lot       4   1.456    0.36412    6.34     0.001
Error    35   2.011    0.05746
Total    39   3.468

Model Summary

       S    R-sq   R-sq(adj)   R-sq(pred)
0.239717   42.00%    35.37%       24.25%
```

b. The estimate of σ_w^2 is $\hat{\sigma}_w^2 = MSE = 0.05746$.

The estimate of σ_B^2 is $\hat{\sigma}_B^2 = \dfrac{MS(A) - MS(B \text{ in } A)}{n_2} = \dfrac{0.36412 - 0.05746}{8} = 0.038333$.

c. To determine if the between lots variation exceeds the within lots variation, we test:

$$H_0 : \sigma_B^2 = 0$$
$$H_a : \sigma_B^2 > 0$$

The test statistic is $F = \dfrac{MS(A)}{MS(B \text{ in } A)} = \dfrac{0.36412}{0.05746} = 6.34$ and the p-value is $p = 0.001$.

Since the p-value is less than α($p = 0.001 < 0.05$), H_0 is rejected. There is sufficient evidence to indicate that the between lots variation exceeds the within lots variation at $\alpha = 0.05$.

14.53 a. There are $g = \dfrac{p(p-1)}{2} = \dfrac{4(4-1)}{2} = 6$ pairwise comparisons.

b. There are no significant differences in the mean energy expended among the colony sizes of 3, 6, and 9. However, the mean energy expended for the colony containing 12 robots was significantly less than that for all other colony sizes.

14.55 The confidence interval for ($\mu_1 - \mu_2$) contains only positive numbers. Therefore, the mean for Depot 1 is greater than the mean for Depot 2.

The confidence interval for ($\mu_1 - \mu_3$) contains only positive numbers. Therefore, the mean for Depot 1 is greater than the mean for Depot 3.

The confidence interval for ($\mu_2 - \mu_3$) contains only positive numbers. Therefore, the mean for Depot 2 is greater than the mean for Depot 3.

Thus, the mean for Depot 3 is the smallest, then the mean for Depot 2 and the mean for Depot 1 is the largest.

14.57 a.   There are $g = \dfrac{p(p-1)}{2} = \dfrac{4(4-1)}{2} = 6$ pairwise comparisons.

b.   The mean soluble magnesium for Sourdough was significantly larger than the means for the other three types of bread. The mean soluble magnesium for Lactate was significantly larger than the means for Control and Yeast. No significant difference could be detected in the mea soluble magnesium between the Control and Yeast. Therefore, the bread that yielded the largest mean is Sourdough and the bread that yielded the lowest mean is either Control or Yeast.

c.   When we consider all 6 comparisons made, collectively the chance of making a Type I error is 0.05.

14.59   For this problem, $\alpha = 0.05$. The critical values are $\omega_{ij} = q_\alpha(p,v)\dfrac{s}{\sqrt{2}}\sqrt{\dfrac{1}{n_i}+\dfrac{1}{n_j}}$

where $p = 5$, $v = 21$, and $s = 0.448565$. From Table 13, Appendix B, $q_{0.05}(5,21) \approx 4.23$. The critical values are:

$$\omega_{12} = \omega_{23} = 4.23\frac{0.448565}{\sqrt{2}}\sqrt{\frac{1}{7}+\frac{1}{6}} = 0.7464$$

$$\omega_{13} = 4.23\frac{0.448565}{\sqrt{2}}\sqrt{\frac{1}{7}+\frac{1}{7}} = 0.7172$$

$$\omega_{14} = \omega_{15} = \omega_{34} = \omega_{45} = 4.23\frac{0.448565}{\sqrt{2}}\sqrt{\frac{1}{7}+\frac{1}{3}} = 0.9259$$

$$\omega_{24} = \omega_{25} = 4.23\frac{0.448565}{\sqrt{2}}\sqrt{\frac{1}{6}+\frac{1}{3}} = 0.9487$$

$$\omega_{45} = 4.23\frac{0.448565}{\sqrt{2}}\sqrt{\frac{1}{3}+\frac{1}{3}} = 1.0955$$

Arranging the five sample means in order from the smallest to the largest, we get:

		Mean Ai/Be Ratios		
SWRA	SD	UMRB-1	UMRB-3	UMRB-2
2.643	2.780	3.497	3.761	4.017

Compare UMRB-2 to SWRA:
$4.017 - 2.643 = 1.374 > 0.9487$ Thus, the means are significantly different.

Compare UMRB-2 to SD:
$4.017 - 2.780 = 1.237 > 0.9487$ Thus, the means are significantly different.

Compare UMRB-2 to UMRB-1:
$4.017 - 3.497 = 0.520 \ngtr 0.7464$ Thus, the means are not significantly different.

Compare UMRB-3 to SWRA:
$3.761 - 2.643 = 1.118 > 0.9259$ Thus, the means are significantly different.

Compare UMRB-3 to SD:
$3.761 - 2.780 = 0.981 > 0.9259$ Thus, the means are not significantly different.

Compare UMRB-1 to SWRA:

$3.497 - 2.643 = 0.854 \ngtr 0.9259$ Thus, the means are significantly different.

From the above, the mean AI/Be ratios for boreholes UMRB-2 and UMRB-3 are significantly greater than the mean AI/Be ratios for boreholes SWRA and SD. No other means are significantly different.

14.61 First, compute the sample mean crack widths for the 4 wetting periods and arrange them in order from the smallest to the largest:

	Mean Crack Widths		
14 weeks	6 weeks	2 weeks	0 weeks
0.0750	0.0833	0.2617	0.6583

From Exercise 4.19, $p = 4$, $v = 33$, $n_i = 12$, $\alpha = 0.05$, and $s = 0.1242$. From Table 13, Appendix B, $q_{0.05}(4,33) \approx 3.85$. Using $\omega = q_\alpha(p,v)s\sqrt{\dfrac{1}{n_i}}$, the critical value is:

$$\omega = 3.85(0.1242)\sqrt{\frac{1}{12}} = 0.13804$$

Compare 0 to 14 weeks:

$0.6583 - 0.0750 = 0.5833 > 0.13804$ Thus, the means are significantly different.

Compare 0 to 6 weeks:

$0.6583 - 0.0833 = 0.5740 > 0.13804$ Thus, the means are significantly different.

Compare 0 to 2 weeks:

$0.6583 - 0.2617 = 0.3956 > 0.13804$ Thus, the means are significantly different.

Compare 2 to 14 weeks:

$0.2617 - 0.0750 = 0.1867 > 0.13804$ Thus, the means are significantly different.

Compare 2 to 6 weeks:

$0.2617 - 0.0833 = 0.1784 > 0.13804$ Thus, the means are significantly different.

Compare 6 to 14 weeks:

$0.0833 - 0.0750 = 0.0083 \ngtr 0.13804$ Thus, the means are not significantly different.

All population mean crack widths are significantly different except the means for 6 and 14 weeks.

14.63 First, compute the mean shear strength for the four antimony amounts and arrange in order from smallest to largest:

	Mean Shear Strengths		
10	0	3	5
17.033	20.175	20.408	20.617

From Exercise 14.34, $p = 4$, $v = 32$, $n_i = 12$, $\alpha = 0.01$, and $s = 1.3139476$. From Table 14, Appendix B, $q_{0.01}(4,32) \approx 4.80$. Using $\omega = q_\alpha(p,v)s\sqrt{\dfrac{1}{n_i}}$, the critical value is:

$$\omega = 4.80(1.3139476)\sqrt{\frac{1}{12}} = 1.82$$

Compare 5 to 10 weeks:

$20.617 - 17.033 = 3.584 > 1.82$ Thus, the means are significantly different.

Compare 5 to 0 weeks:

$20.617 - 20.175 = 0.442 \not> 1.82$ Thus, the means are not significantly different.

Compare 3 to 10 weeks:

$20.408 - 17.033 = 3.375 > 1.82$ Thus, the means are significantly different.

Compare 0 to 10 weeks:

$20.175 - 17.033 = 3.142 > 1.82$ Thus, the means are significantly different.

The mean shear strengths for 0%, 3%, and 5% amounts of antimony are significantly greater than the mean shear strength of 10% of antimony. No other significant differences exist.

14.65   To check for normal assumption, we use MINITAB to construct stem-and-leaf displays of the heat rate variable for each level of the gas turbine variable.

```
Stem-and-leaf of HEATRATE   ENGINE = Advanced    N = 21
Leaf Unit = 100

   8    9    11112224
  (8)   9    56677899
   5   10    124
   2   10    8
   1   11
   1   11    5

Stem-and-leaf of HEATRATE   ENGINE = Aurodiv     N = 7
Leaf Unit = 1000

   2    0    89
   3    1    1
  (2)   1    22
   2    1    4
   1    1    6

Stem-and-leaf of HEATRATE   ENGINE = Traditiona  N = 39
Leaf Unit = 100

   6   10    012334
  17   10    55556667899
  (7)  11    0112223
  15   11    567899
   9   12    24
   7   12    9
   6   13    13
   4   13    5
   3   14
   3   14    667
```

We see no great departures from normality with these plots. Therefore, we believe the assumption is satisfied.

To detect unequal variances, we conducted Levene's test of equal variances below:

Test for Equal Variances: HEATRATE versus Coded ENGINE

```
Method

Null hypothesis          All variances are equal
Alternative hypothesis  At least one variance is different
Significance level       α = 0.05

95% Bonferroni Confidence Intervals for Standard Deviations

ENGINE          N       StDev           CI
Advanced     1  21     638.51   ( 355.49, 1294.40)
Aeroderiv    2   7    2651.85   (1358.99, 7864.19)
Traditional  3  39    1279.28   ( 901.01, 1935.15)

Individual confidence level = 98.3333%

Tests
                                Test
Method                    Statistic   P-Value
Multiple comparisons          —        0.005
Levene                      6.88       0.002
```

We test:

$$H_0 : \sigma_1^2 = \sigma_2^2 = \sigma_3^2$$
$$H_a : \text{At least two } \sigma_i^2\text{'s differ}$$

The test statistic is $F = 6.88$ and the p-value is $p = 0.002$. Since the p-value is less than $\alpha(p = 0.002 < 0.01)$, H_0 is rejected. There is sufficient evidence to indicate that the variances differ. Thus, we have reason to believe that the equal variance assumption is violated.

14.67 To check the normal assumption, we use MINITAB to construct stem-and-leaf displays of the word pairs recalled for each group.

Stem-and-Leaf Display: Scopolamine, Placebo, None

```
Stem-and-leaf of Scopolamine   N = 12
Leaf Unit = 0.10

  1     4   0
  3     5   00
 (6)    6   000000
  3     7
  3     8   000

Stem-and-leaf of Placebo   N = 8
Leaf Unit = 0.10

  1     7   0
  2     8   0
  4     9   00
  4    10   000
  1    11
  1    12   0

Stem-and-leaf of None   N = 8
Leaf Unit = 0.10

  1     8   0
  2     9   0
  3    10   0
 (2)   11   00
  3    12   000
```

We see no great departures from normality with these plots. Therefore, we believe the assumption is satisfied.

To detect unequal variances, we conducted Levene's test of equal variances below.

Test for Equal Variances: Scopolamine, Placebo, None

Method

Null hypothesis   All variances are equal
Alternative hypothesis At least one variance is different
Significance level α = 0.05

95% Bonferroni Confidence Intervals for Standard Deviations

Sample	N	StDev	CI
Scopolamine	12	1.26730	(0.801296, 2.50384)
Placebo	8	1.50594	(0.717733, 4.50908)
None	8	1.50594	(0.637608, 5.07572)

Individual confidence level = 98.3333%

Tests

Method	Test Statistic	P-Value
Multiple comparisons	—	0.844
Levene	0.33	0.725

We test:

$$H_0 : \sigma_1^2 = \sigma_2^2 = \sigma_3^2$$
$$H_a : \text{At least two } \sigma_i^2\text{'s differ}$$

The test statistic is $F = 0.33$ and the p-value is $p = 0.725$. Since the p-value is not less than $\alpha(p = 0.725 \not< 0.05)$, H_0 is not rejected. There is insufficient evidence to indicate that the variances differ. Thus, we have no reason to believe that the equal variance assumption is violated.

14.69   To check for normality, MINITAB was used to construct histograms for each of the treatment combinations. The histograms are:

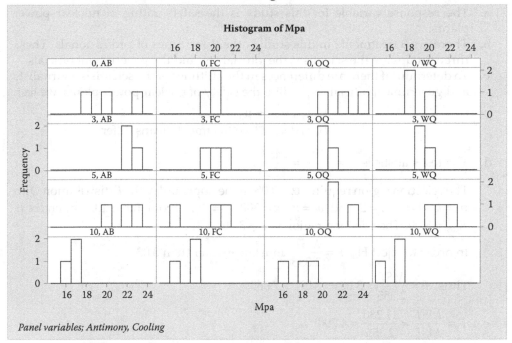

Panel variables; Antimony, Cooling

Since there are only 3 observations for each of the treatments, it is very difficult to assess normality. There is no strong evidence to indicate non-normality.
To check for equal variance, MINITAB was used to create boxplots for each of the treatments:

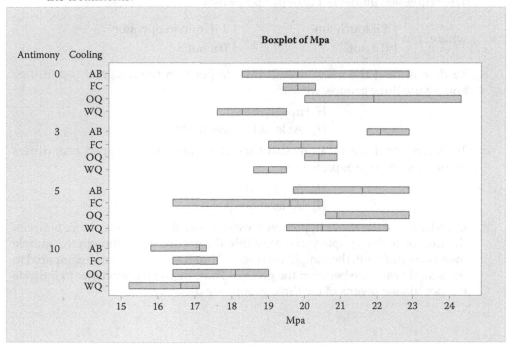

Since the spread of the values appears to vary from treatment to treatment, it appears that the assumption of equal variance may be violated.

14.71 a. The response variable for this study is the safety rating of nuclear power plants.

b. There are 3 treatments in this study – the three types of professionals. These three groups are the scientists, the journalists, and the government officials.

c. To determine if there are differences in the attitudes of the scientists, journalists and government officials regarding the safety of nuclear power plants, we test:

$$H_0 : \mu_1 = \mu_2 = \mu_3$$
$$H_a : \text{At least two treatment means differ}$$

d. The test statistic is $F = \dfrac{MST}{MSE} = \dfrac{MST}{2.355}$.

The rejection region requires $\alpha = 0.05$ in the upper tail of the F distribution with $v_1 = k - 1 = 3 - 1 = 2$ and $v_2 = n - k = 300 - 3 = 297$. From Table 10, Appendix B, $F_{0.05} \approx 3.00$. The rejection region is $F > 3.00$.

In order to reject H_0, $F = \dfrac{MST}{2.355}$ must be greater than 3.03.

Thus, $F = \dfrac{MST}{2.355} > 3.00 \Rightarrow MST > 7.065$.

e. $F = \dfrac{MST}{MSE} = \dfrac{11.280}{2.355} = 4.790$.

Using MINITAB, with $v_1 = 2$ and $v_2 = 297$, the p-value is $p = P(F \geq 4.790) = 0.009$.

14.73 a. This is a completely randomized design. The 45 subjects were randomly divided into three groups.

b. The regression model is $E(y) = \beta_0 + \beta_1 x_1 + \beta_2 x_2$

where $x_1 = \begin{cases} 1 \text{ if touch tone} \\ 0 \text{ if not} \end{cases}$ $x_2 = \begin{cases} 1 \text{ if human operator} \\ 0 \text{ if not} \end{cases}$

c. To determine if the mean overall time to perform the assigned task differs among the three groups, we test:

$$H_0 : \mu_1 = \mu_2 = \mu_3$$
$$H_a : \text{At least two means differ}$$

d. To determine if the mean overall time to perform the assigned task differs among the three groups, we test:

$$H_0 : \beta_1 = \beta_2 = 0$$
$$H_a : \text{At least one } \beta_i \neq 0$$

e. In order to test the above hypotheses, we compare the sample variance between the groups to the sample variance within the groups. Even though the sample means are different, the sample variance within the groups is large compared to the sample variance between the groups. Thus, there is no evidence to indicate the population means of the three groups are different.

14.75   The complete ANOVA table is:

Source	df	SS	MS	F
Agent	2	131.90	65.95	6.36
Ingot	6	268.29	44.715	4.31
Error	12	124.46	10.372	
Total	20	524.65		

a. The treatments in this experiment are the three agents (nickel, iron, and copper).
b. The blocks in this experiment are the 7 ingots.
c. To determine if there is a difference in the mean pressure required to separate the components among the three bonding agents, we test:

$$H_0 : \mu_1 = \mu_2 = \mu_3$$
$$H_a : \text{At least two means differ}$$

The test statistic is $F = 6.36$.

The rejection region requires $\alpha = 0.05$ in the upper tail of the F distribution with $v_1 = p - 1 = 3 - 1 = 2$ and $v_2 = n - b - p + 1 = 21 - 7 - 3 + 1 = 12$. Using Table 10, Appendix B, $F_{0.05} = 2.81$. The rejection region is $F > 2.81$.

d. Since the observed value of the test statistic falls in the rejection region ($F = 6.36 > 2.81$), H_0 is rejected. There is sufficient evidence to indicate a difference in the mean pressure required to separate the components among the three bonding agents at $\alpha = 0.05$.

14.77 a.   Using MINITAB, the results are:

ANOVA: SoilAcid versus Time, pH, Depth

Analysis of Variance for SoilAcid

Source	DF	SS	MS	F	P
Time	2	0.381478	0.190739	19.17	0.000
pH	1	0.030422	0.030422	3.06	0.111
Depth	2	0.067144	0.033572	3.37	0.076
pH*Depth	2	0.007811	0.003906	0.39	0.685
Error	10	0.099522	0.009952		
Total	17	0.586378			

S = 0.0997608   R-Sq = 83.03%   R-Sq(adj) = 71.15%

To determine if interaction between pH and soil depth exists, we test:

H_0: There is no interaction between pH level and soil depth
H_a: There is interaction between pH level and soil depth

The test statistic is $F = 0.39$ and the p-value is $p = 0.685$. Since the p-value is not less than $\alpha(p = 0.685 \not< 0.05)$, H_0 is not rejected. There is insufficient evidence to indicate interaction between pH and soil depth exists at $\alpha = 0.05$.

b. To determine if blocking over time was effective, we test:

H_0: There are no differences among the block means
H_a: There are differences among the block means

The test statistic is $F = 19.17$ and the p-value is $p = 0.000$. Since the p-value is less than $\alpha(p = 0.000 < 0.05)$, H_0 is rejected. There is sufficient evidence to indicate if blocking over time was effective at $\alpha = 0.05$.

14.79   In a two factor factorial design, the first thing we test for in the interaction between the two factors.
To determine if aid type and room order interact, we test:

H_0: Aid type and room order do not interact
H_a: Aid type and room order interact

The test statistic is $F = 1.29$ and the p-value is $p > 010$. Since the p-value is not small, H_0 is not rejected. There is insufficient evidence to indicate aid type and room order interact at $\alpha = 0.05$.
Since the interaction is not significant, tests on the main effects are conducted. To determine if there is a difference in mean travel time among the three aid types, we test:

$H_0 : \mu_S = \mu_M = \mu_N$
$H_a :$ At least two means differ

The test statistic is $F = 76.67$ and the p-value is $p < 0.0001$. Since the p-value is small, H_0 is rejected. There is sufficient evidence to indicate there is a difference in mean travel time among the three aid types at $\alpha = 0.05$.
To determine if there is a difference in mean travel time between the two orders, we test:

$H_0 : \mu_{E/W} = \mu_{W/E}$
$H_a : \mu_{E/W} \neq \mu_{W/E}$

The test statistic is $F = 1.95$ and the p-value is $p > 0.10$. Since the p-value is not small, H_0 is not rejected. There is insufficient evidence to indicate there is a difference in mean travel time between the two orders at $\alpha = 0.05$.

14.81   Using MINITAB, the results are:

```
One-way ANOVA: PD-1, IADC-1-2-6, IADC-5-1-7

Source      DF        SS        MS        F        P
Factor       2     366.6     183.3     9.50    0.006
Error        9     173.7      19.3
Total       11     540.2

S = 4.393      R-Sq = 67.85%      R-Sq(adj) = 60.71%

Level           N      Mean     StDev
PD-1            4    34.300     3.127
IADC-1-2-6      4    28.050     2.167
IADC-5-1-7      4    20.775     6.589
```

To determine if the mean RoP differs among the three drill bits, we test:

$H_0 : \mu_1 = \mu_1 = \mu_3$
$H_a :$ At least two means differ

The test statistic is $F = 9.50$ and the p-value is $p = 0.006$. Since the p-value is less than $\alpha (p = 0.006 < 0.05)$, H_0 is rejected. There is sufficient evidence to indicate there is a difference in the mean RoP values among the three drill bits at $\alpha = 0.05$.

To determine where the differences exist, MINITAB is used to run Tukey's multiple comparison procedure.

```
Tukey 95% Simultaneous Confidence Intervals
All Pairwise Comparisons

Individual confidence level = 97.91%

PD-1 subtracted from:

              Lower    Center   Upper   --+-----+-----+-----+------
IADC-1-2-6   -14.925   -6.250   2.425         (-----*------)
IADC-5-1-7   -22.200  -13.525  -4.850   (-----*-----)
                                        --+------+-----+-----+------
                                        -20     -10    0     10


IADC-1-2-6 subtracted from:

              Lower    Center   Upper   --+------+------+------+-------
IADC-5-1-7   -15.950   -7.275   1.400        (---------*-------)
                                        --+------+-------+------+-----
                                        -20     -10     0      10
```

The mean RoP for PD-1 is significantly greater than the mean RoP of IADC 5-1-7. No other significant differences exist.

14.83 a. Using MINITAB, the results are:

Two-way ANOVA: PERCENT versus TEMP, TIME

```
Source        DF      SS        MS        F       P
TEMP          2     4376.7   2188.36   457.14   0.000
TIME          2     8200.4   4100.19   856.52   0.000
Interaction   4      103.3     25.82     5.39   0.003
Error        27      129.3      4.79
Total        35    12809.6

S = 2.188   R-Sq = 98.99%   R-Sq(adj) = 98.69%
```

The complete ANOVA table is:

Source	df	SS	MS	F	P
TEMP	2	4376.7	2188.36	457.14	0.000
TIME	2	8200.4	4100.19	856.52	0.000
TxT	4	103.3	25.82	5.39	0.003
Error	27	129.3	4.79		
Total	35	12809.6			

b. To determine if time and temperature interact, we test:

H_0: Time and temperature do not interact
H_a: Time and temperature interact

The test statistic is $F = 5.39$ and the p-value is $p = 0.003$. Since the p-value is less than $\alpha(p = 0.003 < 0.05)$, H_0 is rejected. There is sufficient evidence to indicate time and temperature interact to affect percentage of water removed at $\alpha = 0.05$.

c. Using MINITAB, the results are:

Regression Analysis: PERCENT versus T, E, TxE, Tsq, Esq

Analysis of Variance

Source	DF	Adj SS	AdjMS	F-Value	P-Value
Regression	5	12658.1	2531.62	501.22	0.000
E	1	0.8	0.82	0.16	0.690
T	1	0.2	0.22	0.04	0.837
TxE	1	81.0	81.00	16.04	0.000
Esq	1	23.3	23.35	4.62	0.040
Tsq	1	2.7	2.72	0.54	0.469
Error	30	151.5	5.05		
Lack-of-Fit	3	22.3	7.43	1.55	0.224
Pure Error	27	129.2	4.79		
Total	35	12809.6			

Model Summary

S	R-sq	R-sq(adj)	R-sq(pred)
2.24743	98.82%	98.62%	98.29%

Coefficients

Term	Coef	SECoef	T-Value	P-Value	VIF
Constant	-12.3	29.1	-0.42	0.676	
E	-0.188	0.466	-0.40	0.690	103.00
T	0.100	0.481	0.21	0.837	439.00
TxE	0.01125	0.00281	4.00	0.000	61.00
Esq	0.01708	0.00795	2.15	0.040	49.00
Tsq	0.00146	0.00199	0.73	0.469	433.00

Regression Equation

PERCENT = -12.3 - 0.188 E + 0.100 T + 0.01125 TxE
          + 0.01708 Esq+ 0.00146 Tsq

The least squares prediction equation is

$$\hat{y} = -12.3 - 0.188E + 0100T + 0.01125TxE + 0.01708E^2 + 0.00146T^2$$

d. For $T = 120$ and $E = 20$,

$$\hat{y} = -12.3 - 0.188(20) + 0.100(120) + 0.01125(20)(120) + 0.01708(20)^2 + 0.00146(120)^2$$
$$= 50.796$$

This prediction is based on the model that uses all the data, not just the data when $T = 120$ and $E = 20$. We do not expect predictions to equal the values of the sample means.

e. Using MINITAB, the results are:

Prediction for PERCENT

Regression Equation

PERCENT = -12.3 - 0.188 E + 0.100 T + 0.01125 TxE + 0.01708
          Esq+ 0.00146 Tsq

Variable	Setting
E	30
T	140
TxE	4200

```
Esq                900
Tsq              19600

    Fit   SE Fit          95% CI              95% PI
87.2778  1.00856  (85.2180, 89.3375)  (82.2469, 92.3086)
```

The 95% confidence interval for the mean percentage of water removed when temperature is 140 and exposure time is 30 is between 85.2180% and 89.3375%.

14.85 a. Using MINITAB, the results are:

ANOVA: TEMP versus TREAT, BATCH

```
Factor  Type    Levels  Values
TREAT   fixed        4  1, 2, 3, 4
BATCH   fixed        3  1, 2, 3

Analysis of Variance for TEMP

Source  DF      SS      MS      F      P
TREAT    3  1549.7   516.6   2.32  0.175
BATCH    2  2082.2  1041.1   4.68  0.060
Error    6  1335.8   222.6
Total   11  4967.7

S = 14.9211   R-Sq = 73.11%   R-Sq(adj) = 50.70%
```

To determine if there is a difference in mean temperature among the four treatments, we test:

$$H_0 : \mu_1 = \mu_2 = \mu_3 = \mu_4$$
$$H_a : \text{At least two means differ}$$

The test statistic is $F = 2.32$ and the p-value is $p = 0.175$. Since the p-value is not less than $\alpha (p = 0.175 \not< 0.05)$, H_0 is not rejected. There is insufficient evidence to indicate there is a difference in the mean temperature among the four treatments at $\alpha = 0.05$.

b. To determine if there is a difference in mean temperature among the three batches, we test:

$$H_0 : \mu_{B1} = \mu_{B2} = \mu_{B3}$$
$$H_a : \text{At least two means differ}$$

The test statistic is $F = 4.68$ and the p-value is $p = 0.060$. Since the p-value is not less than $\alpha (p = 0.060 \not< 0.05)$, H_0 is not rejected. There is insufficient evidence to indicate there is a difference in the mean temperature among the three batches at $\alpha = 0.05$.

c. Yes. Since no batch differences were detected, it is recommended that no blocks be used in the future. A completely randomized design is recommended to increase degrees of freedom for error.

14.87 a.   Using MINITAB, the results are:

Regression Analysis: LIGHT versus x1, x2, x3, x4, x1x2, x1x3, x1x4, x2x3, ...

Analysis of Variance

Source	DF	Adj SS	Ad jMS	F-Value	P-Value
Regression	15	745.469	49.698	40.78	0.000
x1	1	712.531	712.531	584.64	0.000
x2	1	3.781	3.781	3.10	0.097
x3	1	1.531	1.531	1.26	0.279
x4	1	0.781	0.781	0.64	0.435
x1x2	1	0.281	0.281	0.23	0.637
x1x3	1	0.781	0.781	0.64	0.435
x1x4	1	2.531	2.531	2.08	0.169
x2x3	1	0.781	0.781	0.64	0.435
x2x4	1	0.031	0.031	0.03	0.875
x3x4	1	19.531	19.531	16.03	0.001
x1x2x3	1	1.531	1.531	1.26	0.279
x1x2x4	1	0.281	0.281	0.23	0.637
x1x3x4	1	0.031	0.031	0.03	0.875
x2x3x4	1	0.781	0.781	0.64	0.435
x1x2x3x4	1	0.281	0.281	0.23	0.637
Error	16	19.500	1.219		
Total	31	764.969			

Model Summary

S	R-sq	R-sq(adj)	R-sq(pred)
1.10397	97.45%	95.06%	89.80%

Coefficients

Term	Coef	SE Coef	T-Value	P-Value	VIF
Constant	10.031	0.195	51.40	0.000	
x1	4.719	0.195	24.18	0.000	1.00
x2	-0.344	0.195	-1.76	0.097	1.00
x3	0.219	0.195	1.12	0.279	1.00
x4	-0.156	0.195	-0.80	0.435	1.00
x1x2	0.094	0.195	0.48	0.637	1.00
x1x3	0.156	0.195	0.80	0.435	1.00
x1x4	0.281	0.195	1.44	0.169	1.00
x2x3	-0.156	0.195	-0.80	0.435	1.00
x2x4	-0.031	0.195	-0.16	0.875	1.00
x3x4	0.781	0.195	4.00	0.001	1.00
x1x2x3	-0.219	0.195	-1.12	0.279	1.00
x1x2x4	-0.094	0.195	-0.48	0.637	1.00
x1x3x4	-0.031	0.195	-0.16	0.875	1.00
x2x3x4	0.156	0.195	0.80	0.435	1.00
x1x2x3x4	0.094	0.195	0.48	0.637	1.00

Regression Equation

```
LIGHT = 10.031 + 4.719 x1 - 0.344 x2 + 0.219 x3 - 0.156 x4
        + 0.094 x1x2 + 0.156 x1x3 + 0.281 x1x4 - 0.156 x2x3
        - 0.031 x2x4 + 0.781 x3x4 - 0.219 x1x2x3 - 0.094 x1x2x4
        - 0.031 x1x3x4 + 0.156 x2x3x4 + 0.094 x1x2x3x4
```

To determine if any of the factors contribute to the model, we test:

$$H_0 : \beta_1 = \beta_2 = \beta_3 = \cdots = \beta_{15} = 0$$
$$H_a : \text{At least one } \beta_i \neq 0$$

The test statistic is $F = 40.78$ and the p-value is $p = 0.000$. Since the p-value so small, H_0 is rejected. There is sufficient evidence to indicate at least one of the factors contributes information for the prediction of y for $\alpha > 0.000$.

b. It appears that the interaction of shift and operator (x_3 and x_4) is significant ($t = 4.00, p = 0.001$) and the main effect amount (x_1) is significant ($t = 24.18, p = 0.000$).

c. The complete model is:

$$E(y) = \beta_0 + \beta_1 x_1 + \beta_2 x_2 + \beta_3 x_3 + \beta_4 x_4 + \beta_5 x_1 x_2 + \beta_6 x_1 x_3 + \beta_7 x_1 x_4 + \beta_8 x_2 x_3 + \beta_9 x_2 x_4$$

$$+ \beta_{10} x_3 x_4 + \beta_{11} x_1 x_2 x_3 + \beta_{12} x_1 x_2 x_4 + \beta_{13} x_1 x_3 x_4 + \beta_{14} x_2 x_3 x_4 + \beta_{15} x_1 x_2 x_3 x_4$$

d. There will be 16 degrees of freedom for estimating σ^2.

14.89 a. To find $\sum y$, recall that $\bar{y} = \dfrac{\sum y}{n} \Rightarrow \sum y = n\bar{y}$

$$\sum y_1 = n_1 \bar{y}_1 = 50(8,477) = 423,850$$

$$\sum y_2 = n_2 \bar{y}_2 = 50(10,404) = 520,200$$

$$\sum y = 423,850 + 520,200 = 944,050$$

$$CM = \frac{\left(\sum y\right)^2}{n} = \frac{944,050^2}{100} = 8,912,304,025$$

$$SST = \frac{T_1^2}{n_1} + \frac{T_2^2}{n_2} - CM = \frac{423,850^2}{50} + \frac{520,200^2}{50} - 8,912,304,025$$

$$= 3,592,976,450 + 5,412,160,800 - 8,912,304,025 = 92,833,225$$

b. Recall from Exercise 14.34, $SSE = (n_1 - 1)s_1^2 + (n_2 - 1)s_2^2 + \cdots + (n_p - 1)s_p^2$.

Thus, $SSE = (50 - 1)820^2 + (50 - 1)928^2 = 75,145,616$

c. $SS(Total) = SST + SSE = 92,833,225 + 75,145,616 = 167,978,841$

d. The ANOVA table is:

Source	df	SS	MS	F
Treatment	1	92,833,225	92,833,225	121.07
Error	98	75,145,616	766,792	
Total	99	167,978,841		

e. Since there would only be two x values represented in the simple linear regression, the least squares line would pass through the sample mean y values for these two x values, i.e., $(7, \bar{y}_1)$ and $(28, \bar{y}_2)$.

f. For $x = 7$ days, $\bar{y}_1 = 8,477$. For $x = 28$ days, $\bar{y}_2 = 10,404$.
We know the line passes through the points $(7, 8,477)$ and $(28, 10,404)$. The slope of the line will be $\dfrac{10,404 - 8,477}{28 - 7} = 91.762$. At $x = 0$, the line will cross the y-axis at $8,477 - 7(91.762) = 7,834.666$. Thus, the least squares line will be $\hat{y} = 7,834.666 + 91.762x$.

g. $s = \sqrt{MSE} = \sqrt{766,792} = 875.6666$

$$SS_{xx} = \sum x^2 - \frac{\left(\sum x\right)^2}{n} = 50(7)^2 + 50(28)^2 - \frac{[50(7) + 50(28)]^2}{100} = 11,025$$

For $x = 20$, $\hat{y} = 7,834.666 + 91.762(20) = 9,669.906$
For confidence coefficient 0.95, $\alpha = 0.05$ and $\alpha/2 = 0.05/2 = 0.025$. Using MINITAB with $df = 98$, $t_{0.025} = 1.98$. The 95% confidence interval is:

$$\hat{y} \pm t_{\alpha/2}s\sqrt{\frac{1}{n} + \frac{(x - \bar{x})^2}{SS_{xx}}} \Rightarrow 9,669.906 \pm 1.98(875.6666)\sqrt{\frac{1}{100} + \frac{(20 - 17.5)^2}{11,025}}$$

$$\Rightarrow 9,669.906 \pm 178.229 \Rightarrow (9,491.677, 9,848.1347)$$

h. $r^2 = 1 - \dfrac{SSE}{SS(Total)} = 1 - \dfrac{75,145,616}{167,978,841} = 0.553$

55.3% of the sample variation about the sample mean compressive strength values can be explained by the linear relationship between compressive strength and curing time.

14.91   The complete model is $E(y) = \beta_0 + \beta_1 x_1 + \beta_2 x_2 + \beta_3 x_3 + \beta_4 x_4 + \beta_5 x_1 x_4 + \beta_6 x_2 x_4 + \beta_7 x_3 x_4$

where $x_1 = \begin{cases} 1 \text{ if Adult female} \\ 0 \text{ if not} \end{cases}$ $x_2 = \begin{cases} 1 \text{ if Gravid female} \\ 0 \text{ if not} \end{cases}$ $x_3 = \begin{cases} 1 \text{ if Adult male} \\ 0 \text{ if not} \end{cases}$

$x_4 = \begin{cases} 1 \text{ if Extract trail} \\ 0 \text{ if not} \end{cases}$

To test for interaction between age-sex group and trail, we compare the complete model to the reduced model $E(y) = \beta_0 + \beta_1 x_1 + \beta_2 x_2 + \beta_3 x_3 + \beta_4 x_4$.
Using MINITAB, the results of fitting the complete model are:

```
Regression Analysis: DEVIATION versus x1, x2, x3, x4, x1x4, x2x4, x3x4

Analysis of Variance
Source         DF   Adj SS    Adj MS   F-Value   P-Value
Regression      7    64962    9280.3     12.66     0.000
x1              1      560     559.7      0.76     0.384
x2              1        2       1.7      0.00     0.961
x3              1     3320    3320.5      4.53     0.036
x4              1    20196   20196.0     27.54     0.000
x1x4            1      556     556.4      0.76     0.386
x2x4            1     2234    2234.3      3.05     0.084
x3x4            1      693     692.6      0.94     0.333
Error         112    82132     733.3
Total         119   147094
```

```
Model Summary

       S    R-sq   R-sq(adj)   R-sq(pred)
27.0799   44.16%      40.67%       33.73%
```

$SSE_c = 82,132$

The results of fitting the reduced model are:

Regression Analysis: DEVIATION versus x1, x2, x3, x4

```
Analysis of Variance

Source        DF   Adj SS   Adj MS   F-Value   P-Value
Regression     4    62717   15679.2     21.37     0.000
  x1           1       58      58.0      0.08     0.779
  x2           1     4168    4168.3      5.68     0.019
  x3           1     3917    3917.2      5.34     0.023
  x4           1    46445   46445.5     63.30     0.000
Error        115    84377     733.7
Total        119   147094
```

```
Model Summary

       S    R-sq   R-sq(adj)   R-sq(pred)
27.0871   42.64%      40.64%       37.06%
```

$SSE_R = 84,377$

To determine if age-sex group and trail interact to affect average trail deviation, we test:

H_0: Age-sex group and trail do not interact
H_a: Age-sex group and trail interact

The test statistic is $F = \dfrac{(SSE_R - SSE_C)/(k-g)}{SSE_C/[n-(k=1)]} = \dfrac{(84,377-82,132)/(7-4)}{82,132/[120-(7+1)]} = 1.020.$

The rejection region requires $\alpha = 0.05$ in the upper tail of the F distribution with $v_1 = k - g = 7 - 4 = 3$ and $v_2 = n - (k+1) = 120 - (7+1) = 112$. Using MINITAB, $F_{0.05} = 2.686$. The rejection region is $F > 2.686$.

Since the observed value of the test statistic does not fall in the rejection region ($F = 1.020 \not> 2.686$), H_0 is not rejected. There is insufficient evidence to indicate age-sex group and trail interact to affect average trail deviation at $\alpha = 0.05$.

Since the interaction is not significant, we can test for main effects.

To test for differences in the mean average trail deviations among the 4 age-sex groups, we use the complete model $E(y) = \beta_0 + \beta_1 x_1 + \beta_2 x_2 + \beta_3 x_3 + \beta_4 x_4$ and the reduced mode

$E(y) = \beta_0 + \beta_4 x_4.$

The new complete model is the reduced model from above. Thus, $SSE_C = 84,377$. Using MINITAB, the new results of fitting the new reduced model are:

Regression Analysis: DEVIATION versus x4

```
Analysis of Variance

Source        DF   Adj SS   Adj MS   F-Value   P-Value
Regression     1    46445   46445.5     54.45     0.000
  x4           1    46445   46445.5     54.45     0.000
Error        118   100648     852.9
```

```
Total          119   147094
```

Model Summary

```
       S    R-sq    R-sq(adj)   R-sq(pred)
29.2053   31.58%     31.00%       28.83%
```

$SSE_R = 10,648$

To determine if there are differences in the mean average trail deviation among the 4 age-sex groups, we test:

$$H_0 : \beta_1 = \beta_2 = \beta_3 = 0$$

$$H_a : \text{At least one } \beta_i \neq 0$$

The test statistic is $F = \dfrac{(SSE_R - SSE_C)/(k - g)}{SSE_C /[n - (k = 1)]} = \dfrac{(100,648 - 84,377)/(4 - 1)}{84,377/[120 - (4 + 1)]} = 7.392.$

The rejection region requires $\alpha = 0.05$ in the upper tail of the F distribution with $v_1 = k - g = 4 - 1 = 3$ and $v_2 = n - (k + 1) = 120 - (4 + 1) = 115$. Using MINITAB, $F_{0.05} = 2.684$. The rejection region is $F > 2.684$.

Since the observed value of the test statistic falls in the rejection region ($F = 7.392 > 2.684$), H_0 is rejected. There is sufficient evidence to indicate differences in the mean average trail deviation among the 4 age-sex groups at $\alpha = 0.05$.

To test for differences in the mean average trail deviations between the two types of trails, we use the complete model $E(y) = \beta_0 + \beta_1 x_1 + \beta_2 x_2 + \beta_3 x_3 + \beta_4 x_4$ and the reduced model $E(y) = \beta_0 + \beta_1 x_1 + \beta_2 x_2 + \beta_3 x_3$.

The new complete model is the same as the complete model above. Thus, $SSE_C = 84,377$.

Using MINITAB, the new results of fitting the new reduced model are:

Regression Analysis: DEVIATION versus x1, x2, x3

Analysis of Variance

Source	DF	Adj SS	Adj MS	F-Value	P-Value
Regression	3	16271	5423.72	4.81	0.003
x1	1	58	58.02	0.05	0.821
x2	1	4168	4168.33	3.70	0.057
x3	1	3917	3917.18	3.47	0.065
Error	116	130822	1127.78		
Total	119	147094			

Model Summary

```
       S    R-sq    R-sq(adj)   R-sq(pred)
33.5824   11.06%      8.76%        4.82%
```

$SSE_R = 130,822$

To determine if there are differences in the mean average trail deviation between the two types of trails, we test:

$$H_0 : \beta_4 = 0$$

$$H_a : \beta_4 \neq 0$$

The test statistic is $F = \dfrac{(SSE_R - SSE_C)/(k-g)}{SSE_C/[n-(k=1)]} = \dfrac{(130{,}822 - 84{,}377)/(4-3)}{84{,}377/[120-(4+1)]} = 63.301.$

The rejection region requires $\alpha = 0.05$ in the upper tail of the F distribution with $v_1 = k - g = 4 - 3 = 1$ and $v_2 = n - (k+1) = 120 - (4+1) = 115$. Using MINITAB, $F_{0.05} = 3.924$. The rejection region is $F > 3.924$.

Since the observed value of the test statistic falls in the rejection region ($F = 63.301 > 3.924$), H_0 is rejected. There is sufficient evidence to indicate differences in the mean average trail deviation between the two types of trails at $\alpha = 0.05$.

15

Nonparametric Statistics

15.1 a. To determine if the median amount of caffeine in Breakfast Blend coffee exceeds 300 mg, we test:

$$H_0 : \tau = 300$$

$$H_a : \tau > 300$$

b. Four of the cups in the sample have caffeine contents that exceed 300. Therefore, the test statistic is $S = 4$.

c. Since one of the observations equals 300, this observation is omitted. Using Table 2, Appendix B, with $n = 5$ and $p = 0.5$, $p = P(x \geq 4) = 1 - P(x \leq 3) = 1 - 0.8125 = 0.1875$.

d. Since the p-value is not less than $\alpha(p = 0.1875 \nless 0.05)$, H_0 is not rejected. There is insufficient evidence to indicate the median amount of caffeine in Breakfast Blend coffee exceeds 300 mg at $\alpha = 0.05$.

15.3 a. If the data are not normally distributed, the t-test is unreliable.

b. A possible alternative test would be the sign test.

c. The test statistic is the number of observations that exceed 3 which is $S = 9$.

d. First, eliminate all observations that equal 3. Thus, $n = 17$. Using MINITAB, with $n = 17$ and $p = 0.5$, $p = P(x \geq 9) = 1 - P(x \leq 8) = 1 - 0.5 = 0.5$.

e. Since the p-value is not less than $\alpha(p = 0.5 \nless 0.05)$, H_0 is not rejected. There is insufficient evidence to indicate the median number of wheels exceeds 3 at $\alpha = 0.05$.

15.5 To determine if the median surface roughness of coated interior pipe differs from 2 micrometers, we test:

$$H_0 : \tau = 2$$

$$H_a : \tau \neq 2$$

The number of observations less than 2 is $S_1 = 8$. The number of observations greater than 2 is $S_2 = 11$. The test statistic is the larger of S_1 and S_2 which is $S = 11$.
Using Table 2, Appendix B, with $n = 20$ and $p = 0.5$,

$$p = 2P(x \geq 11) = 2[1 - P(x \leq 10)] = 2(1 - 0.5881) = 0.8238.$$

Since the p-value is not less than $\alpha(p = 0.8238 \nless 0.05)$, H_0 is not rejected. There is insufficient evidence to indicate the median surface roughness of coated interior pipe differs from 2 micrometers at $\alpha = 0.05$.

15.7 a. To determine if the median daily ammonia concentration for all afternoon drive-time days exceeds 1.5 ppm, we test:

$$H_0 : \tau = 1.5$$

$$H_a : \tau > 1.5$$

b. The test statistic is the number of observations greater than 1.5 which is $S = 3$.

c. Since one of the observations equals 1.5, this observation is omitted. Using Table 2, Appendix B, with $n = 7$ and $p = 0.5$, $p = P(x \geq 3) = 1 - P(x \leq 2) = 1 - 0.2266 = 0.7734$.

d. Since the p-value is not less than $\alpha(p = 0.7734 \nless 0.05)$, H_0 is not rejected. There is insufficient evidence to indicate the median daily ammonia concentration for all afternoon drive-time days exceeds 1.5 ppm at $\alpha = 0.05$.

15.9   To determine if the true median level of radon exposure in the tombs is less than 6,000 Bq/m3, we test:

$$H_0 : \tau = 6,000$$

$$H_a : \tau < 6,000$$

The test statistic is the number of observations less than 6,000 which is $S = 9$. Using MINITAB, with $n = 12$ and $p = 0.5$, $p = P(x \geq 9) = 1 - P(x \leq 8) = 1 - 0.9270 = 0.0730$.

Since the p-value is less than $\alpha(p = 0.0730 < 0.10)$, H_0 is rejected. There is sufficient evidence to indicate the true median level of radon exposure in the tombs is less than 6,000 Bq/m3 at $\alpha = 0.10$. The tombs do not need to be closed.

15.11   From Exercise 15.10, we showed that $P(S_1 \geq c) = P(S_2 \leq n - c)$.

For a two-tailed test, the p-value will be $P(S_1 \geq c) + P(S_2 \leq n - c) = 2P(S_1 \geq c)$.

15.13 a. The ranks of the 20 observations are shown below:

Old Design	Rank	New Design	Rank
210	9	216	16.5
212	13.5	217	18.5
211	11	162	4
211	11	137	1
190	7	219	20
213	15	216	16.5
212	13.5	179	6
211	11	153	3
164	5	152	2
209	8	217	18.5
	$T_1 = 104$		$T_2 = 106$

b. The sum of the ranks of the old design is $T_1 = 104$.

c. The sum of the ranks of the new design is $T_2 = 106$.

d. Since $n_1 = n_2$, we can use either T_1 or T_2 can be used as the test statistic. We will use $T = T_1 = 104$.

e. To compare the distributions of bursting strengths for the two designs, we test:

H_0: The two designs have identical probability distributions

H_a: The distributions for the bursting strengths differ for the two designs

Using Table 15, Appendix B, with $n_1 = n_2 = 10$ and $\alpha = 0.05$, $T_L = 79$ and $T_U = 131$. The rejection region is $T \leq 79$ or $T \geq 131$.

Since the observed value of the of the test statistic does not fall in the rejection region ($T = 104 \not\leq 79$ and $T = 104 \not\geq 131$), H_0 is not rejected. There is insufficient evidence to indicate that the distribution for the bursting strengths differ for the two designs at $\alpha = 0.05$.

15.15 First, rank the observations and then sum the ranks:

70-cm	Rank	100-cm	Rank
6.00	1	6.80	2
7.20	3	9.20	5.5
10.20	7.5	8.80	4
13.20	13.5	13.20	13.5
11.40	11	11.20	9.5
13.60	15	14.90	16
9.20	5.5	10.20	7.5
11.20	9.5	11.80	12
$n_1 = 8$	$T_1 = 66$	$n_2 = 8$	$T_2 = 70$

To determine if there is a difference in the locations of the cracking torsion moment distributions for the two types of T-beams, we test:

H_0: The two populations have identical probability distributions

H_a: The probability distribution for the 70-cm slab width is shifted to the right or left of that for the 100-cm slab width

Since $n_1 = n_2$, we can use either T_1 or T_2 can be used as the test statistic. We will use $T = T_1 = 66$.

Using Table 15, Appendix B, with $n_1 = n_2 = 8$ and $\alpha = 0.10$, $T_L = 52$ and $T_U = 84$. The rejection region is $T \leq 52$ or $T \geq 84$.

Since the observed value of the of the test statistic does not fall in the rejection region ($T = 66 \not\leq 52$ and $T = 66 \not\geq 84$), H_0 is not rejected. There is insufficient evidence to indicate a difference in the locations of the cracking torsion moment distributions for the two types of T-beams at $\alpha = 0.10$.

15.17   First, rank the observations and then sum the ranks:

Group 1	Rank	Group 2	Rank
104.1	8	96.7	7
34.0	1	53.6	2
62.5	3	64.4	4
73.8	6	69.7	5
$n_1 = 4$	$T_1 = 18$	$n_2 = 4$	$T_2 = 18$

To compare the distributions of new bone formation for the two groups, we test:

H_0: The two populations have identical probability distributions

H_a: The distributions for the new bone formation differ for the two groups

Since $n_1 = n_2$, we can use either T_1 or T_2 can be used as the test statistic. We will use $T = T_1 = 18$.

Using Table 15, Appendix B, with $n_1 = n_2 = 4$ and $\alpha = 0.10$, $T_L = 12$ and $T_U = 24$. The rejection region is $T \le 12$ or $T \ge 24$.

Since the observed value of the test statistic does not fall in the rejection region ($T = 18 \not\le 12$ and $T = 18 \not\ge 24$), H_0 is not rejected. There is insufficient evidence to indicate the distributions of new bone formation differ for the two groups at $\alpha = 0.10$.

15.19   First, rank the observations and then sum the ranks. If you notice, all of the observations without calcium/gypsum are smaller than all of the observations with calcium/gypsum. Thus, $T_1 = 1275$ and $T_2 = 3775$.

To determine if the addition of calcium/gypsum to the solution impact water quality, we test:

H_0: The two populations have identical probability distributions

H_a: The distribution of water quality without calcium/gypsum is shifted to the right or left of that with calcium/gypsum

The test statistic is

$$Z = \frac{T_1 - \left[\dfrac{n_1 n_2 + n_1(n_1+1)}{2} \right]}{\sqrt{\dfrac{n_1 n_2 (n_1 + n_2 + 1)}{12}}} = \frac{1275 - \left[\dfrac{50(50) + 50(50+1)}{2} \right]}{\sqrt{\dfrac{50(50)(50 + 50 + 1)}{12}}} = -8.617$$

The rejection region requires $\alpha/2 = 0.05/2 = 0.025$ in each tail of the z-distribution. Using Table 5, Appendix B, $z_{0.025} = 1.96$. The rejection region is $z < -1.96$ or $z > 1.96$.

Since the observed value of the test statistic falls in the rejection region ($z = -8.617 < -1.96$), H_0 is rejected. There is sufficient evidence to indicate the addition of calcium/gypsum to the solution impact water quality at $\alpha = 0.05$.

15.21 The $4! = 24$ different ways the ranks can be assigned and the corresponding values of T_2 are:

A	B		A	B		A	B		A	B		A	B		A	B
1	3		1	4		1	2		1	4		1	2		1	3
2	4		2	3		3	4		3	2		4	3		4	2
$T_2 = 7$			$T_2 = 7$			$T_2 = 6$			$T_2 = 6$			$T_2 = 5$			$T_2 = 5$		

A	B		A	B		A	B		A	B		A	B		A	B
2	3		2	4		2	1		2	4		2	1		2	3
1	4		1	3		3	4		3	1		4	3		4	1
$T_2 = 7$			$T_2 = 7$			$T_2 = 5$			$T_2 = 5$			$T_2 = 4$			$T_2 = 4$		

A	B		A	B		A	B		A	B		A	B		A	B
3	2		3	4		3	1		3	4		3	1		3	2
1	4		1	2		2	4		2	1		4	2		4	1
$T_2 = 6$			$T_2 = 6$			$T_2 = 5$			$T_2 = 5$			$T_2 = 3$			$T_2 = 3$		

A	B		A	B		A	B		A	B		A	B		A	B
4	2		4	3		4	1		4	3		4	1		4	2
1	3		1	2		2	3		2	1		3	2		3	1
$T_2 = 5$			$T_2 = 5$			$T_2 = 4$			$T_2 = 4$			$T_2 = 3$			$T_2 = 3$		

If we assume that all outcomes are equally likely, then each has a probability of 1/24.

$$E(T_2) = \frac{7}{24} + \frac{7}{24} + \frac{6}{24} + \frac{6}{24} + \frac{5}{24} + \frac{5}{24} + \frac{7}{24} + \frac{7}{24} + \frac{5}{24} + \frac{5}{24} + \frac{4}{24} + \frac{4}{24}$$

$$+ \frac{6}{24} + \frac{6}{24} + \frac{5}{24} + \frac{5}{24} + \frac{3}{24} + \frac{3}{24} + \frac{5}{24} + \frac{5}{24} + \frac{4}{24} + \frac{4}{24} + \frac{3}{24} + \frac{3}{24}$$

$$= \frac{120}{24} = 5$$

From the formula, $E(T_2) = \frac{n_1 n_2 + n_2(n_2 + 1)}{2} = \frac{2(2) + 2(2+1)}{2} = \frac{10}{2} = 5.$

15.23 a. The results of the t-test may not be valid because the differences may not be normally distributed.

  b. The appropriate nonparametric test is the Wilcoxon Signed Rank test. The hypotheses are:

  H_0: The two populations have identical probability distributions
  H_a: The distributions of the total amount of heavy minerals differ for the twin holes

c & d. The differences and the ranks of the absolute differences are:

Location	1st Hole	2nd Hole	Diff	Abs Diff	Rank
1	5.5	5.7	−0.2	0.2	3.5
2	11	11.2	−0.2	0.2	3.5
3	5.9	6	−0.1	0.1	1.5
4	8.2	5.6	2.6	2.6	15
5	10	9.3	0.7	0.7	6
6	7.9	7	0.9	0.9	7
7	10.1	8.4	1.7	1.7	13.5
8	7.4	9	−1.6	1.6	12
9	7	6	1	1	8
10	9.2	8.1	1.1	1.1	9
11	8.3	10	−1.7	1.7	13.5
12	8.6	8.1	0.5	0.5	5
13	10.5	10.4	0.1	0.1	1.5
14	5.5	7	−1.5	1.5	11
15	10	11.2	−1.2	1.2	10

e. The sum of the ranks for the negative differences is $T_- = 55$ and the sum of the ranks of the positive differences is $T_+ = 65$

f. To determine if there is a difference in the THM distributions of all original holes and their twin holes, we test:

H_0: The two populations have identical probability distributions
H_a: The distributions of the total amount of heavy minerals differ for the twin holes

The test statistic is the smaller of T_- and T_+ which is $T = T_- = 55$.
Using Table 16, Appendix B, with $n = 15$ and $\alpha = 0.05$, $T_0 = 25$. The rejection region is $T \leq 25$.
Since the observed value of the test statistic does not fall in the rejection region ($T = 55 \nleq 25$), H_0 is not rejected. There is insufficient evidence to indicate that there is a difference in the THM distributions of all original holes and their twin holes at $\alpha = 0.05$.

15.25 a. To determine if the distributions of chest injury ratings of drivers and front seat passengers differ, we test:

H_0: The two populations have identical probability distributions
H_a: The distributions of the chest injury ratings differ for drivers and front seat Passengers

b. Using MINITAB, the results are:
Wilcoxon Signed Rank Test: Diff

```
Test of median = 0.000000 versus median ≠ 0.000000

              N for    Wilcoxon           Estimated
        N     Test     Statistic     P     Median
Diff 18        16         23.0    0.021    -4.000
```

The test statistic is $T = T_+ = 23$.

c. Using Table 16, Appendix B, with $n = 16$ and, $\alpha = 0.01$, $T_0 = 19$. The rejection region is $T \le 19$.

d. Since the observed value of the test statistic does not fall in the rejection region ($T = 23 \not< 19$), H_0 is not rejected. There is insufficient evidence to indicate that the distributions of chest injury ratings of drivers and front seat passengers differ at $a = 0.01$.

The two-tailed p-value is $p = 0.021$.

15.27 Some preliminary calculations are:

Day	Field	3D Model	Diff	Abs Diff	Rank
Oct. 24	−58.00	−52.00	−6.00	6.00	3
Dec. 3	69.00	59.00	10.00	10.00	5
Dec. 15	35.00	32.00	3.00	3.00	2
Feb. 2	−32.00	−24.00	−8.00	8.00	4
Mar. 25	−40.00	−39.00	−1.00	1.00	1
24-May	−83.00	−71.00	−12.00	12.00	6

The sum of the ranks for the negative differences is $T_- = 14$ and the sum of the ranks of the positive differences is $T_+ = 7$. The test statistic is the smaller of T_- and T_+ which is $T = T_+ = 7$.

To determine if there is a shift in the change in transverse strain distribution between field measurements and the 3D model, we test:

H_0: The two populations have identical probability distributions

H_a: The distributions for the change in transverse strains differ for the field measurements and the 3D model

The test statistic is $T = 7$.

Using Table 16, Appendix B, with $n = 6$ and $\alpha = 0.05$, $T_0 = 1$. The rejection region is $T \le 1$.

Since the observed value of the test statistic does not fall in the rejection region ($T = 7 \not\le 1$), H_0 is not rejected. There is insufficient evidence to indicate there is a shift in the change in transverse strain distribution between field measurements and the 3D model at $\alpha = 0.05$.

15.29   Some preliminary calculations are:

Zone	Before	After	Diff	Abs Diff	Rank
401	0.000000	0.003595	−0.003595	0.003595	6
402	0.001456	0.007278	−0.005822	0.005822	7
403	0.000000	0.003297	−0.003297	0.003297	5
404	0.002868	0.003824	−0.000956	0.000956	3
405	0.000000	0.002198	−0.002198	0.002198	4
406	0.000000	0.000898	−0.000898	0.000898	2
407	0.000626	0.000000	0.000626	0.000626	1
408	0.000000	0.000000	0.000000	0.000000	-

The sum of the ranks for the negative differences is $T_- = 27$ and the sum of the ranks of the positive differences is $T_+ = 1$. The test statistic is the smaller of T_- and T_+ which is $T = T_+ = 1$.

To compare the sea turtle nesting densities before and after beach nourishing, we test:

H_0: The two populations have identical probability distributions
H_a: The distributions for the two methods differ in location

The test statistic is $T = 1$.

Using Table 16, Appendix B, with $n = 7$ and $\alpha = 0.05$ $T_0 = 2$. The rejection region is $(T \le 2)$.

Since the observed value of the test statistic falls in the rejection region ($T = 1 \le 2$), H_0 is rejected. There is sufficient evidence to indicate the sea turtle nesting densities before and after beach nourishing differ in location at $\alpha = 0.05$.

15.31   For the Wilcoxon signed ranks test, show that $T_+ + T_- = \dfrac{n(n+1)}{2}$ where n is the number of non-zero differences that are ranked.

There are n non-zero differences. These are ranked by their magnitude, ignoring their signs.

Thus, the sum of the ranks is $\dfrac{n(n+1)}{2}$.

T_+ is the sum of the ranks associated with the positive differences and is the sum of the ranks associated with the negative differences. Together, $T_+ + T_-$ must sum to the sum of all the ranks or $\dfrac{n(n+1)}{2}$.

Thus, $T_+ + T_- = \dfrac{n(n+1)}{2}$.

15.33 The 8 different rankings are as follows, where (−) indicates the difference is negative and (+) indicates the difference is positive:

	Rankings							
Observation	1	2	3	4	5	6	7	8
A	1 (+)	2 (+)	1 (−)	1 (+)	2 (−)	2 (+)	1 (−)	2 (−)
B	2 (+)	1 (+)	2 (+)	2 (−)	1 (+)	1 (−)	2 (−)	1 (−)
T_+	3	3	2	1	1	2	0	0
T_-	0	0	1	2	2	1	3	3

If each case is equally likely, then each has a probability of occurring of 1/8.

$$E(T_+) = 3\left(\frac{1}{8}\right) + 3\left(\frac{1}{8}\right) + 2\left(\frac{1}{8}\right) + 1\left(\frac{1}{8}\right) + 1\left(\frac{1}{8}\right) + 2\left(\frac{1}{8}\right) + 0\left(\frac{1}{8}\right) + 0\left(\frac{1}{8}\right) = \frac{12}{8} = 1.5$$

For $n = 2$, $E(T_+) = \dfrac{n(n+1)}{4} = \dfrac{2(2+1)}{4} = \dfrac{6}{14} = 1.5$

15.35 a. The table below shows the ranks for all the observations:

Scopolamine	Rank	Placebo	Rank	No Drug	Rank
5	2.5	8	13	8	13
8	13	10	20.5	9	17
8	13	12	26.5	11	23.5
6	6.5	10	20.5	12	26.5
6	6.5	9	17	11	23.5
6	6.5	7	10	10	20.5
6	6.5	9	17	12	26.5
8	13	10	20.5	12	26.5
6	6.5				
4	1				
5	2.5				
6	6.5				
	$T_1 = 84$		$T_2 = 145$		$T_3 = 177$

b. The sum of the ranks of the observations in group 1 is $T_1 = 84$.
c. The sum of the ranks of the observations in group 2 is $T_2 = 145$.
d. The sum of the ranks of the observations in group 3 is $T_3 = 177$.
e. The test statistic is

$$H = \frac{12}{n(n+1)} \sum \frac{T_i^2}{n_i} - 3(n+1) = \frac{12}{28(28+1)}\left(\frac{84^2}{12} + \frac{145^2}{8} + \frac{177^2}{8}\right) - 3(28+1) = 18.403$$

f. To determine if the distributions for the number of word pairs recalled differ by group, we test:

H_0: The three population probability distributions are identical
H_a: At least two of the three population probability distributions differ in location

The test statistic is $H = 18.403$.

The rejection region requires $\alpha = 0.05$ in the upper tail of the χ^2 distribution with $df = k - 1 = 3 - 1 = 2$. From Table 8, Appendix B, $\chi^2_{0.05} = 5.99147$. The rejection region is $H > 5.99147$.

f. Since the observed value of the test statistic falls in the rejection region ($H = 18.403 > 5.99147$), H_0 is rejected. There is sufficient evidence to indicate that at least two of the distributions of word pairs recalled differ in location at $\alpha = 0.05$.

g. Using MINITAB, the results are:

Mann-Whitney Test and CI: Scopolamine, Placebo

```
             N    Median
Scopolamine  12    6.000
Placebo       8    9.500

Point estimate for η1 - η2 is -3.000
95.1 Percent CI for η1 - η2 is (-4.000,-1.999)
W = 82.5
Test of η1 = η2 vs η1 < η2 is significant at 0.0005
The test is significant at 0.0004 (adjusted for ties)
```

To determine if the distribution of the number of word pairs recalled by group 1 is shifted to the left of the distribution of the number of word pairs recalled by group 2, we test:

H_0: The two population probability distributions are identical
H_a: The distribution of the number of word pairs recalled by group 1 is shifted to the left of the distribution of the number of word pairs recalled by group 2

The test statistic is $T = 82.5$ and the p-value is $p = 0.0005$.
Since the p-value is less than $\alpha (p = 0.0005 < 0.05)$, H_0 is rejected. There is sufficient evidence to indicate the distribution of the number of word pairs recalled by group 1 is shifted to the left of the distribution of the number of word pairs recalled by group 2 at $\alpha = 0.05$.

15.37 a. To determine if the probability distributions of the number of collisions over a 3-year period differ among the five communities, we test:

H_0: The five population probability distributions are identical
H_a: At least two of the five population distributions differ in location

b. The test statistic is:

$$H = \frac{12}{n(n+1)} \sum \frac{T_i^2}{n_i} - 3(n+1)$$

$$= \frac{12}{150(150+1)} \left(\frac{3398^2}{30} + \frac{2249.5^2}{30} + \frac{3144^2}{30} + \frac{1288.5^2}{30} + \frac{1245^2}{30} \right) - 3(150+1) = 71.53$$

c. The rejection region requires $\alpha = 0.05$ in the upper tail of the χ^2 distribution with $df = k - 1 = 5 - 1 = 4$. From Table 8, Appendix B, $\chi^2_{0.05} = 9.48773$. The rejection region is $H > 9.48773$.

Since the observed value of the test statistic falls in the rejection region ($H = 71.53 > 9.48773$), H_0 is rejected. There is sufficient evidence to indicate the probability distributions of the number of collisions over a 3-year period differ among the five communities at $\alpha = 0.05$.

15.39 Using MINITAB, the results are:

Kruskal-Wallis Test: Strength versus Housing

```
Kruskal-Wallis Test on Strength
```

Housing	N	Median	Ave Rank	Z
1	10	37.65	12.9	-0.74
2	6	38.75	15.1	0.20
3	6	39.80	20.5	2.02
4	6	35.65	10.5	-1.34
Overall	28		14.5	

```
H = 5.00  DF = 3  P = 0.172
H = 5.00  DF = 3  P = 0.172 (adjusted for ties)
```

To determine if the strength distributions of the four housing systems differ in location, we test:

H_0: The four population probability distributions are identical
H_a: At least two of the four population distributions differ in location

The test statistic is $H = 5.00$ and the p-value is $p = 0.172$. Since the p-value is not less than $\alpha (p = 0.172 \not< 0.05)$, H_0 is not rejected. There is insufficient evidence to indicate the strength distributions of the four housing systems differ in location at $\alpha = 0.05$.

15.41 Using MINITAB, the results are:
Kruskal-Wallis Test: Ratio versus Group

```
Kruskal-Wallis Test on Ratio
```

Group	N	Median	Ave Rank	Z
SD	3	2.730	3.8	-2.33
SWRA	3	2.730	3.2	-2.49
UMRB-1	7	3.300	12.7	-0.32
UMRB-2	6	3.890	19.8	2.28
UMRB-3	7	3.600	17.5	1.62
Overall	26		13.5	

```
H = 16.26  DF = 4  P = 0.003
H = 16.27  DF = 4  P = 0.003 (adjusted for ties)
```

To determine if the distributions for the ratios differ for the five different boreholes, we test:

H_0: The five population probability distributions are identical
H_a: At least two of the five population probability distributions differ in location

The test statistic is $H = 16.26$ and the p-value is $p = 0.003$. Since the p-value is less than $\alpha (p = 0.003 < 0.10)$, H_0 is rejected. There is sufficient evidence to indicate that the distributions for the ratios differ in location among the five different boreholes at $\alpha = 0.10$.

15.43   There are a total of n observations which are all ranked. Thus, the sum of the ranks of all the observations is $\dfrac{n(n+1)}{2}$.

The ranks within each group are summed to get $T_1, T_2, ..., T_k$.

Thus, $T_1 + T_2 + \cdots + T_3 = \displaystyle\sum_{i=1}^{k} T_i =$ sum of ranks of all the observations which is $\dfrac{n(n+1)}{2}$.

15.45 a. The rejection region requires $\alpha = 0.01$ in the upper tail of the χ^2 distribution with $df = k - 1 = 3 - 1 = 2$. From Table 8, Appendix B, $\chi^2_{0.01} = 9.21034$. The rejection region is $F_r > 9.21034$.

   b. Since the observed value of the test statistic falls in the rejection region $(F_r = 19.16 > 9.21034)$, H_0 is rejected. There is sufficient evidence to indicate that the distributions for the proportion of eye fixations on the interior mirror differ among the three tasks at $\alpha = 0.01$.

   c. Since the observed value of the test statistic does not fall in the rejection region $(F_r = 7.80 \not> 9.21034)$, H_0 is not rejected. There is insufficient evidence to indicate that the distributions for the proportion of eye fixations on the off-side mirror differ among the three tasks at $\alpha = 0.01$.

   d. Since the observed value of the test statistic falls in the rejection region $(F_r = 20.67 > 9.21034)$, H_0 is rejected. There is sufficient evidence to indicate that the distributions for the proportion of eye fixations on the speedometer differ among the three tasks at $\alpha = 0.01$.

15.47   Some preliminary calculations are:

	Ranks			
Temperature	Tin-Lead	Tin-Silver	Tin-Copper	Tin-Silver-Copper
23°C	4	2	1	3
50°C	3	4	1	2
75°C	4	2	1	3
100°C	1	4	3	2
125°C	1	2	3	4
150°C	1	2	4	3
	$T_1 = 14$	$T_2 = 16$	$T_3 = 13$	$T_4 = 17$

To determine if the distributions for the four solder types are identical, we test:

H_0: The probability distributions for the four solder types are identical

H_a: At least two of the probability distributions differ in location

The test statistic is

$$F_r = \frac{12}{bk(k+1)}\sum T_i^2 - 3b(k+1) = \frac{12}{6(4)(4+1)}(14^2 + 16^2 + 13^2 + 17^2) - 3(6)(4+1) = 1$$

The rejection region requires $\alpha = 0.10$ in the upper tail of the χ^2 distribution with $df = k-1 = 4-1 = 3$. From Table 8, Appendix B, $\chi^2_{0.10} = 6.25139$. The rejection region is $F_r > 6.25139$.

Since the observed value of the test statistic does not fall in the rejection region ($F_r = 1 \ngtr 6.25139$), H_0 is not rejected. There is insufficient evidence to indicate that the distributions of plastic hardening measurements differ in location for the four solder types.

15.49 Some preliminary calculations are:

Week	Ranks				
	Monday	Tuesday	Wednesday	Thursday	Friday
1	5	1	4	2	3
2	5	4	3	1	2
3	2.5	2.5	5	1	4
4	2	1	3.5	5	3.5
5	5	1	2	3	4
6	4	2	3	1	5
7	5	3.5	1.5	3.5	1.5
8	4	2	1	3	5
9	1	2	5	3	4
	$T_1 = 33.5$	$T_2 = 19$	$T_3 = 28$	$T_4 = 22.5$	$T_5 = 32$

To determine if the distributions of absentee rates differ for the five days, we test:

H_0: The probability distributions for the five days are identical

H_a: At least two of the probability distributions differ in location

The test statistic is

$$F_r = \frac{12}{bk(k+1)}\sum T_i^2 - 3b(k+1)$$

$$= \frac{12}{9(5)(5+1)}(33.5^2 + 19^2 + 28^2 + 22.5^2 + 32^2) - 3(9)(5+1) = 6.778$$

The rejection region requires $\alpha = 0.10$ in the upper tail of the χ^2 distribution with $df = k - 1 = 5 - 1 = 4$. From Table 8, Appendix B, $\chi^2_{0.10} = 7.77944$. The rejection region is $F_r > 7.77944$.

Since the observed value of the test statistic does not fall in the rejection region ($F_r = 6.778 \not> 7.77944$), H_0 is not rejected. There is insufficient evidence to indicate the distributions of absentee rates differ for the five days at $\alpha = 0.10$.

15.51   Some preliminary calculations are:

Droplet	Length, y	y-Rank	Concentration, x	x-Rank	Difference, d_i	d_i^2
1	22.50	6	0.0	1	5	25
2	16.00	5	0.2	2	3	9
3	13.50	2	0.4	3	−1	1
4	14.00	4	0.6	4	0	0
5	13.75	3	0.8	5	−2	4
6	12.50	1	1.0	6	−5	25

$$\sum d_i^2 = 64$$

a. The ranks of the wicking length values are shown in the table.
b. The ranks of the antibody concentration values are shown in the table.

c. $r_s = 1 - \dfrac{6\sum d_i^2}{n(n^2 - 1)} = 1 - \dfrac{6(64)}{6(6^2 - 1)} = -0.8286$

d. To determine if the wicking length is negatively rank correlated with antibody concentration, we test:

$$H_0 : \rho = 0$$
$$H_a : \rho < 0$$

The test statistic is $r_s = -0.8286$.

Using Table 17, Appendix B, with $\alpha = 0.05$ and $n = 6$, $r_{0.05} = 0.829$. The rejection region is $r_s < -0.829$.

Since the observed value of the test statistic falls in the rejection region ($r_s = -0.8286 < -0.829$), H_0 is rejected. There is sufficient evidence to indicate the wicking length is negatively rank correlated with antibody concentration at $\alpha = 0.05$.

15.53 a. $r_s = 0.179$. There is a weak positive rank correlation between navigability and OIU level.

$r_s = 0.334$. There is a weak positive rank correlation between transactions and OIU level.

$r_s = 0.590$. There is a moderate positive rank correlation between locatability and OIU level.

$r_s = -.115$. There is a weak negative rank correlation between information richness and OIU level.

$r_s = 0.114$. There is a weak positive rank correlation between number of files and OIU level.

b. Whenever the observed p-value is less than $\alpha = 0.10$, H_0 can be rejected.

There is insufficient evidence $(p = 0.148 \not< 0.10)$ to indicate that navigability and OIU level are rank correlated.

There is sufficient evidence $(p = 0.023 \not< 0.10)$ to indicate that transactions and OIU level are rank correlated.

There is sufficient evidence $(p = 0.000 < 0.10)$ to indicate that locatability and OIU level are rank correlated.

There is insufficient evidence $(p = 0.252 \not< 0.10)$ to indicate that information richness and OIU level are rank correlated.

There is insufficient evidence $(p = 0.255 \not< 0.10)$ to indicate that number of files and OIU level are rank correlated.

15.55 a. Some preliminary calculations are:

		Differences, $y_j - y_i (i < j)$	
VOF, y	Batches, x	# Negatives	# Positives
86.68	3		
232.87	4	0	1
372.36	5	0	2
496.51	6	0	3
838.82	7	0	4
1183	8	0	5
	Totals	0	15

To determine if a positive relationship exists between VOF and the number of batches, we test:

$$H_0 : \beta_1 = 0$$

$$H_2 : \beta_1 > 0$$

The test statistic is $C = (-1)(\# \text{negatives}) + (1)(\# \text{positives}) = (-1)(0) + (1)(15) = 15$.

Using Table 18, Appendix B, with $n = 6$ and $x = 15$, the p-value is $P(x \geq C) = P(x \geq 15) = 0.001$.

Since the p-value is less than $\alpha(p = 0.001 < 0.05)$, H_0 is rejected. There is sufficient evidence to indicate there is a positive relationship between VOF and the number of batches at $\alpha = 0.05$.

b. Some preliminary calculations are:

VOF, y	RunTime, x	Differences, $y_j - y_i (i < j)$ # Negatives	# Positives
372.36	12		
232.87	14	1	0
496.51	18	0	2
86.68	27	3	0
1183.00	33	0	4
838.82	42	1	4
Totals		5	10

To determine if a positive relationship exists between VOF and run time, we test:
$$H_0 : \beta_1 = 0$$

$$H_a : \beta_1 > 0$$

The test statistic is $C = (-1)(\#\text{negatives}) + (1)(\#\text{positives}) = (-1)(5) + (1)(10) = 5$.
Using Table 18, Appendix B, with $n = 6$ and $x = 5$, the p-value is $P(x \geq C) = P(x \geq 5) = 0.235$.
Since the p-value is not less than α ($p = 0.235 \not< 0.05$), H_0 is not rejected. There is insufficient evidence to indicate there is a positive relationship between VOF and run time at $\alpha = 0.05$.

15.57 a. The ranks are shown below:

Test	Y1	Rank Y1	X	Rank X	Difference, d_i	d_i^2
A1	4.63	8	12.03	8	0	0
A2	4.32	3	11.32	7	−4	16
A3	4.54	6	9.51	5	1	1
A4	4.09	1	8.25	2	−1	1
A5	4.56	7	9.02	4	3	9
A6	4.48	5	9.97	6	−1	1
A7	4.35	4	8.42	3	1	1
A8	4.23	2	7.53	1	−1	1
					$\sum d_i^2 = 30$	

$$r_s = 1 - \frac{6 \sum d_i^2}{n(n^2 + 1)} = 1 - \frac{6(30)}{8(8^2 - 1)} = 0.643$$

The ranks are shown below:

Test	Y2	Rank Y2	X	Rank X	Difference, d_i	d_i^2
A1	7.17	8	12.03	8	0	0
A2	6.52	4	11.32	7	-3	9
A3	6.31	2	9.51	5	-3	9
A4	6.19	1	8.25	2	-1	1
A5	6.81	6	9.02	4	2	4
A6	6.98	7	9.97	6	1	1
A7	6.45	3	8.42	3	0	0
A8	6.69	5	7.53	1	4	16
						$\sum d_i^2 = 40$

$$r_s = 1 - \frac{6 \sum d_i^2}{n(n^2 - 1)} = 1 - \frac{6(40)}{8(8^2 - 1)} = 0.524$$

The ranks are shown below:

Test	Y3	Rank Y3	X	Rank X	Difference, d_i	d_i^2
A1	385.81	8	12.03	8	0	0
A2	358.44	7	11.32	7	0	0
A3	292.71	5	9.51	5	0	0
A4	253.16	2	8.25	2	0	0
A5	279.82	4	9.02	4	0	0
A6	318.74	6	9.97	6	0	0
A7	262.14	3	8.42	3	0	0
A8	244.97	1	7.53	1	0	0
						$\sum d_i^2 = 0$

$$r_s = 1 - \frac{6 \sum d_i^2}{n(n^2 - 1)} = 1 - \frac{6(0)}{8(8^2 - 1)} = 1.000$$

Thus, Y_1, initial setting time, is the most strongly positively associated with pressure stabilization time with $r_s = 0.643$.

b. Some preliminary calculations are:

Y1	X	Differences, $y_j - y_i (i < j)$	
		# Negatives	# Positives
4.23	7.53		
4.09	8.25	1	0
4.35	8.42	0	2
4.56	9.02	0	3
4.54	9.51	1	3
4.48	9.97	2	3
4.32	11.32	4	2
4.63	12.03	0	7
	Totals	8	20

To determine if a positive relationship exists between initial setting time and pressure stabilization time, we test:

$$H_0 : \beta_1 = 0$$

$$H_a : \beta_1 > 0$$

The test statistic is $C = (-1)(\#\,\text{negatives}) + (1)(\#\,\text{positives}) = (-1)(8) + (1)(20) = 12$. Using Table 18, Appendix B, with $n = 8$ and $x = 12$, the p-value is $P(x \geq C) = P(x \geq 12) = 0.089$.

Since the p-value is not less than $\alpha (p = 0.089 \nless 0.05)$, H_0 is not rejected. There is insufficient evidence to indicate there is a positive relationship between initial setting time and pressure stabilization time at $\alpha = 0.05$.

Some preliminary calculations are:

Y1	X	Differences, $y_j - y_i (i < j)$	
		# Negatives	# Positives
6.69	7.53		
6.19	8.25	1	0
6.45	8.42	1	1
6.81	9.02	0	3
6.31	9.51	3	1
6.98	9.97	0	5
6.52	11.32	3	3
7.17	12.03	0	7
	Totals	8	20

To determine if a positive relationship exists between final setting time and pressure stabilization time, we test:

$$H_0 : \beta_1 = 0$$

$$H_a : \beta_1 > 0$$

The test statistic is $C = (-1)(\#\,negatives) + (1)(\#\,positives) = (-1)(8) + (1)(20) = 12$. Using Table 18, Appendix B, with $n = 8$ and $x = 12$, the p-value is $P(x \geq C) = P(x \geq 12) = 0.089$.
Since the p-value is not less than $\alpha(p = 0.089 \not< 0.05)$, H_0 is not rejected. There is insufficient evidence to indicate there is a positive relationship between final setting time and pressure stabilization time at $\alpha = 0.05$.

Some preliminary calculations are:

		Differences, $y_j - y_i (i < j)$	
Y1	X	# Negatives	# Positives
244.97	7.53		
253.16	8.25	0	1
262.14	8.42	0	2
279.82	9.02	0	3
292.71	9.51	0	4
318.74	9.97	0	5
358.44	11.32	0	6
385.81	12.03	0	7
	Totals	0	28

To determine if a positive relationship exists between maturity index and pressure stabilization time, we test:

$$H_0 : \beta_1 = 0$$

$$H_a : \beta_1 > 0$$

The test statistic is. $C = (-1)(\#\,negatives) + (1)(\#\,positives) = (-1)(0) + (1)(28) = 28$. Using Table 18, Appendix B, with $n = 8$ and $x = 28$, the p-value is $P(x \geq C) = P(x \geq 28) \approx 0.000$.
Since the p-value is less than $\alpha(p = 0.000 \not< 0.05)$, H_0 is rejected. There is sufficient evidence to indicate there is a positive relationship between maturity index and pressure stabilization time at $\alpha = 0.05$.

15.59   Some preliminary calculations are:

Transfer Enhancement, y	Unflooded Area Ratio, x	Differences, $y_j - y_i (i < j)$	
		# Negatives	# Positives
2.9	1.21		
3.2	1.26	0	1
2.8	1.32	2	0
3.5	1.32	0	3
3.7	1.54	0	4
4.1	1.58	0	5
4.2	1.62	0	6
4.5	1.64	0	7
4.9	1.70	0	8
4.7	1.77	1	8
4.5	1.78	2	7
4.6	1.88	2	9
4.9	1.88	0	11
4.4	1.93	6	7
5.3	1.95	0	14
5.2	2.00	1	14
5.1	2.04	2	14
6.1	2.12	0	17
5.2	2.24	2	16
5.3	2.26	1	17
6.7	2.37	0	20
5.8	2.47	2	19
7.0	2.47	0	22
6.0	2.77	3	20
	Totals:	24	249

To determine if a positive relationship exists between heat transfer enhancement and unflooded area ratio, we test:

$$H_0 : \beta_1 = 0$$

$$H_a : \beta_1 > 0$$

The test statistic is $C = (-1)(\# \text{negatives}) + (1)(\# \text{positives}) = (-1)(24) + (1)(249) = 225.$

Using Table 18, Appendix B, with $n = 24$ and $x = 225$, the p-value is $P(x \geq C) = P(x \geq 225) = 0.000.$

Since the p-value is less than $\alpha (p = 0.000 < 0.10)$, H_0 is rejected. There is sufficient evidence to indicate there is a positive relationship between heat transfer enhancement and unflooded area ratio at $\alpha = 0.10$.

15.61 The 36 arrangements along with the corresponding r_s values are:

u	v	u	v	u	v	u	v	u	v	u	v
1	1	1	1	1	2	1	2	1	3	1	3
2	2	2	3	2	1	2	3	2	1	2	2
3	3	3	2	3	3	3	1	3	2	3	1
$r_s = 1$		$r_s = 0.5$		$r_s = 0.5$		$r_s = -0.5$		$r_s = -0.5$		$r_s = -1$	

u	v	u	v	u	v	u	v	u	v	u	v
1	1	1	1	1	2	1	2	1	3	1	3
3	2	3	3	3	1	3	3	3	1	3	2
2	3	2	2	2	3	2	1	2	2	2	1
$r_s = 0.5$		$r_s = 1$		$r_s = -0.5$		$r_s = 0.5$		$r_s = -1$		$r_s = -0.5$	

u	v	u	v	u	v	u	v	u	v	u	v
2	1	2	1	2	2	2	2	2	3	2	3
1	2	1	3	1	1	1	3	1	1	1	2
3	3	3	2	3	3	3	1	3	2	3	1
$r_s = 0.5$		$r_s = -0.5$		$r_s = 1$		$r_s = -1$		$r_s = 0.5$		$r_s = -0.5$	

u	v	u	v	u	v	u	v	u	v	u	v
2	1	2	1	2	2	2	2	2	3	2	3
3	2	3	3	3	1	3	3	3	1	3	2
1	3	1	2	1	3	1	1	1	2	1	1
$r_s = -0.5$		$r_s = 0.5$		$r_s = -1$		$r_s = 1$		$r_s = -0.5$		$r_s = 0.5$	

u	v	u	v	u	v	u	v	u	v	u	v
3	1	3	1	3	2	3	2	3	3	3	3
1	2	1	3	1	1	1	3	1	1	1	2
2	3	2	2	2	3	2	1	2	2	2	1
$r_s = -0.5$		$r_s = -1$		$r_s = 0.5$		$r_s = -0.5$		$r_s = 1$		$r_s = 0.5$	

u	v	u	v	u	v	u	v	u	v	u	v
3	1	3	1	3	2	3	2	3	3	3	3
2	2	2	3	2	1	2	3	2	1	2	2
1	3	1	2	1	3	1	1	1	2	1	1
$r_s = -1$		$r_s = -0.5$		$r_s = -0.5$		$r_s = 0.5$		$r_s = 0.5$		$r_s = 1$	

Since each of the 36 arrangements are equally likely, each has a probability of 1/36.

$$E(r_s) = \sum r_s p(r_s) = 1\left(\frac{1}{36}\right) + 0.5\left(\frac{1}{36}\right) + 0.5\left(\frac{1}{36}\right) - 0.5\left(\frac{1}{36}\right) + \cdots + 1\left(\frac{1}{36}\right) = 0$$

15.63 a. To determine whether the median biting rate is higher in bright sunny weather, we test:

$$H_0 : \tau = 5$$

$$H_a : \tau > 5$$

b. $S = 95$ where S is the number of observations greater than 5. The test statistic is

$$Z = \frac{(S - 0.5) - 0.5n}{0.5\sqrt{n}} = \frac{(95 - 0.5) - 0.5(122)}{0.5\sqrt{122}} = 6.07$$

Using Table 5, Appendix B, the p-value is $p = P(Z > 6.07) \approx 0.0000$.

c. Since the observed p-value is less than $\alpha (p = 0.0000 < 0.01)$, H_0 is rejected. There is sufficient evidence to indicate that the median biting rate in bright, sunny weather is greater than 5 at $\alpha = 0.01$.

15.65   Some preliminary calculations are:

Task	Human	Automated	Diff	Abs Diff	Rank
1	185.4	180.4	5.0	5.0	2
2	146.3	248.5	−102.2	102.2	8
3	174.4	185.5	−11.1	11.1	3
4	184.9	216.4	−31.5	31.5	5
5	240.0	269.3	−29.3	29.3	4
6	253.8	249.6	4.2	4.2	1
7	238.8	282.0	−43.2	43.2	6
8	263.5	315.9	−52.4	52.4	7

The sum of the ranks for the negative differences is $T_- = 33$ and the sum of the ranks of the positive differences is $T_+ = 3$. The test statistic is the smaller of T_- and T_+ which is $T = T_+ = 3$.

To determine if the throughput rates of tasks scheduled by a human differ from those of the automated method, we test:

H_0: The probability distributions of throughput rates of humans and the automated method are identical

H_a: The probability distributions of throughput rates of humans is shifted to the right or left of that for the automated method

The test statistic is $T = 3$.

Using Table 16, Appendix B, with $n = 8$ and $\alpha = 0.01$, $T_0 = 0$. The rejection region is $T \leq 0$.

Since the observed value of the test statistic does not fall in the rejection region $(T = 3 \nleq 0)$, H_0 is not rejected. There is insufficient evidence to indicate that throughput rates of tasks scheduled by a human differ from those of the automated method at $\alpha = 0.01$.

15.67 a. Stem-and-leaf displays for the mercury levels of the three lake types are shown below:

Stem-and-Leaf Display: MERCURY

```
Stem-and-leaf of MERCURY Coded TYPE = 1 N = 21
Leaf Unit = 0.10
```

```
    4    0      0111
   (8)   0      22222233
    9    0      444445
    3    0      7
    2    0      89
```

Stem-and-leaf of MERCURY Coded TYPE = 2 N = 53
Leaf Unit = 0.10

```
   13    0      0001111111111
   24    0      22222233333
   (8)   0      44455555
   21    0      6677777777
   11    0      888999
    5    1      01
    3    1      22
    1    1
    1    1
    1    1
    1    2
    1    2
    1    2      5
```

Stem-and-leaf of MERCURY Coded TYPE = 3 N = 45
Leaf Unit = 0.10

```
    3    0      011
   18    0      222222233333333
   (17)  0      44444444444445555
   10    0      666667
    4    0      89
    2    1      01
```

We see that all of the mercury level distributions exhibit skewed right distributions.

b. The normal assumption necessary for the ANOVA analysis appears to be violated.

c. Using MINITAB, the results are:

Kruskal-Wallis Test: MERCURY versus TYPE

```
Kruskal-Wallis Test on MERCURY

TYPE       N    Median    Ave Rank      Z
1         21    0.3600    48.6       -1.67
2         53    0.4800    63.8        1.08
3         45    0.4100    60.8        0.21
Overall  119              60.0

H = 2.97 DF = 2 P = 0.227
H = 2.97 DF = 2 P = 0.227 (adjusted for ties)
```

To determine if the distributions for the mercury levels differ for the three lake types, we test:

H_0: The three population probability distributions are identical
H_a: At least two of the three population probability distributions differ in location

The test statistic is $H = 2.97$ and the p-value is $p = 0.227$. Since the p-value is not less than $\alpha(p = 0.227 \not< 0.05)$, H_0 is not rejected. There is insufficient evidence to indicate that the locations of the probability distributions for the mercury levels differ for the three lake types at $\alpha = 0.05$.

15.69   Some preliminary calculations are:

		Ranks - Sprays	
Ear	Spray A	Spray B	Spray C
1	2	3	1
2	2	3	1
3	1	3	2
4	3	2	1
5	2	1	3
6	1	3	2
7	2.5	2.5	1
8	2	3	1
9	2	3	1
10	2	3	1
	$T_1 = 19.5$	$T_2 = 26.5$	$T_3 = 14$

To determine if there are differences among the probability distributions of the amounts of aflatoxin present for the three sprays, we test:

H_0: The probability distributions for the three sprays are identical
H_a: At least two of the probability distributions differ in location

The test statistic is

$$F_r = \frac{12}{bk(k+1)} \sum T_i^2 - 3b(k+1)$$

$$= \frac{12}{10(3)(3+1)}(19.5^2 + 26.5^2 + 14^2) - 3(10)(3+1) = 7.85$$

The rejection region requires $\alpha = 0.05$ in the upper tail of the χ^2 distribution with $df = k-1 = 3-1 = 2$. From Table 8, Appendix B, $\chi^2_{0.05} = 5.99147$. The rejection region is $F_r > 5.99147$.

Since the observed value of the test statistic falls in the rejection region ($F_r = 7.85 > 5.99147$), H_0 is rejected. There is sufficient evidence to indicate there are differences among the probability distributions of the amounts of aflatoxin present for the three sprays at $\alpha = 0.05$.

15.71   Some preliminary calculations are:

		Ranks - Computer Programs	
Level	STAAD-III(1)	STAAD-III(2)	Drift
1	3	1.5	1.5
2	3	1	2
3	3	1	2
4	3	1	2
5	3	1	2
	$T_1 = 15$	$T_2 = 5.5$	$T_3 = 9.5$

To determine if the distributions of lateral displacement estimated by the three computer programs differ, we test:

H_0: The probability distributions for the three computer programs are identical

H_a: At least two of the probability distributions differ in location

The test statistic is

$$F_r = \frac{12}{bk(k+1)} \sum T_i^2 - 3b(k+1) = \frac{12}{5(3)(3+1)}(15^2 + 5.5^2 + 9.5^2) - 3(5)(3+1) = 9.1$$

The rejection region requires $\alpha = 0.05$ in the upper tail of the χ^2 distribution with $df = k - 1 = 3 - 1 = 2$. From Table 8, Appendix B, $\chi^2_{0.05} = 5.99147$. The rejection region is $F_r > 5.99147$.

Since the observed value of the test statistic falls in the rejection region ($F_r = 9.1 > 5.99147$), H_0 is rejected. There is sufficient evidence to indicate the distributions of lateral displacement estimated by the three computer programs differ at $\alpha = 0.05$.

15.73   Since both $n_1 > 10$ and $n_2 > 10$, it is appropriate to use the large sample approximation. The data are ranked and the ranks are summed:

Rural	Rank	Urban	Rank
3.5	5	24.0	24
8.1	7	29.0	25
1.8	4	16.0	18
9.0	9	21.0	21
1.6	3	107.0	28
23.0	23	94.0	27
1.5	2	141.0	29
5.3	6	11.0	12.5
9.8	11	11.0	12.5
15.0	17	49.0	26
12.0	14.5	22.0	22
8.2	8	13.0	16
9.7	10	18.0	19.5
1.0	1	12.0	14.5
		18.0	19.5
$n_1 = 14$	$T_1 = 120.5$	$n_2 = 15$	$T_2 = 314.5$

To determine if there is a difference in the PCB levels between rural and urban areas, we test:

H_0: The probability distributions for the rural and urban areas are identical

H_a: The probability distributions for the rural and urban areas differ in location

The test statistic is

$$z = \frac{T_2 - \left[\frac{n_1 n_2 + n_2(n_2 + 1)}{2}\right]}{\sqrt{\frac{n_1 n_2(n_1 + n_2 + 1)}{12}}} = \frac{314.5 - \left[\frac{14(15) + 15(15 + 1)}{2}\right]}{\sqrt{\frac{14(15)(14 + 15 + 1)}{12}}} = 3.906$$

The rejection region requires $\alpha/2 = 0.05/2 = 0.025$ in each tail of the z distribution. From Table 5, Appendix B, $z_{0.025} = 1.96$. The rejection region is $z < -1.96$ or $z < 1.96$.

Since the observed value of the test statistic falls in the rejection region $(Z = 3.906 > 1.96)$, H_0 is rejected. There is sufficient evidence to indicate a difference exists in the PCB levels between the rural and urban areas at $\alpha = 0.05$.

15.75 a. Some preliminary calculations are:

Annealing Time, x	Passivation Potential, y	x-Rank	y-Rank	Difference, di	d_i^2
10	−408	1	1	0	0
20	−400	2	2	0	0
45	−392	3	3	0	0
90	−379	4	5	−1	1
120	−385	5	4	1	1

$$\sum d_i^2 = 2$$

$$r_s = 1 - \frac{6 \sum d_i^2}{n(n^2 - 1)} = 1 - \frac{6(2)}{5(5^2 - 1)} 0.9$$

Since r_s is close to 1, there is a strong positive rank correlation between annealing time and passivation potential.

b. To determine if there is a significant correlation between annealing time and passivation potential, we test:

$$H_0 : \rho = 0$$

$$H_a : \rho \neq 0$$

The test statistic is $r_s = 0.9$.

Using Table 17, Appendix B, with $\alpha = 0.10$ and $n = 5$, $r_{0.10} = 0.9$. The rejection region is $r_s < -0.9$ or $r_s < 0.9$.

Since the observed value of the test statistic does not fall in the rejection region $(r_s = 0.9 \not> 0.9)$, H_0 is not rejected. There is insufficient evidence to indicate that there is a significant correlation between annealing time and passivation potential at $\alpha = 0.10$.

15.77 Some preliminary calculations are:

38°F	Rank	42°F	Rank	46°F	Rank	50°F	Rank
22	16	15	3	14	2	17	6.5
24	18.5	21	14	28	22	18	8.5
16	4.5	26	21	21	14	13	1
18	8.5	16	4.5	19	10.5	20	12
19	10.5	25	20	24	18.5	21	14
		17	6.5	23	17		
$n_1 = 5$	$T_1 = 58$	$n_2 = 6$	$T_2 = 69$	$n_3 = 6$	$T_3 = 84$	$n_4 = 5$	$T_4 = 42$

To determine if the weight increases produced by the four temperatures differ, we test:

H_0: The four population probability distributions are identical

H_a: At least two of the four population probability distributions differ

The test statistic is

$$H = \frac{12}{n(n+1)} \sum \frac{T_i^2}{n_i} - 3(n+1)$$

$$= \frac{12}{22(22+1)} \left(\frac{58^2}{5} + \frac{69^2}{6} + \frac{84^2}{6} + \frac{42^2}{5} \right) - 3(22+1) = 2.03$$

The rejection region requires $\alpha = 0.10$ in the upper tail of the χ^2 distribution with $df = k - 1 = 4 - 1 = 3$. From Table 8, Appendix B, $\chi^2_{0.10} = 6.25139$. The rejection region is $H > 6.25139$.

Since the observed value of the test statistic does not fall in the rejection region ($H = 2.03 \not> 6.25139$), H_0 is not rejected. There is insufficient evidence to indicate that the weight increases produced by the four temperatures differ at $\alpha = 0.10$.

15.79 a. To determine if the median TCDD level in fat tissue exceeds 3 ppt, we test:

$$H_0 : \tau = 3$$

$$H_a : \tau > 3$$

The test statistic is then number of observations greater than 3 which is $S = 14$. Using Table 2, Appendix B, with $n = 20$ and $p = 0.5$, $p = P(x \geq 14) = 1 - P(x \leq 13) = 1 - 0.9423 = 0.0577$.

Since the p-value is not less than $\alpha(p = 0.0577 \not< 0.05)$, H_0 is not rejected. There is insufficient evidence to indicate the median TCDD level in fat tissue exceeds 3 ppt at $\alpha = 0.05$.

b. To determine if the median TCDD level in plasma exceeds 3 ppt, we test:

$$H_0 : \tau = 3$$

$$H_a : \tau > 3$$

The test statistic is the number of observations greater than 3 which is S = 12. Using MINITAB, with $n = 19$ and $p = 0.5, p = P(x \geq 12) = 1 - P(x \leq 11) = 1 - 0.8204 = 0.1796 \cdot$ Since the *p*-value is not less than $\alpha(p = 0.1796 \not< 0.05)$, H_0 is not rejected. There is insufficient evidence to indicate the median TCDD level in plasma exceeds 3 ppt at $\alpha = 0.05$.

c. Some preliminary calculations are:

Vet	Fat	Plasma	Diff	ABS	Rank
1	4.9	2.5	2.4	2.4	11.5
2	6.9	3.5	3.4	3.4	16
3	10.0	6.8	3.2	3.2	15
4	4.4	4.7	-0.3	0.3	4
5	4.6	4.6	0.0	0.0	–
6	1.1	1.8	-0.7	0.7	8
7	2.3	2.5	-0.2	0.2	2.5
8	5.9	3.1	2.8	2.8	14
9	7.0	3.1	3.9	3.9	17
10	5.5	3.0	2.5	2.5	13
11	7.0	6.9	0.1	0.1	1
12	1.4	1.6	-0.2	0.2	2.5
13	11.0	20.0	-9.0	9.0	19
14	2.5	4.1	-1.6	1.6	9
15	4.4	2.1	2.3	2.3	10
16	4.2	1.8	2.4	2.4	11.5
17	41.0	36.0	5.0	5.0	18
18	2.9	3.3	-0.4	0.4	5
19	7.7	7.2	0.5	0.5	6.5
20	2.5	2.0	0.5	0.5	6.5

The sum of the ranks for the negative differences is $T_- = 50$ and the sum of the ranks of the positive differences is $T_+ = 14$. The test statistic is the smaller of T_- and T_+ which is $T = T_- = 50$.

To determine the relationship of the distributions TCDD in fat tissue and plasma differ in location, we test:

H_0: The two populations have identical probability distributions
H_a: The distribution of TCDD in fat tissue is shifted to the right or left of that for plasma
The test statistic is $T = 50$.

Using Table 16, Appendix B, with $n = 19$ and $\alpha = 0.05$, $T_0 = 46$. The rejection region is $T \leq 46$.

Since the observed value of the test statistic does not fall in the rejection region $(T = 50 \not\leq 46)$, H_0 is not rejected. There is insufficient evidence to indicate that the distribution of TCDD in fat tissue is shifted to the right or left of that for plasma at $\alpha = 0.05$.

d. Using MINITAB, the Spearman rank correlation on the ranks of the variables is:

Correlation: R-Fat, R-Plasma

```
Pearson correlation of R-Fat and R-Plasma = 0.774
```

To determine if the TCDD levels in fat tissue and plasma are positively rank correlated, we test:

$$H_0 : \rho = 0$$

$$H_a : \rho > 0$$

The test statistic is $r_s = 0.774$.

Using Table 17, Appendix B, with $\alpha = 0.05$ and $n = 20$, $r_{0.05} = 0.377$. The rejection region is $r_s > 0.377$.

Since the observed value of the test statistic falls in the rejection region ($r_s = 0.774 > 0.377$), H_0 is rejected. There is sufficient evidence to indicate at $\alpha = 0.05$.

16

Statistical Process and Quality Control

16.1   a.   The center line is $\bar{x} = \dfrac{\sum x}{n} = \dfrac{2,744}{12} = 228.67.$

    b. $s^2 = \dfrac{\sum x^2 - \dfrac{\left(\sum x\right)^2}{n}}{n-1} = \dfrac{822,542 - \dfrac{2,744^2}{12}}{12-1} = 17,734.60606$

       $s = \sqrt{17,734.60606} = 133.17$

       The lower and upper control limits are:

       $\text{LCL} = \bar{x} - 3s = 228.67 - 3(133.17) = -170.84$

       $\text{UCL} = \bar{x} + 3s = 228.67 + 3(133.17) = 628.18$

  c.   Since all of the number updates fall between the lower and upper control limits, the process appears to be in control.

16.3   **Site 1**

    The center line is $\bar{x} = \dfrac{\sum x}{n} = \dfrac{2,238.7}{25} = 89.548.$

    $s^2 = \dfrac{\sum x^2 - \dfrac{\left(\sum x\right)^2}{n}}{n-1} = \dfrac{200,696.8552 - \dfrac{2,238.7^2}{25}}{25-1} = 9,40615$

    $s = \sqrt{9.40615} = 3.0669$

    The lower and upper control limits are:

    $\text{LCL} = \bar{x} - 3s = 89.548 - 3(3.0669) = 80.3473$

    $\text{UCL} = \bar{x} + 3s = 89.548 + 3(0.0669) = 98.7487$

The variable control chart is:

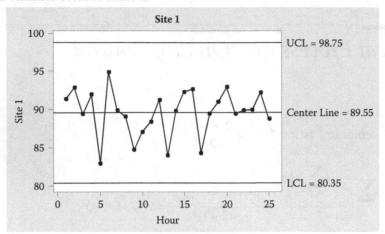

Since all of the drug concentrations fall between the lower and upper control limits, the process appears in control.

Site 2

The center line is $\bar{x} = \dfrac{\sum x}{n} = \dfrac{2,225.83}{25} = 89.0332$.

$$s^2 = \frac{\sum x^2 - \dfrac{\left(\sum x\right)^2}{n}}{n-1} = \frac{198,440.2889 - \dfrac{2,225.83^2}{25}}{25-1} = 11.1467$$

$s = \sqrt{11.1467} = 3.3387$

The lower and upper control limits are:

$\text{LCL} = \bar{x} - 3s = 89.0332 - 3(3.3387) = 79.0171$

$\text{UCL} = \bar{x} + 3s = 89.0332 + 3(3.3387) = 99.0493$

The variable control chart is:

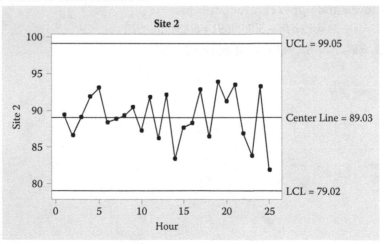

Since all of the drug concentrations fall between the lower and upper control limits, the process appears in control.

16.5 a. The center line is $\bar{x} = \dfrac{\sum x}{n} = \dfrac{117.9}{20} = 5.895$.

$$s^2 = \frac{\sum x^2 - \dfrac{\left(\sum x\right)^2}{n}}{n-1} = \frac{697.39 - \dfrac{117.9^2}{20}}{20-1} = 0.12471$$

$s = \sqrt{0.12471} = 0.3531$

The lower and upper control limits are:

LCL $= \bar{x} - 3s = 5.895 - 3(0.3531) = 4.8357$

UCL $= \bar{x} + 3s = 5.895 + 3(0.3531) = 6.9543$

The variable control chart is:

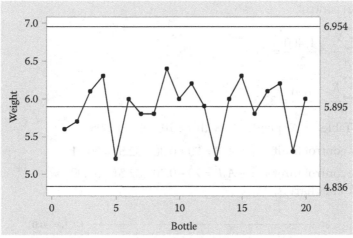

Since all weights fall between the lower and upper control limits, the process appears in control.

16.7 a&b. The center line is $\bar{x} = \dfrac{\sum x}{n} = \dfrac{3.773}{27} = 0.13974$.

$$s^2 = \frac{\sum x^2 - \dfrac{\left(\sum x\right)^2}{n}}{n-1} = \frac{0.527447 - \dfrac{3.773^2}{27}}{27-1} = 0.0000079$$

$s = \sqrt{0.0000079} = 0.0028092$

The lower and upper control limits are:

LCL $= \bar{x} - 3s = 0.13974 - 3(0.00281) = 0.1313$

UCL $= \bar{x} + 3s = 0.13974 + 3(0.00281) = 0.1482$

The variable control chart is:

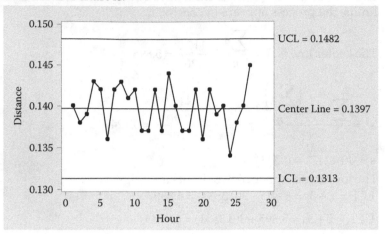

c. Since all of the measurements fall inside the control limits, the process appears to be in control.

16.9 a. $\bar{\bar{x}} = \dfrac{\sum \bar{x}_i}{n} = \dfrac{1,400}{20} = 70$

b. $\bar{R} = \dfrac{\sum R_i}{n} = \dfrac{650}{20} = 32.5$

c. From Table 19, Appendix B, with $n = 10$, $A_2 = 0.308$.

Upper control limit $= \bar{\bar{x}} + A_2 \bar{R} = 70 + 0.308(32.5) = 80.01$

Lower control limit $= \bar{\bar{x}} - A_2 \bar{R} = 70 - 0.308(32.5) = 59.99$

d. The $\bar{x}$-bar chart is:

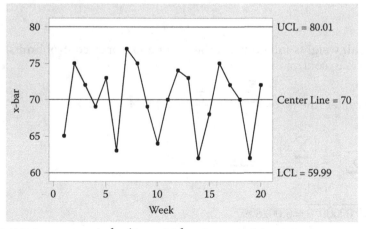

The process appears to be in control.

e. The new plot is:

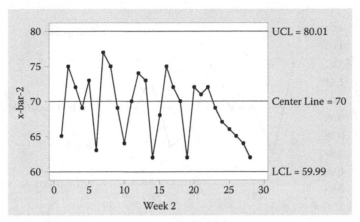

No. All of the new measurements are within the control limits.

16.11   For each of the sampling times, we will compute the sample mean pin lengths and the sample ranges.

30-minute Interval	Sample Mean $\bar{x}_i$	Range R_i
1	1.020	0.06
2	0.970	0.08
3	0.996	0.04
4	0.994	0.05
5	1.020	0.09
6	0.984	0.14
7	0.974	0.09
8	1.012	0.08
9	0.988	0.04
10	1.000	0.07

a.   The center line is $\bar{\bar{x}} = \dfrac{\sum \bar{x}_i}{n} = \dfrac{9.958}{10} = 0.9958$.

b. $\bar{R} = \dfrac{\sum R_i}{n} = \dfrac{0.740}{10} = 0.074$

From Table 19, Appendix B, with $n = 5$, $A_2 = 0.577$.

Upper control limit $= \bar{\bar{x}} + A_2\bar{R} = 0.9958 + 0.577(0.074) = 1.0385$

Lower control limit $= \bar{\bar{x}} - A_2\bar{R} = 0.9958 - 0.577(0.074) = 0.9531$

c.   The $\bar{x}$-chart is:

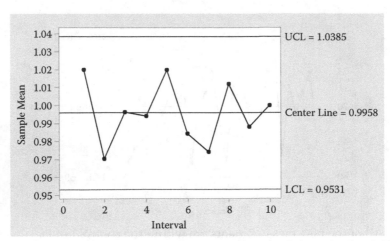

d. The Defense Department's specification is that the firing pins be 1.00 ± 0.08 or from 0.92 to 1.08. Since all of the sample means are well within this range, the manufacturing process appears to be in control.

16.13 a. $P(\bar{x} > UCL) = P(\bar{x} > \mu + 3\sigma_{\bar{x}}) = P(Z > 3) = 0.5 - 0.4987 = 0.0013$ (Using Table 5, Appendix B)

b. $P(Y = 3) = \left(\begin{array}{c} 67 \\ 3 \end{array} \right) 0.0013^3 (0.9987)^{67-3} = 0.0000968$

16.15 a&b. For each hour, the sample mean and sample range are computed:

Hour	Sample Mean $\bar{x}_i$	Range R_i	Hour	Sample Mean $\bar{x}_i$	Range R_i
1	0.1378	0.009	15	0.1418	0.010
2	0.1430	0.008	16	0.1404	0.013
3	0.1412	0.015	17	0.1400	0.006
4	0.1398	0.006	18	0.1418	0.008
5	0.1400	0.010	19	0.1404	0.008
6	0.1392	0.008	20	0.1388	0.006
7	0.1412	0.010	21	0.1414	0.006
8	0.1400	0.008	22	0.1414	0.007
9	0.1420	0.007	23	0.1406	0.008
10	0.1392	0.013	24	0.1414	0.013
11	0.1396	0.012	25	0.1404	0.008
12	0.1414	0.009	26	0.1420	0.007
13	0.1412	0.003	27	0.1410	0.008
14	0.1406	0.008			

The center line is $\bar{\bar{x}} = \dfrac{\sum \bar{x}_i}{n} = \dfrac{3.7976}{27} = 0.14065.$

$\bar{R} = \dfrac{\sum R_i}{n} = \dfrac{0.234}{27} = 0.00867$

From Table 19, Appendix B, with $n = 5$, $A_2 = 0.577$.

Upper control limit $= \bar{\bar{x}} + A_2\bar{R} = 0.14065 + 0.577(0.00867) = 0.14565$

Lower control limit $= \bar{\bar{x}} - A_2\bar{R} = 0.14065 - 0.577(0.00867) = 0.13565$

The $\bar{x}$-chart is:

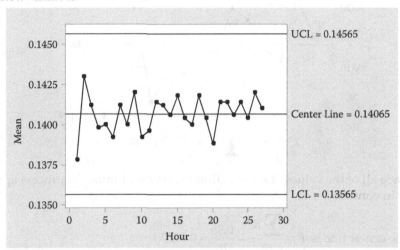

c. The process appears to be in control because all sample means are between the upper and lower control limits.

16.17 a. From Table 19, Appendix B, with $n = 4$, $D_3 = 0$ and $D_4 = 2.282$.

Upper control limit $= D_4\bar{R} = 2.282(0.335) = 0.7645$

Lower control limit $= D_3\bar{R} = 0(0.335) = 0$

b. All of the values of R are within the upper and lower control limits, so the process appears to be in control. There are no trends in the data to indicate the presence of special causes of variation.

c. Yes. This process appears to be in control. Therefore, these control limits could be used to monitor future output.

d. Of the 30 R values plotted, there are only 6 different values. Most of the R values take on one of three values. This indicates that the data must be discrete (take on a countable number of values), or that the path widths are multiples of each other.

16.19 a. The center line is $\bar{R} = 14.87$

From Table 19, Appendix B, with $n = 5$, $D_3 = 0$ and $D_4 = 2.115$.

Upper control limit $= D_4\bar{R} = 2.115(14.87) = 31.45$

Lower control limit $= D_3\bar{R} = 0(14.87) = 0$

b.   The range would be $R = 24.7 - 2.2 = 22.5$. Since this value is within the control limits, then the station manager should do nothing.

16.21   From Exercise 6.12, $\bar{R} = 0.1392$ and $n = 8$.

From Table 19, Appendix B, with $n = 8$, $D_3 = 0.136$ and $D_4 = 1.864$.

Upper control limit $= D_4\bar{R} = 1.864(0.1392) = 0.2596$

Lower control limit $= D_3\bar{R} = 0.136(0.1392) = 0.0189$

The R-chart is:

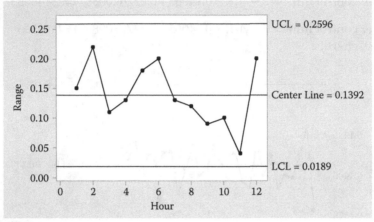

Since all of the values of R are within the control limits, the process appears to be in control.

16.23 a.   The center line is $\bar{R} = \dfrac{\sum R_i}{n} = \dfrac{0.38}{16} = 0.02375$.

From Table 19, Appendix B, with $n = 2$, $D_3 = 0$ and $D_4 = 3.276$.

Upper control limit $= D_4\bar{R} = 3.276(0.02375) = 0.0778$

Lower control limit $= D_3\bar{R} = 0(0.07958) = 0$

The R-chart is:

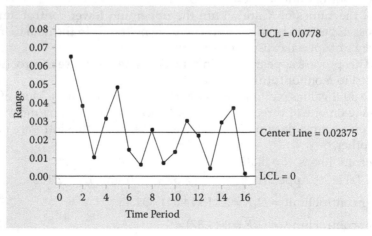

b. The center line is $\bar{x} = \dfrac{\sum \bar{x}_i}{n} = \dfrac{3.561}{16} = 0.22256$.

From Table 19, Appendix B, with $n = 2$, $A_2 = 1.880$.

Upper control limit $= \bar{x} + A_2\bar{R} = 0.22256 + 1.880(0.02375) = 0.26721$

Lower control limit $= \bar{x} - A_2\bar{R} = 0.22256 - 1.880(0.02375) = 0.17791$

The $\bar{x}$ -chart is:

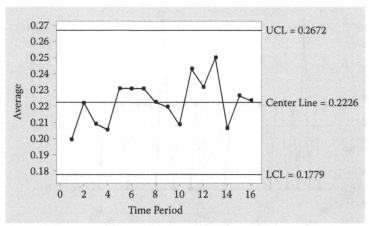

c. For the R-chart, all values of R are within the control limits. Thus, the process appears to be in control. For the $\bar{x}$ -chart, all of the averages are within the control limits. Thus, the process appears to be in control. An estimate of the true average thickness of the expensive layer is $\bar{x} = 0.22256$. .

16.25 a. From Exercise 16.16 **a**, $\bar{R} = 0.8065$.

From Table 19, Appendix B, with $n = 3$, $D_3 = 0$ and $D_4 = 2.575$.

Upper control limit $= D_4\bar{R} = 2.575(0.8065) = 2.07674$

Lower control limit $= D_3\bar{R} = 0(0.8065) = 0$

The R-chart is:

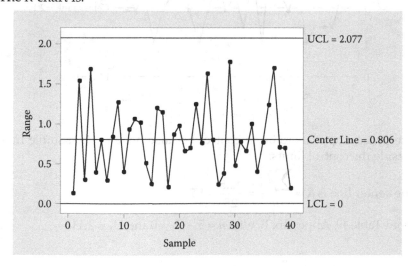

The process appears to be in control because none of the sample ranges are outside the control limits.

b.   From Exercise 16.16 **b**, $\bar{R} = 0.75$.

From Table 19, Appendix B, with $n = 3$, $D_3 = 0$ and $D_4 = 2.575$.

Upper control limit $= D_4\bar{R} = 2.575(0.75) = 1.93125$

Lower control limit $= D_3\bar{R} = 0(0.75) = 0$

The R-chart is:

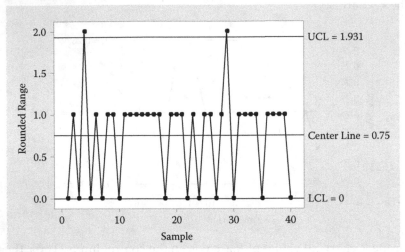

The process does not appear to be in control because some of the sample ranges are outside the control limits.

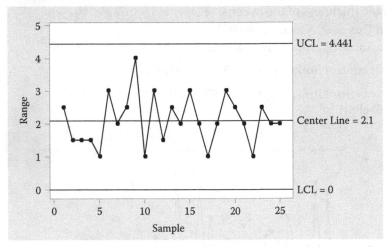

The process appears to be in control because none of the sample ranges are outside the control limits.

b.   The center line is $\bar{R} = \dfrac{\sum R_i}{n} = \dfrac{42.5}{25} = 1.7$.

From Table 19, Appendix B, with $n = 5$, $D_3 = 0$ and $D_4 = 2.115$.

Upper control limit $= D_4\bar{R} = 2.115(1.7) = 3.5955$

Lower control limit $= D_3\bar{R} = 0(1.7) = 0$

The R-chart is:

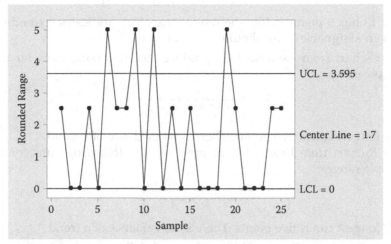

The process does not appear to be in control because some of the sample ranges are outside the control limits.

d. We get two different answers as to whether this process is in control, depending on theaccuracy of the data. When the data were measured to an accuracy of .5 gram, theprocess appears to be in control. However, when the data were measured to an accuracy of only 2.5 grams, the process appears to be out of control. The same data were used for each chart – just measured to different accuracies.

16.27 a. There are 3 runs of lengths 2, 5, and 4. There are no apparent trends.

    b. There are 6 runs of lengths 1, 1, 3, 2, 1, and 4. There are no apparent trends.

    c. There are 3 runs of lengths 5, 5, and 2. There are no apparent trends.

    d. There are 4 runs of lengths 1, 5, 1, and 8. There are 8 consecutive points on the same side of the center line. This indicates a trend is present.

    e. There are 6 runs of lengths 1, 1, 3, 2, 2, and 1. There are no apparent trends.

    f. There are 3 runs of lengths 1, 9, and 1. There are 9 consecutive points on the same side of the center line. This indicates a trend is present.

16.29 The $\bar{x}$-chart from Exercise 16.9 produces the following runs for the $k = 20$ sample means:

$$\underbrace{-}\underbrace{++}\underbrace{-}\underbrace{+}\underbrace{-}\underbrace{++}\underbrace{--}0\underbrace{++}\underbrace{--}\underbrace{++}0\underbrace{-}\underbrace{+}$$
$$\;1\;\;\;\;2\;\;\;3\;4\;5\;\;\;6\;\;\;\;\;7\;\;\;8\;\;9\;\;\;\;10\;\;\;11\;\;\;12\,13\,14$$

The longest run is two points. There is no evidence of a trend.

The original R-chart from Exercise 16.18 produces the following runs for the $k = 20$ sample ranges:

$$\underbrace{-}\underbrace{+}\underbrace{-}\underbrace{+++}\underbrace{+}\underbrace{--}\underbrace{+++}\underbrace{-}\underbrace{+}\underbrace{-}\underbrace{+}\underbrace{--}\underbrace{+}\underbrace{-}$$
$$\;1\;2\;3\;\;\;\;\;4\;\;\;\;\;\;5\;\;\;\;\;\;6\;\;\;\;7\;8\;9\;10\;11\;\;12\,13$$

The longest run is four points. There is no evidence of a trend.

The extended R-chart from Exercise 16.18 produces the following runs for the $k = 28$ sample ranges:

$$-+-+++++--+++-+-+-+---+---------$$
1 2 3   4   5   6   7 8 9 10 11 12   13

Run 13 has 9 points below the range mean. This indicates a trend is present and an assignable cause should be searched for.

16.31   The $\bar{x}$ -chart from Exercise 16.12 produces the following runs for the $k = 12$ sample means:

$$+-+-+-++-++$$
1 3 3 4 5 6 7 9 10 11

The longest run is two points. There is no evidence of a trend.
The R-chart from Exercise 16.21 produces the following runs for the $k = 12$ sample ranges:

$$++--++------+$$
1   2   3   4   5

The longest run is five points. There is no evidence of a trend.

16.33   The $\bar{x}$-chart from Exercise 16.16 produces the following runs for the $k = 40$ sample means:

$$---+----++-+++++-++--+-+--++-+--+--+++-$$
1   2   3   4   5   6   7 8   9 10 11 12 13   14 15 16 17 18 19   20   21

The longest run is five points. There is no evidence of a trend.
The R-chart from Exercise 16.25 produces the following runs for the $k = 40$ sample ranges:

$$-+-+---+-+-++-+-++--+-+-+---+-+---+---++---$$
1 2 3 4   5   6 7   8   9   10 11 12   13 14 15 16   17   18   19   20 21 22   23

The longest run is three points. There is no evidence of a trend.

16.35 a.   Upper control limit $= \bar{p} + 3\sqrt{\dfrac{\bar{p}(1-\bar{p})}{n}} = 0.0105 + 3\sqrt{\dfrac{0.0105(1-0105)}{1000}} = 0.0202$

Lower control limit $= \bar{p} + 3\sqrt{\dfrac{\bar{p}(1-\bar{p})}{n}} = 0.0105 - 3\sqrt{\dfrac{0.0105(1-0.0105)}{1000}} = 0.0008$

b.   Since all of the observations are within the control limits, the process appears to be in control.

c.   The rational subgrouping strategy says that samples should be chosen so that it gives the maximum chance for the measurements in each sample to be similar and so that it gives the maximum chance for the samples to differ. By selecting 1000 consecutive chips each time, this gives the maximum chance for the measurements in the sample to be similar. By selecting the samples every other day, there is a relatively large chance that the samples differ.

16.37   The sample proportions of leaks is computed for each week and are shown in the table:

Week	Number	Leaks	$\hat{p}_j$
1	500	36	0.072
2	500	28	0.056
3	500	24	0.048
4	500	26	0.052
5	500	20	0.04
6	500	56	0.112
7	500	26	0.052
8	500	28	0.056
9	500	31	0.062
10	500	26	0.052
11	500	34	0.068
12	500	26	0.052
13	500	32	0.064

$$\bar{p} = \frac{\text{Total number of leaks}}{\text{Total number of pumps sampled}} = \frac{393}{6,500} = 0.06046$$

Upper control limit

$$= \bar{p} + 3\sqrt{\frac{\bar{p}(1-\bar{p})}{n}} = 0.06046 + 3\sqrt{\frac{0.06046(1-0.06046)}{500}} = 0.09244$$

Lower control limit

$$= \bar{p} - 3\sqrt{\frac{\bar{p}(1-\bar{p})}{n}} = 0.06046 + 3\sqrt{\frac{0.06046(1-0.06046)}{500}} = 0.02848$$

The *p*-chart is:

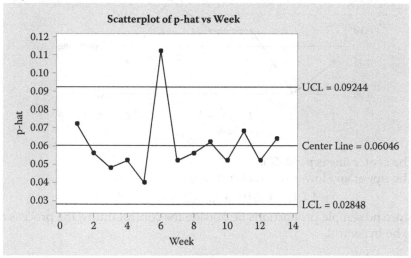

Since one sample proportion falls outside the upper control limit, it appears that the process is out of control.

16.39 a. To compute the proportion of defectives in each sample, divide the number of defectives by the number in the sample, 20: $\hat{p}_i = \frac{Y}{n}$. The sample proportions are listed in the table:

Week	Number	Defectives	$\hat{p}_j$
1	20	1	0.0500
2	20	0	0.0000
3	20	2	0.1000
4	20	2	0.1000
5	20	3	0.1500
6	20	1	0.0500

$$\bar{p} = \frac{\text{Total number of defectives}}{\text{Total number of units sampled}} = \frac{9}{120} = 0.075$$

$$\text{Upper control limit} = \bar{p} + 3\sqrt{\frac{\bar{p}(1-\bar{p})}{n}} = 0.075 + 3\sqrt{\frac{0.075(1-0.075)}{20}} = 0.25169$$

$$\text{Lower control limit} = \bar{p} - 3\sqrt{\frac{\bar{p}(1-\bar{p})}{n}} = 0.075 - 3\sqrt{\frac{0.075(1-0.075)}{20}} = -0.10169 \approx 0$$

The *p*-chart is:

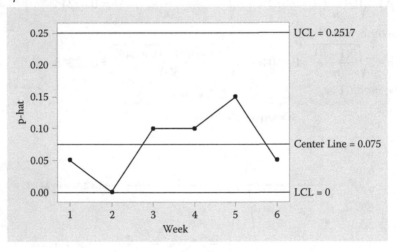

b. The center line is $\bar{p} = 0.075$.

c. The upper and lower control limits are:

$$\text{UCL} = 0.25169 \text{ and } \text{LCL} = -0.10169 \approx 0$$

Since no sample proportions lie outside the control limits, the process appears to be in control.

16.41 a. To compute the proportion of defectives in each sample, divide the number of defectives by the number in the sample, 50: $\hat{p}_i = \frac{Y}{n}$, The sample proportions are listed in the table:

Day	Number	Defectives	$\hat{p}_i$	Day	Number	Defectives	$\hat{p}_i$
1	50	11	0.22	12	50	23	0.46
2	50	15	0.30	13	50	15	0.30
3	50	12	0.24	14	50	12	0.24
4	50	10	0.20	15	50	11	0.22
5	50	9	0.18	16	50	11	0.22
6	50	12	0.24	17	50	16	0.32
7	50	12	0.24	18	50	15	0.30
8	50	14	0.28	19	50	10	0.20
9	50	9	0.18	20	50	13	0.26
10	50	13	0.26	21	50	12	0.24
11	50	15	0.30				

$$\bar{p} = \frac{\text{Total number of defectives}}{\text{Total number of units sampled}} = \frac{270}{1,050} = 0.2571$$

$$\text{Upper control limit} = \bar{p} + 3\sqrt{\frac{\bar{p}(1-\bar{p})}{n}} = 0.2571 + 3\sqrt{\frac{0.2571(1-0.2571)}{50}} = 0.4425$$

$$\text{Lower control limit} = \bar{p} + 3\sqrt{\frac{\bar{p}(1-\bar{p})}{n}} = 0.2571 - 3\sqrt{\frac{0.2571(1-0.2571)}{50}} = 0.0717$$

The *p*-chart is:

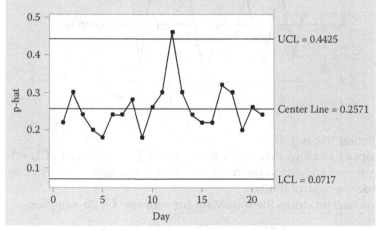

b. The center line is $\bar{p} = 0.2571$.
c. The upper and lower control limits are:

$$\text{UCL} = 0.4425 \text{ and LCL} = 0.0717$$

d. The process does not appear to be in control as sample proportion 12 falls out-
    side the control limits. Dropping this sample out and recalculating the control
    limits, we get:

$$\bar{p} = \frac{\text{Total number of defectives}}{\text{Total number of units sampled}} = \frac{270}{1,050} = 0.247$$

$$\text{Upper control limit} = \bar{p} + 3\sqrt{\frac{\bar{p}(1-\bar{p})}{n}} = 0.247 + 3\sqrt{\frac{0.247(1-0.247)}{50}} = 0.430$$

$$\text{Lower control limit} = \bar{p} - 3\sqrt{\frac{\bar{p}(1-\bar{p})}{n}} = 0.247 - 3\sqrt{\frac{0.247(1-0.247)}{50}} = 0.064$$

e. The p-chart produces the following runs for the $k = 21$ sample proportions:

    $$-+-----+-+++++---++-+-$$
    $$1\ 2 \quad\quad 3 \quad\quad 4\ 5 \quad 6 \quad\quad 7 \quad 8\ 9\ 10\ 11$$

    The longest run is five points. There is no evidence of a trend.

16.43 a.   The center line is $\bar{c} = \dfrac{\sum c_i}{n} = \dfrac{130}{20} = 6.5$.

Upper control limit $= \bar{c} + 3\sqrt{\bar{c}} = 6.5 + 3\sqrt{6.5} = 14.149$

Lower control limit $= \bar{c} + 3\sqrt{\bar{c}} = 6.5 - 3\sqrt{6.5} = -1.149 \approx 0$
The c-chart is:

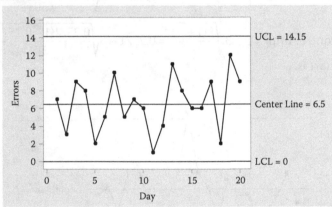

b. The center line is $\bar{c} = 6.5$.
c. The upper and lower control limits are: UCL $= 14.149$ and LCL $= 0$
    The process appears to be in control since no sample number of defectives lie
    outside the control limits.
d. The c-chart produces the following runs for the $k = 20$ samples:

    $$+-++--+-+-+---++--+-++$$
    $$1\ 2\ 3 \quad 4\ 5\ 6\ 7 \quad 8 \quad 9 \quad 10\ 11\ 12\ 13$$

    The longest run is three points. There is no evidence of a trend.

16.45 a.   The center line is $\bar{c} = \dfrac{\sum c_i}{n} = \dfrac{436}{50} = 8.72$.

Upper control limit $= \bar{c} + 3\sqrt{\bar{c}} = 8.72 + 3\sqrt{8.72} = 17.579$

Lower control limit $= \bar{c} - 3\sqrt{\bar{c}} = 8.72 - 3\sqrt{8.72} = -0.139 \approx 0$
The c-chart is:

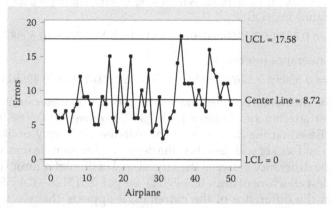

Since the number of errors for aircraft 36 falls outside the control limits, the process appears to be out of control.

b.   The c-chart produces the following runs for the $k = 50$ samples:

The longest run is seven points. There is evidence of a trend.

16.47 a. $\bar{x} = \dfrac{\sum x_i}{n} = \dfrac{4,340.9}{80} = 54.26125$

$s^2 = \dfrac{\sum x^2 - \dfrac{\left(\sum x\right)^2}{n}}{n-1} = \dfrac{235,679.03 - \dfrac{4,340.9^2}{80}}{80-1} = 1.7262$

$s = \sqrt{1.7262} = 1.3138$
The tolerance interval is $\bar{x} \pm Ks$.
From Table 20, Appendix B, with $n = 80, \alpha = 0.05,$ and $\gamma = 0.99, K = 2.986$. The tolerance interval is:

$\bar{x} \pm Ks \Rightarrow 54.26125 \pm 2.986(1.3138) \Rightarrow 54.26125 \pm 3.9230 \Rightarrow (50.3382, 58.1543)$

b.   The specification interval is $54 \pm 4 \Rightarrow (50, 58)$. The lower limit of the specification interval is just smaller than the lower tolerance limit. The upper limit of the specification interval is just lower than the upper tolerance limit. The specifications are not being met.

c.   From Table 21, Appendix B, with $\alpha = 0.05$ and $\gamma = 0.99, n = 473$. Since $n = 80$ in this example, we cannot form the tolerance interval.

16.49 a.   Since μ and σ are known, the tolerance interval for 99% of the complaint rates is:

$$\mu \pm z_{0.005}\sigma \Rightarrow 26 \pm 2.576(11.3) \Rightarrow 26 \pm 29.1088 \Rightarrow (-3.1088, 55.1088)$$

This is a 100% tolerance interval. Since μ and σ are known and the population is assumed normal, we know that 99% of the data will fall within 2.576 standard deviations of the mean.

16.51 a.   The 95% tolerance interval for 99% of the weights is $\bar{x} \pm Ks$.
From Table 20, Appendix B, with $n = 50, \alpha = 0.05$ and $\gamma = 0.99, K = 3.126$. The 95% tolerance interval is:

$$\bar{x} \pm Ks \Rightarrow 0.5120 \pm 3.126(0.0010) \Rightarrow 0.5120 \pm 0.003126 \Rightarrow (0.508874, 0.515126)$$

b.   The 95% tolerance interval is:

$$\bar{x} \pm Ks \Rightarrow 0.5005 \pm 3.126(0.0015) \Rightarrow 0.5005 \pm 0.004689 \Rightarrow (0.495811, 0.505189)$$

c.   Taking the difference between the smallest socket diameter limit (0.508874) and the largest attachment diameter limit, (0.505189), we get $0.508874 - 0.505189 = 0.003685$. Based on the difference of the extremes, it is very unlikely to find an extension and socket with less than the desired minimum clearance of 0.004 inch.

d.   Taking the difference between the largest socket diameter limit (0.515126) and the smallest attachment limit (0.495811), we get $0.515126 - 0.495811 = 0.019315$. Based on the difference of the extremes, it appears that we will find some attachment and socket pairs that have a maximum clearance of more than the required 0.015.

e.   Let $y_1 = $ socket diameter and $y_2 = $ attachment diameter. The socket diameters are approximately normally distributed with a mean μ_1 estimated to be 0.512 and standard deviation σ_1 estimated to be 0.001. The attachment diameters are approximately normally distributed with a mean μ_2 estimated to be 0.5005 and a standard deviation σ_2 estimated to be 0.0015. Therefore, the distribution of $y_1 - y_2$ is approximately normally distributed with mean $\mu_1 = \mu_2$ estimated to be $0.512 - 0.5005 = 0.0115$ and variance $\sigma_1^2 + \sigma_2^2$ estimated to be $0.001^2 + 0.0015^2 = 0.00000325$.

$$P(y_1 - y_2 > 0.015) = P\left(Z > \frac{0.015 - 0.0115}{\sqrt{0.00000325}}\right) = P(Z > 1.94)$$

$$= 0.5 - P(0 < Z < 1.94) = 0.5 - 0.4738 = 0.$$

(Using Table 5, Appendix B)

16.53 a.   If the output distribution is normal with a mean of 1000 and a standard deviation of 100, then the proportion of the output that is unacceptable is:

$$P(X < 980) + P(X > 1,020) = P\left(Z < \frac{980 - 1,000}{100}\right) + P\left(Z > \frac{1,020 - 1,000}{100}\right)$$

$$= P(Z < -0.2) + P(Z > 0.2) = (0.5 - 0.0793) + (0.5 - 0.0793) = 0.8414$$

(Using Table 5, Appendix B)

b. $C_p = \dfrac{\text{USL-LSL}}{6\sigma} \approx \dfrac{1,020 - 980}{6(100)} = \dfrac{40}{600} = 0.0667$

Since the value of C_p is less than 1, the process is not capable.

16.55 a. Using MINITAB, the capability analysis diagram is:

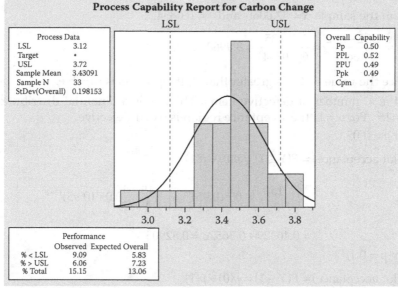

b. Five of the observations are outside the specification limits. Thus, the percentage is $(5/33) \times 100\% = 15.15\%$.

c. From the sample, $s = 0.1982$.

$$C_p = \frac{\text{USL} - \text{LSL}}{6\sigma} = \frac{3.72 - 3.12}{6(0.1982)} = 0.5045$$

Since the value of C_p is less than 1, the process is not capable.

16.57 a. Using MINITAB, the capability analysis diagram is:

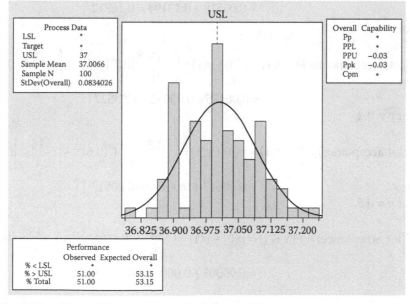

The LSL = 35 is off the chart to the left.

b. Fifty-one of the observations are above the upper specification limit. Thus, the percentage is $(51/100) \times 100\% = 51\%$.

c. From the sample $\bar{x} = 37.0066$ and $s = 0.0834$.

$$C_p = \frac{USL - LSL}{6\sigma} = \frac{37 - 35}{6(0.0834)} = 3.9968$$

d. Since the value of C_p is greater than 1, the process is capable.

16.59 a. Let $y =$ number of defective items. Then y is a binomial distribution with $n = 15$. For $\alpha = 1$, the acceptance region is 0 or 1 defective.

For $p = 0.05$,

$$P(\text{lot acceptance}) = P(Y \le 1) = p(0) + p(1)$$

$$= \binom{15}{0} 0.05^0 (0.95)^{15} + \binom{15}{1} 0.05^1 (0.95)^{15-1}$$

$$= 0.46329 + 0.36576 = 0.82905$$

For $p = 0.1$,

$$P(\text{lot acceptance}) = P(Y \le 1) = p(0) + p(1)$$

$$= \binom{15}{0} 0.1^0 (0.9)^{15} + \binom{15}{1} 0.1^1 (0.9)^{15-1}$$

$$= 0.20589 + 0.34315 = 0.54904$$

For $P = 0.2$,

$$P(\text{lot acceptance}) = P(Y \le 1) = p(0) + p(1) = \binom{15}{0} 0.2^0 (0.8)^{15} + \binom{15}{1} 0.2^1 (0.8)^{15-1}$$

$$= 0.03518 + 0.13194 = 0.16712$$

For $p = 0.3$,

$$P(\text{lot acceptance}) = P(Y \le 1) = p(0) + p(1) = \binom{15}{0} 0.3^0 (0.7)^{15} + \binom{15}{1} 0.3^1 (0.7)^{15-1}$$

$$= 0.00475 + 0.03052 = 0.03527$$

For $p = 0.4$,

$$P(\text{lot acceptance}) = P(Y \le 1) = p(0) + p(1) = \binom{15}{0} 0.4^0 (0.6)^{15} + \binom{15}{1} 0.4^1 (0.6)^{15-1}$$

$$= 0.00047 + 0.00470 = 0.00517$$

For $p = 0.5$,

$$P(\text{lot acceptance}) = P(Y \le 1) = p(0) + p(1) = \binom{15}{0} 0.5^0 (0.5)^{15} + \binom{15}{1} 0.5^1 (0.5)^{15-1}$$

$$= 0.00003 + 0.00046 = 0.00049$$

The operating characteristic curve is:

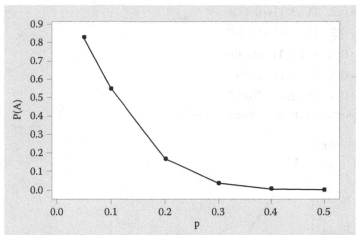

b. For AQL = 0.05, the producer's risk is α, the probability of rejecting the lot if the lot fraction is 0.05.

$$P(Y > 1) = 1 - P(Y \le 1) = 1 - p(0) - p(1)$$

$$= 1 - \binom{15}{0} 0.05^0 (0.95)^{15} - \binom{15}{1} 0.05^1 (0.95)^{15-1}$$

$$= 1 - 0.46329 - 0.36576 = 0.17095$$

c. For $p_1 = 0.20$, the consumer's risk is β, the probability of accepting the lot if the lot fraction is 0.20.

$$P(Y \le 1) = p(0) + p(1) = \binom{15}{0} 0.2^0 (0.8)^{15} + \binom{15}{1} 0.2^1 (0.8)^{15-1}$$

$$= 0.03518 + 0.13194 = 0.16712$$

16.61 a. Let y = number of defective items. Then y has a binomial distribution with $n = 10$. For $\alpha = 1$ and AQL = 0.025, the producer's risk is:

$$P(Y > 1) = 1 - P(Y \le 1) = 1 - p(0) - p(1)$$

$$= 1 - \binom{10}{0} 0.025^0 (0.975)^{10} - \binom{10}{1} 0.025^1 (0.975)^{10-1}$$

$$= 1 - 0.77633 - 0.19906 = 0.02461$$

b. For $p_1 = 0.15$, the consumer's risk is β, the probability of accepting the lot if the lot fraction is 0.15.

$$P(Y \le 1) = p(0) + p(1) = \binom{10}{0} 0.15^0 (0.85)^{10} + \binom{10}{1} 0.15^1 (0.85)^{10-1}$$

$$= 0.19687 + 0.34743 = 0.54430$$

c. To compute the operating characteristic curve, we compute $P(Y \le 1)$ for various values of p. Using Table 2, Appendix B, with $n = 10$, we get:

For $p = 0.1$, $P(Y \leq 1) = 0.7361$

For $p = 0.2$, $P(Y \leq 1) = 0.3758$

For $p = 0.3$, $P(Y \leq 1) = 0.1493$

For $p = 0.4$, $P(Y \leq 1) = 0.0464$

For $p = 0.5$, $P(Y \leq 1) = 0.0107$

For $p = 0.6$, $P(Y \leq 1) = 0.0017$

The operating characteristic curve is:

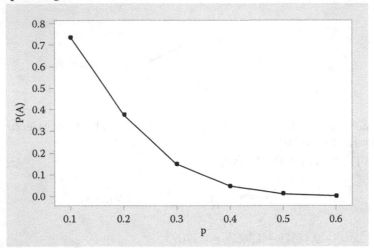

The consumer risks accepting lots containing a lot fraction defective equal to $p = 0.20$ approximately 37.6% of the time. The fact that β is so large for $p = 0.20$ indicates that this sampling plan would be of little value in practice. The plan needs to be based on a larger sample size.

16.63 a.   Using Table 22 of Appendix B, for a lot of 5000 items, the code for a reduced inspection level is **J**.

Now using Table 23, we find the row labeled **J** and see the recommended size sample of $n = 80$. We follow this row across to the 4% AQL and see we will accept the lot if 7or fewer are defective.

b.   Using Table 22 of Appendix B, for a lot of 5000 items, the code for a normal inspection level is **L**.

Now using Table 23, we find the row labeled **L**and see the recommended size sample of $n = 200$. We follow this row across to the 4% AQL and see we will accept the lot if 14 or fewer are defective.

c.   Using Table 22 of Appendix B, for a lot of 5000 items, the code for a tightened inspection level is **M**.

Now using Table 23, we find the row labeled **M** and see the recommended size sample of $n = 315$. We follow this row across to the 4% AQL and see we will accept the lot if 21 or fewer are defective.

16.65 The center line is $\bar{x} = \dfrac{\sum x}{n} = \dfrac{29.97}{20} = 1.4985$.

$$s^2 = \dfrac{\sum x^2 - \dfrac{\left(\sum x\right)^2}{n}}{n-1} = \dfrac{44.911408 - \dfrac{29.97^2}{20}}{20-1} = 0.000071736$$

$s = \sqrt{0.000071736} = 0.0085$

The lower and upper control limits are:

$$\text{LCL} = \bar{x} - 3s = 1.4985 - 3(0.0085) = 1.4730$$
$$\text{UCL} = \bar{x} + 3s = 1.4985 + 3(0.0085) = 1.524$$

The variable control chart is:

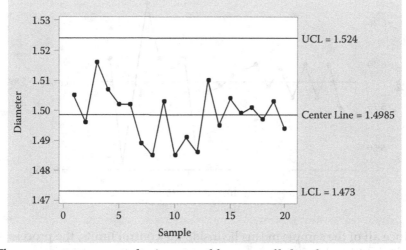

The process appears to be in control because all the observations are within the control limits.

16.67 To contract an $\bar{x}$–chart, we first compute the sample mean and sample range for each hour.

Hour	Sample Mean $\bar{\bar{x}}_i$	Range R_i	Hour	Sample Mean $\bar{x}_i$	Range R_i
1	5.73333	0.2	11	5.86667	0.6
2	6.00000	0.6	12	5.83333	0.2
3	5.80000	0.8	13	5.46667	0.5
4	6.00000	0.5	14	6.03333	0.1
5	5.80000	1.1	15	5.96667	0.6
6	5.96667	1.5	16	6.03333	0.4
7	5.86667	0.4	17	6.36667	0.5
8	6.00000	0.4	18	5.86667	0.5
9	5.96667	0.8	19	5.40000	0.2
10	5.93333	0.4	20	6.03333	0.1

The center line is $\bar{\bar{x}} = \dfrac{\sum \bar{x}_i}{n} = \dfrac{117.93333}{20} = 5.89667$.

$\bar{R} = \dfrac{\sum R_i}{n} = \dfrac{10.4}{20} = 0.52$

From Table 19, Appendix B, with $n = 3$, $A_2 = 1.023$.

Upper control limit $= \bar{\bar{x}} + A_2\bar{R} = 5.89667 + 1.023(0.52) = 6.42863$

Lower control limit $= \bar{\bar{x}} + A_2\bar{R} = 5.89667 - 1.023(0.52) = 5.36471$

The $\bar{x}$-chart is:

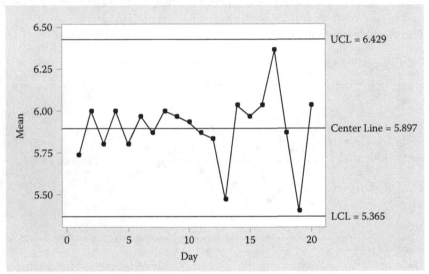

Since all of the sample means lie inside the control limits, the process appears to be in control.

The center line is $\bar{R} = 0.52$.

From Table 19, Appendix B, with $n = 3$, $D_3 = 0$ and $D_4 = 2.575$.

Upper control limit $= D_4\bar{R} = 2.575(0.52) = 1.339$

Lower control limit $= D_3\bar{R} = 0(0.52) = 0$

The *R*-chart is:

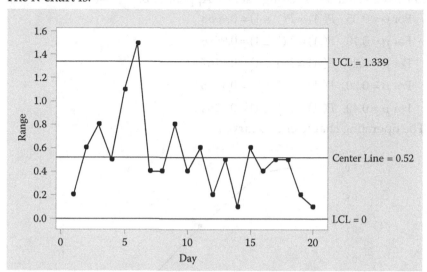

The range from day 6 is outside the control limits. Therefore, the process appears to be out of control.

16.69 a. To sketch operating characteristic curves, we need to compute $P(A) = P$ (accepting lot) for different values of p. Let y = number of defective items in n trials. Then y has a binomial distribution.

Let $n = 5$ and $a = 1$. Using Table 2, Appendix B,

For $p = 0.05$, $P(A) = P(Y \leq 1) = 0.9774$

For $p = 0.10$, $P(A) = P(Y \leq 1) = 0.9185$

For $p = 0.20$, $P(A) = P(Y \leq 1) = 0.7373$

For $p = 0.30$, $P(A) = P(Y \leq 1) = 0.5282$

For $p = 0.40$, $P(A) = P(Y \leq 1) = 0.3370$

The operating characteristic curve is:

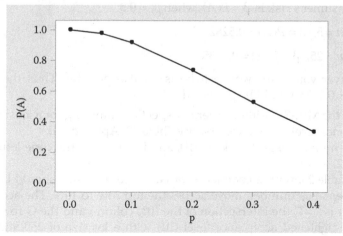

Let $n = 25$ and $a = 5$. Using Table 2, Appendix B,

For $p = 0.05$, $P(A) = P(Y \leq 1) = 0.9988$

For $p = 0.10$, $P(A) = P(Y \leq 1) = 0.9666$

For $p = 0.20$, $P(A) = P(Y \leq 1) = 0.6167$

For $p = 0.30$, $P(A) = P(Y \leq 1) = 0.1935$

For $p = 0.40$, $P(A) = P(Y \leq 1) = 0.0294$

The operating characteristic curve is:

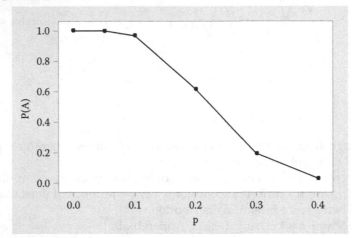

b. As a seller, one would want the producer's risk, α, to be as small as possible. The producer's risk is $\alpha = 1 - P(A)$ when $p = AQL = 0.10$.

When $n = 5$, $\alpha = 1 - P(A) = 1 - 0.9816 = 0.0814$

When $n = 25$, $\alpha = 1 - P(A) = 1 - 0.9666 = 0.0334$

As a seller, you would want α to be as small as possible. Thus, the second plan with $\alpha = 0.0334$ would be preferred.

c. As a buyer, one would want the consumer's risk, β, to be as small as possible. The consumer's risk is $\beta = P(A)$ when $p_1 = 0.3$.

When $n = 5$, $\beta = P(A) = 0.5282$

When $n = 25$, $\beta = P(A) = 0.1935$

As a buyer, you would want β to be as small as possible. Thus, the second plan with $\beta = 0.1935$ would be preferred.

16.71 a. To find the MIL-STD-105D general inspection sampling for a lot size of 250 and acceptance level of 10%, we first use Table 22, Appendix B.
For the normal inspection level (II), and lot size 250, the code letter is **G** from Table 22.
From Table 23, with a code letter of **G**, the sample size should be 32. To find the acceptance number, move across the top row to 10%. The acceptance (Ac) number is $a = 7$, the intersection of the 10% column and the **G** row.

b. For the tightened acceptance level (III) with a lot size of 250 and acceptance level of 10%, we again start with Table 22. For $n = 250$ and tightened acceptance level III, the code letter is **H** from Table 22.

From Table 23, with a code letter of **H**, the sample size should be 50. To find the acceptance number, move across the top row to 10%. The acceptance (Ac) number is $a = 10$, the intersection of the 10% column and the **H** row.

16.73 a. Let y = number of defectives in $n = 20$ trials. The distribution of y is binomial. For $a = 2$ and AQL $= 0.05$, the producer's risk is $\alpha = 1 - P(A) = 1 - P(Y \le 2) = 1 - 0.8245 = 0.0755$ (from Table 2, Appendix B).

b. The consumer's risk is $\beta = P(A)$. For $p_1 = 0.10$,

$$\beta = P(A) = P(Y \le 2) = 0.6769 \text{ (from Table 2, Appendix B)}.$$

c. To find the operating Characteristic curve, we must find $P(A)$ for several values of p.

For $p = 0.1$, $P(A) = P(Y \le 2) = 0.6769$ (from Table 2, Appendix B)

For $p = 0.2$, $P(A) = P(Y \le 2) = 0.2061$

For $p = 0.3$, $P(A) = P(Y \le 2) = 0.0355$

For $p = 0.4$, $P(A) = P(Y \le 2) = 0.0036$

For $p = 0.5$, $P(A) = P(Y \le 2) = 0.0002$

The operating curve is:

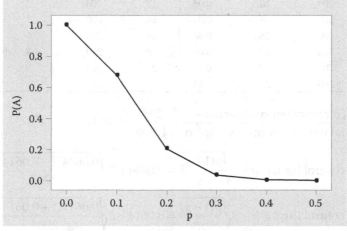

d. Using Table 22 in Appendix B, the lot code for a lot size of 1500 is **K**. Using Table 23, we find a recommended sample size of $n = 125$. Using AQL $= 0.05$, we should accept the lot if 10 or fewer are defective (use AQL $= 0.04$ for a conservative value).

e. For $n = 125$ and $a = 10$.

Let y = number of defective items in $n = 125$ trials.

Then y has a binomial distribution with $n = 125$ and $p = $ AQL $= 0.05$. The producer's risk is $\alpha = 1 - P(A) = 1 - P(Y \le 10)$.

Since $n = 125$, we will use the normal approximation to the binomial with $\mu = np = 125(0.05) = 6.25$ and $\sigma = \sqrt{npq} = \sqrt{125(0.05)(0.95)} = 2.4367$.

Thus, $\alpha = 1 - P(A) = 1 - P(Y \le 10) \approx 1 - P\left(Z \le \dfrac{10.5 - 6.25}{2.4367}\right) = 1 - P(Z \le 1.74)$

$$= 1 - (0.5 + P(0 \le Z \le 1.74)) = 1 - 0.5 - 0.4591 = 0.0409$$

f. For $n = 125$ and $p_1 = 0.08$, the consumer's risk is $\beta = P(Y \leq 10)$. Using the normal approximation to the binomial with $\mu = np = 125(0.08) = 10$ and $\sigma = \sqrt{npq} = \sqrt{125(0.08)(0.92)} = 3.03315$.

Thus, $\beta = P(A) = P(Y \leq 10) \approx P\left(Z \leq \dfrac{10.5 - 10}{3.03315}\right) = P(Z \leq 0.16)$

$= 0.5 + P(0 \leq Z \leq 0.16) = 0.5 + 0.0636 = 0.5636$

16.75 a. To compute the proportion of defectives in each sample, divide the number of defectives by the number in the sample, 200: $\hat{p}_i = \frac{Y}{n}$. The sample proportions are listed in the table:

Sample	Number	Defectives	$\bar{p}_i$	Sample	Number	Defectives	$\hat{p}_i$
1	200	4	0.02	12	200	20	0.10
2	200	6	0.03	13	200	20	0.10
3	200	11	0.06	14	200	17	0.09
4	200	12	0.06	15	200	13	0.07
5	200	5	0.03	16	200	10	0.05
6	200	10	0.05	17	200	11	0.06
7	200	8	0.04	18	200	7	0.04
8	200	16	0.08	19	200	6	0.03
9	200	17	0.09	20	200	8	0.04
10	200	20	0.10	21	200	9	0.05
11	200	28	0.14				

$$\bar{p} = \frac{\text{Total number of defectives}}{\text{Total number of units sampled}} = \frac{258}{4,200} = 0.0614$$

$$\text{Upper control limit} = \bar{p} + 3\sqrt{\frac{\bar{p}(1 - \bar{p})}{n}} = 0.0614 + 3\sqrt{\frac{0.0614(1 - 0.0614)}{200}} = 0.1123$$

$$\text{Lower control limit} = \bar{p} - 3\sqrt{\frac{\bar{p}(1 - \bar{p})}{n}} = 0.0614 - 3\sqrt{\frac{0.0614(1 - 0.0614)}{200}} = 0.0105$$

The *p*-chart is:

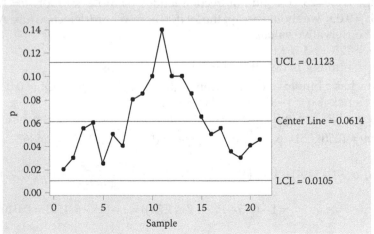

b. No, the control limits should not be used to monitor future process output. One observation lies outside the control limits. The process appears to be out of control.

16.77 a. Using MINITAB, the capability analysis diagram is:

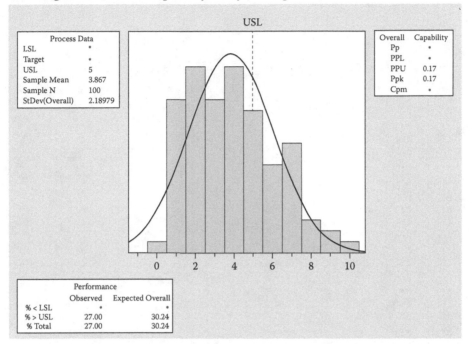

b. For an upper specification limit of 5, there are 27 observations above this limit. Thus, $(27/100) \times 100\% = 27\%$ of the observations are unacceptable. It does not appear that the process is capable.

c. From Exercise 16.76, the process appears to be in control. Thus, it is appropriate to estimate C_p.

From the sample, $\bar{x} = 3.867$ and $s = 2.190$.

$$C_p = \frac{\text{USL} - \text{LSL}}{6\sigma} \approx \frac{5 - 0}{6(2.190)} = 0.381$$

Since the C_p is less than 1, the process is not capable.

d. There is no lower specification limit because management has no time limit below which is unacceptable. The variable being measured is time customers wait in line. The actual lower limit would be 0.

17

Product and System Reliability

17.1   The density function for the exponential distribution is

$$f(t) = \begin{cases} \dfrac{e^{-t/\beta}}{\beta} & 0 \le t \le \infty \\ 0 & \text{otherwise} \end{cases}$$

The cumulative distribution function of the exponential distribution is

$$F(t) = 1 - e^{-t/\beta}$$

The hazard rate is $z(t) = \dfrac{f(t)}{1 - F(t)} = \dfrac{\dfrac{e^{-t/\beta}}{\beta}}{1 - (1 - e^{-t/\beta})} = \dfrac{\dfrac{e^{-t/\beta}}{\beta}}{e^{-t/\beta}} = \dfrac{1}{\beta}$

For $\beta = \dfrac{1}{\lambda}$, $z(t) = \dfrac{1}{(1/\lambda)} = \lambda$

The graph of the hazard rate is:

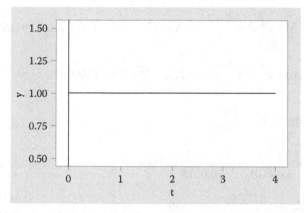

17.3 a.   The density function for t is: $f(t) = \dfrac{1}{\sqrt{2\pi}} e^{-\frac{(t-3)^2}{2}} \quad -\infty < t < \infty$

For $t = 0, 1, 2, \ldots, 6$,

$$f(0) = \frac{1}{\sqrt{2\pi}} e^{-\frac{(0-3)^2}{2}} = \frac{1}{\sqrt{2\pi}} e^{-\frac{9}{2}} = 0.00443 \quad f(1) = \frac{1}{\sqrt{2\pi}} e^{-\frac{(1-3)^2}{2}} = \frac{1}{\sqrt{2\pi}} e^{-\frac{4}{2}} = 0.05399$$

$$f(2) = \frac{1}{\sqrt{2\pi}} e^{-\frac{(2-3)^2}{2}} = \frac{1}{\sqrt{2\pi}} e^{-\frac{1}{2}} = 0.24197 \quad f(3) = \frac{1}{\sqrt{2\pi}} e^{-\frac{(3-3)^2}{2}} = \frac{1}{\sqrt{2\pi}} e^0 = 0.39894$$

$$f(4) = \frac{1}{\sqrt{2\pi}} e^{-\frac{(4-3)^2}{2}} = \frac{1}{\sqrt{2\pi}} e^{-\frac{1}{2}} = 0.24197 \quad f(5) = \frac{1}{\sqrt{2\pi}} e^{-\frac{(5-3)^2}{2}} = \frac{1}{\sqrt{2\pi}} e^{-\frac{4}{2}} = 0.05399$$

$$f(6) = \frac{1}{\sqrt{2\pi}} e^{-\frac{(6-3)^2}{2}} = \frac{1}{\sqrt{2\pi}} e^{-\frac{9}{2}} = 0.00443$$

$F(t_0) = P(t \le t_0)$ where t has a normal distribution with $\mu = 3$ and $\sigma = 1$. Using Table 5, Appendix B:

$$F(0) = P(t \le 0) = P\left(Z \le \frac{0-3}{1}\right) = P(Z \le -3) = 0.5 - P(0 \le Z \le 3) = 0.5 - 0.4987 = 0.0013$$

$$F(1) = P(t \le 1) = P\left(Z \le \frac{1-3}{1}\right) = P(Z \le -2) = 0.5 - P(0 \le Z \le 2) = 0.5 - 0.4772 = 0.0228$$

$$F(2) = P(t \le 2) = P\left(Z \le \frac{2-3}{1}\right) = P(Z \le -1) = 0.5 - P(0 \le Z \le 1) = 0.5 - 0.3413 = 0.1587$$

$$F(3) = P(t \le 3) = P\left(Z \le \frac{3-3}{1}\right) = P(Z \le 0) = 0.5$$

$$F(4) = P(t \le 4) = P\left(Z \le \frac{4-3}{1}\right) = P(Z \le 1) = 0.5 + P(0 \le Z \le 1) = 0.5 + 0.3413 = 0.8413$$

$$F(5) = P(t \le 5) = P\left(Z \le \frac{5-3}{1}\right) = P(Z \le 2) = 0.5 + P(0 \le Z \le 2) = 0.5 + 0.4772 = 0.9772$$

$$F(6) = P(t \le 6) = P\left(Z \le \frac{6-3}{1}\right) = P(Z \le 3) = 0.5 + P(0 \le Z \le 3) = 0.5 + 0.4987 = 0.9987$$

We know that $z(t) = \frac{f(t)}{1 - F(t)}$.

Thus, $z(0) = \frac{f(0)}{1 - F(0)} = \frac{0.00443}{1 - 0.0013} = 0.0044$

$$z(1) = \frac{f(1)}{1 - F(1)} = \frac{0.05399}{1 - 0.0228} = 0.0552 \qquad z(2) = \frac{f(2)}{1 - F(2)} = \frac{0.24197}{1 - 0.1587} = 0.2876$$

$$z(3) = \frac{f(3)}{1 - F(3)} = \frac{0.39894}{1 - 0.5} = 0.7979 \qquad z(4) = \frac{f(4)}{1 - F(4)} = \frac{0.24197}{1 - 0.8413} = 1.5247$$

$$z(5) = \frac{f(5)}{1 - F(5)} = \frac{0.05399}{1 - 0.9772} = 2.3680 \qquad z(6) = \frac{f(6)}{1 - F(6)} = \frac{0.00443}{1 - 0.9987} = 3.4077$$

b. The plot of $z(t)$ is:

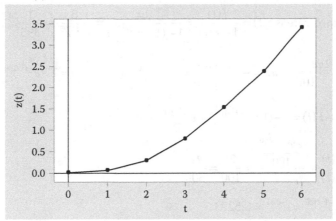

17.5 a. The density function for the exponential distribution is

$$f(t) = \begin{cases} \dfrac{e^{-t/460}}{460} & 0 \le t \le \infty \\ 0 & \text{otherwise} \end{cases}$$

The cumulative distribution function of the exponential distribution is
$F(t) = 1 - e^{-t/460}$

The hazard rate is $z(t) = \dfrac{f(t)}{1 - F(t)} = \dfrac{\dfrac{e^{-t/460}}{460}}{1 - \left(1 - e^{-t/460}\right)} = \dfrac{\dfrac{e^{-t/460}}{\beta}}{e^{-t/460}} = \dfrac{1}{460}$

b. The density function for the exponential distribution is

$$f(t) = \begin{cases} \dfrac{e^{-t/2880}}{2880} & 0 \le t \le \infty \\ 0 & \text{otherwise} \end{cases}$$

The cumulative distribution function of the exponential distribution is
$F(t) = 1 - e^{-t/2880}$

The hazard rate is $z(t) = \dfrac{f(t)}{1 - F(t)} = \dfrac{\dfrac{e^{-t/2880}}{2880}}{1 - (1 - e^{-t/2880})} = \dfrac{\dfrac{e^{-t/2880}}{\beta}}{e^{-t/2880}} = \dfrac{1}{2880}$

c. The density function for the exponential distribution is

$$f(t) = \begin{cases} \dfrac{e^{-t/395}}{395} & 0 \le t \le \infty \\ 0 & \text{otherwise} \end{cases}$$

The cumulative distribution function of the exponential distribution is
$F(t) = 1 - e^{-t/395}$

The hazard rate is $z(t) = \dfrac{f(t)}{1 - F(t)} = \dfrac{\dfrac{e^{-t/395}}{395}}{1 - (1 - e^{-t/395})} = \dfrac{\dfrac{e^{-t/395}}{\beta}}{e^{-t/395}} = \dfrac{1}{395}$

17.7 a. $F(t) = \displaystyle\int_0^t \dfrac{2ye^{-y^2/100}}{100}\, dy = -e^{-y^2/100}\Big]_0^t = -e^{-t^2/100} - (-e^{-0^2/100}) = 1 - e^{-t^2/100}$

b. $R(t) = 1 - F(t) = 1 - (1 - e^{-t^2/100}) = e^{-t^2/100}$

$z(t) = \dfrac{f(t)}{R(t)} = \dfrac{\dfrac{2te^{-t^2/100}}{100}}{e^{-t^2/100}} = \dfrac{2t}{100} = \dfrac{t}{50}$

c. $R(8) = e^{-8^2/100} = e^{-0.64} = 0.5273$

$z(8) = \dfrac{8}{50} = 0.16$

17.9 a. The mean of the Weibull distribution is found using $\mu = \beta^{1/\alpha}\Gamma\!\left(\dfrac{\alpha+1}{\alpha}\right)$.

From Table 6, Appendix B, $\Gamma\!\left(\dfrac{\alpha+1}{\alpha}\right) = \Gamma\!\left(\dfrac{3.5+1}{3.5}\right) = \Gamma(1.2857) \approx 0.89904$

Solving for β, we get

$\mu = \beta^{1/\alpha}\Gamma\!\left(\dfrac{\alpha+1}{\alpha}\right) \Rightarrow 2{,}370 = \beta^{1/3.5}(0.89904) \Rightarrow 2636.145222 = \beta^{1/3.5}$

$\Rightarrow \beta \approx 9.4057 \times 10^{11}$

b. The density function of the Weibull distribution is $f(t) = \dfrac{\alpha}{\beta}t^{\alpha-1}e^{-t^\alpha/\beta}$.

The cumulative distribution function is $F(t) = 1 - e^{-t^\alpha/\beta}$.

Thus, $R(t) = 1 - F(t) = 1 - (1 - e^{-t^\alpha/\beta}) = e^{-t^\alpha/\beta}$

The hazard rate for the Weibull distribution is

$z(t) = \dfrac{f(t)}{R(t)} = \dfrac{\dfrac{\alpha}{\beta}t^{\alpha-1}e^{-t^\alpha/\beta}}{e^{-t^\alpha/\beta}} = \dfrac{\alpha}{\beta}t^{\alpha-1} = \dfrac{3.5}{9.4057 \times 10^{11}}t^{2.5} = 3.7211 \times 10^{-12}t^{2.5}$

c. $z(5{,}000) = 3.7211 \times 10^{-12}(5{,}000)^{2.5} = 0.006578$

17.11 a. If Y has an exponential distribution with parameter β, then $\dfrac{2Y}{\beta}$ has a chi-square distribution with 2 degrees of freedom. Thus, $\sum \dfrac{2Y}{\beta} = \dfrac{2\sum Y}{\beta}$ has a chi-square distribution with $2n$ degrees of freedom. Using the pivotal statistic, $\dfrac{2\sum Y_i}{\beta}$, the confidence interval is:

$$P\!\left(\chi^2_{1-\alpha/2} \le \dfrac{2\displaystyle\sum y_i}{\beta} \le \chi^2_{\alpha/2}\right) = 1 - \alpha$$

$$P\left(\chi^2_{1-\alpha/2} \le \frac{2\sum y_i}{\beta} \le \chi^2_{\alpha/2}\right) = P\left(\frac{1}{\chi^2_{1-\alpha/2}} \ge \frac{\beta}{2\sum y_i} \ge \frac{1}{\chi^2_{\alpha/2}}\right) = P\left(\frac{2\sum y_i}{\chi^2_{1-\alpha/2}} \ge \beta \ge \frac{2\sum y_i}{\chi^2_{\alpha/2}}\right)$$

$$\text{or } P\left(\frac{2\sum y_i}{\chi^2_{\alpha/2}} \le \beta \le \frac{2\sum y_i}{\chi^2_{1-\alpha/2}}\right) = 1-\alpha$$

For confidence coefficient 0.95, $\alpha = 0.05$ and $\alpha/2 = 0.05/2 = 0.025$. From Table 8, Appendix B, with $df = 2n = 2(20) = 40$, $\chi^2_{0.025} = 59.3417$ and $\chi^2_{0.975} = 24.4331$.

From the data, $\sum y_i = 12,323$. The 95% confidence interval is

$$\left(\frac{2(12,323)}{59.3417} \le \beta \le \frac{2(12,323)}{24.4331}\right) \Rightarrow (415.32, 1008.71)$$

We are 95% confident that the mean number of cycles to failure is between 415.32 and 1008.71.

b. For the exponential distribution the hazard rate is $1/\beta$. Substituting the 95% confidence limits for β, we get the following 95% confidence interval for the hazard rate:

$$\left(\frac{1}{1008.71} \le \frac{1}{\beta} \le \frac{1}{415.32}\right) \Rightarrow (0.00099, 0.00241)$$

We are 95% confident that the true hazard rate is between 0.00099 and 0.00241.

17.13 a. If Y has an exponential distribution with parameter β, then $\frac{2Y}{\beta}$ has a chi-square distribution with 2 degrees of freedom. Thus, $\sum \frac{2Y}{\beta} = \frac{2\sum Y}{\beta}$ has a chi-square distribution with 2n degrees of freedom. Using the pivotal statistic, $\frac{2\sum Y_i}{\beta}$, the confidence interval is:

$$P\left(\chi^2_{1-\alpha/2} \le \frac{2\sum y_i}{\beta} \le \chi^2_{\alpha/2}\right) = 1-\alpha$$

$$P\left(\chi^2_{1-\alpha/2} \le \frac{2\sum y_i}{\beta} \le \chi^2_{\alpha/2}\right) = P\left(\frac{1}{\chi^2_{1-\alpha/2}} \ge \frac{\beta}{2\sum y_i} \ge \frac{1}{\chi^2_{\alpha/2}}\right) = P\left(\frac{2\sum y_i}{\chi^2_{1-\alpha/2}} \ge \beta \ge \frac{2\sum y_i}{\chi^2_{\alpha/2}}\right)$$

$$\text{or } P\left(\frac{2\sum y_i}{\chi^2_{\alpha/2}} \le \beta \le \frac{2\sum y_i}{\chi^2_{1-\alpha/2}}\right) = 1-\alpha$$

For confidence coefficient 0.90, $\alpha = 0.10$ and $\alpha/2 = 0.10/2 = 0.05$. Using MINITAB, with $df = 2n = 2(29) = 58$, $\chi^2_{0.05} = 76.7778$ and $\chi^2_{0.95} = 41.4920$.

From the data, $\sum y_i = 3599.75$. The 90% confidence interval is

$$\left(\frac{2(3599.75)}{76.7778} \le \beta \le \frac{2(3599.75)}{41.4920}\right) \Rightarrow (93.7706, 173.5154)$$

b. For confidence coefficient 0.90, $\alpha = 0.10$ and $\alpha/2 = 0.10/2 = 0.05$. Using MINITAB, with $df = 2n = 2(22) = 44$, $\chi^2_{0.05} = 60.4809$ and $\chi^2_{0.95} = 29.7875$.

From the data, $\Sigma\, y_i = 3599.75$. The 90% confidence interval is

$$\left(\frac{2(4351.25)}{60.4809} \le \beta \le \frac{2(4351.25)}{\chi^2_{0.95} = 29.7875} \right) \Rightarrow (143.8884,\ 292.1527)$$

17.15   The approximate confidence interval for the mean time between failures is:

$$\frac{2(\text{Total life})}{\chi^2_{\alpha/2}} \le \beta \le \frac{2(\text{Total life})}{\chi^2_{1-\alpha/2}}$$

where total life is $\displaystyle\sum_{i=2}^{r} t_i + (n-r)T = 4048 + (100-3)2000 = 198,048$

For confidence coefficient 0.99, $\alpha = 0.01$ and $\alpha/2 = 0.01/2 = 0.005$. From Table 8, Appendix B, with $df = 2r+2 = 2(3)+2 = 8$, $\chi^2_{0.005} = 21.9550$ and $\chi^2_{0.995} = 1.344419$. The 99% confidence interval is:

$$\frac{2(\text{Total life})}{\chi^2_{\alpha/2}} \le \beta \le \frac{2(\text{Total life})}{\chi^2_{1-\alpha/2}} \Rightarrow \frac{2(198,048)}{21.9550} \le \beta \le \frac{2(198,048)}{1.344419}$$

$$\Rightarrow (18,041.27,\ 294,622.44)$$

We are 99% confident that the mean time between failures of the capacitors is between 18,041.27 and 294,622.44 hours.

17.17   The approximate confidence interval for the mean time between failures is:

$$\frac{2(\text{Total life})}{\chi^2_{\alpha/2}} \le \beta \le \frac{2(\text{Total life})}{\chi^2_{1-\alpha/2}}$$

where total life is $\displaystyle\sum_{i=2}^{r} t_i + (n-r)T = 3307.5 + (96-37)135 = 11,272.5$

For confidence coefficient 0.95, $\alpha = 0.05$ and $\alpha/2 = 0.05/2 = 0.025$. From Table 8, Appendix B, with $df = 2r+2 = 2(37)+2 = 76$, $\chi^2_{0.025} \approx 106.629$ and $\chi^2_{0.975} \approx 57.1532$. The 99% confidence interval is:

$$\frac{2(\text{Total life})}{\chi^2_{\alpha/2}} \le \beta \le \frac{2(\text{Total life})}{\chi^2_{1-\alpha/2}} \Rightarrow \frac{2(11,272.5)}{106.629} \le \beta \le \frac{2(11,272.5)}{57.1352} \Rightarrow (211.4,\ 394.5)$$

17.19   Some preliminary calculations are:

Time (i)	$x_i = \ln i$	# Survivors (n_i)	$\hat{R}(i) = \dfrac{n_i}{n}$	$-\ln \hat{R}(i)$	$y_i = \ln[-\ln \hat{R}(i)]$
1	0.00000	47	0.94	0.061875	-2.782633
2	0.69315	39	0.78	0.248461	-1.392468
3	1.09861	29	0.58	0.544727	-0.607470
4	1.38629	18	0.36	1.021651	0.021420
5	1.60944	11	0.22	1.514128	0.414840
6	1.79176	5	0.10	2.302585	0.834032
7	1.94591	3	0.06	2.813411	1.034398
8	2.07944	1	0.02	3.912023	1.364055

For the model $y_i = \beta_0 + \beta_1 x_i + \varepsilon_i$, we need to estimate β_0 and β_1 by the method of least squares. Using MINITAB, the results are:

```
Regression Analysis: y versus x

Analysis of Variance

Source        DF    Adj SS    Adj MS    F-Value   P-Value
Regression     1   13.6869   13.6869   10517.64     0.000
   x           1   13.6869   13.6869   10517.64     0.000
Error          6    0.0078    0.0013
Total          7   13.6947

Model Summary

        S    R-sq   R-sq(adj)   R-sq(pred)
0.0360740   99.94%     99.93%       99.91%

Coefficients

Term         Coef   SE Coef   T-Value   P-Value   VIF
Constant  -2.7744    0.0287    -96.72     0.000
x          1.9879    0.0194    102.56     0.000   1.00


Regression Equation

y = -2.7744 + 1.9879 x
```

a.   Using the method of least squares:

$$\hat{\alpha} = \hat{\beta}_1 = 1.9879$$

$$\hat{\beta} = e^{-\hat{\beta}_0} = e^{-(-2.7744)} = 16.029$$

b.   The confidence interval for α is the same as the confidence interval for β_1 which is $\hat{\beta}_1 \pm t_{\alpha/2} s_{\hat{\beta}_1}$

For confidence coefficient 0.95, $\alpha = 0.05$ and $\alpha/2 = 0.05/2 = 0.025$. From Table 7, Appendix B, with $df = n - 2 = 8 - 2 = 6$, $t_{0.025} = 2.447$. The 95% confidence interval is:

$$\hat{\beta}_1 \pm t_{\alpha/2} s_{\hat{\beta}_1} \Rightarrow 1.9879 \pm 2.447(0.0194) \Rightarrow 1.9879 \pm 0.0475 \Rightarrow (1.9404,\ 2.0354)$$

c. To find the confidence interval for β, we first find a confidence interval for β_0. Then the confidence interval for β is found by raising e to the negative endpoints of the confidence interval for β_0.

   The form of the confidence interval for β_0 is $\hat{\beta}_0 \pm t_{\alpha/2} s_{\hat{\beta}_0}$
   From part **b**, $t_{0.025} = 2.447$. The confidence interval for β_0 is:

$$\hat{\beta}_0 \pm t_{\alpha/2} s_{\hat{\beta}_0} \Rightarrow -2.7744 \pm 2.447(0.0287) \Rightarrow -2.7744 \pm 0.0702$$

$$\Rightarrow (-2.8446,\ -2.7042)$$

   The confidence interval for β is: $(e^{-(-2.7042)},\ e^{-(-2.8446)}) \Rightarrow (14.9424,\ 17.1947)$

17.21 a. From Exercise 17.19, $\hat{\alpha} = 1.987949$ and $\hat{\beta} = 16.029$.
   For the Weibull distribution, the density function is

$$f(t) = \begin{cases} \dfrac{\alpha}{\beta} t^{\alpha-1} e^{-t^{\alpha}/\beta} & 0 \le t < \infty,\ \alpha > 0,\ \beta > 0 \\ 0 & \text{otherwise} \end{cases}$$

And the cumulative distribution function is $F(t) = 1 - e^{-t^{\alpha}/\beta}$.
The hazard rate is

$$z(t) = \frac{f(t)}{1 - F(t)} = \frac{\dfrac{\alpha}{\beta} t^{\alpha-1} e^{-t^{\alpha}/\beta}}{1 - (1 - e^{-t^{\alpha}/\beta})} = \frac{\alpha}{\beta} t^{\alpha-1} = \frac{1.987949}{16.029} t^{0.987949} = 0.124022 t^{0.987949}$$

The graph is:

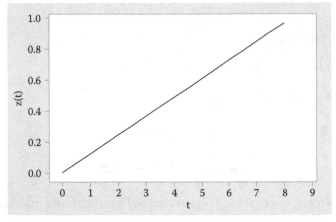

b. $z(4) = 0.124022(4)^{0.987949} = 0.4879$

17.23 a. The table is:

Time (*i*)	# Survivors (n_i)
1	$12 - 3 = 9$
2	$12 - 8 = 4$
3	$12 - 10 = 2$

b. Expanding the table in part **a**, we get:

Time (*i*)	$x_i = \ln i$	# Survivors (n_i)	$\hat{R}(i) = \frac{n_i}{n}$	$-\ln \hat{R}(i)$	$y_i = \ln[-\ln \hat{R}(i)]$
1	0.00000	9	0.750	0.28768	−1.24590
2	0.69315	4	0.333	1.09861	0.09405
3	1.09861	2	0.167	1.79176	0.58320

Using MINITAB, the results are:

```
Regression Analysis: y versus x

Analysis of Variance

Source       DF   Adj SS   Adj MS   F-Value   P-Value
Regression    1  1.77092  1.77092     78.61     0.071
  x           1  1.77092  1.77092     78.61     0.071
Error         1  0.02253  0.02253
Total         2  1.79345

Model Summary

       S   R-sq   R-sq(adj)   R-sq(pred)
0.150089  98.74%     97.49%       79.08%

Coefficients

Term        Coef  SE Coef  T-Value  P-Value   VIF
Constant  -1.201    0.143    -8.38    0.076
x          1.694    0.191     8.87    0.071  1.00

Regression Equation

y = -1.201 + 1.694 x
```

Using the method of least squares:

$$\hat{\alpha} = \hat{\beta}_1 = 1.694 \text{ and } \hat{\beta} = e^{-\hat{\beta}_0} = e^{-(-1.201)} = 3.323$$

c. The confidence interval for α is the same as the confidence interval for β_1
For confidence coefficient 0.95, $\alpha = 0.05$ and $\alpha/2 = 0.05/2 = 0.025$. From Table 7, Appendix B, with $df = n - 2 = 3 - 2 = 1$, $t_{0.025} = 12.706$. The 95% confidence interval is:

$$\hat{\beta}_1 \pm t_{\alpha/2} s_{\hat{\beta}_1} \Rightarrow 1.694 \pm 12.706(0.191) \Rightarrow 1.694 \pm 0.441 \Rightarrow (-0.733, 4.121)$$

Using $t_{0.025} = 12.706$, the confidence interval for β_0 is

$$\hat{\beta}_0 \pm t_{\alpha/2} s_{\hat{\beta}_0} \Rightarrow -1.201 \pm 12.706(0.143) \Rightarrow -1.201 \pm 1.817 \Rightarrow (-3.018, 0.616)$$

The confidence interval for β is: $(e^{-(0.616)}, e^{-(-3.018)}) \Rightarrow (0.540, 20.450)$

    d. Let Y = time until repair.

$$F(y) = 1 - e^{-y^\alpha/\beta}$$

Substituting the estimates of α and β,

$$P(Y \le 2) = F(2) = 1 - e^{-2^{1.694}/3.323} = 1 - 0.3777 = 0.6223$$

17.25   For a series system consisting of 4 independently operating components, A, B, C, and D, the reliability of the system is:

$$P(\text{series system functions}) = p_A p_B p_C p_D = 0.88(0.95)(0.90)(0.80) = 0.60192$$

17.27 a. Since the system software will fail if at least one of the software code statement fails, the system operates as a series.

      b. The reliability of the system is:

$$P(\text{series system functions}) = P(\text{first component functions})$$
$$\times (\text{second component functions})$$
$$\times \cdots \times (k^{th} \text{ component functions}) = (1 - p_1)(1 - p_2)\cdots(1 - p_k)$$

where p_i = probability i^{th} component fails.

17.29   To find the reliability of the system, we must find the reliability of several subsystems:

The reliability of the subsystem (parallel)

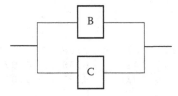

is $P(\text{subsystem B and C functions}) = p_{BC} = 1 - (1 - p_B)(1 - p_C)$

$$= 1 - (1 - 0.95)(1 - 0.85)$$

$$= 1 - 0.0075 = 0.9925$$

The reliability of the subsystem (parallel)

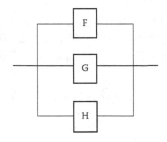

is $P(\text{subsystem F, G, and H functions}) = p_{FGH} = 1 - (1 - p_F)(1 - p_G)(1 - p_H)$

$$= 1 - (1 - 0.80)(1 - 0.95)(1 - 0.95)$$

$$= 1 - 0.0005 = 0.9995$$

The reliability of the subsystem (series)

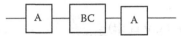

is $p_{ABCD} = p_A p_{BC} p_D = 0.90(0.9925)(0.85) = 0.7592625$

The reliability of the subsystem (series)

is $p_{EFGH} = p_E p_{FGH} = 0.98(0.9995) = 0.97951$

The reliability of the subsystem (parallel)

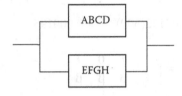

is $P(\text{system functions}) = 1 - (1 - p_{ABCD})(1 - p_{EFGH})$

$$= 1 - (1 - 0.7592625)(1 - 0.97951)$$

$$= 1 - 0.00493 = 0.99507$$

17.31 a. $P(\text{system functions}) = 0.95$

If the system has 3 identical components connected in series,

$$P(\text{system functions}) = p_A p_B p_C = p^3 = 0.95$$

Thus, $p = \sqrt[3]{0.95} = 0.983$

b. If the components are connected in parallel,

$$P(\text{system functions}) = 1 - (1 - p_A)(1 - p_B)(1 - p_C)$$

$$= 1 - (1 - p)^3 = 0.95$$

$$\Rightarrow (1 - p)^3 = 1 - 0.95 = 0.05$$

$$\Rightarrow 1 - p = \sqrt[3]{0.05} = 0.3684$$

$$\Rightarrow p = 0.6316$$

17.33 a. For a Weibull distribution, the cumulative distribution function is $F(t) = 1 - e^{-t^\alpha/\beta}$.

If $\alpha = 0.05$ and $\beta = 0.70$, then $F(t) = 1 - e^{-t^{0.05}/0.70}$.

The probability the light fails before time $t = 1000$ is

$$P(Y \le 1000) = F(1000) = 1 - e^{-1000^{0.05}/0.70} = 1 - 0.1329 = 0.8671$$

b. The reliability of the light is $R(t) = 1 - F(t) = 1 - (1 - e^{-t^{0.05}/0.70}) = e^{-t^{0.05}/0.70}$

Thus, $R(500) = e^{-500^{0.05}/0.70} = 0.1424$

c. The hazard rate of the light is $z(t) = \dfrac{f(t)}{R(t)}$.

For the Weibull distribution, $f(t) = \dfrac{\alpha}{\beta} t^{\alpha-1} e^{-t^{\alpha}/\beta}$.

With $\alpha = 0.05$ and $\beta = 0.70$, $f(t) = \dfrac{0.05}{0.7} t^{0.05-1} e^{-t^{0.05}/0.7} = 0.07142 t^{-0.95} e^{-t^{0.05}/0.7}$

Thus, $z(t) = \dfrac{f(t)}{R(t)} = \dfrac{0.07142 t^{-0.95} e^{-t^{0.05}/0.7}}{e^{-t^{0.05}/0.7}} = 0.07142 t^{-0.95}$.

For $t = 500$, $z(500) = 0.07142(500)^{-0.95} = 0.000195$

17.35 a. $F(t) = P(Y \le t) = \displaystyle\int_0^t \frac{1}{\beta} dy = \frac{y}{\beta}\Big]_0^t = \frac{t}{\beta} - \frac{0}{\beta} = \frac{t}{\beta}$

$R(t) = 1 - F(t) = 1 - \dfrac{t}{\beta}$

$z(t) = \dfrac{f(t)}{R(t)} = \dfrac{\dfrac{1}{\beta}}{1 - \dfrac{t}{\beta}} = \dfrac{1}{\beta - t}$

b. When $\beta = 10$, $z(t) = \dfrac{1}{10 - t}$

For $t = 0$, $z(0) = \dfrac{1}{10 - 0} = 0.10$ For $t = 1$, $z(1) = \dfrac{1}{10 - 1} = 0.1111$

For $t = 2$, $z(2) = \dfrac{1}{10 - 2} = 0.125$ For $t = 3$, $z(3) = \dfrac{1}{10 - 3} = 0.1429$

For $t = 4$, $z(4) = \dfrac{1}{10 - 4} = 0.1667$ For $t = 5$, $z(5) = \dfrac{1}{10 - 5} = 0.2$

The graph of the hazard rate is:

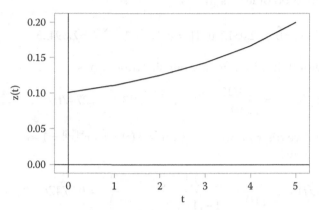

c. The reliability of the system when $t = 4$ and $\beta = 10$ is $R(4) = 1 - \dfrac{4}{10} = 0.6$

17.37 a. Some preliminary calculations are:

Hour (i)	$x_i = \ln i$	# Survivors (n_i)	$\hat{R}(i) = \frac{n_i}{n}$	$-\ln \hat{R}(i)$	$y_i = \ln[-\ln \hat{R}(i)]$
1	0.00000	438	0.876	0.132389	−2.022009
2	0.69315	280	0.560	0.579818	−0.545040
3	1.09861	146	0.292	1.231001	0.207828
4	1.38629	51	0.102	2.282782	0.825395
5	1.60944	15	0.030	3.506558	1.254635

Using MINITAB, the results are:

```
Regression Analysis: y versus x

Analysis of Variance

Source        DF    Adj SS    Adj MS   F-Value   P-Value
Regression     1   6.66502   6.66502   5624.83     0.000
  x            1   6.66502   6.66502   5624.83     0.000
Error          3   0.00355   0.00118
Total          4   6.66858

Model Summary

        S    R-sq   R-sq(adj)   R-sq(pred)
0.0344228   99.95%     99.93%       99.79%

Coefficients

Term         Coef   SE Coef   T-Value   P-Value   VIF
Constant  -2.0007    0.0302    -66.34     0.000
x          2.0312    0.0271     75.00     0.000   1.00

Regression Equation
y = -2.0007 + 2.0312 x
```

Using the method of least squares:

$$\hat{\alpha} = \hat{\beta}_1 = 2.0312 \text{ and } \hat{\beta} = e^{-\hat{\beta}_0} = e^{-(-2.0007)} = 7.39423$$

b. For the Weibull distribution, the density function is:

$$f(t) = \frac{\alpha}{\beta} t^{\alpha-1} e^{-t^\alpha/\beta} = \frac{2.0312}{7.39423} t^{2.0312-1} e^{-t^{2.0312}/7.39423} = 0.2747 t^{1.0312} e^{-t^{2.0312}/7.39423}$$

The cumulative distribution function is $F(t) = 1 - e^{-t^\alpha/\beta} = 1 - e^{-t^{2.0312}/7.39423}$.
The hazard rate is:

$$z(t) = \frac{f(t)}{1 - F(t)} = \frac{0.2747 t^{1.0312} e^{-t^{2.0312}/7.39423}}{1 - \left(1 - e^{-t^{2.0312}/7.39423}\right)} = 0.2747 t^{1.0312}$$

The reliability $R(t)$ is:

$$R(t) = 1 - F(t) = 1 - (1 - e^{-t^{2.0312}/7.39423}) = e^{-t^{2.0312}/7.39423}$$

c. $P(t \geq 1) = R(1) = e^{-1^{2.0312}/7.39423} = 0.8735$

17.39   The density function for the exponential distribution is $f(t) = \frac{1}{\beta} e^{-t/\beta}$.

For $\beta = 1000$, $f(t) = \frac{1}{1000} e^{-t/1000}$

The cumulative distribution function is $F(t) = 1 - e^{-t/\beta}$
For $\beta = 1000$, $F(t) = 1 - e^{-t/1000}$
The reliability of a single component is $R(t) = 1 - F(t) = 1 - (1 - e^{-t/1000}) = e^{-t/1000}$

At $t = 1400$, the reliability of a component is $R(1400) = e^{-1400/1000} = 0.2466$
The reliability of a system composed of two resistors connected in series is:

$$P(\text{system functions}) = p_A p_B = 0.2466(0.2466) = 0.0608$$

17.41 a. The approximate confidence interval for the mean time between failures is:

$$\frac{2(\text{Total life})}{\chi^2_{\alpha/2}} \leq \beta \leq \frac{2(\text{Total life})}{\chi^2_{1-\alpha/2}}$$

where total life is $\displaystyle\sum_{i=1}^{r} t_i + (n-r)t_r = 5,283 + (10-4)2,266 = 18,879$

For confidence coefficient 0.95, $\alpha = 0.05$ and $\alpha/2 = 0.05/2 = 0.025$. From Table 8, Appendix B, with $df = 2r = 2(4) = 8$, $\chi^2_{0.025} = 17.5346$ and $\chi^2_{0.975} = 2.17973$. The 95% confidence interval is:

$$\frac{2(\text{Total life})}{\chi^2_{\alpha/2}} \leq \beta \leq \frac{2(\text{Total life})}{\chi^2_{1-\alpha/2}} \Rightarrow \frac{2(18,879)}{17.5346} \leq \beta \leq \frac{2(18,879)}{2.17973}$$

$$\Rightarrow (2,153.343, \ 17,322.329)$$

b. The cumulative distribution function for the exponential distribution is $F(t) = 1 - e^{-t/\beta}$.

The reliability function is $R(t) = 1 - F(t) = 1 - (1 - e^{-t/\beta}) = e^{-t/\beta}$.

The point estimator for β is $\hat{\beta} = \dfrac{\displaystyle\sum_{i=1}^{r} t_i + (n-r)t_r}{r} = \dfrac{18,879}{4} = 4,719.75$.

Thus, the estimate of the reliability function is $\hat{R}(t) = e^{-t/4,719.75}$

The probability that the semiconductor will still be in operation after 4,000 hours is:

$$\hat{R}(4,000) = e^{-4,000/4,719.75} = 0.4285$$

The 95% confidence interval for the reliability is found by substituting the endpoints of the confidence interval for β into the equation for the reliability. The 95% confidence interval is:

$$e^{-4,000/2,153.343} \leq e^{-4000/\beta} \leq e^{-4,000/17,322.329} \Rightarrow (0.1561, \ 0.7938)$$

c. The hazard rate is $z(t) = \dfrac{f(t)}{R(t)} = \dfrac{\dfrac{1}{\beta}e^{-t/\beta}}{e^{-t/\beta}} = \dfrac{1}{\beta}$.

For $\hat{\beta} = 4,719.75$, the hazard rate is estimated by $z(t) = \dfrac{1}{4,719.75} = 0.000212$.

Because the hazard rate is constant, it is just as likely that the component will fail in one unit interval as in any other.

The 95% confidence interval for the hazard rate is found by substituting the endpoints of the confidence interval for β into the equation for the hazard function. The 95% confidence interval is:

$$\left(\frac{1}{17,322.329}, \ \frac{1}{2,153.343} \right) \Rightarrow (0.0000577, \ 0.0004644)$$

17.43   To find the reliability of the system, we must first find the reliability of the subsystems.
The parallel subsystem with components A and B has reliability probability:

$$P_{AB} = 1 - (1 - p_A)(1 - p_B) = 1 - (1 - 0.85)(1 - 0.75) = 1 - 0.0375 = 0.9625$$

The parallel subsystem with components D and E has reliability probability:

$$P_{DE} = 1 - (1 - p_D)(1 - p_E) = 1 - (1 - 0.90)(1 - 0.95) = 1 - 0.005 = 0.995$$

The parallel subsystem with components C and DE has reliability probability:

$$P_{CDE} = 1 - (1 - p_C)(1 - p_{DE}) = 1 - (1 - 0.75)(1 - 0.995) = 1 - 0.00125 = 0.99875$$

The parallel subsystem made up of components C, D, E, and F has reliability probability:

$$P_{CDEF} = 1 - (1 - p_{CD})(1 - p_{EF}) = 1 - (1 - 0.7225)(1 - 0.93752) = 0.98266$$

The complete series system made up of two subsystems, AB, and CDE, has reliability probability:

$$P(\text{system}) = p_{AB}p_{CDE} = 0.9625(0.99875) = 0.9613$$

17.45 $$f(t) = z(t)R(t) = z(t)e^{-\int_0^t z(y)\,dy}$$

Appendix A: Matrix Algebra

A.1 a. $\mathbf{AB} = \begin{bmatrix} 3 & 0 \\ -1 & 4 \end{bmatrix} \begin{bmatrix} 2 & 1 \\ 0 & -1 \end{bmatrix} = \begin{bmatrix} 3(2)+0(0) & 3(1)+0(-1) \\ -1(2)+4(0) & -1(1)+4(-1) \end{bmatrix} = \begin{bmatrix} 6 & 3 \\ -2 & -5 \end{bmatrix}$

b. $\mathbf{AB} = \begin{bmatrix} 3 & 0 \\ -1 & 4 \end{bmatrix} \begin{bmatrix} 1 & 0 & 3 \\ -2 & 1 & 2 \end{bmatrix} = \begin{bmatrix} 3(1)+0(-2) & 3(0)+0(1) & 3(3)+0(2) \\ -1(1)+4(-2) & -1(0)+4(1) & -1(3)+4(2) \end{bmatrix} = \begin{bmatrix} 3 & 0 & 9 \\ -9 & 4 & 5 \end{bmatrix}$

c. $\mathbf{BA} = \begin{bmatrix} 2 & 1 \\ 0 & -1 \end{bmatrix} \begin{bmatrix} 3 & 0 \\ -1 & 4 \end{bmatrix} = \begin{bmatrix} 2(3)+1(-1) & 2(0)+1(4) \\ 0(3)-1(-1) & 0(0)-1(4) \end{bmatrix} = \begin{bmatrix} 5 & 4 \\ 1 & -4 \end{bmatrix}$

A.3 a. $\mathbf{AB}$ is a 3×3 matrix $(3 \times 2)(2 \times 4) \Rightarrow 3 \times 4$

b. No, it is not possible to find $\mathbf{BA}$. In order to multiply two matrices, the inner dimension numbers must be equal $\Rightarrow (2 \times 4)(3 \times 2) \Rightarrow 4 \neq 3$.

A.5 a. $\mathbf{AB} = \begin{bmatrix} 1 & 0 & 0 \\ 0 & 3 & 0 \\ 0 & 0 & 2 \end{bmatrix} \begin{bmatrix} 2 & 3 \\ -3 & 0 \\ 4 & -1 \end{bmatrix} = \begin{bmatrix} 1(2)+0(-3)+0(4) & 1(3)+0(0)+0(-1) \\ 0(2)+3(-3)+0(4) & 0(3)+3(0)+0(-1) \\ 0(2)+0(-3)+2(4) & 0(3)+0(0)+2(-1) \end{bmatrix} = \begin{bmatrix} 2 & 3 \\ -9 & 0 \\ 8 & -2 \end{bmatrix}$

b. $\mathbf{CA} = \begin{bmatrix} 3 & 0 & 2 \end{bmatrix} \begin{bmatrix} 1 & 0 & 0 \\ 0 & 3 & 0 \\ 0 & 0 & 2 \end{bmatrix} = \begin{bmatrix} 3(1)+0(0)+2(0) & 3(0)+0(3)+2(0) & 3(0)+0(0)+2(2) \end{bmatrix} = \begin{bmatrix} 3 & 0 & 4 \end{bmatrix}$

c. $\mathbf{CB} = \begin{bmatrix} 3 & 0 & 2 \end{bmatrix} \begin{bmatrix} 2 & 3 \\ -3 & 0 \\ 4 & -1 \end{bmatrix} = \begin{bmatrix} 3(2)+0(-3)+2(4) & 3(3)+0(0)+2(-1) \end{bmatrix} = \begin{bmatrix} 14 & 7 \end{bmatrix}$

A.7 a. $\begin{bmatrix} 1 & 0 \\ 0 & 1 \end{bmatrix}$

b. $\mathbf{IA} = \begin{bmatrix} 1 & 0 \\ 0 & 1 \end{bmatrix} \begin{bmatrix} 3 & 0 & 2 \\ -1 & 1 & 4 \end{bmatrix} = \begin{bmatrix} 1(3)+0(-1) & 1(0)+0(1) & 1(2)+0(4) \\ 0(3)+1(-1) & 0(0)+1(1) & 0(2)+1(4) \end{bmatrix} = \begin{bmatrix} 3 & 0 & 2 \\ -1 & 1 & 4 \end{bmatrix} = \mathbf{A}$

c. $\begin{bmatrix} 1 & 0 & 0 \\ 0 & 1 & 0 \\ 0 & 0 & 1 \end{bmatrix}$

d. $\mathbf{AI} = \begin{bmatrix} 3 & 0 & 2 \\ -1 & 1 & 4 \end{bmatrix} \begin{bmatrix} 1 & 0 & 0 \\ 0 & 1 & 0 \\ 0 & 0 & 1 \end{bmatrix} = \begin{bmatrix} 3(1)+0(0)+2(0) & 3(0)+0(1)+2(0) & 3(0)+0(0)+2(1) \\ -1(1)+1(0)+4(0) & -1(0)+1(1)+4(0) & -1(0)+1(0)+4(1) \end{bmatrix}$

$= \begin{bmatrix} 3 & 0 & 2 \\ -1 & 1 & 4 \end{bmatrix} = \mathbf{A}$

A.9 $\mathbf{AA}^{-1} = \mathbf{A}^{-1}\mathbf{A} = \mathbf{I}$

$$\mathbf{AA}^{-1} = \begin{bmatrix} 12 & 0 & 0 & 8 \\ 0 & 12 & 0 & 0 \\ 0 & 0 & 8 & 0 \\ 8 & 0 & 0 & 8 \end{bmatrix} \begin{bmatrix} 1/4 & 0 & 0 & -1/4 \\ 0 & 1/12 & 0 & 0 \\ 0 & 0 & 1/8 & 0 \\ -1/4 & 0 & 0 & 3/8 \end{bmatrix} = \begin{bmatrix} 1 & 0 & 0 & 1 \\ 0 & 1 & 0 & 0 \\ 0 & 0 & 1 & 0 \\ 0 & 0 & 0 & 1 \end{bmatrix}$$

$$\mathbf{A}^{-1}\mathbf{A} = \begin{bmatrix} 1/4 & 0 & 0 & -1/4 \\ 0 & 1/12 & 0 & 0 \\ 0 & 0 & 1/8 & 0 \\ -1/4 & 0 & 0 & 3/8 \end{bmatrix} \begin{bmatrix} 12 & 0 & 0 & 8 \\ 0 & 12 & 0 & 0 \\ 0 & 0 & 8 & 0 \\ 8 & 0 & 0 & 8 \end{bmatrix} = \begin{bmatrix} 1 & 0 & 0 & 1 \\ 0 & 1 & 0 & 0 \\ 0 & 0 & 1 & 0 \\ 0 & 0 & 0 & 1 \end{bmatrix}$$

A.11   To verify Theorem 1.1 show $\mathbf{DD}^{-1} = \mathbf{D}^{-1}\mathbf{D} = \mathbf{I}$

$$\mathbf{DD}^{-1} = \begin{bmatrix} d_{11} & 0 & 0 & \cdots & 0 \\ 0 & d_{22} & 0 & \cdots & 0 \\ 0 & 0 & d_{33} & \cdots & 0 \\ \cdot & \cdot & \cdot & \cdots & 0 \\ \cdot & \cdot & \cdot & \cdots & 0 \\ \cdot & \cdot & \cdot & \cdots & 0 \\ 0 & 0 & 0 & \cdots & d_{nn} \end{bmatrix} \begin{bmatrix} 1/d_{11} & 0 & 0 & \cdots & 0 \\ 0 & 1/d_{22} & 0 & \cdots & 0 \\ 0 & 0 & 1/d_{33} & \cdots & 0 \\ \cdot & \cdot & \cdot & \cdots & 0 \\ \cdot & \cdot & \cdot & \cdots & 0 \\ \cdot & \cdot & \cdot & \cdots & 0 \\ 0 & 0 & 0 & \cdots & 1/d_{nn} \end{bmatrix} = \begin{bmatrix} 1 & 0 & 0 & \cdots & 0 \\ 0 & 1 & 0 & \cdots & 0 \\ 0 & 0 & 1 & \cdots & 0 \\ \cdot & \cdot & \cdot & \cdots & 0 \\ \cdot & \cdot & \cdot & \cdots & 0 \\ \cdot & \cdot & \cdot & \cdots & 0 \\ 0 & 0 & 0 & \cdots & 1 \end{bmatrix}$$

$$\mathbf{D}^{-1}\mathbf{D} = \begin{bmatrix} 1/d_{11} & 0 & 0 & \cdots & 0 \\ 0 & 1/d_{22} & 0 & \cdots & 0 \\ 0 & 0 & 1/d_{33} & \cdots & 0 \\ \cdot & \cdot & \cdot & \cdots & 0 \\ \cdot & \cdot & \cdot & \cdots & 0 \\ \cdot & \cdot & \cdot & \cdots & 0 \\ 0 & 0 & 0 & \cdots & 1/d_{nn} \end{bmatrix} \begin{bmatrix} d_{11} & 0 & 0 & \cdots & 0 \\ 0 & d_{22} & 0 & \cdots & 0 \\ 0 & 0 & d_{33} & \cdots & 0 \\ \cdot & \cdot & \cdot & \cdots & 0 \\ \cdot & \cdot & \cdot & \cdots & 0 \\ \cdot & \cdot & \cdot & \cdots & 0 \\ 0 & 0 & 0 & \cdots & d_{nn} \end{bmatrix} = \begin{bmatrix} 1 & 0 & 0 & \cdots & 0 \\ 0 & 1 & 0 & \cdots & 0 \\ 0 & 0 & 1 & \cdots & 0 \\ \cdot & \cdot & \cdot & \cdots & 0 \\ \cdot & \cdot & \cdot & \cdots & 0 \\ \cdot & \cdot & \cdot & \cdots & 0 \\ 0 & 0 & 0 & \cdots & 1 \end{bmatrix}$$

A.13 a. Rewrite the linear equations as:

$$10v_1 + 0v_2 + 20v_3 = 60$$

$$0v_1 + 20v_2 + 0v_3 = 60$$

$$20v_1 + 0v_2 + 68v_3 = 176$$

$$\mathbf{A} = \begin{bmatrix} 10 & 0 & 20 \\ 0 & 20 & 0 \\ 20 & 0 & 68 \end{bmatrix} \quad \mathbf{V} = \begin{bmatrix} v_1 \\ v_2 \\ v_3 \end{bmatrix} \quad \mathbf{G} = \begin{bmatrix} 60 \\ 60 \\ 176 \end{bmatrix}$$

b. Show $\mathbf{A}^{-1}\mathbf{A} = \mathbf{A}\mathbf{A}^{-1} = \mathbf{I}$

$$\mathbf{A}^{-1}\mathbf{A} = \begin{bmatrix} 17/70 & 0 & -1/14 \\ 0 & 1/20 & 0 \\ -1/14 & 0 & 1/28 \end{bmatrix} \begin{bmatrix} 10 & 0 & 20 \\ 0 & 20 & 0 \\ 20 & 0 & 68 \end{bmatrix} = \begin{bmatrix} 1 & 0 & 0 \\ 0 & 1 & 0 \\ 0 & 0 & 1 \end{bmatrix}$$

$$\mathbf{A}\mathbf{A}^{-1} = \begin{bmatrix} 10 & 0 & 20 \\ 0 & 20 & 0 \\ 20 & 0 & 68 \end{bmatrix} \begin{bmatrix} 17/70 & 0 & -1/14 \\ 0 & 1/20 & 0 \\ -1/14 & 0 & 1/28 \end{bmatrix} = \begin{bmatrix} 1 & 0 & 0 \\ 0 & 1 & 0 \\ 0 & 0 & 1 \end{bmatrix}$$

c. $\mathbf{V} = \mathbf{A}^{-1}\mathbf{G} = \begin{bmatrix} 17/70 & 0 & -1/14 \\ 0 & 1/20 & 0 \\ -1/14 & 0 & 1/28 \end{bmatrix} \begin{bmatrix} 60 \\ 60 \\ 176 \end{bmatrix} = \begin{bmatrix} 2 \\ 3 \\ 2 \end{bmatrix}$

Printed in the United States
by Baker & Taylor Publisher Services

Printed in the United States
by Baker & Taylor Publisher Services